Novel Innovation Design for the Future of Health

Michael Friebe
Editor

Novel Innovation Design for the Future of Health

Entrepreneurial Concepts for Patient Empowerment and Health Democratization

 Springer

Editor
Michael Friebe
IDTM Recklinghausen,
Otto-von-Guericke University
Magdeburg, Germany

ISBN 978-3-031-08193-4   ISBN 978-3-031-08191-0   (eBook)
https://doi.org/10.1007/978-3-031-08191-0

This Springer imprint is published by the registered company Springer Nature Switzerland AG
The registered company address is: Gewerbestrasse 11, 6330 Cham, Switzerland

This book is dedicated to my parents and YOU

"You know the greatest danger facing us is ourselves, an irrational fear of the unknown. But there's no such thing as the unknown — only things temporarily hidden, temporarily not understood."
—Captain James T. Kirk

This is a short but also difficult page. Nobody typically reads the DEDICATION and nobody really remembers who was listed here.

But it is a page that the author writes last and—considering the amount of text—spends the most time on.

This book in essence is mainly dedicated to YOU, who holds it right now as a real physical book or as a digital version. The reason? Well, YOU must be someone who wants to create health-related innovation and has a purpose. And YOU see that there are a lot of global challenges related to WELL BEING and GOOD HEALTH that cannot be solved with the current focus of healthcare provision and the currently applied methodologies and mindsets.

And of course this book is also dedicated to the ones that are close to me, that have helped and supported me for decades, that have stimulated me, and that have challenged me in my private and professional life.
I really appreciate and love you ... I may not have told you yet or not often enough and I am sorry for that.
You know me and you accept me that way.
Please stick around and help me!
Thank You!
*—**Michael Friebe***

Foreword

> Let's put the power of exponential technologies into patient's hands and revolution-ize how we live.
> *We are edging closer towards a dramatically extended healthspan.*
> *Where "100 years old can become the new 60"*
> —Peter Diamandis, MD

Rethink Health Innovation Generation with Exponential Technology Convergence

Accelerating advancement of exponential technologies is actually *old news*.

So, what's the *new* news in 2022 and useful for health innovators?

That formerly independent waves of exponentially accelerating technology are beginning to *converge* with other independent waves of exponentially accelerating technology.

In other words, these waves are starting to overlap—stacking atop one another, producing tsunami-sized behemoths that threaten to wash away (read: "reinvent") almost every industry in their path.

For example, the speed of drug development is accelerating—not *only* because biotechnology (sequencing, CRISPR, etc.) is progressing at an exponential rate, but because AI, quantum computing, and other exponentials are converging on the field.

When an innovation creates a new market and washes away an existing one, we use the term "disruptive innovation," such as when silicon chips replaced vacuum tubes at the beginning of the digital age. That was a disruptive innovation.

It is pretty clear that no industry is immune to these developments.

Why would that therefore not be true for the healthcare business? Big data + machine and federated learning, novel sensors, 5G, gene information, and 3D printing will undoubtedly cause a disruption of how healthcare is seen and performed and also where it takes place. And with that disruption we will most likely see an adapted and completely changed health business model. More home care, more prevention and prediction, a higher health personalization, and an increased participation of the individual will help to identify and fix health problems long before they are noticeable or even diagnosable. Do we then still need the amount of hospital beds that we have today? Do we still need the same sick-care financing that covers the treatment and therapy cost?

And ultimately that should lead to developing new ways to fight diseases associated with aging leading to an increased health longevity.

As an entrepreneur, being able to see around the corner of tomorrow and being agile enough to adapt is critical to your success.

What you should remember as you think about how to leverage these new business models is that the rate at which technology is accelerating *is itself accelerating*.

Yet even in this dynamic environment, countless businesses are still anchored by a mentality of maintaining—competing *solely* on operational execution.

But as an innovator and entrepreneur, it's more vital than ever that you leverage these business models for success in the decades ahead.

Each one is a revolutionary way of creating value—each is a force for acceleration.

All of this requires a novel look on education and educating, and this book will possibly help to solve some of the big challenges in and around global health through creating awareness, helping to develop an exponential mind, and providing a valuable toolset for innovators.

Peter Diamandis, MD Recently named by Fortune as one of the "World's 50 Greatest Leaders," Peter H. Diamandis is the founder and executive chairman of the XPRIZE Foundation, which leads the world in designing and operating large-scale incentive competitions. He is also the executive founder of Singularity University, a graduate-level Silicon Valley institution that counsels the world's leaders on exponentially growing technologies. As an entrepreneur, Diamandis has started over 20 companies in the areas of longevity, space, venture capital and education. He is co-founder of BOLD Capital Partners, a venture fund with $250M investing in exponential technologies, and co-

(continued)

founder and Vice Chairman of Celularity, Inc., a cellular therapeutics company. He is also a New York Times Bestselling author of three books: Abundance—The Future Is Better Than You Think, BOLD—How to go Big, Create Wealth & Impact the World and The Future is Faster Than You Think. Contact: info@diamandis.com

XPRIZE Foundation   Peter Diamandis
Culver City, CA, USA

SINGULARITY University
Santa Clara, CA, USA

Foreword

We have to challenge, disrupt, and find new solutions to make healthcare more affordable, more accessible, and ultimately more equitable.
—*Prof. Shafi Ahmed, MD*

The Exciting Future of Surgery ... but that Is Not the Primary Concern of the Future of Health

I started writing this foreword with a perspective on the future of surgery, as I am a practicing surgeon. The new concept of digital surgery follows on from the longer journey of digital health, and it is incredible and reassuring to see the recent developments in this field globally with international teams solving important problems.

It is safe to assume that some of these will actually be translated to the patient's bedside over the next few years. Augmented and virtual reality, combined with artificial intelligence, will help a surgeon plan and navigate a complex procedure more safely. Semi-autonomous surgery is almost at a touching distance and becoming a reality. The rapid expansion of surgical robots and improvements in connectivity may guide us to the point of the surgical singularity and to the Shangri La of a fully autonomous operation.

Is that good or bad?

Well, if you argue from a patient's perspective—and that is the point of view that one should always have—with a proven better accuracy and efficacy, that is fairly easy to answer. However, we are not quite there yet, but I am of the firm belief that it will not take much longer. From a perspective of a surgeon, there may be a concern that it will remove the need for surgeons, but I am convinced that these technologies will actually enhance and augment our work and remove variation and improve outcomes that are driven by data. There will always be a need for surgeons in decision making and judgment which are essential and require a myriad of skills including empathy and communication. Surgeons will be vital in emergency cases, or for very complicated procedures that the AI-powered robot is not yet trained for and to importantly manage complications. This brings me neatly to the global health workforce shortages and to highlight that clinical education has not yet completely embraced immersive technologies and simulation. We will naturally repurpose our roles as surgeons similar to the radiologist embracing AI diagnostic tools.

I am also someone who has close ties to many lower income nations. What I see as health needs there has nothing to do with the incremental developments that we are working on in the countries with established healthcare systems. I am deeply concerned about the lack of access to advanced technologies there and the associated cost of providing them. Only a small percentage of the population can actually afford the health offerings. We need to rethink our current approach of making incremental developments and move toward a more disruptive approach that would lead to a democratization of health. I am also encouraged by those same countries that understand frugal innovation as well as having the foresight and determination to leapfrog Western health systems by thinking outside conventional attitudes to health and care delivery.

We should constantly question the current healthcare business model as we can and need to do better. We need to shift much more toward health monitoring and prevention.

The future of health must be working from a patient's perspective, addressing the big challenges, using the power of exponential technologies, and developing needed tools in interdisciplinary and global teams. The book is a great starting point for rethinking the current process of health innovation.

Prof. Shafi Ahmed, MD, is a world renowned, multi award winning surgeon, teacher, futurist, innovator, and entrepreneur. He is a 3x TEDx speaker and is faculty at Singularity University. He has delivered over 250 keynotes in 30 countries.

After studying medicine at Kings College Hospital Medical School London, he completed his surgical training in London. Ahmed is currently a

(continued)

Laparoscopic Colorectal surgeon specialising in colorectal cancer at The Royal London and St Bartholomew's Hospitals.

His mission is to merge the world of medicine, global education, and virtual and augmented reality to democratize and scale surgical education to make it affordable and accessible to everyone using the power of connectivity to allow equitable surgical care. Contact: info@medicalrealities.com

Barts Health NHS Trust   Shafi Ahmed
London, UK

Royal College of Surgeons
London, UK

Medical Realities
London, UK

Foreword

Longevity medicine is the epitome of a sustainable disruption in healthcare innovation, characterized by symbiotically inclusive interdisciplinarity of clinical medicine, gerosciences, AI, and computational science, potentiated by collaborative ecosystem with the public health, industry, and governance.

—*Prof. Evelyne Yehudit Bischof, MD*

Longevity Medicine as a Symbiosis and Catalysator of Health Innovation and Technology

The health sector's core demands shifted thanks to the massive advancements in reactive medicine, facilitated and accelerated by progressive novel technologies, and a resulting lifespan extension. The silver tsunami arrived in most of the demographics; the global health burden is dominated by chronic diseases of old age.

Health innovations in sick care and early prevention have already reached a solid niveau, and further developments will now focus on striving excellence, such as in interventional robotics, targeted therapies, and enhanced imaging. AI is virtually ubiquitous in all health sectors.

With the rapid ascent of AI and ML, especially deep learning, as well as federal and transfer learning, fundamental tools were given to enrich the nascent field of longevity medicine: AI-based precision medicine that aims at expanding the healthy lifespan by identifying the risks of diseases and mitigating and eliminating them. Uniquely inclusive, multimodal, and interdisciplinary, this field is ultimately the future of health innovation – and the future is now. Artificial and human intelligence will need to converge more efficiently toward optimization of individual biological age.

Various stakeholders, including the industry, are as essential and partaking in shaping healthcare and implementing strategic applications as now. This is due to the granularity and permeability of key arenas: computational and AI science, gerosciences, medicine, and engineering. Combined with social AI maturity and increasing demand for personalization in diagnostics and therapies, innovators will embrace the enchanting challenge to forge on the existing data collection modalities, e.g., POC, continuous monitoring, NLP, and multiomics, and create integration strategies towards new trajectories and improved AI-developed algorithms.

Innovation requires new responsibilities for all, especially to clinicians like myself who also engage as scientists, academic tutors, and public health advisors. In the present technological perpetuum mobile, it is imperative to work along executors and assure that sustainability, credibility, and safety is assured for what we all share: the will to achieve healthy longevity and lifespan with minimal to no age-related diseases.

As a longevity physician, leading my patients and educating peers and stakeholders such as investors, insurances, or policy makers often includes illustrating the opportunities, limitations, and perspective of healthcare innovations. This book offers a compilation of valuable themes, equipping the reader with information, trends, skills, and resources necessary to navigate and excel in health innovation.

Prof. Evelyne Yehudit Bischof, MD, is an Associate Professor at Shanghai University of Medicine and Health Sciences and a visiting associate professor at the Tel Aviv University School of Medicine. Other assignments are delegate doctor of the Israeli Consulate China, Shanghai; Chief physician Internal medicine University Hospital of Jiatong School of Medicine Renji, Shanghai; Longevity concierge physician Human Longevity Inc., San Diego; Scientific physician at International Center for Multimorbidity and Complexity in Medicine (ICMC), Universität Zürich, Schweiz. Evelyne is an internal medicine specialist and longevity physician, trained a.e. at Harvard Medical School affiliated hospitals (Mass General Hospital, Beth Israel MD, Dana Farber) and Columbia University NYC. She is a specialist in internal medicine, with research focus on Artificial Intelligence (AI) and digital health, especially in the fields of oncology, preventative and precision medicine, biogerontology, and geronto-oncology. She published over 90 peer-reviewed papers and is a

(continued)

frequent speaker at scientific and medical conferences in Asia and in Europe. Contact: bischofevelyne@gmail.com

Shanghai University of Medicine
and Health Sciences, International
Center for Multimorbidity and Complexity
in Medicine (ICMC), Universität Zürich
Zurich, Switzerland

Evelyne Yehudit Bischof

Preface

Innovation Design for the Future of Health

Entrepreneurial Concepts for Patient Empowerment and Health Democratization

If you believe that global healthcare and health delivery is fine, that we have no real challenges to solve in that space, and that we should continue to exclusively focus on diagnosing and subsequently treating sick people with more and more complicated and expensive devices and processes . . . then thank you for buying this book, but it is not really meant for you!

For everyone else "Thank You" for allowing me to take you on a journey from current healthcare services and delivery processes to a future of health vision.

In a high-income country you may say that everything is actually working well when we are sick. You are right and that service should continue to become better, more efficient, personalized, and more patient centric. The digital health transformation process will for sure provide tools to clinicians and patients that will likely empower the patient significantly and create a changed and more equal patient–doctor relation. That is good!

Rethink, if you are not from one of the high-income nations and empathize with the possible issues of low-income nations if you are situated in a rather wealthy environment.

We likely will have technologies available that will allow us to predict and prevent many diseases before they are noticeable, we will be able to increase our healthy lifespan (longevity), and we will be able to benefit from novel prevention and treatment strategies. This in turn will trigger new segments and new business models, especially since these may be rather inexpensive and not require a doctor input anymore. So we may need a new "*INNOVATION DESIGN for the FUTURE of HEALTH*."

The best recipe to achieve meaningful regional and global solutions is to stimulate entrepreneurship. A core element of the book is therefore to present "**Entrepreneurial Concepts for Patient Empowerment and Health Democratization.**"

Fig. 1 Left: The needs and the determinants of personal health are not really aligned with the actual healthcare delivery processes and goals, and the great challenges around health are also not effectively dealt with. Right: A core concept of the book is to stimulate entrepreneurial activities

Figure 1 on the left shows the main determinants of personal health, how healthcare delivery looks like, and the major challenges that we need to address, and on the right the different entrepreneurial (or intrapreneurial) starting points and the issues that this venture has to deal with. The book attempts to provide methodologies that should prepare for the needed future thinking and for an exponential mindset. This will likely increase your odds of developing successful and impactful products and services.

The book is complemented by articles of leading experts that will provide some relevant health innovation-related insights that are provided as chapters in the following main parts:

- *General Introduction about "What is Wrong with Health?" and "What Should the Future of Health be?" (Part I: Chaps. 1–5).*
- *Followed by "Exponential Medicine + Technologies + Mindset" (Part II: Chaps. 6–10) and "Future Health Value Propositions" (Part III: Chaps. 11–17).*
- *After that we start with "Innovation Methodology Basics" (Part IV: Chaps. 18–22) and discuss "Ethical Design Considerations for Health Innovations" (Part V: Chap. 23).*
- *The next two parts follow with a more generic title of "Health Innovation Design" (Part VI: Chaps. 24–26) and a more specific presentation of the "Purpose Launchpad Health" (Part VII: Chap. 27).*
- *"Health Leadership, Skills and other Methodologies" is covered in the succeeding part (Part VIII: Chaps. 28–31), followed by*
- *The important part on "Health Entrepreneurship" (Part IX: Chaps. 32–39).*
- *Several European and Corporate "Health Innovation Education and Incubation" (Part X: Chaps. 40–48) setups are then presented as case examples and the book closes with*
- *A summary presentation and explanation of the "Purpose Launchpad Health: Toolset Templates and Principles" (Part XI: Chap. 49).*

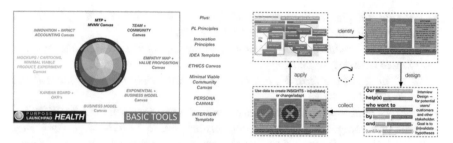

Fig. 2 Left: The main innovation method presented—Purpose Launchpad Health (based on Purpose Launchpad developed by Francisco Palao and the Purpose Alliance (www. purposealliance.com)). 8 Segments and many tools—all of which will be presented and discussed. Right: a typical agile iteration process based on the Lean Startup principle that is a core element of understanding the problem and creating something that actually has a validated need

I was very much inspired by the innovation laws of Peter Diamandis (see instructions and download link at https://www.diamandis.com/blog/how-you-can-use-peters-laws), especially these three, while all other ones listed are awesome as well and make you think about the future and your current role in shaping the future.

1. *The best way to predict the future is to create it yourself!*
2. *When given a choice... take both!!*
3. *Do it by the BOOK, but be the AUTHOR (or in my case the EDITOR and main AUTHOR).*

And as a former university professor I also like another one very much "*Bureaucracy is an obstacle to be conquered with persistence, confidence and a bulldozer when necessary.*"

I firmly believe that we also need to adjust/rethink our current education system that is often too siloed and does not focus on creating impact. So I hope that the book is also used for teaching health technology innovation design in a classroom setup. For details on how to do that (also hybrid or completely online), see Chaps. 40–42.

Figure 2 shows the core innovation framework used in this book, the Purpose Launchpad Health (PLH) based on the great work of Francisco Palao, and many international contributors (the editor being one of them).

This framework will be explained and discussed in detail, with examples and by providing all the used templates (Part XI).

I finish this preface with a list of the key innovation principles of this book:

- *Use agile and iterative approaches to validate hypotheses and assumptions in all aspects.*
- *Whenever possible think 10x not 10%—question the current setup!*
- *Experimentation (product, customer, business model, ...) is key to a successful implementation of a novel product or process idea.*

- *Put yourself often in the position of the core user / patient and try to see the situation from their point of view (EMPATHIZE).*
- *Do not start building before defining and evaluating the problem that you are trying to address.*
- *Good questions are most important—if you ask the right questions you will find the answers!*
- *Embrace failure or invalidating ideas and with that focus on the LEARNING.*
- *Write everything down for a learning history and to be able to go back when needed. Especially important for the regulatory approval process of health products.*
- *Come up with many solution ideas—do not limit yourself to the first one that you fall in love with.*
- *Use minimal viable prototypes (MVP) for validation checks and customer experiments.*
- *Evaluate alternative solutions based on their DESIRABILITY (Is there a verifiable customer need?), FEASIBILITY (Can we build solutions that will actually be working and satisfying these needs?), and VIABILITY/SUSTAINABILITY (Can we do that within a business model? . . . or do we envision a new one? . . . and can we do that within a sustainable future business model?).*
- *Have a (massive transformative) purpose combined with a longer-term vision, a shorter term mission, and a definition of the organizations core values. Or in other words: have a great futuristic and far-reaching plan, which is too big for you at the moment and too far away, but also come up with realistic milestones on how you want to reach that.*

The book will address all these points in the different parts and chapters.

Enjoy the reading and learning . . .
. . . **and a big Thank You to all the Contributors!**

Recklinghausen and Magdeburg   Michael Friebe
Germany

Acknowledgments

A big "thank you" to all the contributors—you all did a great job, you said very quickly "yes" after I asked you, and you are all pretty much in line with the core principles of the book.

Prof. Dr. Nassir Navab from the Technical University München (TUM), Chair of Computer Aided Medical Procedures, needs to be mentioned and acknowledged. An incredible innovative researcher who showed a great openness towards the positive impact of nonprofessional academics, who also has been a decade-long supporter for me in research (TUM-IAS fellowship), and who has been part of several entrepreneurial ventures with me.

The lecture that I started there is still taught every semester with over 500 graduates in the meantime (thanks to Dr. Jörg Traub for continuing it and for being a friend) and over 25 startups. He stimulated me to apply for a full-time professorship, and the Otto-von-Guericke-University in Magdeburg eventually appointed me as research professor. The 5 years there were extremely productive, I learned a lot, still have many friends and supporters there, and am still closely connected to the medical faculty as honorary professor. It opened my eyes however with respect to needed reforms and adjustments for research perspectives (only talking about healthtech related), value propositions, and education of health innovators. To me it is more clear than ever that meaningful transformation and innovations require a multidisciplinary environment with low bureaucratic burden. It is very difficult however—maybe even impossible—for normal universities to be agile enough to adapt to the changing environments. But while my time at the university—coming from two decades of being an entrepreneur and CEO—was a rather frustrating reality check, it was also one of the main triggers for this book.

For stimulating foresight, for introducing novel concepts, and for generally influencing my thinking and acting I want to acknowledge Singularity University, A360, Exponential Medicine, OpenExO, Purpose Alliance, Syntropic Enterprise, Growth Institute, and the Healthcaptains—many of the book contributors are part of at least one of these organizations.

I need to mention the influence of friends and collaborators in academia from IIT Kharagpur India, Misr University of Science and Technology Egypt, Queensland University of Technology Brisbane/Australia, and since 2022 the AGH University of Science and Technology in Krakow Poland ... and equally important the connections and ongoing discussions with industry and here specifically Siemens Healthineers (Innovation Think Tank and the MRI and Advanced Therapies Business Units), GE Healthcare (Edison Accelerator and the Chief Medical Office EMEA), Brainlab, Visus, Olympus, BBraun Aesculap, Bracco, and many startup companies with their great ideas and motivated founders ... and finally the professional organizations and networking setups MedEcon Ruhr, IEEE EMBS, SPIE, RSNA, MedTec Pharma, Business Angel Network Deutschland (BAND), CARS, and EWG.

Isabella Geiger and Stefan Hellwig, my coworkers, and co-shareholders have helped and supported me and my crazy ideas for close to 30 years now. Thanks a lot for the trust and confidence.

And finally I have to acknowledge the contributions of my wife Peggy. You will not find any article from her, but you can be ensured that without her this book would not have been finished.

Contents

List of Abbreviations

10X	10 times (thinking)
3D	Three-dimensional . . .e.g., 3D printer
4PH	Prediction, prevention, personalization, participation health
6D	Digital, disappointment, disrupt, dematerialize, demonetize, democratize
ABC	Acibadem Biodesign Center
ACTH	Adrenocorticotropic hormone
AI	Artificial Intelligence
AT	Assistive Technologies
AUC	Area under the curve
BCAA	Branched-chain amino acids
BD2K	Big data to knowledge
BLSS	Bioregenerative life support system
BMC	Business Model Canvas
BME	Biomedical Engineering
BME IDEA	Biomedical Engineering, Innovation, Design, and Entrepreneurship Alliance
BOSI	Builder, Opportunist, Specialist, Innovator (bosidna.com)
CAGR	Compounded annual growth rate
CB	Certified Body (regulatory)
CCC	China Compulsory Certification
CDMO	Contract design and manufacturing organization
CDS	Clinical decision support
CE	Conformité Européenne
CER	Clinical evaluation report
CoG	Cost of goods
CPT	Current Procedural Terminology (for Medical Reimbursement)
CQC	China Quality Certification
CQC	Care Quality Commission
CRISPR	Clustered regularly interspaced short palindromic repeats
CRO	Clinical Research Organization
CT	Computed Tomography
CTP	Cell Therapy Product

CV-19	COVID-19
CVD	Cardiovascular Diseases
DACH	Germany (D), Austria (A), Switzerland (CH)
DiGa	Digitale Gesundheitsanwendung (Digital Health App)
DIY	Do it yourself
DL	Deep Learning
DNA	Desoxyribonucleic Acid
DR	Dietary Restrictions
DRG	Diagnostic Related Case Groups (for Medical Reimbursement)
ECG	Electrocardiogram
ECTS	European Credit Transfer and Accumulation System
EDV	Expertise development Program
EHR	Electronic Health Record
ENT	Ear, Nose, Throat
ePA	Elektronische Patientenakte (electronic medical record)
EPO	European Patent Office
EuGH	European Court
EVP	Ethically viable Product
EXO	Exponential Organisations (www.openexo.com)
FCC	Federal Communications Commission
FDA	Food and Drug Administration
FDR	False Discovery Rate
GDP	Gross Domestic Product
GE	General Electric Corporation
GM	General Management
HALE	Health adjusted life expectancy
HCP	Healthcare Provider
HIN	High Income Nation
HMW	How might we
I3-EME	Identify Invent Implement with Engineers Medical Staff and Economics
ICER	Institute of Clinical and Economic Review
ICT	Information and Communications Technology
IEEE	Institute of Electrical and Electronics Engineers
IGT	Image guided Therapies
IoT	Internet of Things
IP + IPR	Intellectual Property Rights
IPO	Initial Public Offering
ISO	International Organization for Standards
IT	Information Ttechnology
ITT	Innovation Think Tank
IUD	Intrauterine Device
KOL	Key Opinion Leader
LDH	Lactate Dehydrogenase
LEO	Low Earth Orbit

LIN	Low-income Nation
MDR	Medical Device Regulation
mHealth	Mobile Health
MIC	Ministry of Internal Affairs (Regulatory Body in Japan)
ML	Machine Learning
MRI	Magnetic Resonance Imaging
MTP	Massive Transformative Purpose
MVMV	Moonshot, Vision, Mission, Values
MVP	Minimum Viable Prototype (sometimes also product)
NASA	North American Space Agency
NB	Notified Body (regulatory)
NCD	Noncommunicable Diseases
NHS	National Health Service (UK)
NICE	National Institute of Health and Care Excellence
NIH	National Institutes of Health (US)
OECD	Organisation for Economic Co-operation and Development
OEM	Original Equipment Manufacturer
OKR	Objectives and Key Results
OPSP	One-Page Strategic Plan (Growth Institute)
P4	Prediction, Prevention, Personalization, Participation
PCR	Polymerase Chain reaction
PCT	Patent Cooperation Treaty
PL	Purpose Launchpad
PLB	Purpose Launchpad Board
PLH	Purpose Launchpad Health
PLM	Purpose Launchpad Mood
PLP	Purpose Launchpad Points
PLP	Purpose Launchpad Progress
POC	Proof of concept
QALY	Quality-adjusted life year
QM	Quality Management
QMH	Quality Management Handbook
QMS	Quality Management System
R&D	Research and Development
RA	Regulatory Affairs
RIH	Responsible Innovation in Health
RM	Risk Management
ROC	Receiver Operating Characteristics
ROS	Reactive Oxygen Species
RRI	Responsible Research and Innovation
SDG	Sustainable Development Goals (United Nations)
SOD	Superoxide dismutase
SPE	Solar particle event
T2I2	Technological Innovations in Therapy and Imaging
TD	Technical documentation

TRISH	Translation Research Institute for Space Health
UN	United Nations
USD	United States Dollar
VDI	Verein deutscher Ingenieure (Association of German Engineers)
VPC	Value Proposition Canvas
WHO	World Health Organization
WIPO	World Intellectual Property Organization
WMH	Wearable Health Monitor

Part I

What Is Wrong with Health? What Should the Future of Health Be?

INNOVATION DESIGN for the FUTURE of HEALTH

1

Michael Friebe

Abstract

This chapter describes in further detail the motivations to write this book. It also shows how the author, with a background as a medical technology entrepreneur and also as a university professor, developed his belief that we need a different approach to health innovation. A new and agile approach powered by a set of human skills that are not taught at universities is required. So one of the intents for this book is to accompany teaching on Healthtech Innovation Generation at universities, but also to show novel innovation methods for entrepreneurs and also intrapreneurial activities. And finally to stimulate rethinking the current health innovation processes and goals.

Keywords

Exponential medicine · Global health challenges · Biomedical engineering curriculum · Planetary health · Individual health

M. Friebe (✉)
AGH University of Science and Technology, Krakow, Poland

Otto-von-Guericke University, Magdeburg, Germany

IDTM GmbH, Recklinghausen, Germany

FOM University of Applied Science, Center for Innovation and Business Development, Essen, Germany
e-mail: info@friebelab.org

M. Friebe (ed.), *Novel Innovation Design for the Future of Health*,
https://doi.org/10.1007/978-3-031-08191-0_1

What will you learn in this chapter?
- A personal journey towards rethinking Health Innovation goals and towards health democratization
- An overview of the Health Challenges, Individual Health needs, and likely developments
- The definition of Innovation in the Health context
- The problem with the need for disruption and the lack of business model to support that
- Exponential technology developments and their impact on health innovations

Let me start by summing up the last 30 years of my professional life.

After completing a Bachelor in Engineering in 1988, I took the opportunity to leave Germany for the Bay Area and eventually stayed for 5 years.

On my return, I started a company that was operating mobile diagnostic equipment, combined with the sale of accessories. Eventually we invested in diagnostic imaging centres and in radiology start-ups in Germany and neighbouring countries. So, we were providing a service and invested in companies that were reimbursed for their products and services!

From 2010, I became more involved in teaching and health research, which led to 5 years as a full-time research professor. Over the last 10 years, I have been involved in more than 35 start-ups and have also been an investor. I would certainly say that this has not been a normal, one-dimensional life!

However, it was in my role of a distinguished lecturer for the Engineering in Medicine and Biology Society (EMBS) of the IEEE [1] that my mindset towards health innovation changed and with that, my motivations and personal goals.

As part of that assignment I was teaching and, indirectly through the teamwork in these lectures, experienced the healthcare systems and problems that existed in Jordan, Turkey, India, Tunisia, Australia, Egypt, and several other countries. What I learned and experienced was that: in many countries, healthcare is a privilege that only few can really afford; frequently, in areas that are disconnected from an urban metropolis, no decent service is available; and finally, the needs for innovation differ from country to country and even among regions in the same country.

Whereas before I was exclusively seeking out unmet clinical needs in Germany or the USA, I now became more interested in trying to identify problem areas in the health delivery process whose resolution would benefit individuals, rather than only look for business opportunities related to health.

In rural India, the question of whether an artificial intelligence and sensor-based homecare device is better than a doctor is irrelevant, if no doctor is available or a personal visit to a doctor is unaffordable. In such health populations, low-cost, portable, and connected devices could make a huge difference. In "developed" nations, the value propositions of health innovations are often not primarily concerned with the needs of the patient and society, but emphasize on whether and how a new device or process fits into existing business models of health service reimbursements.

Exposure to the exponential organization concept (OpenExO—www.openexo.com; Singularity University—www.su.org; Exponential Medicine—https://singularityhub.com/exponential-medicine/) and to a novel purpose-oriented innovation methodology (www.purposelaunchpad.com) had a significant influence on my thinking—and doing!

And the final stimulus came from my time as a full-time university professor teaching biomedical engineering students.

Their ideas for innovations had to be based on previously identified and validated health-related problems and their mission was to translate outputs from the research lab into entrepreneurial ventures or to connect and work with industry.

In this process, I was shocked to see and realize how far degree-based education falls short in equipping students to anticipate and address future needs: the research system and funding in the medtech/healthtech area are almost exclusively focused on incremental innovation.

Future Health requires a new and agile approach to innovation, powered by a set of human skills that are not taught at universities. So one of the intents for this book is to accompany teaching on Healthtech Innovation Generation [2].

In first place, the book is intended as a HOW-TO resource and guide for potential or active Innovators in this space, whether they work in a start-up or are entrepreneurs in a larger organization.

I believe the book's content could also provide helpful guidance for health politicians. And, of course, I would be thrilled if it could stimulate needed change in the education sector.

Today's education system is just like the healthcare segment, unfortunately: very siloed and in need of deep reform, not only to match current needs, but to become agile in its ability to adjust to future developments.

My hope is that the book will stimulate a change in teaching, towards future- and purpose-oriented problem-solving, in generating needs-driven innovation, and, by engaging students more actively in these more effective processes, in bringing about that change in behaviour.

The book will NOT provide deep science, the objective is to stimulate innovation that actually benefits the individual, including as a patient. We will be discussing "purpose" a lot, will be talking about exponential mindsets, as well as the application of exponential or deep technologies in enabling and delivering novel health approaches that not only treat you well when you are sick, but focus maximum attention, effort, and investment in developing tools and processes that act before you actually get sick [3].

In the current approach to healthcare most of the available resources are focused on building up care facilities that diagnose, treat, and help you recover after therapy.

Prevention-based medicine will need to be organized entirely differently.

Its infrastructure will be improved continuously and clinicians will have the best tools available to ensure accurate, precise treatment with the fastest recovery potential and the least side effects.

We know that, through early detection, a majority of all health problems could have been prevented entirely, or at least the severity of a treatment would have been reduced.

Let us assume that only a portion of these negative events could be avoided in the next decades. What would that mean for health infrastructure (e.g. hospitals, hospital beds, clinical staff needs, health-related costs), for health insurance, for primary and secondary care?

The technology developments of the last decade, and especially, **combinations** of different digital technologies, e.g. artificial intelligence, miniaturized sensors, genetic information, gene editing, 3D printing, more, and more autonomous robotics (we call these combinations "convergence") are making rapidly growing sources and volumes of information more accessible and available, very often at very little cost [4–6].

Well, here we have a problem!

Innovation has an invention component and a commercial component.

If these technology-driven innovations with a high potential for disruption (bringing radical change to existing approaches and business models) are not providing a return on investment, why do them?

First, it is clear that there is room for the development of new and attractive business models.

But the existing system needs to be flexible enough to allow these developments to land. The current healthcare systems are far from being agile because existing business models are based on diagnosing and treating a sick person. You go to your primary care physician typically when you are sick (sitting with many other sick patients in the same room does not make you healthier!). When a diagnosis is unclear you are sent to a specialist, perhaps via a radiologist who provides diagnostic imaging to confirm or invalidate certain potential causes.

After that, you receive treatment (exercise, dietary, a minor surgical procedure, pharmaceuticals, physical therapy, . . .) or are sent to a hospital for a more extensive therapy, which could include surgery (hopefully using minimally-invasive approaches). Once you are released from the hospital you may receive the further treatment as outlined initially.

In this system, the hospital infrastructure **needs** sick patients, to sustain itself.

Healthy people provide currently no or little revenue—from the patient, or health insurers, or government health agencies—for healthcare providers

In many countries you can—or are forced to—sign up for health insurance that covers the high costs in case of you get sick and need care, precisely for the same reason that you take up other insurance policies. You do not really expect your house to burn down or your car to be damaged, but in the unlikely event that this happens you are covered. So insurance mitigates the effects of a potential negative event. And health insurance works in the same way: your hope and intention is that you stay healthy and do not need to go to a doctor or a hospital. Healthy people paying into the system subsidize the costs incurred by sick patients. In general this a good and empathetic approach to balancing a portfolio of health risk but does not result in better population health.

This book and the author is NOT against innovation to improve diagnosis, therapy, and treatments. Of course, we should work to continuously improve the quality of healthcare services and the efficiency and effectiveness of health delivery including patients and clinician experiences within that system.

But what is proposed is to put the **patient,** not the delivery system, at the centre of all activity.

The most important goal should be to significantly lower the cost of healthcare services. Not by 3% or 5%, but by 75% or 90%! Only with a dramatic reduction in cost disruptive approaches will be implemented. Buckminster Fuller said *"You never change things by fighting the existing reality. To change something build a new model that makes the existing one obsolete"!*

Unfortunately, the medtech and healthtech industry has no real interest or incentive to do this. Understandably: commercial organizations need to produce a gross margin (revenue minus direct cost of materials and manufacturing) that is sufficiently high to cover total operational expenses for engineering, sales, buildings, infrastructure, and other costs. With a significant cost reduction more systems can potentially be placed, but it is unlikely that this will produce the same gross margin in absolute value.

"*You never change things by fighting the existing reality. To change something, **build a new model that makes the** existing model obsolete.*"

~ R. Buckminster Fuller

Disrupting Health and Healthcare with system change is revolutionary; many would resist changing what is currently in place. But that is always the case when existing stakeholders are concerned that they could lose out. Nor do Investors typically like Disruption because it brings business models that are unknown, novel, and unproven [2, 7–9].

While disruption is needed and should be a goal to work towards, a step approach is recommended: implementing incremental innovations but based on a clearly defined purpose and a vision to which all subscribe. And the purpose of our activities should not only be to create new innovations for billable health services that benefit the health providers and other involved stakeholders, but to do this in ways that are based on what a patient wants and what creates benefit for that patient.

The subtitle of the book *"Entrepreneurial Strategies and Development Concepts towards Patient Empowerment and Health Democratisation"* already indicates what I believe is needed (entrepreneurship) to transform the current system and what the goals of that transformation should be. Both Health Entrepreneurship and Digital Health will be addressed in several chapters.

We will also present concepts and important aspects of planetary and sustainable health plus other determinants of individual health.

Figure 1.1 shows these Health Determinants, spanning the current problem spaces in the 2020s healthcare systems in most western nations and some of the open global challenges.

Noting, from that figure, that Individual Health depends only to 15% on the quality of the healthcare system. What would happen if we spent more attention and money on the 85%?

Future opportunity to innovate will undoubtedly also come from research in advanced understanding of molecular and biological processes [3], and, while the innovation methodologies are also applicable to this science as well, this book will not cover them explicitly (Fig. 1.2).

We are very aware that, for many, the vision of the book may seem far out in the future and that realities do not support all of the innovation ideas presented [8]. However, we believe in the imperative of educating future health innovators about novel, purpose-oriented thinking and innovation generation resulting in ways that allow people to manage their own health [9].

In taking the path of applying exponential technologies to address open health challenges and future needs, we cannot predict WHEN WILL IT HAPPEN and WHAT WILL IT LEAD TO [2]. But the past has shown that disruption quickly turns industries around, creates new players, and eliminates existing organizations that were not willing or prepared to adjust for emerging change.

The hope is that we will reach the goal of HEALTH DEMOCRATIZATION, where everyone on this planet has access to tools, devices, and deep-tech that allow personalized prevention and prediction.

Not every step is easy, but the results give even small teams the power to change the world in a faster and more impactful way than traditional business ever could [10].

The book will provide several individual parts,

- *an overview of Health Innovation Needs*
- *insights into exponential technologies and deep-tech for health applications*
- *an introduction to relevant novel innovation methodologies dedicated or adapted to the health segment*
- *perspectives on future-oriented healthtech education, ethical considerations*
- *entrepreneurial issues, with examples of global innovation setups and*
- *an outlook on digital health, exponential medicine, space health and its transfer back to earth*

Every chapter provides 3–4 points at the beginning on what will be covered and concludes with 5–10 key learning points and take-home messages at the end.

Depending on your previous knowledge, innovation situation, and environment (student, health innovator, start-up entrepreneur, intrapreneur) you may also be

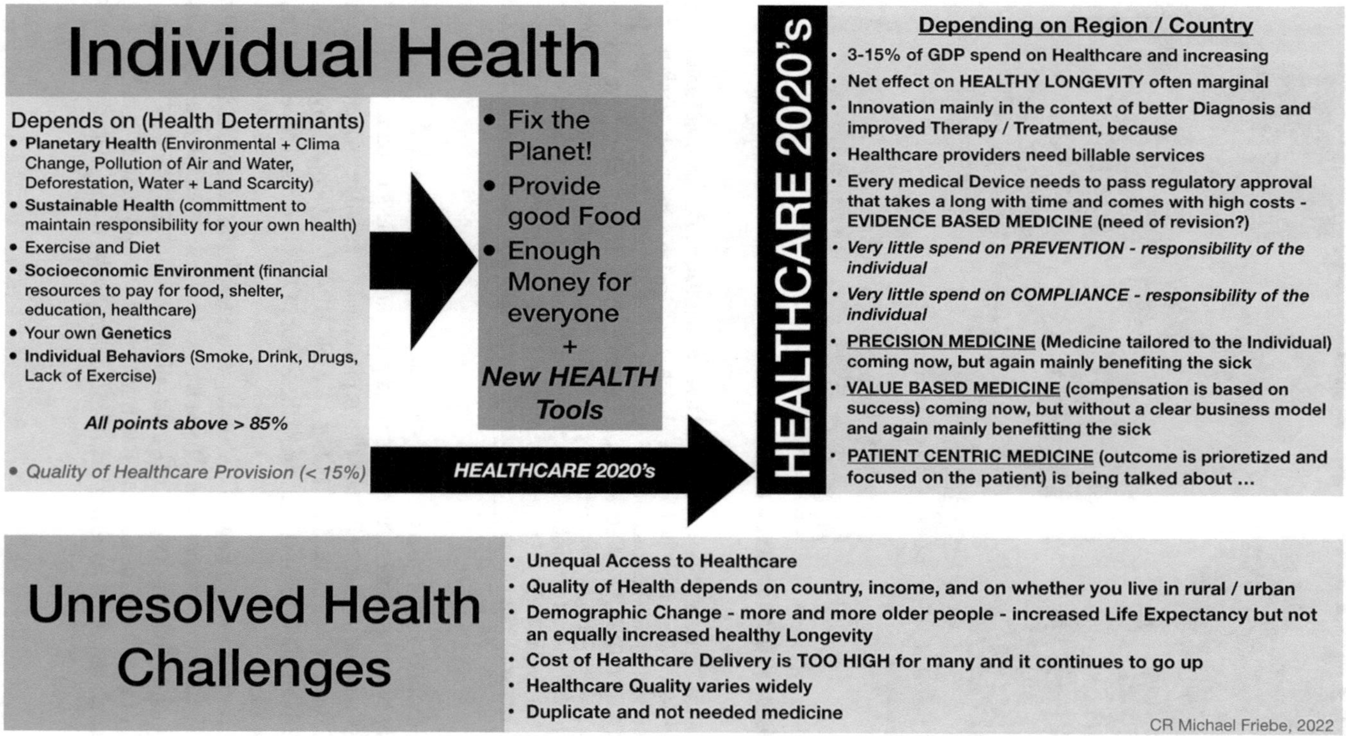

Individual Health

Depends on (Health Determinants)

- Planetary Health (Environmental + Clima Change, Pollution of Air and Water, Deforestation, Water + Land Scarcity)
- Sustainable Health (committment to maintain responsibility for your own health)
- Exercise and Diet
- Socioeconomic Environment (financial resources to pay for food, shelter, education, healthcare)
- Your own Genetics
- Individual Behaviors (Smoke, Drink, Drugs, Lack of Exercise)

All points above > 85%

- *Quality of Healthcare Provision (< 15%)*

- Fix the Planet!
- Provide good Food
- Enough Money for everyone

+ *New HEALTH Tools*

HEALTHCARE 2020's

HEALTHCARE 2020's

Depending on Region / Country

- 3-15% of GDP spend on Healthcare and increasing
- Net effect on HEALTHY LONGEVITY often marginal
- Innovation mainly in the context of better Diagnosis and improved Therapy / Treatment, because
- Healthcare providers need billable services
- Every medical Device needs to pass regulatory approval that takes a long with time and comes with high costs - EVIDENCE BASED MEDICINE (need of revision?)
- *Very little spend on PREVENTION - responsibility of the individual*
- *Very little spend on COMPLIANCE - responsibility of the individual*
- PRECISION MEDICINE (Medicine tailored to the Individual) coming now, but again mainly benefiting the sick
- VALUE BASED MEDICINE (compensation is based on success) coming now, but without a clear business model and again mainly benefitting the sick
- PATIENT CENTRIC MEDICINE (outcome is prioretized and focused on the patient) is being talked about ....

CR Michael Friebe, 2022

Unresolved Health Challenges

- Unequal Access to Healthcare
- Quality of Health depends on country, income, and on whether you live in rural / urban
- Demographic Change - more and more older people - increased Life Expectancy but not an equally increased healthy Longevity
- Cost of Healthcare Delivery is TOO HIGH for many and it continues to go up
- Healthcare Quality varies widely
- Duplicate and not needed medicine

Fig. 1.1 Individual Health depends on several Health Determinants, but only to 15% on the quality of the health system (top left). And the systems of most developed nations come with many issues that future innovation should address (top right). However, many health challenges remain unsolved especially on a global level (bottom)

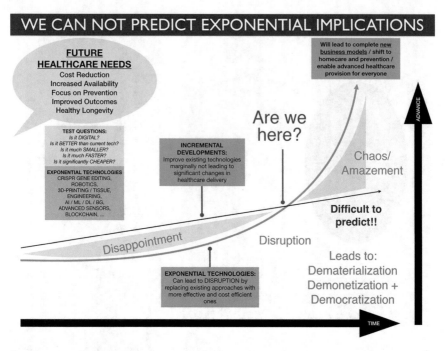

Fig. 1.2 [from 2] Exponential Technologies and their implications on the Future of Health cannot be exactly predicted (as anything in the future), but they will lead to a disruption, which in effect will bring new delivery and business models. The figure shows the 6D's (Peter Diamandis—[10]) applied to FUTURE HEALTHCARE NEEDS. Exponential technologies (that all have to have a digitally base and are digitally scalable), going through a disappointment phase (underperforming based on expectations and predictions), eventually leading to a disruption replacing existing approaches and technologies. Dematerialization (making things smaller), Demonetization (affordability increase) then lead to Democratization (availability for Everyone on this planet) of HEALTH

interested in only certain parts and/or chapters, which also means that there will occasionally be some content repetition from previous chapters or parts.

The main innovation methodology that is presented and used in an adapted health innovation form is based on PURPOSE LAUNCHPAD, developed by Francisco Palao and other collaborator and authors (the editor of this book and several authors are some of them).

These awesome groups have also developed a PURPOSE MANIFESTO with a massive transformative purpose (MTP), beliefs, values, and principles as a possible guiding principle for our actions and innovation work in the future. This is truly inspiring for me and my future work and maybe it is for you as well (see Fig. 1.3).

I very much thank all contributors of this book for their support.

Fig. 1.3 The PURPOSE MANIFESTO with Massive Transformative Purpose, beliefs, values, and principles (www.purposemanifesto.org)

Take-Home Messages

- Planetary Health is one important determinant of individual health and the health of society.
- Socio-economics is another.
- The quality of the current healthcare system has a relatively small influence on individual health (on average!).
- In most developed (apologies for that qualifying statement) countries the healthcare business model is primarily based on billing for medical or related health services—for people who are not well (also known as "sickcare").
- Currently, novel devices and systems are developed mainly for quality improvements and to enable novel and more advanced or more accurate diagnosis and therapies.
- The big challenges of future healthcare (very high cost and therefore not affordable for a large part of the global population, unequal access, changes in demographics, uneven quality, ...) can only be addressed with system changes and rethinking of the entire health paradigm (dramatically lowering costs, increasing homecare, becoming more data-based).
- A novel Health Innovation methodology is proposed as a core part of this book. This is purpose based and oriented towards patient benefit.
- Diagnosis and treatment will continue to be improved, but Medicine needs to implement and be driven by new metrics and new areas of focus: Precision and Personalized I Patient Centric I Value Based.

References

1. Friebe M (2019) What is it like to be an IEEE EMBS distinguished lecturer? https://www.embs. org/pulse/articles/what-is-it-like-to-be-an-ieee-embs-distinguished-lecturer/
2. Friebe M (2020) Healthcare in need of innovation: exponential technology and biomedical entrepreneurship as solution providers (Keynote Paper). Proc. SPIE 11315, Medical Imaging 2020: Image-Guided Procedures, Robotic Interventions, and Modeling, 113150T (16 March 2020). https://doi.org/10.1117/12.2556776
3. Joshi I (2017) Waiting for deep medicine. Lancet 2019(393):1193–1194. https://doi.org/10. 1016/S0140-6736(19)30579-3
4. Dhavan A et al (2015) Current and future challenges in point-of-care technologies: a paradigm-shift in affordable global healthcare with personalized and preventive medicine. IEEE J Transl Eng Health Med 3:2800110
5. Kraft D (2016) The future of healthcare is arriving — 8 exciting areas to watch. https:// singularityhub.com/2016/08/22/exponential-medicine-2016-the-future-of-healthcare-is-com ing-faster-than-you-think/
6. Diamandis P (2016) Disrupting todays healthcare system. http://www.diamandis.com/blog/ disruptingtodays-healthcare-system
7. Christensen C, Bohmer R, Kenagy J (2000) Will disruptive innovations cure health care? Harv Bus Rev Sept–Oct 2000 issue. https://hbr.org/2000/09/will-disruptiveinnovations-cure-health-care
8. Friebe M (2017) International Healthcare Vision 2037. New technologies, educational goals and entrepreneurial challenges. Otto-von-Guericke-Universität, Magdeburg. ISBN: 978-3-944722-59-7
9. Christensen C, Waldeck A, Fogg R (2017) The innovation health care really needs: help people manage their own health. Harv Bus Rev. https://hbr.org/2017/10/the-innovation-health-care-really-needs-help-people-manage-their-own-health?autocomplete=true
10. Diamandis P (2016) The 6 D's. https://www.diamandis.com/blog/the-6ds

Michael Friebe received the B.Sc. degree in electrical engineering, the M.Sc. degree in technology management from Golden Gate University, San Francisco, and the Ph.D. degree in medical physics in Germany. He spent 5 years in San Francisco, as a Research and Design Engineer with an MRI and ultrasound device manufacturer. He is a German citizen with expertise in diagnostic imaging and image-guided therapies, as a/an Founder/Innovator/CEO/Investor and a Scientist. He is also a Research Fellow with the Technical University of Munich, Munich; an Adjunct Professor with the Queensland University of Technology, Brisbane; and a honorary Professor of image-guided therapies with Otto von Guericke University, Magdeburg, Germany. Since 2022, he has been a Professor of biomedical engineering innovation with the AGH University of Science and Technology, Krakow, Poland. He is a listed inventor of more than 80 patents and has authored over 200 papers, and was part of over 35 Medical Technology Start-Ups. He is a Board Member of four medical technology start-up companies and an investment partner of a MedTec-fund. From 2016 to 2018, he was a Distinguished Lecturer of the IEEE EMBC teaching innovation generation and MedTec entrepreneurship. He is also a coach and trainer for OpenExO, and a master Launchpad mentor for the Purpose Alliance.

Health Innovations from an Innovators' Perspective

2

Michael Friebe

Abstract

Innovation is the translation of technologies and inventions to a market. There are different starting and obviously different ending points. And with exponential and converging technologies, huge challenges in health delivery to solve, and the need for a novel approach to health, there are a lot of opportunities for health innovators. This paper will discuss some of the innovation starting points (individual turned entrepreneur, innovator, intrapreneur, and entrepreneur turned intrapreneur), the differences and areas where the health innovation process differs significantly from other industries. And it will present the core principles of the book. Having a purpose is an important starting point for most innovation activities, but even more when you deal with health-related issues and identifying solutions. The economic scaling and growth for health-related innovations is typically more expensive, takes more time, and requires more patience from all involved, especially from the innovators and investors. And with the expected massive changes in all aspects of health (service, delivery, move to prevention, needed cost reduction, ageing population, and many more) you need to have a disruptive/radical vision (10 years) towards reaching your purpose, with intermediate more traditional and incremental intermediaries (2–3 years).

M. Friebe (✉)
AGH University of Science and Technology, Krakow, Poland

Otto-von-Guericke University, Magdeburg, Germany

IDTM GmbH, Recklinghausen, Germany

FOM University of Applied Science, Center for Innovation and Business Development, Essen, Germany
e-mail: info@friebelab.org

Keywords

Health entrepreneur · Health intrapreneur · Health innovation process · Health disruption · Edge business · Core business · Innovation principles · Minimal viable prototype · Massive transformative purpose

> **What will you learn in this chapter?**
> - The organization of the book
> - The core principles of this book with respect to health innovation
> - Who should read it

2.1 Introduction

There are several differences—at least they currently are, maybe changing in the near future—when you compare the translation of inventions/ideas/concepts in the health space with other industries. Health is the ultimate good and important to everyone, health is also a big business, and it deals with humans that could be harmed or even die because of wrong diagnosis, monitoring, and therapeutical treatments.

Our current system of evidence-based medicine and regulatory approval has the goal of maximizing the benefits with ensuring least harm. This approach however is based on the current way of delivering health services and using medical technology products, when you are sick. That will hopefully change!

2.2 Motivation

Your motivation, to read this book and apply the concepts, could be (actually they should be anyway) based on a purpose, but they could also initially be stimulated by your experiences in the health system, because you were working on novel concepts and were thinking about making something bigger out of it, or because you are an insider working for a medical technology/pharmtech/biotech company or are employed in healthcare operations.

Everyone's goal (economically, and with respect to solving open and unmet clinical/health issues) is to create a PAINKILLER for a BURNING PROBLEM. The experience has shown that especially technologically oriented potential innovators are often not checking the problem in depth and immediately apply their technological expertise to build something based on not validated assumptions. In the education and research "industry" (universities) this is typically even worse. Technical departments at universities have a specific technical expertise and clinical departments a specific clinical expertise. Putting these two together will create a solution for the narrow clinical problem identified by the clinicians with the dedicated technological expertise of the engineers.

What is rarely asked is:

- How could this described/identified problem be prevented all together?
- Is there a global need for the solution?
- What are alternative solutions?
- And many more that could lead to solutions outside the current healthcare setup

What we, therefore, propose is to have a global look and dare to have a vision that moves diagnosis and treatment/therapy from the current place of health delivery to a less expensive and more patient-friendly one (e.g. from secondary care to homecare—much cheaper with a higher quality of life).

Also, what is a problem in the USA, for example, could not be a problem at all in other countries or vice versa? Do not limit yourself!

Figure 2.1 shows some of the different starting points. You may have gotten exposed to a health-related problem or issue because you believe you have found a health problem that needs to be urgently solved. It could also be that you have a personal motivation to address health challenges or issues. You find yourself in the HEALTH INNOVATOR/ENTREPRENEUR part on the left of Fig. 2.1. You would dive deep into the problem in a PATIENT/USER EMPATHIZE phase and come up with hypotheses and many ideas that need to be (in)validated. You would run many customer experiments with crude prototypes. That process goes relatively fast, is focused on iterative learning using agile methodologies, and does not cost much. Many project ideas and initiatives do not survive that phase because the problem is not really a problem, the motivation has dropped, or there is not enough support to continue. The next phase requires a team, a good enough project plan and value propositions packaged in a future vision to convince early investors. The goal would be to further validate the ideas and create solutions that allow you to find early "paying" customers/users. "Crossing the chasm" is the title of a book that describes the need and presents concept on how to move from a very small number of "crazy" early users to create an attractive product and offering for the majority customers [1]. This phase requires much more capital, a larger team and many start-ups and projects will not reach that phase. And then you need to scale your operations and maybe even create new markets and additional products.

The difference for an intrapreneur/innovator within an existing, most likely successful medical technology or health delivery organization is that it comes with an established business and customer base. The new idea needs to fit into this existing operation (CORE). The innovation is probably with respect to improving the current offering or by extending the product to new markets or customers. These companies have existing expertise in developing, producing, marketing and will likely follow a typical waterfall approach to translate the ideas to their existing customer base. For many product improvement ideas this is the most reasonable method.

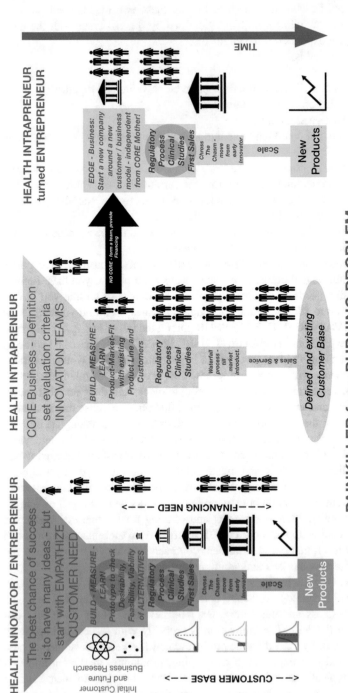

PAINKILLER for a BURNING PROBLEM

based on existing healthcare issues (incremental / traditional innovation); with completely new technology (radical innovation) - based on novel and not yet developed Health Business Models (architectural and disruptive + frugal innovations)

Fig. 2.1 Three possible scenarios of health innovation starting points. An innovator (person/team) that will become an entrepreneur on the left (RED). Many ideas are reduced in agile and iterative learning and validation processes, further reduced in a subsequent start-up phase. At the same time the staff and financing needs to go up to eventually achieve a sustainable operation. You could also come from within an existing organization (INTRAPRENEUR (ORANGE) with a goal to develop a new product or product improvement that fits to the existing offering (CORE business compatibility). The development and translation process is typically following a waterfall principle. And in case the new concepts require a new business model or address a customer that is not being served by the organization (EDGE business) this may then lead to a spin-out with much the same problems and issues as the ENTREPRENEUR

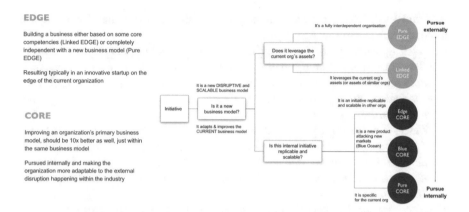

Fig. 2.2 We use the differentiation CORE and EDGE particularly when speaking about innovation that comes from within established companies or organizations. Some of them (EDGE) are not related to the current operation and business model and should either be discarded, sold, or externally pursued in a new entity

And in case that the intrapreneurial activity turns out innovation approaches that are not in line with the existing business model and operational setup of the organization (EDGE) then there is still the option of spinning a company out. Important difference here, this new company needs to act and develop like a real start-up and needs to be detached from the mother organization. If that is not the case, it will act, behave, and argue just like the mother. It could be a little easier for these companies to get started, as they may get initial funding from the mother. We typically distinguish here between LINKED EDGE (disruptive and scalable business model that uses the mother companies assets—e.g. manufacturing capabilities, some engineering, research cooperations) and PURE EDGE (completely interdependent with respect to the mother company). Also the CORE opportunities can be further classified into EDGE-CORE, BLUE CORE, and PURE CORE depending on whether it is unique to the mother company, can be used in other organizations, or is a product that highlights one or few specific features while reducing others. Something that is called VALUE INNOVATION in a BLUE OCEAN strategy (Value Innovation = simultaneous pursuit of differentiation and low cost leading to increased value for both the buyers and the company). See Fig. 2.2 for a short explanation and some criteria.

But all concepts/methods that we present should be applicable for all these scenarios as they are all based on some important core principles.

2.3 Core Principles of the Book

The core principles of the book are presented in somewhat sequential order, but this does not mean that the first ones are the most important ones. All are important! Please check compliance and ensure that you revisit them regularly. The bold and underlined word combinations may help in the future to remember. They are also listed again in the following chapters.

- *Use* of *agile and iterative approaches to validate hypotheses and assumptions* in all aspects—means that if you use a tool that is presented, check and (in)validate, apply the learnings, change or adapt your plan and concepts, and define new experiments to help you advance. Create feedback loops with your team, customers/users, partners.
- *Whenever possible **think 10×** (how can we improve, reduce, increase by a factor of 10 or more). With this approach you are forced to rethink existing solutions. If your goal is to improve, reduce, increase something by 10%, you will automatically look at which parameter can be tweaked to achieve that and may never discover impactful concepts.*
- ***Experimentation is key***. *With experimentation we do not mean a science experiment (even though that could also be valuable especially for checking the technical feasibility), but rather a customer or user related verification one. Do not do experiments that you already know the results off and not experiments that will not change anything on whether they are validated or not.*
- ***Put yourself often in the position of the core user/patient and try to see the situation from their point of view***. *Of course that needs to be validated in experiments. But it is the starting point of virtually all innovation methodologies presented in this book (**EMPATHIZE** in Design Thinking, IDENTIFY in BIODESIGN and I3-EME, CUSTOMER segment in Purpose Launchpad Health, MEASURE in LEAN STARTUP)* [2–8].
- ***Do not start building before*** *you have gone through some initial steps of **defining and evaluating the problem that you are trying to address** (What is the real problem? Who are the customers? Who are the decision makers? What will happen with the problem solution with the introduction of novel technologies?)*
- *Good questions are most important—**if you ask the right questions you will find the answers!** This is also the base for the insightful customer interviews. Do not seriously (you can though for opening the conversation) ask questions for which you already know (not assume) the answers and do not seriously ask questions (again, you can for conversational purposes) that do not change anything.*
- ***Embrace failure or invalidating ideas and with that focus on LEARNING***. *Every time you falsify a hypotheses you should be happy.*
- ***Write everything down for a learning history and to be able to go back*** *at a later point to revisit and recheck. Things change and what was considered not a feasible idea could be 3 or 5 years later with changed environmental conditions, different foci, and advanced technologies.*
- ***Come up with many solution ideas—do not limit yourself to the first one that you fell in love with.*** *Always ask and find out for yourself, your team, your supervisor, your professor on whether an idea actually solves a patients' or health stakeholders' problems and on whether it can create an impact.*
- ***Use Minimal Viable Prototypes (MVP) for validation checks, customer experiments,*** ... *The purpose of these MVP is to find out what the customer values the most or likes the best. They do not have to be functional, but they*

should show and allow to understand a particular feature or value proposition. They could be physical (paper/carton or 3D printed models) or picture based (cartoons or drawings, movie animations, product and software simulations).

- *Prototyping (MVP and functional prototypes) is the process of building quick, cheap, and rough study models. The goal is to **evaluate alternative solutions based on their DESIRABILITY** (Is there a verifiable customer need?), **FEASIBILITY** (Can we build solutions that will actually be working and satisfying these needs?), **and VIABILITY** (Can we do that within an existing business model? . . . or do we need a new one?).*
- ***Have a** guiding (**massive transformative**) **purpose** for the company/organizations' work combined with a **longer-term vision, a shorter term mission and** a definition of the organizations' **core values** is an essential motivator.*

2.4   What Are the Book's Limitations?

While we believe that entrepreneurial activities will probably be most impactful with respect to transforming the healthcare setup, we will not be able to cover the aspect of actually starting a company and with that to discuss financing and other project and company management options that are important for that. We will provide some sub-chapters that cover individual elements however.

Steve Blank described a *start-up as a temporary setup in search of a sustainable business model.* In the Health space that typically requires a regulatory approved and certified product.

Depending on the product, its intended use (e.g. EASY: health style product measuring certain physiological parameters; COMPLICATED: sensor based Digital Health product/software for diagnosis and monitoring; VERY COMPLICATED: anything that is placed for a longer period of time inside the human body for diagnosis/monitoring/therapy) this can take several years, and will cost many millions (USD/€). What we want to provide you with is to define such a business model and get enough insights to make you somewhat confident, excited, and motivated to take that step. We also want to provide you a toolset that allows you to analyse and summarize the needs and the opportunities with your approach, formulate a future vision, and with that have a good base to communicate with potential investors.

And, we believe that the current education with respect to the above-mentioned core principles and health innovation needs requires a reset.

Take-Home Messages
- The intention to create a PAINKILLER for a BURNING PROBLEM is a key motivation for Health Innovation activities.

(continued)

- Unfortunately the PAINKILLER that is developed is too often not solving a large enough problem/need.
- Agile and iterative methodologies are key to finding and defining the problem.
- $10\times$ thinking is a core principle of disruptive and radical innovation.
- Entrepreneur/Intrapreneur (CORE and EDGE) needs will be covered in the book.
- Start-up financing and presentations or other important aspects beyond the Exploration and Evaluation/Validation phase will not be covered in depth.

References

1. Moore C, McKenna R (2006) Crossing the chasm: marketing and selling high-tech products to mainstream customers. ISBN-13: 978-0062353948
2. d.school, Hasso Plattner Institute of Design at Stanford: An Introduction to Design Thinking Process Guide (2013). https://web.stanford.edu/~mshanks/MichaelShanks/files/509554.pdf. Retrieved 3 March 2022
3. Yock P, Zenios S, Makower J (2015) Biodesign: the process of innovating medical technologies, 2nd edn. Cambridge University Press, Cambridge. ISBN-13: 978-1107087354
4. Fritzsche H, Boese A, Friebe M (2021) From 'bench to bedside and back': rethinking MedTec innovation and technology transfer through a dedicated Makerlab. J Health Des 6(2):382–390
5. Palao F (2021) Purpose Launchpad Guide V 2.0. https://www.purposelaunchpad.com/methodology. Retrieved 10 March 2022. V 1.0 also available at https://www.scribd.com/document/486477523/Purpose-Launchpad. Retrieved 10 March 2022
6. Friebe M, Fritzsche H, Morbach O, Heryan K (2022) The PLH – Purpose Launchpad Health – Meta-methodology to explore problems and evaluate solutions for biomedical engineering impact creation. IEEE EMBC Conference Paper, July 2022 (not published yet)
7. Ries E (2011) The lean startup: how today's entrepreneurs use continuous innovation to create radically successful businesses. ISBN-13: 978-0307887894
8. Blank S, Dorf B (2020) The startup owner's manual: the step-by-step guide for building a great company. Wiley. ISBN-13: 978-1119690689

Michael Friebe received the B.Sc. degree in electrical engineering, the M.Sc. degree in technology management from Golden Gate University, San Francisco, and the Ph.D. degree in medical physics in Germany. He spent 5 years in San Francisco, as a Research and Design Engineer with an MRI and ultrasound device manufacturer. He is a German citizen with expertise in diagnostic imaging and image-guided therapies, as a/an Founder/Innovator/CEO/Investor and a Scientist. He is also a Research Fellow with the Technical University of Munich, Munich; an Adjunct Professor with the Queensland University of Technology, Brisbane; and a honorary Professor of image-guided therapies with Otto von Guericke University, Magdeburg, Germany. Since 2022, he has been a Professor of biomedical engineering innovation with the AGH University of Science and Technology, Krakow, Poland. He is a listed inventor of more than 80 patents and has authored over 200 papers, and was part of over 35 Medical Technology Start-Ups. He is a Board Member of four medical technology start-up companies and an investment partner of a MedTec-fund. From 2016 to 2018, he was a Distinguished Lecturer of the IEEE EMBC teaching innovation generation and MedTec entrepreneurship. He is also a coach and trainer for OpenExO, and a master Launchpad mentor for the Purpose Alliance.

From SICKCARE to HEALTHCARE to HEALTH

Michael Friebe

Abstract

There are significant challenges in global healthcare. Some countries have abundant services, but are stuck with a rather nimble and expensive system that focuses on incremental innovations. Other geographies are still in need of basic tools, infrastructure and require completely different, inexpensive, and with that more disruptive solutions to satisfy their healthcare needs.

Next-Generation Healthcare systems will need to focus on prevention/early detection and pro-active therapy will employ exponential technologies. This likely will lead to significant changes in the way we experience, think about, and deliver healthcare. A digitally empowered patient will play a much more important role.

This article will help to set the stage for a more methodological approach to health innovation that is not based on incremental thinking, but rather on defining a future-oriented purpose with a goal to eliminate—or at least significantly reduce—global health delivery inequalities by developing intelligent, data based, inexpensive, and portable devices and tools plus embed them into different care and monitoring environments (e.g., Home instead of Practice/Hospital). Important in that context is to also rethink the current health business models that are based on reimbursing and paying for services of patients rather than on preventing people from becoming patients.

M. Friebe (✉)
AGH University of Science and Technology, Krakow, Poland

Otto-von-Guericke University, Magdeburg, Germany

IDTM GmbH, Recklinghausen, Germany

FOM University of Applied Science, Center for Innovation and Business Development, Essen, Germany
e-mail: info@friebelab.org

Keywords

Healthcare innovation · Exponential technologies · Prevention · Health disruption · Reverse innovation · Future healthcare

What will you learn in this chapter?
- The need to change the way we deal with HEALTH Innovation
- That Health is too expensive, in-equal, inaccessible for many and that we need to rethink device/tool/process developments based on their impact and purpose—it is up for Disruption!
- We should work toward a system that helps the individual to optimize their own health rather than to focus on income generation for the health providers

3.1 SICKCARE

In the coming years/decades, we will hopefully experience a shift from the current SICK-CARE provision (you mainly go to see a doctor/hospital in case of a health problem; a doctor/hospital/pharmacy is only getting paid when you are sick and need to be diagnosed, a therapy ordered or otherwise treated) to a system that embraces technologies and activities that help you maintain to stay healthy.

Education and training of Health Innovators needs to be adjusted to these developments focussing on solving unmet clinical needs of the patient, which requires a solid understanding of current health problems (regional, global), future technologies, economic realities, and global health markets. At the same time the education and focus of the clinical professions needs to be adapted. Not the diagnosis and therapy needs to be in the focus, but prevention, prediction, and personalized medicine. We need to see health innovations designed by multidisciplinary groups with the purpose of equalizing access to health tools and focussing on healthy longevity for everyone.

The paper will present and discuss some of these future global innovation needs and the subsequent need for a change in teaching health innovation generation that needs to include learning—(creativity, critical thinking, collaboration) and life skills (flexibility, leadership, social), as well as basic economic and entrepreneurial training—all of which are currently not taught or emphasized.

3.2   With Disruption Toward Patient-Centered HEALTHCARE

The World Health Organization (WHO) explains that "health innovation" improves the efficiency, effectiveness, quality, sustainability, safety, and/or affordability of healthcare. This definition also includes "new or improved" health policies, practices, systems, products and technologies, services, and delivery methods that result in improved healthcare [1, 2].

If you are actively involved in Healthcare delivery or in developing innovations in and for that segment, you continuously hear about the upcoming and necessary changes that will come in the next years. Phrases like "disruption is needed," "healthcare is sickcare," "rude staff and high cost," "empathy is missing in healthcare," "uneven availability," and many more are frequently used to describe the current system of providing health treatments. And most of the ones listed are correctly describing the situation in the developed world.

Health Delivery Systems have wrong incentives, are very expensive, and come with an unequal level of access. All these are indicators for an industry that needs disruption [1, 2].

Different problems exists in the less developed world however, where proper healthcare services are often not available, unaffordable for a large part of the population, and where diagnostic and treatment quality is not up to the standards that we are used to living in the richer countries. But that may actually be advantageous in the coming years, because these countries may be able to adapt health improvements based on disruptive low-cost devices much faster and with a higher willingness.

We are in the fourth industrial revolution with the evolution and application of artificial intelligence/machine learning/big data in virtually every segment, widespread use of robotics, additive manufacturing, and many more dubbed as Exponential Technologies, because of their digital—exponential—base and other attributes that can lead to more democratic access for anyone. Healthcare has been very rigid in adopting these technologies and is still using the fax machine and handwritten reports as the standard of communication in many places.

One reason for that has been the application of scientific methods for healthcare decision-making—evidence-based medicine—that requires extensive clinical trials with statistic evaluations for regulatory approval and subsequent consideration as standard of care. This is a very expensive and time-consuming approach. But this gold standard is also an average approach that does not always consider the individual health issues effectively and in certain cases also leads to detrimental effects.

The presented case study is a very typical example in the current healthcare system of many nations and many may say that this is OK. The responsibility for change is transferred to the patient. In case of another undiagnosed problem or non-compliance by the patient that could however easily lead to a chronic condition that is significantly more difficult to treat and massively more expensive.

Michael, 57 years, Researcher and Serial Entrepreneur
Michael regularly goes to a health-check with his primary care physician (Internal Medicine). He is physically very fit, but has a high stress level. During the recent visit he complained about increased tiredness. The check-up does not reveal any problems with the exception of a slightly increased thyroid. Blood and urine tests show values slightly outside a normal range for several liver and thyroid markers. The doctor orders an additional thyroid scan (turns out negative), a change in diet, and a new blood test in 6 month.

It could also be that the values that are considered normal for the general population are very high or low for Michael. They do not allow a personalized look. Other problems are the diagnostic "noise," the fact that different doctors would draw different conclusions and that one and the same would decide differently based on changed environmental or personal circumstances (e.g., family issues, soccer team lost, does not feel well, was at a party last night, and drank too much, . . .) [3, 4].

Doctors are trained for many years to become experts in their particular field. They see patients that have symptoms that they are trained to diagnose and to provide recommendations on a treatment/therapy. They get only involved when the patient already has the symptoms.

The hospitals provide excellent care for patients that need to undergo a more extensive, often invasive, therapy. After successful treatment, possibly even before, they may have to take medication and/or go to a physical therapy to help with the recovery.

Current healthcare delivery—through medical practices, pharmacies, medical technology industry, hospitals, insurers, and others—and the current health business model is dependent on a sick person being diagnosed and subsequently treated. That is the reason why many call the current system actually SICK-CARE provision. You pay for a health insurance to cover expenses when you are sick and the providers get reimbursed for treatments, but keeping you healthy is not incentivized properly.

So here for the first time the question to you:

Do you not believe that the convergence of technologies that are either already available or that may soon find its way into healthcare will provide novel ways of employing health sensors and machine learning to provide personalized health assessments that can be used to identify changes in the health status before any symptoms can be observed?

Or maybe another example of a typical scenario:

It is winter season and you have a cough, your nose runs, and you have a slightly increased temperature. You already got a negative CV-19 (or other virus test in the future) test via a drive through test center. Normal procedure would be to either do nothing and wait or to go see a doctor in your condition, to wait for an hour in a packed waiting room with many sick people. Then to eventually see a doctor that likely spends very little time with you to quickly diagnose a flue. You leave the practice with a prescription and go the next pharmacy to pick up the cold medicine. This is not a scenario 50 years ago. This is 21st century Germany and probably reality in many other countries as well.

What about a novel telehealth system at home with advanced sensors measuring temperature, evaluating face expressions, analyzing coughing sounds, ... combined with machine learning algorithms that compare an assessed health status with a previous individual base. The evaluated data is then send as a result file with a confidence statement (e.g., 95% flue) to the primary care physician, who evaluates, has the option to video connect to the patient, issues an electronic prescription that is automatically send to the closest pharmacy with the medication delivered to the home. Science-Fiction? No, the technology exist already ... everything could already be implemented! But why is it not?

And why actually take the step via the doctor?
What is stopping or hindering the implementation of such tools?
Several reasons why western healthcare systems are reluctant to change [1, 5]:

- The fee per service model in Healthcare has just lead to offering more services.
- An example like the one above would make the diagnosis more affordable, but this is not in the interest of the stakeholders. The oversight and final decision by the doctor is used as an argument, just as the mistrust in technology. The reluctance to share medical data is another reason for the implementation delays.
- A majority of health-related Investments by venture capitalists (>90%) are spend on devices, tools, pharmaceutical developments that make healthcare more expensive.
- More competition in healthcare delivery and provision leads to more expenses not to less because the contenders want to offer better and more advanced services than their competitor.
- Value-based healthcare needs metrics that have not been properly defined yet.

Maybe the best chances are in nations with less developed health systems that have issues with availability and access to health services, and where western healthcare is not affordable for most people. And maybe these are also the countries which health innovators should address, something we call reverse innovation. Introduce novel, portable, intelligent, data-driven, and inexpensive systems in countries with a very high need. With a positive outcome and high patient acceptance and use it would then be difficult to argue against use in Germany, Europe, Japan, or the USA.

But one big problem remains.

If there are no business models that could create a decent return on investment for investors, these disruptive ideas will not be supported—at least by the majority of investors. Disruption, also frugal disruptions in healthtech (e.g., reducing the system cost for diagnosis or therapy by a factor of 10 or 100 while maintaining the needed functionality), comes with novel business models that are not existing yet and that cannot even be defined yet.

Difficult to convince someone to invest in a novel idea for which the revenue stream cannot be described yet. Politics will likely have to play a more important role here to make FUTURE HEALTHCARE a reality in the near future!

3.3   From HEALTHCARE to HEALTH

We will need to move toward a real HEALTHCARE system that keeps you healthy and in case you are actually sick diagnoses and treats you (Patient Centric Healthcare) customized and individually tailored (Precision + Personalized Medicine). Reimbursement of these HEALTH "maintenance" services should then be based on actual outcome and patient experience (Value Based Healthcare) and no longer merely on services that were rendered.

See Fig. 3.1 for some issues with a move from SICK-CARE to HEALTHCARE and on to personal HEALTH comparing future directions with the current setup [6, 9–12].

A lot of healthcare professions, regulators, payers, and politicians have opinions on how to change and what to do, but the healthtech/medtech innovator is rarely involved in the discussion.

The healthtech innovator—defined for this paper as a technical domain expert working for healthcare improvements—is often considered as the one that can

CURRENT SICK-CARE SETUP	FUTURE HEALTH DELIVERY
Treating SICKNESS	Focussing on PREVENTION
Procedure based REIMBURSEMENT (no procedure, no reimbursement!)	Value (Outcome) based REIMBURSEMENT
Evidence based Medicine with a one-size fits all approach	PERSONALIZED MEDICINE
Lots of INVASIVE Therapies	Focus on MINIMAL-INVASIVE THERAPIES
Treatment / Healthcare provision is CENTERED around the PROVIDER	PATIENT CENTRIC MEDICINE

PRECISION + PERSONALIZED MEDICINE

Customization of healthcare with medical decisions, treatments, practices or products being tailored to the individual patient

VALUE BASED HEALTHCARE

Healthcare providers are compensated for the health and well-being of their patient population rather than for services rendered

PATIENT CENTRIC HEALTHCARE

 Prioretize patient outcomes and meet consumer expectations for on-demand care delivered on THEIR terms.

Adapted from: Dhavan AR et al (2015) Current and Future Challenges in Point-of-Care Technologies: A Paradigm-Shift in Affordable Global Healthcare with Personalized and Preventive Medicine. IEEE Journal of Translational Engineering in Health and Medicine 2015 Mar 5;3:2800110. doi: 10.1109/JTEHM.2015.2400019; Kumar S. (2018). Patient-Centric Approach. https://www.businesspoc.com/patient-centric-approach/; Drumond, N. J Pharm Innov (2019). https://doi.org/10.1007/s12247-019-09407-2

Fig. 3.1 The table on the left shows the current setup of healthcare provision (SICK-CARE) and future developments. On the right you find some definitions of the PATIENT CENTRIC, PERSONALIZED, and VALUE BASED HEALTHCARE. From [6–8]

research and develop in an area that is considered by others as a future innovation segment.

Reasons are that many of these innovators and future health entrepreneurs lack the non-technical skills needed to determine and shape the future of HEALTHCARE and personal HEALTH. Another reason is that the current system of healthcare provision is setup in disconnected (data/communication) silos that are only reluctantly communicate with each other and share information, and that also are very rigid with respect to changing provision routes and institutional responsibilities (Outpatient/Inpatient/Regulators/Insurers/Pharmaceutical- and Medtec providers/ Health Economy/Politics).

3.4 Conclusion

The main intent of this paper is to highlight the need for, and subsequently present an innovation methodology to design future-oriented twenty-first century healthtech systems that are based on discovering and evaluating the actual needs of the individuals rather than basing it on the needs of the healthcare providers. And a methodology that is

- purpose driven, that should
- benefit the individual, should try to
- resolve some of the big issues in global healthcare, namely inequalities, inaccessibilities, focus on prevention rather than treatment, healthy longevity, but it should also help to
- reduce health-associated costs and the focus on generating revenue through billable services

This does not only involve a mind shift change with respect to how we approach health innovation, but it also requires novel education approaches involving multidisciplinary teams, rethinking the health-related financing and investments, motivations to adapt through changes in health politics, and a push to a more sustainable health for everyone.

Sustainable health can only develop if environmental conditions (e.g., good air and water quality, sufficient supplies of healthy and balanced food, education opportunities), economic stability (personal income, balanced society) are matched with health that treats you when you are sick, but that has the main goal to prevent you from getting sick by employing predictive and participative tools and delivers high-quality care without damaging the environment. It needs to focus on creating health at a population level (ONE HEALTH), while also being conscious about the enormous quantities of energy, waste (toxic and non-toxic), materials, water, and chemicals it uses or produces.

All of these ideas, wishes, and demands need to be embedded in a health business model that embraces and incentivizes initiatives, products, and developments based on new health metrics.

Let me close with a poem from Dorothy Oger (created March 30, 2022 during a session on the FUTURE of HEALTH as part of the SYNTROPIC WORLD CONFERENCE with Michael Friebe and Michael Smits as speakers):

My model for living long and healthy
Is to not rely only on healthcare
As it stands today,
Caring for people to care about life
Caring for nature, for our collectives
using our common sense,
Our compassion, our sense of well-being,
Listening to the body
Learning from our body
It is healthier than
The healthcare system we have today.

Take-Home Messages
- Healthcare—in the western world—is getting more expensive while inequalities widen and access gets worse.
- Healthcare has been rather immune against disruption because stakeholders are incentivized offering billable services.
- Disruption—using exponential technologies like machine learning, 3D printing, advanced sensors, genetic engineering, robotics—could lead to much more affordable and at the same time to a health provision that addresses the needs of the individual rather than the needs of the health providers.
- Business models for changed health value propositions (patient centered, personalized, prevention focused, more homecare, and less professional practice based) do not exist in reality yet.
- Evidence-based medicine and the required gold standard testing is not in line with novel exponential technology-based developments.
- Healthcare investors spend the majority of their money on making medicine more expensive.
- More intense competition in health provision is—in contrast to most other industries—leading to an increase in cost.
- A possible solution could be Reverse Innovation—introducing disruptive and patient-centered concepts in countries with a higher need and lower barriers of entry and return them to the established healthcare systems after positive results and high patient acceptance.
- Government and health politics need to play a more prominent role in stimulating disruption.
- Health innovations can only lead to a more sustainable health, if it is accompanied with economic and environmental improvements.
- A new health business model with new metrics is needed.

References

1. Kimble L, Massoud R (2016) WHAT DO WE MEAN BY INNOVATION IN HEALTHCARE? EMJ Innov 1(1):89–91
2. Christensen C, Waldeck A, Fogg R (2017) How disruptive innovation can finally revolutionize healthcare. INNOSIGHT/CHRISTENSEN Institute. https://www.innosight.com/wp-content/uploads/2017/05/How-Disruption-Can-Finally-Revolutionize-Healthcare-final.pdf. Retrieved 28 February 2022
3. Christensen C, Bohmer R, Kenagy J (2000) Will disruptive innovations cure health care? HARV BUS REV, Sept–Oct issue. https://hbr.org/2000/09/will-disruptiveinnovations-cure-health-care
4. Fierro J et al (2004) Diagnostic agreement in the evaluation of image-guided breast core needle biopsies. Am J Surg Pathol 28:126
5. Shakurada S et al (2012) Inter-rater agreement in the assessment of abnormal chest X-ray findings for tuberculosis between two Asian countries. BMC Infect Dis 12:article31
6. Christensen C, Waldeck A, Fogg R (2017) The innovation health care really needs: help people manage their own health. Harv Bus Rev. https://hbr.org/2017/10/the-innovation-health-care-really-needs-help-people-manage-their-own-health?autocomplete=true
7. Diamandis P (2016) Disrupting todays healthcare system. http://www.diamandis.com/blog/disruptingtodays-healthcare-system
8. Friebe M (2020) Healthcare in need of innovation: exponential technology and biomedical entrepreneurship as solution providers. Proc. SPIE 11315, Medical Imaging 2020: Image-Guided Procedures, Robotic Interventions, and Modeling, 113150T. https://doi.org/10.1117/12.2556776
9. Schroeder S (2007) We can do better — Improving the health of the American people. N Engl J Med 2007(357):1221–1228
10. Dhavan A et al (2015) Current and future challenges in point-of-care technologies: a paradigm-shift in affordable global healthcare with personalized and preventive medicine. IEEE J Transnatl Eng Health Med 3:2800110
11. Kraft D (2016) The future of healthcare is arriving — 8 exciting areas to watch. https://singularityhub.com/2016/08/22/exponential-medicine-2016-the-future-of-healthcare-is-coming-faster-than-you-think/
12. Friebe M, International Healthcare Vision 2037 (2017) New technologies, educational goals and entrepreneurial challenges. Otto-von-Guericke-Universität, Magdeburg. ISBN: 978-3-944722-59-7

Michael Friebe received the B.Sc. degree in electrical engineering, the M.Sc. degree in technology management from Golden Gate University, San Francisco, and the Ph.D. degree in medical physics in Germany. He spent 5 years in San Francisco, as a Research and Design Engineer with an MRI and ultrasound device manufacturer. He is a German citizen with expertise in diagnostic imaging and image-guided therapies, as a/an Founder/Innovator/CEO/Investor and a Scientist. He is also a Research Fellow with the Technical University of Munich, Munich; an Adjunct Professor with the Queensland University of Technology, Brisbane; and a honorary Professor of image-guided therapies with Otto von Guericke University, Magdeburg, Germany. Since 2022, he has been a Professor of biomedical engineering innovation with the AGH University of Science and Technology, Krakow, Poland. He is a listed inventor of more than 80 patents and has authored over 200 papers, and was part of over 35 Medical Technology Start-Ups. He is a Board Member of four medical technology start-up companies and an investment partner of a MedTec-fund. From 2016 to 2018, he was a Distinguished Lecturer of the IEEE EMBC teaching innovation generation and MedTec entrepreneurship. He is also a coach and trainer for OpenExO, and a master Launchpad mentor for the Purpose Alliance.

Future Look on Health: Opportunities

4

Michael Friebe

Abstract

A future can be painted, can be wished for, can be reasoned, but best is to actively work toward a future that benefits the individual and the society. It is a good practice—and we do it for this article as well—to analyze the current situation and formulate a future goal.

Healthcare is complicated and many stakeholders are involved that have—also not unexpected—different goals with respect to what future for health delivery should look like.

It is not the intention of this paper to go into all the details and present all the reasons why we are doing, what we are doing, and why there are many difficulties to overcome, but to have and present an optimistic view.

Knowing all the different interest helps you to empathize however. Analyzing the healthcare providers gives you valuable information on current value propositions for innovation, and by looking at technological possibilities and general developments combined with some future predictions will allow you to look a little into the future.

This paper will present some of the issues and some ideas and methodologies that innovators and stakeholders are considering and using.

M. Friebe (✉)
AGH University of Science and Technology, Krakow, Poland

Otto-von-Guericke University, Magdeburg, Germany

IDTM GmbH, Recklinghausen, Germany

FOM University of Applied Science, Center for Innovation and Business Development, Essen, Germany
e-mail: info@friebelab.org

Keywords

Future of health · Sci-Fi stories · Storytelling · Forecasting · Need-based analysis ·
Healthcare issues · Healthcare challenges · Exponential medicine

What will you learn in this chapter?
- Healthcare is a big business
- The stakeholders that pay (Government, Employers, Individuals/Society) do not get what they want and do not determine the future of health … at the moment!
- Future Health Innovations will be based on deep technologies that will eventually lead to a later transformation from a home care health to a professional care health
- Areas that require the attention of health innovators for a patient-centered medicine

4.1   Introduction

Let us start with a short clarification on some terminology. "Healthcare" includes everything that deals with making sure you are diagnosed and treated when you are sick. "Healthcare delivery" describes the processes and the directly involved parties. With the term "Health" I want to refer to the individuals' expectations to stay healthy and maintain that as long as possible (also referred to as healthy longevity). This, of course, requires actions and input from the individual itself like having a healthy diet, regular exercise, avoiding drugs, monitoring their own health, and several more. This also requires that the environment that the individual lives in is "healthy." Unfortunately we are doing a pretty good job at the moment of self-destroying the world that we live in. While the planetary health is not in the focus of this book, it needs to be in the focus when analyzing individual health issues (Fig. 4.1).

But even then you may have an accident or you may get sick or you need to be worked on, body parts will need to be "reworked" or "replaced/exchanged" or you may need clinical support for the final days of your life. You then want and expect an accurate and quick diagnosis that is subsequently followed up by the best possible therapeutical intervention and an effective post-therapy treatment. And you expect that these health services are centered around your needs as patient.

Fig. 4.1 Painting a future vision of what Health could look like or should be capable of is not new. Already in 1926, Fritz Kahn showed a vision of a Doctor of the Future with many patient monitoring tools sent to an observing physician [1]

4.2   Healthcare Is a Big Business

Healthcare is a huge business! In the USA alone the Healthcare sector takes a 18% share of the Gross Domestic Product. That is roughly 4 Trillion USD (that is 4.000 Billion USD!) growing with single digits every year. More and more people cannot afford to pay the insurance premiums, while the ones that can afford it get the best possible treatment. But let's face it, these are still poor people, because they are sick.

Now other countries may officially only spend 3% of their GDP on healthcare, but may not have health insurances and possibly even have larger issues with access and affordability, while at the same time do not have the top equipment needed to provide top diagnosis and top therapies.

Problem in the USA and similar setups? Incredible high cost with increasing competition. Competition in healthcare has not led—like in other industries—to a lower price point. On the contrary, it has led to an increase, because health services are now marketed with having the latest and most advanced equipment, the best doctors, the most luxurious hospital rooms. And individuals want to go to the best place for their health.

Problem in the other countries? Not having enough health offerings that individuals can afford and that provide a good quality.

Shocking however that despite a great healthcare system more than 250,000 people die in the USA alone due to medical errors, which makes it the third leading cause of death.

Most of these medical errors are not caused by bad doctors, they rather represent systemic problems, like poorly coordinated care, fragmented insurance networks, the absence or underuse of safety nets, and other protocols, in addition to unwarranted variation in physician practice patterns that lack accountability [2]. This statement also provided solution ideas already, but none have been implemented so far.

Why? Because there are no incentives to invest in solutions and there is no push yet from political authorities. Also this sentence provides some solution ideas, shift funds and implement guidelines and laws protecting the patient more.

4.3   Healthcare Stakeholders

Let us list some of the stakeholders and their interests (we use the USA and some European Countries as base—may look different in other geographies). This is important to empathize with their reasoning and to provide good arguments and value propositions for your idea/product (empathize with them):

- Hospital: provide excellent (cost-effective) service to sick people and bill for the service provision (a good value proposition would be to provide an expensive process/device at a cheaper price with identical product features, or to add product features—quality improvements, time savings, ...—that can be marketed by the hospital for a similar price).
- Primary care facility: see patients (sick people), diagnose quick, suggest and/or perform a treatment and bill for the service provision (a good value proposition here would be to provide a process/device that allows additional billable diagnosis/treatments).
- Secondary and tertiary care facility: work with primary care facilities to offer (to sick people) highly specialized diagnostic and therapy services and bill for the service provision (Value propositions see above).
- After-care services (Rehabilitation, Physiotherapy): Provide billable services to a diagnosed patient or as a follow-up on an invasive therapy. They are interested in products that allow them to offer highly effective solutions at the lowest possible cost or products that can be uniquely marketed.
- Health Insurance Companies: offer an attractive package for an attractive price to the insured, negotiate with the aforementioned providers good financial terms (a good value proposition would be to identify alternative solutions for expensive therapies, to avoid duplicate diagnosis and treatments, to be able to manage data points, to ensure compliance).
- Pharmaceutical Companies: provide billable medications that treat identified diseases/illnesses or reduce negative effects of that disease/illness. Their goal is to develop—in the future personalized medication—with highest possible treatment success and lowest side effects.
- Medical Technology Providers (established ones): provide improved equipment and solutions that are needed by the first 4 to bill for health-related services and to help distinguish one clinical offering from another. The developed and offered

systems need to be paid for by the health providers and are purchased because they are needed to reduce the cost base or increase the revenue side.

- Start-Up Medical Technology/Health Technology Innovators: Identify an Unmet Clinical Need or Improvement opportunity that leads to improved healthcare by offering novel devices/processes/informations that lead to future revenue for the company. To get there they need to convince investors to fund the development, which they will only do with a high return on investment potential. So it needs to meet the above-mentioned value propositions.
- Health Innovation Investors: Want to invest in novel solutions that come with a business model and relatively low risk of failure, which means it needs to fit into the existing modus operandi.
- Governments (local/regional/federal): Want to ensure that everything needed to guarantee a comprehensive healthcare delivery is present with highest quality. They are also interested (or should be) that healthcare is available for everyone without major waiting time or other restrictions. In many countries healthcare for elderly or individuals with no or little income has to be paid for by the local/regional communities. With that they should be interested in lower cost. These government entities are also responsible for Public Health issues, which is especially in times of viruses a big burden and responsibility.
- Regulatory Agencies, notified bodies, regulatory consultants: Ensure that service and products are solving problems safely and with a high efficacy. No new products, no approvals, and no consultations. Systems and concepts that fall outside evidence-based medicine are difficult to assess and with that it is difficult to obtain approval for. The regulatory processes are directly linked to the current healthcare provision.
- Patients/Individuals: Would like to not having to go to any of the first 4 on this list. They would love to not need to buy a product from number 6 on the list (pharmaceutical companies) and do not like the interaction with number 5. And an interaction with anyone from number 7 and onward is also unlikely. Number 8 could be an exception, if the product is affordable and allows to take, manage, and monitor the own health effectively.
- Employers: Often have to subsidize health insurance expenses and obviously would like to have healthy and fit employees with as little as possible sick time. With that they are interested in a cost-effective health insurance and means that would prevent employees from getting sick, in accurate and quick diagnosis with follow-up therapies that lead to fast recovery.

The above list is likely not complete with respect to the stakeholders and shallow with respect to the described business model and value propositions.

What is clear though is that almost all mentioned are part of the healthcare revenue side that is financed by the employer, patient, and the government. Typically the payer determines the solutions, the offerings, and the future directions. Not the case in healthcare!

So for a future healthcare we need to rethink the current setup!

4.4   Future Health Goals

One of the Sustainable Development Goals (SDGs) is HEALTH and WELLBEING
(SDG 3). Sustainable Health as part of Human Health is in the intersection of
ensuring that our planet and environment are not harmed or altered, that we work
and cooperate within an ethical framework with other individuals and organizations
for the benefit of the community, and that we ensure socioeconomic stability for
everyone.

While this article is not discussing these aspects we want to make sure that they
have been mentioned and will need to be considered as essential requirements for
GOOD HEALTH AND WELLBEING. What we will discuss and point out though
is the need for the individual to actively participate in their own health (Diet,
Exercise, No Drugs, . . .) and by employing predictive, preventive, and personalized
health assessment and monitoring tools—the P4 Medicine [3].

But also the current healthcare delivery needs to be improved with novel
innovations. They should follow the quadruple aim of future healthcare
developments. Outcome of diagnosis/therapy should be improved just as the patient
and clinical staff experience while significantly lowering the cost for devices,
workflows, and processes [4].

Unfortunately we are not working based on the "recipe" as shown in Fig. 4.2:

Fig. 4.2 We require a future focus based on P4-Medicine [3] in combination with the Quadruple
Aims of Healthcare Developments to improve Human Health and work toward one of the 17 Sustainable Development Goals of the United Nations. Human Health improvements require however
also a stable socioeconomic environment, an ethical framework, and a healthy planet. Are we
working on that at the moment?

4.5   Will Digitization, Deep-Tech/Exponential Technologies Help?

Last year we organized a dedicated FUTURE OF HEALTH online Science-Fiction cartoon creation event with over 150 participants. The goal was to tell a story of what healthcare will be able to do or should allow us to do with a 20-year time horizon from an individual perspective.

This daylong event produced seven great and insightful stories (see Fig. 4.3 and [5]). Gene editing CRISPR technology, Artificial Intelligence, 3D Printing, Wearables as well as Prevention and Health Monitoring were mentioned in several of the stories. And interestingly enough the main place for health service and monitoring was seen at home (or anywhere with mobile devices) and not at a hospital or primary care physician.

It reflects the wish of the individual to be in the center of the healthcare provision, but that currently deals with a healthcare setup that provides a complicated and expensive setup for fixing health issues.

We have actually used these type of future looks and storytelling in the past to create initial health device and service solution concepts [5, 6].

Even if these may considered to be far away and too disruptive they could be used to go backward from the future to the present and define incremental innovation steps that should be available in 1 year, 3 years, or 5 years to reach that vision.

The new digital-based technologies—also already mentioned in these cartoons— could provide new solutions and approaches for value creation in the healthcare space, provide new product and service innovations including involving 3D printer, (semi) autonomous robotic systems, self-learning diagnostic systems, data integration and evaluation [7].

Fig. 4.3   A recent Global Future of Health event with 150 participants showed the vision of mainly patients. The seven independently developed stories [5] had several common technologies that are seen as key to the health provision. CRISPR, Artificial Intelligence, 3D Printing, Wearables were mentioned in several of the stories. In most of the stories the main place for health monitoring and prevention is seen at home (or anywhere with mobile devices) and not at a hospital or primary care physician

There is no doubt that health digitization is just at the beginning and will gain significant speed in the coming years.

These technologies will also help the move toward a value-based medicine, but at the same time change the roles of many of the current stakeholders.

The medtech companies will need to change from being hardware providers to intelligent software companies that also provide health sensors with the data outcome being immediately integrated into an electronic health record. They will also need to cooperate much more with competitors and with data companies. And they will need to add products and services for a home care platform with separate devices, sensors, and data analytics. The home care platform data points for health status monitoring will create a personal health record that will be combined (data merge and integration) with the systems and deep-tech information collected in the electronic health record of the professional care system. This will lead to a Digital Twin that can answer the questions

- Diagnostic: Why certain health events happened?
- Predictive: What will possibly/likely happen in the future?
- Prescriptive: What can we do right now or in the near future to prevent that from happening?

4.6 Future Health Innovation Needs

We have discussed the need to move to a 4P medicine already in some depth.

Cost reduction is another that needs to be urgently addressed. Past and present innovations were focused on improving/enhancing existing solutions (10% thinking) vs. focusing on patient's unmet medical needs to radically rethink existing solutions (10 times thinking) [5] (Fig. 4.4).

Figure 4.5 shows such a $10\times$ approach. If we could introduce innovations that allow to shift healthcare delivery from a community hospital to a home setting, we would be able to increase the quality of life and reduce the care cost by $10–100\times$ per day. The above figure is rather qualitative but nicely shows the potential exponential factors.

Other midterm challenges that we need to find solutions for [9]—big challenge (C), fix existing healthcare (F), new medicine (N):

- *Demographic changes and with that increase of age-related diseases (e.g., neurodegenerative) (C)*
- *Increase in life expectancy but not an equal increase in health longevity (see above) (C, N)*
- *Significantly reduce cost of healthcare (C, N)*
- *Data silos that are not connected (fragmented health) (F, N)*
- *Eliminate the clinical noise in diagnosis and therapy (F, N)*
- *Precision and personalized medicine (F, N)*

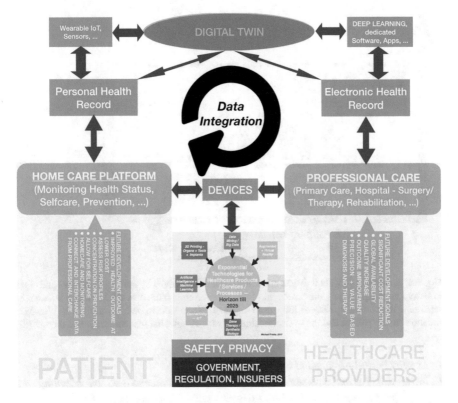

Fig. 4.4 Exponential technologies/deep-tech (yellow) will provide novel tools for professional as well as home care use. They will also have additional effects on the professional care side. The data points generated on both sides will need to be merged and will lead to a predictive, personalized medicine focused on prevention and early detection with the help of a DIGITAL TWIN [8]

- *Increased need in elderly care services, but not enough staff (F, N)*
- *Develop new health business models for data and artificial intelligence based solutions (C, N)*
- *Develop new business models for prevention and health prediction (C, N)*
- *Eliminate the resistance of clinicians toward the adoption of advanced deep-tech based tools (F)*
- *Implement more technology-based approaches for use in universal healthcare and as safeguards to protect privacy, data security, and equality [10] (N)*
- *Ensure that everyone (global) has access to these new tools and services to reach health democratization (C)*
- *Quality of Health depends on country, income, and on whether you live in rural/ urban (C)*
- *Healthcare Quality varies widely (C, F)*

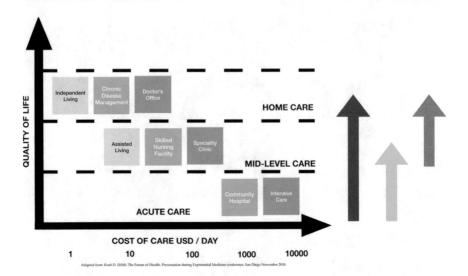

Fig. 4.5 Dramatic (exponential) cost reductions could be achieved by moving treatments and after-care from very expensive acute care to mid-level care and even more moving it to home care. At the same time the patients quality of life would improve. Who wants to be in a hospital? A Win–Win situation for the patient and the overall treatment costs. A potential loss for acute care though [8]

- *Duplicate and not needed medicine (F)*
- *Move away from a disease-specific approach concentrated on counting the number of services provided to a people-centered approach (C, F)*
- *While many parts of healthcare require in-person care the pandemic has shown that virtual care and remote monitoring has proven beneficial and cost-effective—much more can be easily done [11, 12] (N)*
- *Integration of pharmaceutical and medical care management is needed (F)*
- *Innovation is seen as a threat by certain health constituencies, and there have been institutions turning down innovations to protect investments already made (C)*
- *Eliminate Fake Medicine (F)*
- *Provide tools and concepts to move healthcare delivery from acute care to mid-level care or even to home care and move mid-level-care to home care (F, N)*
- *Reduce/Eliminate Medical Errors (F)*
- *Provide solutions for neglected diseases that are not economically attractive enough (F)*
- *Focus on the Patient Needs (C, N)*
- *Make Health Sustainable (C, C, C, N—but only in small parts topic of this book)*
- *Address the Health of our Planet (C, C, C, C—but not the topic of this book)*

4.7 Early Questionnaire Results

In late 2021, we conducted a survey among a total of 50 engineers, data scientists, some clinicians, Medtech industry staff, students, and academic staff. Not meant to create deep scientific conclusions, just to get some insights in what people that are involved in healthcare innovations think themselves (the results of a larger dataset have been submitted for publication and do not vary significantly).

Below you will find some of the condensed results.

Looking back at past predictions with respect to the growth of digital-based technologies we see that typically the 3–5 years horizon was overestimated, meaning people thought we should have been further than what actually happened, but on a 10 year they generally underestimated the progress significantly. The reason is the typical linear thinking process of us humans that analyze what happened in the last couple of years and then linearly project that growth into the future.

But exponential/deep-tech technologies do not develop linearly, but as the descriptor says following an exponential curve with an initial disappointment phase and a very fast performance increase after disruption (better performance than current technologies).

We asked about the most likely implemented Technologies in 3–5 years (Electronics and Sensors, Digital Healthcare, Health Wearables, Data Management), something that already happened or is known. The same question for the 10-year horizon added artificial intelligence and robotics to that (see Fig. 4.6).

We also asked which technologies would reinforce each other and with that have a potential to create a huge impact on future health. Digital Healthcare and Artificial Intelligence were the most often selected combination. Also not surprising, as everyone already sees the effects of these. But, of course, three or more technologies could be put together as well, e.g. sensors + artificial intelligence + virtual reality + robotics to create semi- or fully-autonomous surgical robot systems (see Fig. 4.7).

Another insightful question (Fig. 4.8 top) was on where these technologies will be used (Home care and Hospitals) in the future and which technologies will have the greatest impact on healthcare delivery (Fig. 4.8 bottom) in 5–10 years (AI for prediction and Prevention, Sensors for Monitoring).

And finally we asked about the entity that should be financially responsible for the implementation (Governments and Insurances) and what the biggest issues are with respect to the accuracy of predictions (Higher Longevity and increase of chronic and neurodegenerative diseases)—see Fig. 4.9.

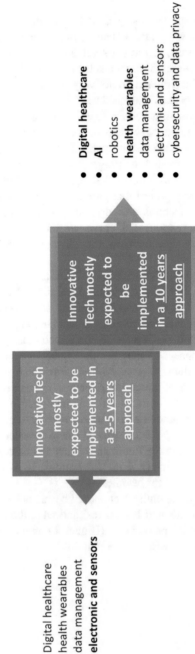

- Digital healthcare
- health wearables
- data management
- **electronic and sensors**

Innovative Tech mostly expected to be implemented in a 3-5 years approach

Innovative Tech mostly expected to be implemented in a 10 years approach

- **Digital healthcare**
- **AI**
- robotics
- **health wearables**
- data management
- electronic and sensors
- cybersecurity and data privacy

Fig. 4.6 Question: Which innovative Technologies would you expect to be implemented in a 3–5-year and 10-year horizon. For the short term the respondents believe in electronics and sensors, digital healthcare, wearables, and data management. Digital healthcare, artificial intelligence, and health wearables are the top three for the 10-year horizon

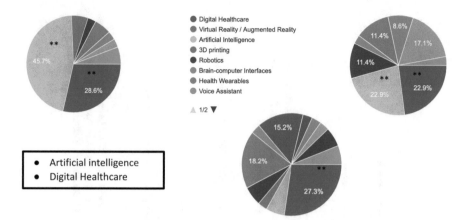

Fig. 4.7 Question: Which of the following innovative technologies do you expect to reinforce each other most and thus impact the future of health? Options were: digital Healthcare, virtual reality/ augmented reality, artificial intelligence, 3D printing, robotics, Brain–Computer Interfaces, health wearables, voice assistant, new touch interfaces, minimal invasive therapy systems, environmental protection and sustainability, data management, electronic and sensors, cybersecurity and data privacy, gene editing, (CRISPR/Cas9). The most selected technologies to mutually reinforce each other were artificial intelligence, digital healthcare, and health wearables

4.8   Conclusion

There are many opportunities and big challenges and problems to solve. Health Innovations are not driven—at the moment—by the payers and by the patient needs. This will change.

P4Medicine will be the new paradigm and that will be based on deep-tech. Existing healthcare services will be continuously improved based on the Quadruple Aim development goals. $10\times$ approaches—especially when applied to lowering diagnostic and treatment costs will disrupt certain health segments.

> **Take-Home Messages**
> - Screw business as usual (title of a book from Sir Richard Branson)—we need to adapt such a mentality for Future of Health Innovations.
> - There are many many many innovation opportunities, but there is also a reluctance by institutions and clinicians to adapt them (does not fit their business model, already invested capital in infrastructure and systems have not been amortized).
> - Value propositions for devices and tools that improve current healthcare delivery need to obviously improve the service quality. They also need to be patient centered and improve the experience (decrease frustrations) of the clinical staff. Most importantly they need to lower the costs significantly!—QUADRUPLE AIM—.

(continued)

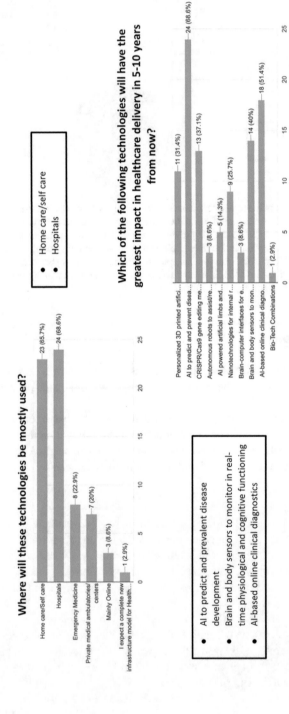

Fig. 4.8 TOP Question: Where will these new technologies primarily/mostly be used—HOME CARE and HOSPITAL. BOTTOM Question: Which of these technologies will have the greatest impact in healthcare delivery in 5–10 years from now? The most three common answers were AI to predict and prevent disease development, brain and body sensors to monitor in real time physiological and cognitive functions, and AI-Based online clinical diagnostics

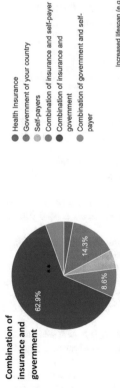

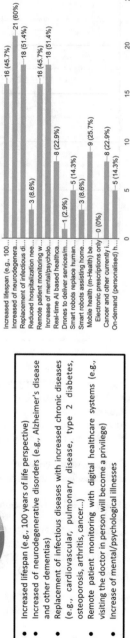

Fig. 4.9 Question for the LEFT: Who should be in charge paying for the implementation of these advanced technologies? Answer: A combination of insurance and government. Question for the RIGHT: What would you expect to be the main ISSUE(s) that we will encounter when forecasting the future of health? Answer: We will likely live a lot longer (comment author: I hope so!), but at the same time will likely encounter increased neurodegenerative and chronic diseases

- Future value will be created if devices/tools will use health data and monitoring for Disease Prediction and with that help Prevention. They will also need to be Personalized to the individual user. But the Participation of the individual in their own health is of course required!
- Individual Health Data will be collected at home (Personal/Home Health Record) and merged with the data points from professional care (Electronic Health Record) to create a comprehensive DIGITAL TWIN.
- Deep-Tech and other exponential health technologies will provide predictive, prescriptive, and diagnostic information toward a prevention medicine.
- Current Innovations are mainly based on 10% (Incremental) approaches, because with that they fit into the existing business models. Future innovations need to embrace a $>10\times$ goal even though a fitting business model is not defined yet.
- The human component is and will be very important in the entire health delivery process, but the reluctance to accept and embrace novel technologies because of fear of changing environments needs to be eased by safeguards.
- Politics will need to be more proactive and act on behalf of the patient needs and to address Planetary and Sustainable Health.

References

1. von Debschitz U, von Debschitz T (2017) FRITZ KAHN Infographics Pioneer with a foreword by Steven Heller TASCHEN. Bibliotheca Universalis Cologne. ISBN 978-3-8365-0493-5) – also see https://www.fritz-kahn.com
2. Johns Hopkins study suggests medical errors are third-leading cause of death in U.S. https://hub.jhu.edu/2016/05/03/medical-errors-third-leading-cause-of-death/. Retrieved 05 March 2022
3. Flores M, Glusman G, Brogaard K, Price ND, Hood L (2013) P4 medicine: how systems medicine will transform the healthcare sector and society. Per Med 10(6):565–576. https://doi.org/10.2217/pme.13.57
4. Sikka R, Morath JM, Leape L (2015) The Quadruple Aim: care, health, cost and meaning in work. BMJ Qual Saf 0:1–3. https://doi.org/10.1136/bmjqs-2015-004160
5. Friebe M, Hitzbleck J, Merkel K (2021) SciFi HIVE – Future of Health – Global Workshop. Cartoonbook available at https://www.researchgate.net/publication/351109123_FUTURE_of_HEALTH_%2D%2D_CARTOONBOOK_%2D%2D_story_results_from_the_global_SciFi_Hive_Event_March_20_2021_%2D%2D_PREDICT_PREVENT_PERSONALIZE_PARTICIPATE
6. Friebe M, International Healthcare Vision 2037 (2017) New Technologies, Educational Goals and Entrepreneurial Challenges. Otto-von-Guericke-Universität, Magdeburg. ISBN: 978-3-944722-59-7
7. Porsche Consulting – Digital Medtech Transformation. www.porsche-consulting.com
8. Friebe M (2020) Healthcare in need of innovation: exponential technology and biomedical entrepreneurship as solution providers. Proc. SPIE 11315, Medical Imaging 2020: Image-

Guided Procedures, Robotic Interventions, and Modeling, 113150T (16 March 2020). https://doi.org/10.1117/12.2556776

9. Gomes-Osman J, Solana-Sánchéz J, Rogers E, Cattaneo G, Souillard-Mandar W, Bates D, Gomez EJ, Tormos-Muñoz JM, Bartrés-Faz D, Pascual-Leone Á (2021) Aging in the digital age: using technology to increase the reach of the clinician expert and close the gap between health span and life span. Front Digit Health 3:755008. https://doi.org/10.3389/fdgth.2021.755008

10. Kulkov I (2021) Next-generation business models for artificial intelligence start-ups in the healthcare industry. Int J Entrepreneurial Behav Res 1355–2554. https://doi.org/10.1108/IJEBR-04-2021-030

11. Wilson D, Sheikh A, Görgens M, Ward K (2021) Technology and universal health. Coverage: Examining the role of digital health. J Glob Health 11:16006

12. Holden RJ, Boustani MA, Azar J (2021) Agile Innovation to transform healthcare: innovating in complex adaptive systems is an everyday process, not a light bulb event. BMJ Innov 7:499–505

Michael Friebe received the B.Sc. degree in electrical engineering, the M.Sc. degree in technology management from Golden Gate University, San Francisco, and the Ph.D. degree in medical physics in Germany. He spent 5 years in San Francisco, as a Research and Design Engineer with an MRI and ultrasound device manufacturer. He is a German citizen with expertise in diagnostic imaging and image-guided therapies, as a/an Founder/Innovator/CEO/Investor and a Scientist. He is also a Research Fellow with the Technical University of Munich, Munich; an Adjunct Professor with the Queensland University of Technology, Brisbane; and a honorary Professor of image-guided therapies with Otto von Guericke University, Magdeburg, Germany. Since 2022, he has been a Professor of biomedical engineering innovation with the AGH University of Science and Technology, Krakow, Poland. He is a listed inventor of more than 80 patents and has authored over 200 papers, and was part of over 35 Medical Technology Start-Ups. He is a Board Member of four medical technology start-up companies and an investment partner of a MedTec-fund. From 2016 to 2018, he was a Distinguished Lecturer of the IEEE EMBC teaching innovation generation and MedTec entrepreneurship. He is also a coach and trainer for OpenExO, and a master Launchpad mentor for the Purpose Alliance.

Navigating Towards a Future of "One Health"

5

Henri Michael von Blanquet and Michael Friebe

Abstract

One Health is built on the simple understanding that the human health, animal health, and the health of our shared environmental nature (Planetary Health) are part of a deeply interconnected system. In addition, the Concept of *One Health* is the unity of multiple practices that work together locally, nationally, and globally to help achieve optimal health for people, animals, and our nature as a whole. When people, animals, and the environment are put together they make up the *One Health Triad* showing how the health of the people, animals, and the nature as whole is linked to one another. *One Health* is a global concept that will help to advance healthcare in the twenty-first century towards sustainability. Global also means that one negative action or incidence (e.g. global warming, climate change) could have a severe effect on the health of everyone and everything. While a lot of the interactions are known, we do not have a unified concept that if deployed properly could help to save all living beings and our environment.

H. M. von Blanquet
The Healthcaptains Club, Nieblum, Germany
e-mail: president@healthcaptains.club

M. Friebe (✉)
AGH University of Science and Technology, Krakow, Poland

Otto-von-Guericke University, Magdeburg, Germany

IDTM GmbH, Recklinghausen, Germany

FOM University of Applied Science, Center for Innovation and Business Development, Essen, Germany
e-mail: info@friebelab.org

Keywords

One Health · Interconnected systems · Transdisciplinary knowledge networks ·
One healthcare innovation · Planetary Health

What will you learn in this chapter?
- The limitations of human medicine to achieve sustainable health outcomes
- The definitions of sustainable, Planetary Health and "One Health"
- How to overcome the barriers between human medicine, veterinary medicine and research and knowledge on our shared environment as a twenty-first-century challenge
- Interconnected System Thinking on Human, Animal, and Environmental Health
- The need of transdisciplinary knowledge networks for One Health
- The Chance for One Health Innovations as a new Market for sustainable Health Solutions

5.1   Introduction

Let us start with some definitions and concepts. **Sustainable Health** is defined as a personal commitment to actively participate and take needed actions to maintain your own health and to act pro-actively preventative, e.g., through healthy food intake, abstain from any drug use, exercise, and generally to keep your mind and body in shape. However, the greatest impact on your personal health is having a stable socioeconomic environment and by ensuring that public health (and with that personal health) is protected from climate change and other related environmental effects.

The healthcare industry is among the most carbon-intensive sectors in the industrialized world. It is responsible for around 4.5% of the global greenhouse gas emissions and a similar percentage of toxic air pollutants. The US healthcare system, for example, is responsible for 25% of the total healthcare-related greenhouse emissions [1].

Sustainability for public health aspects has been defined as the capacity to maintain services at a level that will provide ongoing prevention and treatment for a health problem even after termination of major financial, managerial, and technical assistance [2]. With that the current delivery system and financing structures in the western world need to be considered unsustainable with in-equal health resources distribution and a growing number of people that do not—or have only limited—access to sufficient healthcare, especially for secondary and tertiary services. The only way to solve that is to redesign some—or all—parts of the health delivery and financing together with the stakeholders (individuals/patients, health providers, insurance companies and companies, governments, and others directly or indirectly involved) [3].

5.2 Planetary Health

Planetary health is a complex transdisciplinary field that attempts to provide solutions addressing the impacts of the human disruptions to our planet and all live on it. While we have managed to improve—in average—the health and wealth of the global population we have also caused a sharp decline in the health of Earth. We have caused the global temperatures to rise year after year through increased environmental pollution, exploitation of natural resources, rising population, reduced biodiversity/freshwater supplies/farmable lands, and other related negative activities. Humanity has had and continues to have a dramatic impact on the planet's natural systems and its ability to heal itself.

The quality of the air we need to breathe and the water that we require to survive is decreasing and certain areas on this planet have become uninhabitable causing people to leave and relocate to other areas.

We already see and feel the impact of all these changes and some of the poorest areas are hit the hardest. Everything is interconnected and what we do today has an effect for the coming generations. We all know that, but it is nevertheless important to repeat and make everyone aware of the fact that the human health can only improve or be maintained with good "planetary health". To achieve that we need to work across boundaries and transdisciplinary … and we need to do that fast. Figure 5.1 shows the interconnection between the underlying drivers (consumption, advances, economies), their effect on the ecological drivers and what they cause with respect to the planet's health and eventually to human health [4].

5.3 One Health

One Health combines sustainable and planetary health towards an integrated, unifying approach that aims to balance and optimize the health of people, animals, and ecosystems. It recognizes that the health of humans, domestic and wild animals, plants, and the wider environment (including ecosystems) are closely linked and interdependent.

The approach mobilizes multiple sectors, disciplines, and communities at varying levels of society to work together to foster well-being and tackle threats to health and

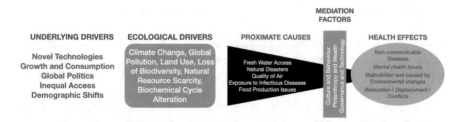

Fig. 5.1 Planetary health: protecting human health on a rapidly changing planet. Adapted from [3]

ecosystems, while addressing the collective need for clean water, energy and air, safe and nutritious food, taking action on climate changes, and contributing to sustainable development.

Today's health problems are frequently complex, transboundary, multifactorial, and across species, and if approached from a purely medical, veterinary, or ecological standpoint, it is unlikely that sustainable mitigation strategies will be produced.

The One Health model has gained momentum in recent years due to the discovery of the multiple interconnections that exist between animal and human disease. Recent estimates place zoonotic diseases as the source for 60% of total human pathogens, and 75% of emerging human pathogens.

There are many more areas that urgently need the One Health approach, at all levels of academia, government, industry, policy and research, because of the inextricable interconnectedness of animal, environmental, human, nature, and planet health. These are:

- Agricultural production and land use
- Animals as Sentinels for Environmental agent and contaminants detection and response
- Antimicrobial resistance mitigation
- Biodiversity/Conservation Medicine
- Climate change and impacts of climate on health of animals, ecosystems, and humans
- Comparative Medicine: commonality of diseases among people and animals such as cancer, obesity, and diabetes
- Disaster preparedness and response
- Disease surveillance, prevention and response, both infectious (zoonotic) and chronic diseases
- Environmental Health
- Food Safety and Security
- Human—Animal bond
- Natural Resources Conservation
- Occupational Health Risks
- Plant/Soil health
- Public policy and regulation
- Research, both basic and translational
- Water Safety and Security
- Welfare/Well-being of animals, humans, ecosystems, and planet

The One Health approach is holistic (see Fig. 5.2). Thinking and acting are transformed across the classic disciplines and sectors into a synergistic integrative approach for the benefit of people, animals, the environment and their common health (= One Health).

These are social tasks that are closely linked to the sustainability goals. In an interdependent world, there is a growing consensus that goals in the health sector

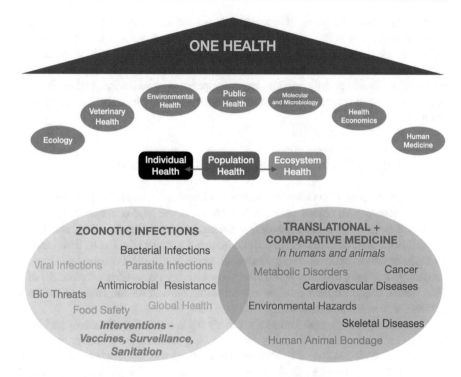

Fig. 5.2 Schematic representation of the One Health Umbrella, developed by the One Health Sweden and the One Health Initiative Autonomous pro bono team adapted from [5]—https://onehealthinitiative.com/

require efficient interdisciplinary collaboration, including technical, economic, and digital competence [5].

Global health threats such as climate change, emerging or re-emerging pathogens from Ebola virus to SARS coronavirus-2 (CoV-2), and the silent epidemic of noncommunicable diseases (including cardiovascular disease, cancer, lung disease, diabetes, and mental health) are phenomena that prove that, for example, environmental protection and species protection, as well as poverty reduction and access to appropriate education, must not be neglected.

It is important to find holistic solutions that go far beyond the treatment of the sick. In this context, the health chances for the different population groups, for animals, plants, and their common environment must be increased preventively. It is important to offer appropriate health protection and at the same time to implement health promotion in the awareness of One Health as a political concept anytime and anywhere (mainstreaming approach). And this must be done in a diversity-sensitive way, i.e. with attention and reference to established cultures, available resources, and the diversity of people, i.e. their differences in gender, age, origin, education, other dimensions of diversity and living conditions.

This can only be done if coordinated and multidisciplinary measures and collaborations across the sectors (public and private) succeed. An informed and evidence-based health policy is trend-setting, as is corporate responsibility, investment in research and education, and good governance and leadership. In order to realize One Health as an overall concept, cooperative, multilateral, and committed democratic action is required in all areas of society, in every country, and on a global level [6].

5.4   Health Through Behaviour and Circumstances

A One Health approach is relevant in almost all healthcare areas. This is also shown by veterinarians who coined the term One Health in the field of zoonoses (60–70% of all infectious diseases are zoonoses) in the last century. They were concerned with the strict exchange of germs in animal and human encounters, including their carriers from the environmental reservoir. Above all, neglected tropical diseases and new or recurring infectious diseases (epidemics) impressively demonstrate the benefit of a One Health approach. But also the more recent research on the environment as a habitat for humans and animals.

Considering the environment as a reservoir for germs and pollution of all kinds, but also as a source of well-being, is a growing research perspective, also with reference to well-being and quality of life studies, which includes the designed environment (i.e. the individual and environmental factors with which the World Health Organization operationalizes in terms of bio-psycho-social health).

Promoting health, for example, by reducing heat and air pollution, building healthy cities (greening, but also energy supply and waste and sewage management), new agricultural concepts (cultivation methods, pest control, etc.), but also new forms of Health behaviour (exercise, nutrition, rest, and integration) for the heterogeneous populations, which must be part of a health for all plan, corresponds to the One Health concept, as does combating the SARS-CoV-2 pandemic.

5.5   One Health as an Academic Convergence Challenge

In the field of 360° Next Generation Healthcare in Human Medicine the professional community is aware that the huge parallelity of innovation areas happening at the same time is interdependent and is therefore creating a "super-convergence" in the transformation of healthcare and health industry as described in many chapters of this book.

In One Health we are still missing convergence strategies because fully different to human healthcare with, e.g., hundreds of university hospitals and thousands of research institutions globally as *360° Next Generation Healthcare Transformation Hubs* "from bench to bedside and bedside to bench" running established innovation and education cycles and thousands of start-ups in every innovation field we are missing to start with real *One Health Research Institutions* and *One Health*

Innovation Technology Parks to support One Health Innovations and to create new Business Models for an real One Health Industry.

One Health is still not established in the academic institutional training of doctors, veterinarians, and the additionally many environmental and planetary health scientific fields as a whole. The missing One Health Education and Research strategies of national and international universities is the missing link towards new One Health Innovations and new One Health Business Platforms. Still we see the investment in One Health Education as the key factor to open this transdisciplinary field towards new Business [7, 8].

To create a first breakthrough in the academic world the authors are interacting directly with the University Hospital Bonn where they are planning to integrate an *One Health Institute* on the University Hospital Campus to start with transdisciplinary education and research in this field. Additionally The Health Captains Club is planning an One Health Summit with the University Hospital Bonn as Co-Host starting 2023 to bring the needed stakeholders together to create the start of an international One Health Knowledge Network. Every first little step counts. This article should inspire the community to take action and we are looking forward to make it happen together [9].

Without the breakthrough in the Convergence Challenge of academic institutions One Health as a scientific and strategic approach remains far too much in the theoretical field. This article is foremost a call of action!

Take-Home Messages
- *Human and Veterinary Health are directly connected to PlanetaryHealth.*
- *The One Health Umbrella represents the transdisciplinary One Health Innovation fields for new Convergence Business Solutions.*
- *Sustainable Health needs to be—among the personal health—concerned with environmental effects and especially with climate change.*
- *Healthcare services are responsible for 4.5% of the total CO_2 emissions and of toxic air pollutants—we need urgent action for more transformation of scientific institutions and healthcare providers towards zero emission buildings—like the J. Craig Venter Institute in La Jolla in California is serving as a model for sustainable research buildings worldwide; source: www.jcvi.org*
- *CO_2 emission measurements and more environmental diagnostic data points need to become metrics for future health and for health stakeholders.*
- *The breakthrough key for One Health is the implementation of One Health Education inside Academia and new One Health Conference Systems and Platforms to build One Health Knowledge Networks as the missing Innovation Hubs.*
- *This article is a call for action—your input is welcome.*

References

1. Eckelman M, Huang K, Lagasse R, Senay E, Dubrow R, Sherman J (2020) Health care pollution and public health damage in The United States: an update. HEALTH AFFAIRS 39(12): 2071–2079. https://doi.org/10.1377/hlthaff.2020.01247
2. https://healthcareglobal.com/top10/top-10-ways-healthcare-industry-can-be-more-sustainable. Retrieved 15 March 2022
3. DeVoe J (2008) The unsustainable US health care system: a blueprint for change. Ann Fam Med 6(3):263–266. https://doi.org/10.1370/afm.837
4. Myers SS (2017) Planetary Health: protecting human health on a rapidly changing planet. The Lancet 390(10114):2860–2868
5. Mackenzie JS, Jeggo M (2019) The one health approach-why is it so important? Trop Med Infect Dis 4(2):88. https://doi.org/10.3390/tropicalmed4020088
6. Frankson R, Hueston W, Christian K, Olson D, Lee M, Valeri L, Hyatt R, Annelli J, Rubin C (2016) One health core competency domains. Front Public Health 4:192. https://www.ncbi.nlm. nih.gov/pmc/articles/PMC5020065/
7. Togami E, Gardy JL, Hansen GR, Poste GH, Rizzo DN, Wilson ME, Mazet JAK (2018) Core competencies in one health education: what are we missing? NAM Perspectives. Discussion Paper, National Academy of Medicine, Washington, DC. https://doi.org/10.31478/201806a
8. Rabinowitz PM, Natterson-Horowitz BJ, Kahn LH, Kock R, Pappaioanou M (2017) Incorporating one health into medical education. BMC Med Educ 17:45. https://bmcmededuc. biomedcentral.com/articles/10.1186/s12909-017-0883-6
9. Binot A, Duboz R, Promburom P, Phimpraphai W, Cappele J, Lajaunie C, Goutard FL, Pinyopommintr T, Fiquie M, Roger FL (2015) A framework to promote collective action within the One Health community of practice: using participatory modelling to enable interdisciplinary, cross-sectoral and multi-level integration One Health 1:44–48. eCollection 2015 Dec. https://www.ncbi.nlm.nih.gov/pubmed/28616464

Henri Michael von Blanquet , MD. "At the latest in the face of the global corona pandemic, we have to completely restructure medicine, health sciences, health industry and the health systems worldwide towards sustainability. It is about human life, the life of our families and friends, our employees and colleagues, the life of those entrusted to us and it is about our life and our nature". Dr. von Blanquet is the inventor and President of THE HEALTH CAPTAINS CLUB and the PRECISION MEDICINE ALLI-ANCE. To combine and scale value-based "Profit- with Non--Profit" Innovations and Business, he is, since 2022, the founder of the business-hub "HEALTHCAPTAINS+COMPANY 360° Next Generation Healthcare"—headquartered in Berlin and on Föhr Island at the Westcoast of North Europe.

Michael Friebe received the B.Sc. degree in electrical engineering, the M.Sc. degree in technology management from Golden Gate University, San Francisco, and the Ph.D. degree in medical physics in Germany. He spent 5 years in San Francisco, as a Research and Design Engineer with an MRI and ultrasound device manufacturer. He is a German citizen with expertise in diagnostic imaging and image-guided therapies, as a/an Founder/Innovator/CEO/Investor and a Scientist. He is also a Research Fellow with the Technical University of Munich, Munich; an Adjunct Professor with the Queensland University of Technology, Brisbane; and a honorary Professor of image-guided therapies with Otto von Guericke University, Magdeburg, Germany. Since 2022, he has been a Professor of biomedical engineering innovation with the AGH University of Science and Technology, Krakow, Poland. He is a listed inventor of more than 80 patents and has authored over 200 papers, and was part of over 35 Medical Technology Start-Ups. He is a Board Member of four medical technology start-up companies and an investment partner of a MedTec-fund. From 2016 to 2018, he was a Distinguished Lecturer of the IEEE EMBC teaching innovation generation and MedTec entrepreneurship. He is also a coach and trainer for OpenExO, and a master Launchpad mentor for the Purpose Alliance.

Part II

Exponential Medicine + Technologies + Mindset

Exponential Technologies for an Exponential Medicine

6

Michael Friebe

Abstract

Digital based and vastly accelerating technologies are called exponential. They lead to new insights, more inventions, more discoveries and when combined to even faster acceleration. They will change existing approaches significantly and will doubtless have a huge impact on health and health delivery eventually leading from current continuous care to a more pro-active and connected care largely at home and on to real-time prediction and prevention medicine. These technologies will also be essential for solving the huge challenges of health and healthcare, like demographic change, increased life expectancy and with that neurodegenerative and chronic diseases, high cost of health services and unequal access and largely varying quality.

Some developments for new digital health tools will be further analyzed based on their potential use for disruption and/or challenge solutions.

Keywords

Exponential technologies · Exponential medicine · 6D's · Digital health · Accelerating return · Health democratization

M. Friebe (✉)
AGH University of Science and Technology, Krakow, Poland

Otto-von-Guericke University, Magdeburg, Germany

IDTM GmbH, Recklinghausen, Germany

FOM University of Applied Science, Center for Innovation and Business Development, Essen, Germany
e-mail: info@friebelab.org

© The Author(s), under exclusive license to Springer Nature Switzerland AG 2022
M. Friebe (ed.), *Novel Innovation Design for the Future of Health*,
https://doi.org/10.1007/978-3-031-08191-0_6

> **What Will You Learn in This Chapter?**
> - How to define exponential technologies
> - The 6D's of exponential technologies
> - How health democratization can be reached
> - That combining several technologies could lead to even faster acceleration and solutions that will have a disruptive/radical effect on healthcare as we know it today
> - The definition of Digital Health and the cultural transformation it requires

6.1 Introduction

We have already talked about the need to change current healthcare, to address the large challenges related to personal health and healthcare delivery. How do we get from the current healthcare delivery of treating sick people in a continuous manner (can take from few days to many months, face to face, in healthcare institutions), to a pro-active and connected care (home care, sensor and analytics based) that only takes minutes, and further toward a real-time health-monitoring setup (wearable sensor systems, predictive multi-sensor analytics). In other words, from SICKCARE to HEALTHCARE and on to (personalized, prevention based) HEALTH.

We have also talked about the fact that current healthcare delivery is a huge business in which the main customer, the individual, is not determining and essentially not getting what they really want and need. Technologies and inventions only become inventions if they find a market and paying customer. If there is a chance that this can happen, then external capital is available. Novel, incremental approaches for the current healthcare setup do have a business model and are therefore supported by venture capitalist, private or strategic investors.

Disruptive (also frugal) and architectural innovations in the health space do not have a convincing and defined business model yet. With that financing is limited. Governments may have to stimulate early developments through grant financing and other support. But there is a big need especially in areas with underdeveloped economic and healthcare systems. Healthcare is much more of a privilege there, clinical staff is missing, and general availability and quality is low. Novel, inexpensive, portable, connected, data- and sensor-based offerings will likely be much quicker accepted there (anything is better than nothing!) [1].

Several already available and continuously improving technologies could help to solve some of the challenges and issues with healthcare. Especially, when they are combined with each other and with that reinforce the effects and outcomes. These digital-based technologies (e.g., artificial intelligence, deep learning, augmented and virtual reality, digital biology/biotech, blockchain, CRISPR, 3D printing, advanced sensors, and several more) that allow change at an accelerated speed are called exponential technologies. The accelerated speed is not only referring to performance improvements but also to exponential cost reductions without performance loss (e.g., number of transistors per area, computational power, data storage, data generated and transferred).

A current robotic surgery system is in effect a telemanipulated physician-controlled multi-tool device that allows more accurate/precise surgical interventions through small incisions. So, still a human-controlled system. Are we moving to a more autonomous setup in the future? Maybe one even without a human surgeon? There is a wide variation among innovators when that will happen. Some say 10 years, others 50 years. If you ask surgeons and clinicians then the time line typically gets pushed further into the future and a high percentage still do believe that such autonomous systems will or at least should not happen. Interesting!

But if we agree that deep-technologies provide huge benefits in the diagnosis and prediction of events. If we also see that multi-sensor systems in combination with signal processing and artificial intelligence can provide real-time information that would otherwise not be available. When these are now fused and combined with the mechanical surgical robot device (camera systems, integrated diagnostic imaging, multi-sensor arrays, predictive data analytics, and automated path and therapy planning to just name a few) would we not believe that this is leading to a better and preferred setup?

Let us go quickly back to exponential technologies. The "Law of Accelerating Returns" postulated by Ray Kurzweil in 2001 states that the more advanced a system becomes the faster it progresses [2, 3].

Exponentially improving technologies lead to new insights, more inventions, more discoveries and then lead jointly to even faster. So we can assume that fusing exponentially accelerating technologies we will see even more accelerating results. Most of them are not predictable and can lead to effects and outcomes that we cannot foresee. These developments always come with a dual nature. They will likely improve certain things significantly or change current setups for the better. But there could and likely will also be negative effects. An agile, and generally followed ethical and moral compass for developers and innovators is getting more and more important.

6.2   6D's Toward Health Democratization

Peter Diamandis introduced the concept of the 6D's of exponential technologies in 2016. *Digital*-based technologies go through an initial *disappointment* phase, a phase of low performance combined with expensive and awkward/complicated setups. The technology accelerates exponentially with respect to price/performance and eventually leads to *disrupting* the linear /incrementally improving current technology. Continuous acceleration *dematerializes* or significantly reduces size, weight, manufacturing complexity. More and more systems are built for a fast growing customer base that in return lead to lower costs or *demonetization*. With lower costs everyone is able to afford and use the system, something we call *democratization* [4].

2030: External sensor systems that measure physiological parameters and additional inputs—that are currently not used for health assessments and diagnosis— packaged in a wearable or as a smartphone attachment. Health platforms will be fed and supplied with the obtained and measured data (digital biomarkers/health tools /therapeutics) that will connect all current stakeholders (In-Patient/Out-Patient care-centers, patients/current non-patients, governments, health insurances and other payors, pharma/biotech/medtech, . . .).

The data sets will characterize a biometric health status baseline. Signal processing, deep learning and big data, platform analysis are then leading to early prediction and prevention and are also able to diagnose certain (maybe almost all) diseases with a > 90% efficacy at home. They are provided for free by the payors or governments. A diagnosis leads to some automatic triggers: a clinical expert may be called to manage the patient-journey with more advanced services that are automatically booked and organized; a tele-therapy will be set up; a digital prescription is sent to the local pharma supplier that delivers to the door; . . . A health-monitoring and healthcare delivery setup that is more patient oriented and centered, more digital in all aspects, significantly more effective and cheaper.

Think for yourself on whether that is real science-fiction and what implementation depends on. And think/anticipate what that means for the number of hospitals, the services and systems offered, for life expectancy, healthy longevity, clinical staff, care staff, emergency medicine, and many more. Exponential technologies will lead to new health, to exponential medicine [5].

6.3   Future Innovation Digital Health Topics

Most of the future innovations will be digital based and therefore I want to spend a paragraph on relevant innovation topics in that field. Important is that we distinguish between a novel approach as previously explained (no business model yet—disruptive or architectural innovation), improvement of existing systems and setups (incremental/traditional innovation), on whether it solves or helps to solve large challenges (for example increased life expectancy, neurodegenerative and chronic diseases, other age-related problems, inequalities, access to good health, significant cost reductions), and on whether it uses deep-technologies (combinations of several exponential technologies). The last one was dropped again, as all the topics listed use two or more accelerating technologies.

Digital Health is not just e(lectronic)-health, or m(obile)-health, or a new health 2.0, but they are all technical solutions that are part of it. Telemedicine, for example, is just a way to perform medical consultations instead of physical via a videoconferencing system. It certainly provides advantages, but the patient–doctor relationship is untouched and unchanged.

The definition of "Digital Health is the cultural transformation of how disruptive technologies that provide digital and objective data accessible to both *caregivers and patients* leads to an equal level doctor–patient relationship with *shared decision-*

making and the democratization of care" (Bertalan Mesko, The Medical Futurist, https://medicalfuturist.com).

This is an important sentence, as it highlights several major future development aspects and goals of Future Health innovations, as well as some of the major issues with the current setup. While we are already producing massive amounts of health data, they are not connected, exchanged, and are certainly not available to the patients to allow the participation in health decision-making ... and often they are not even available to the primary physicians and caretakers. Disruption will lead to transformation, not only economically with respect to business model, but also with respect to responsibilities.

Consider what new health insights are produced through (or lead to)

- patient-generated healthcare data (e.g., established and novel wearables) in combination with the
- technologies and services of big tech companies (e.g., Amazon, Google, Apple, Alibaba, Microsoft)
- do it yourself (DIY) health, wearables, and wellness apps combined with
- novel offerings that provide direct consumer to health models (e.g., preventative healthcare memberships)
- New and next-generation diagnostics (maybe even an at-home medical lab) may possibly lead to doctorless exams and novel treatments (digital therapeutics)

Many other technologies are on the horizon with a potentially massive impact on healthcare. These may be in the DISAPPOINTMENT phase, but they improve exponentially fast and could lead to DISRUPTION soon.

The list below was copied from an IEEE Conference (2022-ICT Solutions for e-health) call for papers [6]. It shows a fairly comprehensive list of narrow topics where digital health solutions could be implemented or could play a role in a future delivery.

It also indicates where novel health sensors for monitoring or diagnostic are combined with advanced signal processing and data collection/evaluation.

A blank in Fig. 6.1 does not indicate that this does not apply at the moment, but that it may not be known or is still unclear.

6.4 Conclusion

Quickly (exponentially) accelerating and combined digital-based technologies will lead to new solutions that cannot be predicted or foreseen with respect to what they will be able to do and when. They will however, and that is clear, lead to a novel healthcare delivery that is real time and predictive leading to prevention-based personal health. Many obstacles till have to be overcome with respect to regulation, acceptance, financing, and reimbursement and especially with respect to a needed cultural transformation.

	Disruptive or architectural	Incremental / Traditional	Demographics (age related diseases)	Cost Reduction / Democrati-zation	Homecare and Personal Health	Deep-Tech
• Cloud computing applications for eHealth	X			X	X	
• Internet of Things (IoT) applications for eHealth	X		X	X	X	X
• Assistive Technology (AT)		X	X		X	X
• Bioinformatics and Computational Biology and Medicine	X	X	X	X	X	X
• Monitoring of Vital Functions with Sensor and ICT Systems	X	X	X	X	X	X
• Biosensors and Sensor Networks		X	X	X		X
• Advanced Biosignal Processing	X	X	X	X	X	X
• Telehealth, Telecare, Telemonitoring, Telediagnostics	X	X	X	X		X
• e-Healthcare, m-Healthcare, x-Health	X		X	X	X	X
• Assisted Living		X	X		X	X
• Smartphones in BME Applications	X	X		X	X	X
• Social Networking, Computing and Education for Health		X	X			
• Computer Aided Diagnostics		X				X
• Improved Therapeutic and Rehabilitation Methods		X	X			X
• Intelligent Bio-signal Interpretation	X	X				X
• Data and Visual Mining for Diagnostics		X				X
• Advanced Medical Visualization Techniques		X				X
• Personalized Medical Devices and Approaches	X	X	X	X	X	X
• Hardware & Software personalized assistive technologies		X	X		X	X
• Assistive systems for users who are blind or visually impaired		X				X
• Integration between home-based assistive technologies and patient health data	X	X	X	X	X	X
• User interfaces for home-based assistive technologies		X	X		X	X
• Use of prescription systems and assistive technologies		X		X	X	
• Healthcare modeling and simulation		X		X		X
• Knowledge discovery and decision support	X	X	X	X	X	X
• Biomedical data processing	X	X				
• Wearable devices	X	X		X		
• Sensor-based mHealth	X		X	X	X	X

Fig. 6.1 A list of digital health topics that are currently in scientific research and their effect with respect to novel health, current healthcare delivery, and for solving some of the big challenges

Several digital health-related technology ideas already have a demonstrated potential to be disruptive.

Established health tech companies and healthcare providers and institutions need to rethink their strategy. A 3–5-year vision may not be enough, especially with technological developments that could lead to disruptions faster than expected. R & D departments need to be allowed to also develop moonshots and based on 10× thinking, rather than to almost exclusively focus on improving quality of existing products.

Take-Home Messages
- Technologies that develop with respect to performance or price/performance very fast are called exponential (artificial intelligence, gene sequencing and editing, virtual/augmented/mixed reality, size/functionality/cost of health sensors, 3D printing and their clinical applications, and many more).
- Fusing/converging several of these technologies will even develop faster and more impactful (sensors plus deep-tech/data analytics, health data fusion and assessments with wearables and diagnostic imaging, surgical robotics with sensors and artificial intelligence based path planning and decision-making).
- Accelerating technologies lead to other unforeseeable events (on society, employment, business models).
- Exponential technologies develop following the 6D's—they are digital based and scalable, are initially quite disappointing compared to existing approaches, improve fast and eventually disrupt the existing technological solutions. Further reduction in size (dematerialization) and cost (demonetization) will then lead to access for everyone (democratization).
- The dual nature of technologies need to be considered—good things always come with negative effects. For example, a shift to home care will lead to a reduction in professional care. Expensive and scarce medical technologies will be replaced with.
- Digital Health will require a cultural transformation that will eventually lead to a doctor–patient relationship with shared decision-making and a democratization of care.
- Start-ups can greatly benefit from these networks to get access in the different domains and phases throughout the creation, start-up, market access, and scale-up.
- Start-Ups also actively contribute to the network with new methods, products, solutions, and approaches. The dynamic of them will actively implement exponential innovation into the ecosystem.
- The joint activities of established players and new players will make the ecosystem persistent and sustainable, adopting to change with a great reach and grounded base.

References

1. Friebe M (2017) Exponential technologies + reverse innovation = solution for future healthcare issues? What does it mean for university education and entrepreneurial opportunities? Open J Bus Manag 5:458–469. https://doi.org/10.4236/ojbm.2017.53039
2. Kurzweil R (2004) The law of accelerating returns. In: Teuscher C (ed) Alan Turing: life and legacy of a great thinker. Springer, Berlin. https://doi.org/10.1007/978-3-662-05642-4_16
3. Buchanan M (2008) The law of accelerating returns. Nat Phys 4:507. https://doi.org/10.1038/nphys1010
4. Diamandis P (2016) The 6 D's. https://www.diamandis.com/blog/the-6ds. Accessed 9 Mar 2022
5. Kraft D (2011) Exponential medicine. https://danielkraftmd.net/exponential-medicine. Accessed 10 Mar 2022
6. IEEE conference on ICT solutions for e-Health (2022). https://www.icts4ehealth.icar.cnr.it. Accessed 10 Mar 2022

Michael Friebe received the B.Sc. degree in electrical engineering, the M.Sc. degree in technology management from Golden Gate University, San Francisco, and the Ph.D. degree in medical physics in Germany. He spent five years in San Francisco, as a Research and Design Engineer with an MRI and ultrasound device manufacturer. He is a German citizen with expertise in diagnostic imaging and image-guided therapies, as a/an Founder/Innovator/CEO/Investor and a Scientist. He is also a Research Fellow with the Technical University of Munich, Munich; an Adjunct Professor with the Queensland University of Technology, Brisbane; and a honorary Professor of image-guided therapies with Otto von Guericke University, Magdeburg, Germany. Since 2022, he has been a Professor of biomedical engineering innovation with the AGH University of Science and Technology, Krakow, Poland. He is a listed inventor of more than 80 patents and has authored over 200 papers, and was part of over 35 Medical Technology Start-Ups. He is a Board Member of four medical technology start-up companies and an investment partner of a MedTec-fund. From 2016 to 2018, he was a Distinguished Lecturer of the IEEE EMBC teaching innovation generation and MedTec entrepreneurship. He is also a coach and trainer for OpenExO, and a master Launchpad mentor for the Purpose Alliance.

Exponential Medicine: Challenges of Human Spaceflight Bringing Innovations for Earth—A Case Study

7

T. Smith, A. Peterman, and D. Donoviel

Abstract

Long-duration deep space human spaceflight poses many health risks that have never been encountered on Earth and will require new technologies to keep astronauts healthy. The distance from Earth presents many communications challenges and the lack of any possible expedient return to Earth presents an opportunity for a new kind of clinical training and decision support. The Translation Research Institute for Space Health (TRISH), partnered with NASA, has identified the need for digital twin technology and contextual artificial intelligence to autonomously support an astronaut diagnosing and even treating an acute condition during a mission. Existing clinical decision support technologies are tightly bound to medical record processing and lack the ability to simulate and model the conditions of space. To meet these novel requirements, a clinical videogame developer was identified to model the space environment and the human body using their gaming medical education engine to enable real-time decision support for astronauts. Innovating for space will likely push the boundaries of conventional medicine and could very well change the way clinicians learn and practice here on Earth in the future.

T. Smith
Center for Healthy Air, Water and Soil, Christina Lee Brown Environment Institute, University of Louisville School of Medicine, Louisville, KY, USA
e-mail: ted.smith@louisville.edu

A. Peterman · D. Donoviel (✉)
Translational Research Institute for Space Health, Center for Space Medicine, Baylor College of Medicine, Houston, TX, USA
e-mail: andrew@innovationscientists.org; donoviel@bcm.edu

© The Author(s), under exclusive license to Springer Nature Switzerland AG 2022
M. Friebe (ed.), *Novel Innovation Design for the Future of Health*,
https://doi.org/10.1007/978-3-031-08191-0_7

Keywords

Computational modeling · Medical simulation · Digital twin · Clinical decision support · Augmented reality · Training

What Will You Learn in This Chapter?
- Space health deals with very unique problems with respect to health diagnosis and treatment.
- A Digital Twin approach and gamification is suggested.
- Tools and processes developed to address the lack of in-person physician support on space missions may also be applicable to health equity and access problems on Earth.

7.1   Introduction

Medical innovations are often inspired by the need to treat clinical pathologies such as traumatic injury or complex disease. However, some innovations are triggered and constrained by the environment of the patient and the clinician. Space travel is one such situation, where there is a wide range of health concerns that may also need to be addressed under challenging conditions [1]. Further complicating matters, the patient may also need to be the clinician. The Translation Institute for Space Health (TRISH) was established through a NASA cooperative agreement with a consortium, led by the Baylor College of Medicine with partners the California Institute of Technology and the Massachusetts Institute for Technology, to help identify and promote breakthrough clinical science and technology solutions to keep space travelers safe and healthy and bring collateral benefit to terrestrial medicine.

Long-duration space flight, whether in the form of an extended stay at an orbiting outpost around the moon, or a deep space mission to Mars, will present health risks that will require timely identification, diagnosis, and treatment. There is clearly a need for robust, personalized clinical decision support (CDS) for these long-duration missions where return to Earth for treatment is not timely, or even possible. The science fiction series Star Trek invented the mythical "Tricorder" device that, with the wave of a wand, could diagnose any health condition and paired with instantaneous transport to a futuristic spaceship clinic for treatment, made fictitious deep space survivable for humans.

Taking inspiration from science fiction, we can begin to identify the necessary components of a smart system capable of addressing known and even unforeseen clinical requirements ideally for circumstances where telemedicine is not practical or possible due to time-delays imposed by the distance from Earth.

At TRISH, we are developing capabilities to address potential gaps in the level of care requirements already identified by NASA, and specifically, those related to "self-care" and "autonomous" scenarios. One capability is the inflight incorporation of multiscale personalized simulations based on the astronaut's actual clinical data. In addition to high-fidelity medical imagery, a next-generation decision support system should also factor in the special environmental conditions that mirror changes in the appearance and function of physiological systems in the space environment. Finally, in the likely eventuality that the operator of this platform is not a clinician, it is critical to have responsive and adaptive artificial intelligence guidance. We also foresee these tools serving as an in silico analog for preflight training and research to prepare for deep space operational circumstances.

Our institute created the Personalized Avatar Responsive Simulation in Virtual AnaLogs (PARSiVAL) Initiative to encourage innovators to develop a Digital Human-Twin (avatar) that would make possible advanced, hyper-personalized clinical decision support and training systems for astronauts on deep space exploration missions, and for simulative use in virtual analog environments. The PARSiVAL acronym comes from the medieval romance "Parzival," written by Wolfram von Eschenbach. In the story, Parzival quests for the Holy Grail. TRISH's PARSiVAL initiative was designed to expand upon the original definition of the TRISH-funded Virtual Human Simulation Framework Project [2], and to address limitations of that project's scope and vision—specifically the limiting of digital avatars to virtual renderings informed primarily by available physiologic data—by expanding the scope to include the development of multiscale computational modeling and simulation frameworks to characterize the in vivo mechanical environment and produce simulative data beyond what is currently available, and a clinical decision support framework to guide the integration of the outputs of these models with useful clinical guidance and training. To put it simply, we are searching for the Holy Grail of clinical decision support and simulation, modeling the cooperation between Parzival and Artemis in the book, "Ready Player One" as well. Their cooperation is a model for how we cooperate with NASA and others in our efforts to develop a space health toolkit supporting Artemis missions to the Moon.

As described earlier, the components of this system would include personalized digital avatar technology and multiscale model development for key physiologic systems, and a framework for developing advanced autonomous clinical decision support capabilities through the integration of multiscale simulation model outputs for both guideline-based evaluation and avatar-centered virtual representation in an analog environment. To determine which physiological systems combined with what environmental circumstances, TRISH's Science Office turned to the NASA Human Research Program requirements documentation (called the Evidence Books) and the Exploration Medical Capability (ExMC) Medical System Concept of Operations for Mars Exploration Mission-11 baseline document [3]. This analysis of medical technologies used on ISS and those planned for Artemis and interviews with Subject Matter Experts in relevant topic areas were integrated with a review of relevant research and technologies, and market analyses to identify drivers of these technologies and companies at the forefront of their development (Fig. 7.1).

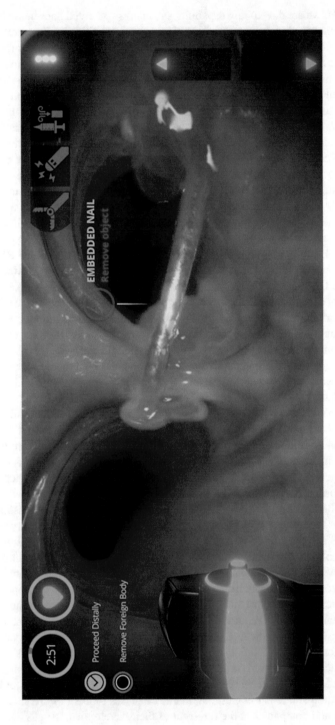

Fig. 7.1 Simulation of laryngoscopic interrogation of airway injury (source: LevelEx Inc)

7.2 Where Will the PARSiVAL Initiative and Other Exponential Space Medicine Innovations Come from?

Sometimes, these advances can come from some surprising sources.

For example, much of clinical decision support technology is tied directly to electronic medical record systems which do not offer simulation or training. The TRISH team identified video game software developer Level Ex Inc. as a promising alternative pathway to achieve a PARSiVAL objective. Level Ex creates video games for doctors that work at the intersection of medicine and entertainment. With titles such as Top Derm, Cardio Ex, Airway Ex, Pulm Ex, and Gastro Ex, they have developed high-fidelity anatomic and physiologic human simulations of disease and treatment.

These solutions are not just games, they are vetted products that earn 750,000 clinicians users required Continuing Medical Education (CME) credits. Importantly, as video game developers, Level Ex designers place a high premium on realism and gamification mechanisms to encourage deep learning and engagement. However, for such a platform to meet all PARSiVAL's requirements, the virtual human simulation for space would need to be enhanced to:

- *Identify and attempt to prioritize medical conditions and spaceflight adaptations of concern—many of which are outside of the current game database.*
- *Incorporate personalized anatomical visualization derived from astronaut clinical images to support long-duration spaceflight mission capabilities and mitigate risks identified in the areas of training, clinical decision support, and autonomous crew health care.*
- *Create applications that include the integration of historical clinical data (personalized when possible), research-driven, environmentally appropriate models of tissues, organs, and organ systems, advanced real-time physiologic monitoring systems, and the environmental systems of the astronaut's spacecraft and habitat, to drive insights and visualizations within the personalized anatomical model.*
- *Incorporate anatomical mapping and surgical planning software features to support the diagnostic visualization of space flight adaptations like microgravity and fractional gravity fluid shifts and related conditions, such as Spaceflight Associated Neuro-ocular Syndrome (SANS), kidney stones, and vascular atrophy.*
- *Incorporate actual astronaut MRI and ultrasound research data.*
- *Create user stories and product backlog items for Digital Twin prototype development based upon information captured from veteran astronauts and aerospace medical team interviews as well as a review of published research reports.*

This is clearly a situation where re-imagining clinical decision support requires a journey from terrestrial clinical simulation and gaming to converge with the unique requirements of space medicine to deliver healthcare very differently to keep astronauts healthy on long missions.

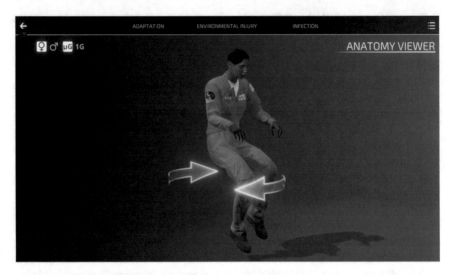

Fig. 7.2 Virtual Human Simulation for Spaceflight (source: LevelEx Inc)

One of the most compelling features of an application like this is in the ability to tune the actual physical laws that are modeled into the simulation to "behave" in accordance with special environmental conditions. Of course, turning off gravity is possible in a virtual environment and the behavior of tissues according to the partial pressures of gasses and fluids in the space environment can be modeled. Simulating the effects of space travel on health is about offering high-fidelity environmental medicine (Fig. 7.2).

It is very possible that all these newly converged technologies for space may find their way back to Earth. For example, here on Earth, medical errors are often the product of an un-converged system. A recent review of 1325 radiology medical malpractice claims revealed the significant role that improper image interpretation plays in subsequent treatment [4]. Many of these common errors could be reduced here on Earth with next-generation systems that are part electronic medical record, part decision support, and part gamified continuing medical education that will improve training and subsequently the quality of care and patient outcomes.

Take-Home Messages for Innovating for Space
1. Be clear about the details of the problem to be solved
2. Determine what adjacent problems have been solved and how
3. Consult subject matter experts to avoid missing obvious parameters
4. Scan the commercial landscape in the category for nascent applications, possibly in adjacent markets
5. Consider other industries that have developed aspects of the solution
6. Find the "win-win" where solving one problem may create a valuable generic capability

Acknowledgments This work was supported by the Translational Research Institute for Space Health through NASA Cooperative Agreement NNX16AO69A. We are extremely grateful to the talented Level Ex team, Victoria Perizes, Erik Funkhouser, Clifton Garner, and Eric Gantwerker, whose work brought the Virtual Human Simulation Framework concept to life.

References

1. HRP Roadmap (2022) Risk of adverse health outcomes & decrements in performance due to medical conditions that occur in mission, as well as long term health outcomes due to mission exposures inflight medical conditions. https://humanresearchroadmap.nasa.gov/Risks/risk.aspx? i=95 . Accessed 27 Feb 2022
2. https://www.bcm.edu/academic-centers/space-medicine/translational-research-institute/ research/trish-innovations/video-games-to-advance-virtual-simulation-in-space-medicine; Accessed 27 Feb 2022
3. https://www.sparknotes.com/lit/ready-player-one/summary/. Accessed 27 Feb 2022
4. https://www.nasa.gov/specials/artemis/. Accessed 27 Feb 2022
5. Medical System Concept of Operations for Mars Exploration Mission-11: Exploration Medical Capability (ExMC) Element - Human Research Program https://ntrs.nasa.gov/ citations/20200001715. Accessed 27 Feb 2022
6. Siegal D, Stratchko LM, DeRoo C (2017) The role of radiology in diagnostic error: a medical malpractice claims review. Diagnosis 4(3):125–131

Dr. Smith co-founded the Envirome Institute in 2018 and joined as a research associate professor in the division of environmental medicine. He is responsible for a wide range of environmental monitoring studies as part of the Institute's interventional Green Heart clinical trial. His environmental medicine research also includes human space health and he serves as a member of the Translation Research Institute for Space Health scientific advisory board. He was a Co-Investigator on the 1998 NASA Neurolab mission. Ted received his B.S. in Biology and Psychology from Allegheny College, his M.S. and Ph.D. in Experimental Psychology from Miami University and completed his post-doctoral studies at the Massachusetts Institute of Technology Human Systems Laboratory.

Andrew Peterman co-founded Innovation Scientists LLC, a software company and consulting firm creating tools for science Program Directors, philanthropic organizations, and private-equity firms who seek to efficiently impact the speed at which radical new technologies are developed. He has a background in Healthcare IT and Enterprise Architecture, and he led the development of the Personalized Avatar Responsive Simulation in Virtual AnaLogs (PARSiVAL) program while with the Translational Research Institute for Space Health (TRISH).

Dr. Dorit Donoviel serves as an Executive Director for the Translational Research Institute for Space Health (TRISH), a NASA-funded institute led by the Baylor College of Medicine (BCM) in Houston, TX with the California Institute of Technology and the Massachusetts Institute of Technology. Dr. Donoviel previously served as a deputy chief scientist and Industry Lead for the National Space Biomedical Research Institute (NSBRI). Before that, Dr. Donoviel worked in the private sector, where she led metabolism drug discovery programs at Lexicon Pharmaceuticals. Dr. Donoviel is an Associate Professor in the Department of Pharmacology and Chemical Biology and the Center for Space Medicine at BCM.

Healthy Longevity

8

Beatrice Barbazzeni

Abstract

Life expectancy has been increasing over time due to better living conditions, advances in medicine, education, and public health concerns. However, this is not associated with increased health longevity due to the increasing rate of chronic diseases, mostly affecting the ageing population. Hence, a negative effect should be expected on the global economy, social systems, and health practices across the most developed countries. The need of translating this biological process into a healthier outcome would benefit the entire healthcare system, but also at the individual and private level. The understating of ageing as a complex event would be necessary when developing innovative approaches to promote health and preventive measures, such as the geroscience, the P4 of medicine, or agetech. To support the growth of a Longevity Industry, the importance of supporting health longevity as a novel medical discipline is recommended when transforming the way in which healthcare is understood and delivered.

Keywords

Health longevity · Chronic diseases · Global economy · Geroscience · The P4 of medicine · Agetech · Longevity industry

B. Barbazzeni (✉)
ESF-GS ABINEP International Graduate School, Otto-von-Guericke-University, Magdeburg, Germany
e-mail: beatrice.barbazzeni@med.ovgu.de

© The Author(s), under exclusive license to Springer Nature Switzerland AG 2022
M. Friebe (ed.), *Novel Innovation Design for the Future of Health*,
https://doi.org/10.1007/978-3-031-08191-0_8

What Will You Learn in This Chapter?
- The increasing rate in life expectancy is not proportional to a parallel increase in health.
- Demographic changes towards the ageing population determine the rise of chronic diseases.
- A negative effect should be expected on the global economy, social and public systems, and health services across developed countries.
- Understanding the ageing process is the key to developing innovative approaches to promote health.
- The growth of the Longevity Industry would be supported by health longevity as a novel discipline in medicine.

Since the nineteenth century, life expectancy has been linearly increasing by almost 2.5 years each decade, particularly in most developed countries [1]. This factor finds its explanation in better public health practices, education, and advances in medicine [1]. Indeed, the discovery of antibiotics and vaccinations positively impacted reducing infectious and parasitic diseases, thus decreasing the rate of deaths while positively influencing longevity [2]. A recent statistic from the World Health Organization found that in the time frame between 2000 and 2019, life expectancy increased by more than 6 years, specifically from 66.8 to 74.3 years. However, although healthy or health adjusted life expectancy (HALE) has also raised in 2019 by 8% from 58.3 to 63.7 years, this fact was mostly related to decreased mortality rates and not reduced age-related disabilities. Hence, increased HALE of 5.4 years does not correspond with a parallel extended life expectancy of 6.6 years [3]. Accordingly, most previous research focused on quantity instead of ameliorating the quality of life with the effect of extending years to ageing and related diseases [1]. Moreover, this evidence translates into having a negative impact on the economy and social system across several emancipated countries; pension, public health, and social costs represent about 25% of the GDP in Europe with an expected rise in the coming years [1]. Being longevity not yet accompanied by HALE, ageing represents one of the main risk factors for developing chronic pathologies and neurodegenerative disease, pressuring the global economy financially, both at the public and private level and stimulating the entire healthcare system of taking proper actions to overcome the burden [2].

Although treatments have been found to be effective in reducing and eliminating certain disease types, these have been replaced by chronic, non-communicable diseases (NCDs); progressively developed in our century. Among age-related chronic diseases, the most common are cardiovascular diseases (CVD), cancer [4], type 2 diabetes, osteoporosis, arthritis, kidney, neurodegenerative, and pulmonary diseases leading to socio-economic challenges due to the increased need for constant care and medical interventions [3]. However, several factors such as a healthy lifestyle, proper nutrition, physical activities, reduced tobacco smoking consumption, socio-economic and work-related status, public health measures, and healthcare accessibility play a central role in determining the rate and intensity of such disease development [4]. Nevertheless, progress in biomedical research results of primary importance when investigating relevant risk factors responsible for morbidity in ageing and disease development, highlighting the understanding of which medical research should be mostly indicated in reducing mortality rates and NCDs incidence while ameliorating HALE, as well as, alleviating the entire healthcare management [1].

In contrast to reduced fertility rates, the need for novel approaches in healthcare to support healthy longevity is particularly high when considering demographic changes towards the ageing population. Indeed, by 2050, approximately 20% of the global population is expected to be represented by elders over 65 years old, reaching almost 2 billion people in a few decades [3]. Thus, the shift towards innovative health programmes and interventions in a preventive, predictive, and proactive medicine perspective would be needed to cope with increasing age-related disabilities while supporting and promoting health longevity [3]. Changes in life expectancy across years in different countries became a central topic consequently in public health communication; an indicator of socio-economic development and available healthcare interventions. Hence, medical research investigating ageing- and disease-related factors of observed mortality rates would become essential because its results would also impact political, social, and public health-informed decisions influencing longevity and life quality [4].

Understanding ageing as a considerable and complex event in constant expansion is necessary when implementing innovative approaches. In the perspective of the economics of ageing, the European Commission predicted that the public health costs would increase between 0.9 and 1.6% of GDP by 2070 compared to 6.8% in 2016 [5]. Thus, the need to develop a unique assessable approach, to foresee and analyse market resources, risks and obstacles gave rise to the Longevity industry. It is a multidimensional sector extended to different domains such as science, technology, as well as political, financial, and social to generate growth and positive outcomes in promoting healthy longevity based on innovative trends and sub-sectors like *geroscience*, the *P4 of medicine,* and assistive technology such as *agetech* that combines digital technology with medicine to ameliorate life quality, independence, cognitive skills monitoring, and social connection [5].

Towards predictive and preventive medicine, data science and artificial intelligence (AI) systems represent important approaches leading to the development of the Longevity industry. Research on regenerative medicine is evolving, raising hopes and promising foresight to enhance HALE and consequently positively transform the healthcare management system [6]. In particular, discoveries in epigenetics, gene therapies, and transcriptomic and the identification of multimodal biomarkers related to specific patterns and mechanisms underlying disease onset would contribute to acquiring insightful information on the ageing process. However, translating such results into a clinical setting still requires more effort and the integration of technological advances [6]. Reversing or delaying disabilities and themselves, the understanding of biological ageing has also been a topic widely examined in molecular biology and genetics related to gene expression mechanism

[7]. In addition, the consideration of genetic and epigenetic factors producing different responses to environmental stimuli throughout life has also been connected to disease development [7]; particularly, these factors are relevant in the perspective of continuously monitoring a patient's behaviour in response to medications.

Therefore, the investigation of basic mechanisms of ageing has influenced different research fields related to biology. Studies on cellular functioning have highlighted those principal biological processes of ageing [7] further to develop proper intervention despite the complexity of age-related mechanisms. Furthermore, several studies investigating structural and functional decline behind ageing were drawn to explore principal factors that determine the severity of diseases or delay their onset leading to extended health-span and autonomy; a novel approach in geriatric medicine called "geroscience" [2]. Indeed, reducing the burden of chronic diseases would generate parallel effects on care management, the interplay between stakeholders, policymakers, and generally the worldwide economy [2]. Demographic changes towards ageing represent the twenty-first-century phenomena characterized by challenges and opportunities to implement innovative solutions supported by science and technologies. Hence, giving value to health and longevity can be achieved by extending lifespan or increasing quality of life [8]. In this regard, advances in science and technologies brought by digitalization and Industry 4.0 generate several advantages in preventing disabilities while promoting autonomy, productivity, and well-being against functional and physiological decline.

Past researchers associated health longevity with improved biomedical innovation throughout history. Recently, studies conducted by Lichtenberg [8] aimed to measure the effect of biomedical development on health longevity across Americans and other sample populations in recent decades. Based on different approaches and methodologies, medical innovations were measured in relation to pharmaceutical, diagnostic imaging innovation (e.g. CT and MRI scans), and the impact of inpatient hospital procedures [8]. Moreover, measures such as mean *vintage* [8] of medical products and services, the number of available medications, as well as, lifespan (e.g. quantity), and functionality of drugs on health (e.g. quality) were taken to investigate the impact of biomedical innovation on health over longitudinal sample data. The first study explored the effect of medical care procedures (e.g. diagnostic imaging, treatments, and care practices) and behavioural risk factors (e.g. obesity, smoking, AIDS, education, income, health insurance reimbursements) on longevity. It was found that improved medical techniques and medical treatment prescriptions positively affected life expectancy.

In contrast, a decrease in the quality of medical education determined a negative impact on lifespan. In addition, reduced incidence of obesity, smoking, and AIDS positively affected life expectancy. Still, no correlations were found with insurance costs or education level, whereas an inverse correlation with income growth was observed. A second study [8] evaluated the effect of newly released drugs on premature mortality rates from rare diseases over longitudinal data collected in the USA and France between 1999 and 2007. It was found that premature mortality was unrelated to the number of drugs approved within 2 years earlier but significantly related to those drugs approved 3–4 years earlier. In a third study [8], pharmaceutical innovation on individuals receiving Social Security Disability Insurance support was investigated, finding that predicted increase in disability was inversely related to increased availability of new medications. Moreover, in a fourth study [8], pharmaceutical innovation was explored in relation to functional limitations in elders based on a nursing home service survey, finding that the increased availability of novel treatments was related to improved independence and autonomy of daily living activities. Lastly, in a fifth study [8], Lichtenberg explored the effect of newly developed drugs in cardiovascular diseases on hospitalization and mortality rates between 1995 and 2003, finding that increased access to specific drugs was positively associated with decreased hospitalization periods and deaths, but further studies would be needed to clarify associated risk factors of these obtained results.

Moreover, to reduce the rate of chronic diseases in promoting a healthy life span, preventive medicine with a geroscience approach would be needed in clinical practice [9]. However, the investigation of mechanisms and processes underlying ageing would require the assessment of a large volume of data acquired on a longitudinal scale, including different populations, socio-economic conditions, and other behavioural characteristics [9]. Thus, the evolution of biomedical research towards AI allowed the implementation of an advanced method of computing data to organize and analyse big datasets over multiple parameters (e.g. biological data, blood tests analysis, gene expression information) to find similar patterns typically observed in ageing. This advanced approach was found to predict the outcome of clinical trials and the discovery of drugs towards personalized medicine [9], as well as powered by deep learning (DL) methods to establish "biological age" based on an individual's biological data [9]. AI technologies would definitely further develop the study on ageing and disease-related processes to define specific and individual interventions based on recurrently observed biomarkers [9]. With this goal in mind, treatments and programmes to stimulate health longevity characteristics will be applied in research and clinical settings, giving rise to the sector of longevity medicine (see Fig. 8.1) and further supporting the evolution of the Longevity industry [9].

In conclusion, to transform longevity into a novel medical discipline, ageing should be viewed as a condition that needs to be treated, a perspective that privileges adopting a preventive over a reactive approach while targeting ageing instead of treating chronic diseases impacting health lifespan. A revised education that prepares clinicians and health professionals is necessary to bring innovation into medicine while embracing new approaches and methods. Longevity medicine is a sub-field of precision medicine that advances diagnostics and interventional methods, mostly based on AI technologies to enhance health and life expectancy [10]. Emerging technologies would be implemented to discover geroprotective and personalized treatments, identify specific biomarkers of ageing and its development, interventions to delay the progress of ageing to arrest or reverse it, and lastly, the speculation and impact on the healthcare system from a social and economic perspective [10]. Advanced AI-based technologies allow the longitudinal and continuous monitoring of the ageing process over larger populations and scaled over different biomarkers to define and identify typical and recurrent ageing patterns.

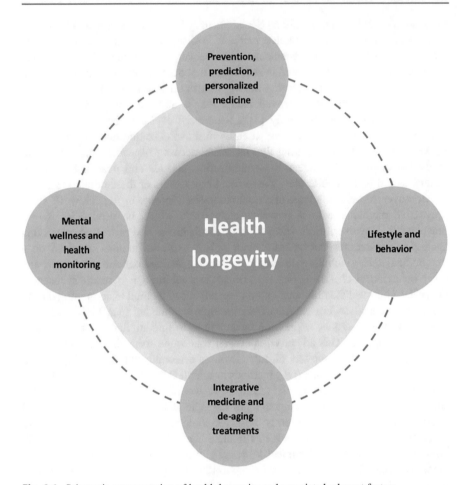

Fig. 8.1 Schematic representation of health longevity and associated relevant factors

Moreover, deep learning methods could also be used as ageing clocks to estimate an individual's biological age based on routine clinical data. Such technologies can be transferred to assess and supervise health and disease risks by adjusting lifestyle and interventional programmes suitable for each individual. These deep learning clocks could become part of the doctor's instruments supporting HALE [9, 10]. This has the effect of promoting personalized and preventive medicine (e.g. drugs, analysis of medical data) in clinical settings while revolutionizing how healthcare processes operate [10].

Take-Home Messages

- *Increased life expectancy should be accompanied with health longevity, thus a novel healthcare approach is needed towards this outcome.*
- *The understanding of ageing, as a complex biological process, is the starting point when developing innovative methods to promote health longevity.*
- *The growth of the Longevity Industry would be facilitated by a revised education in which longevity becomes a novel discipline in medicine generating innovation in healthcare.*
- *Health longevity is the results of mutual contributing factors such as the P4 of medicine, mental wellness and health monitoring, active lifestyle and healthy habits, integrative medicine and de-ageing treatments.*

References

1. Brown GC (2015) Living too long: the current focus of medical research on increasing the quantity, rather than the quality, of life is damaging our health and harming the economy. EMBO Rep 16(2):137–141. https://doi.org/10.15252/embr.201439518
2. Vaiserman A, Lushchak O (2017) Implementation of longevity-promoting supplements and medications in public health practice: achievements, challenges and future perspectives. J Transl Med 15:160. https://doi.org/10.1186/s12967-017-1259-8
3. World Health Organization (WHO) (2020) GHE: life expectancy and healthy life expectancy. https://www.who.int/data/gho/data/themes/mortality-and-global-health-estimates/ghe-life-expectancy-and-healthy-life-expectancy. Accessed 2 Sep 2021
4. Klenk J, Keil U, Jaensch A et al (2016) Changes in life expectancy 1950–2010: contributions from age- and disease-specific mortality in selected countries. Popul Health Metrics 14:20. https://doi.org/10.1186/s12963-016-0089-x
5. Goldsmith C (2019) The longevity industry comes of age. The New Economy. https://www.theneweconomy.com/business/the-longevity-industry-comes-of-age.   Accessed 03 September 2021
6. Moskalev A, Anisimov V, Aliper A, Artemov A, Asadullah K, Belsky D, Baranova A, de Grey A, Dixit VD, Debonneuil E, Dobrovolskaya E, Fedichev P, Fedintsev A, Fraifeld V, Franceschi C, Freer R, Fülöp T, Feige J, Gems D, Gladyshev V et al (2017) A review of the biomedical innovations for healthy longevity. Aging (Albany NY) 9(1):7–25. https://doi.org/10.18632/aging.101163
7. Institute of Medicine (US) Committee on a National Research Agenda on Aging, Lonergan ET (ed) (1991) Extending Life, enhancing life: a national research agenda on aging. National Academies Press (US), Washington, DC. 2, Basic Biomedical Research. https://www.ncbi.nlm.nih.gov/books/NBK234008/
8. Lichtenberg FR (2015) The impact of biomedical innovation on longevity and health. Nordic J Health Econ. https://doi.org/10.5617/njhe.1290. https://www.researchgate.net/publication/267781624_The_Impact_of_Biomedical_Innovation_on_Longevity_and_Health
9. Zhavoronkov A, Bischof E, Lee KF (2021) Artificial intelligence in longevity medicine. Nat Aging 1:5–7. https://doi.org/10.1038/s43587-020-00020-4

10. Colangelo M (2021) AI-enabled doctors will specialize in longevity medicine. Longevity technology. https://www.longevity.technology/ai-enabled-doctors-will-specialise-in-longevity-medicine/. Accessed 03 September 2021

Beatrice Barbazzeni is a proactive scientist and prolific author by *ExO Insight*. With a background in psychobiology, she is currently finishing a Ph.D. in cognitive neuroscience. She likes to challenge herself by joining different research institutes and interdisciplinary teams, with a particular interest in exponential medicine and health longevity. Beatrice is an effective communicator, fluently speaking several languages which allows her to share multicultural experiences and network globally. Last but not least, she has a fantastic passion for fitness, art, and music. The key to her success is discipline and grit; the values that inspire her to become a leader in science.

The Science of Health Longevity

9

Beatrice Barbazzeni

Abstract

An emerging field of science, called longevity medicine, is rapidly evolving aimed at investigating and detecting early biomarkers of age-related diseases based on advanced technologies. Thus, the availability of these advances in medicine would transform healthcare globally while paving the way towards a preventive, predictive, and personalized system. However, a revised education is necessary to prepare future healthcare professionals capable of identifying unmet clinical needs with a patient-centric, outcome-based, and multidisciplinary approach. Moreover, well-being, positive emotions, and dietary restrictions were found to potentially have geroprotective effects against age-related processes, and therefore related to a healthier lifespan. Furthermore, in response to the increasing rate of the ageing population and besides innovative trends in medicine, the promotion of worldwide events would positively influence public health and longevity. Thus, the importance of evolving a longevity mindset, made of healthy habits and a proactive attitude, would reverse the old concept of ageing, usually related to diseases and mortality, into an opportunity to cultivate health with positive, wiser, and forward-looking thinking.

Keywords

Longevity medicine · Age reversal · Artificial intelligence · Biomarkers · Positive emotions · Dietary restrictions · Global events · Longevity mindset

B. Barbazzeni (✉)
ESF-GS ABINEP International Graduate School, Otto-von-Guericke-University, Magdeburg, Germany
e-mail: beatrice.barbazzeni@med.ovgu.de

What Will You Learn in This Chapter?
- Longevity medicine is a novel field of medicine aimed at investigating age-related biomarkers through the implementation of advanced technologies (artificial intelligence and machine learning).
- Positive emotions, dietary restrictions, and healthy habits might have geroprotective effects and thus be related to a healthier lifespan.
- In response to the increasing rate of the ageing populations, investing in worldwide events is needed to promote public health through awareness and a proactive behaviour.
- Evolving a longevity mindset would reverse the concept of ageing from a negative state of being into an opportunity to cultivate health.

9.1 Longevity Medicine

A novel branch in medicine is now quickly evolving towards investigating health longevity based on advanced methods to detect age-related biomarkers. This field is called longevity medicine and involves a few sectors from geroscience, precision, preventive, functional medicine, and biogerontology. Indeed, the focus of longevity medicine is the possibility to detect early signs of age-related (communicable or non-communicable) disease development through artificial intelligence (AI) and machine learning (ML) methods, while preventing its progression [1, 2]. Thus, the education of physicians should be revised to gain expertise in this emerging field of science, especially when facing the challenges brought by this century in healthcare. Indeed, the entire healthcare system is undergoing a transformative process where interdisciplinary teams merge to meet clinical needs with a patient-centric and outcome-based approach [1, 3]. The availability of advanced AI-, ML-based systems, deep learning, and digital technologies represents ground-breaking procedures towards the understanding of ageing and longevity mechanisms, and in which chronological age can be distinguished from biological age. Therefore, their application in preventive, predictive, and precision medicine would support physicians and clinical professionals in their practice, while evolving the entire system from reactive to preventative and proactive. The ultimate goal of healthcare is to promote health longevity and age reversal [1, 2].

In response to the ongoing increase of the ageing population, which in turn will translate into a global social and economic concern, healthcare providers should be prepared in addressing patients' needs with innovative and individualized treatments but also in fostering the shift in the lifespan from sickness to health [1, 2]. Hence, longevity medicine bringing advanced methods, mostly based on AI, in diagnostics and treatments, requires clinical professionals to be well prepared in their implementation. Whether longevity medicine is about remote monitoring of a patient health data, the identification of recurrent age-related patterns, ageing clocks capable of estimating an individual's biological age, or disease-related biomarkers detection, it is the future of health that asks for the acquisition of a novel set of skills and knowledge [1, 2].

Besides medicine and outstanding scientific treatments to reverse age, how can you boost habits for longevity on a daily basis?

9.2   Positive Emotions for a Healthier and Longer Life

A study by Diener and Chan [4] reported that the perception of well-being, measured as life satisfaction, optimism, more positive over negative emotions, is related to a healthier and longer life expectancy. Indeed, several longitudinal studies have demonstrated that individuals who feel happy and satisfied live longer and healthier even when controlling variables such as health and socioeconomic status are at baseline. Thus, experiments on animals and humans, in which the physiological change in perceiving well-being was examined, raised interest. Nevertheless, it is still doubtful whether subjective well-being can positively influence health longevity even in patients affected by diseases (e.g. cancer). Although positive emotions showed potential in improving health and longevity, more studies are needed to enhance these findings, with more robust statistical evidence and experimental designs [1].

9.3   Geroprotective Mechanisms Behind Dietary Restrictions

Differently from malnutrition, dietary restriction (DR) was initially discovered in 1917 by Osborn, although the benefits of DR have been explored later in the 1980s when the first single-gene mutation was found to extend the lifespan of nematode worms [1, 5]. Therefore, more studies have been conducted to investigate how mutations may influence health and longevity, and how these findings would benefit the prevention of age-related diseases in humans while suggesting novel clinical interventions. However, although the majority of experiments were conducted based on invertebrates, rodents, and non-human primates, additional research is needed to prove and confirm the effect of DR in humans to protect from developing cancer, obesity, cardiometabolic and neurodegenerative conditions, and other diseases. In addition, it should also be considered that the effect of DR on longevity is also influenced by genetic and epigenetic factors. Indeed, even in genetically identical rodents, the same type of DR had a diverse effect on the lifespan [1, 5].

In this regard, a review by Green, Lamming, and Fontana [5] discussed evidence about dietary restrictions on mammalian organisms and their effect on stimulating specific mechanisms at the physiological, metabolic, and molecular level underlying age-related disease prevention in humans. Moreover, different types of fasting, protein restriction, or reduction in essential amino acids to promote healthy longevity were also discussed.

Investigating non-mammalian organisms can reveal genetic, epigenetic processes and microbial factors underlying ageing and related diseases, including nutrient sensing, genomics, and protein homeostasis [5]. In several organisms, DR was also found to increase stem cell renewal, physiological, and regeneration of injured tissues, through the inhibition of mTORC1 signalling [1, 5]. Moreover, DR was found to modulate multiple metabolic pathways which typically decline in ageing. Thus, investigated geroprotective mechanisms were the down-regulation of GH and insulin/IGF1 signalling, the reduction of mTORC1 signalling, the activation of GCN2, and reduced protein synthesis, the effect of FGF21 signalling related to nutritional stress and fasting, the activation of sirtuins (a family of nicotinamide adenine dinucleotide (NAD+)-dependent deacetylases), and the reduction of oxidative stress [5].

Nevertheless, recent findings did show how the restriction of specific nutrients and timing of food intake are relevant in modulating age-related processes [1, 5]. A few dietary manipulations were protein restriction, which positively extended lifespan and independently of calorie intake in rodents [1, 5], methionine, threonine, tryptophan, and branched-chain amino acids (BCAAs) restriction was also identified as capable of increasing a healthy lifespan. Moreover, meal frequency and timing might also impact longevity and health. Recently, imposed daily fasting was found to generate metabolic effectables to enhance lifespan in mice [1, 5], but the geroprotective effects of DR were found on cognition and frailty [1, 5]. Furthermore, a 24-h fast every other day or twice/week had the effect of extending lifespan in mice and rats, and prolonged, periodic fasting that lasts more than 24 h and repeated once or twice/month had similar effects. Lastly, time-restricted feeding would regenerate disrupted circadian clock rhythms in mice, while protecting them from developing obesity, insulin resistance, inflammations, and hyperinsulinaemia [1, 5].

Promoting healthy habits with specific eating choices would be crucial in stimulating geroprotective mechanisms to prevent chronic and age-related diseases. However, more studies are needed to better investigate how nutrients (e.g. amino acids, sugar, fats, etc.), DR, regular exercise, and cognitive training may influence longevity and health. Even though genetic and epigenetic factors interact with DR and diverse effects might be expected in humans due to their heterogeneity, future developments should focus on identifying personalized approaches to dietary interventions to promote health longevity at the individual level [5].

9.4 Health Longevity and the Global Grand Challenge

Advances in longevity medicine and public health are going to positively affect health longevity and this event is evolving in parallel to a demographic change towards the ageing population. Today 8.5% of people (617 million) are aged over 65 and by 2050 this estimation is predicted to double reaching 1.6 billion worldwide [1]. Thus, in the next 30 years, individuals aged 80 or even older are expected to more than triple, by growing from 126 million to 447 million [1]. This phenomenon would have a considerable effect on the global economy and care delivery services.

However, why should not we think the reverse? Indeed, developments in regenerative and longevity medicine, characterized by disruptive trends, technologies, and treatments would bring more benefits than burdens, especially when the focus is on promoting a healthier and longer life. Thus, if we look at the other side of the coin, there will be just opportunities [1].

In this regard, the U.S National Academy of Medicine founded a global movement to improve health and well-being from a physical, mental, and social perspective called the "Healthy Longevity Global Challenge". Accordingly, this movement aims to [1]:

- Transform challenges into opportunities in response to a growing ageing population
- Investigating innovative methods to address ageing while promoting healthy longevity
- Support transformative and scalable innovations across the world
- Create a solid ecosystem made of scientists, entrepreneurs, clinical professionals, policymakers, and innovators to foster the future of health

Moreover, the challenge is structured into two main initiatives [1]:

1. *The Global Roadmap for Healthy Longevity*: aims to create an independent International Commission to provide recommendations across worldwide societies regarding strategies, social and behavioural attitudes, technological and scientific advances in the perspective of healthy longevity.
2. *The Healthy longevity Global Competition*: launched in 2019, it is a multi-million-dollar international competition to boost advances in healthy longevity with rewards and prizes. This competition welcomes innovative ideas from science, medicine, public health, engineering, entrepreneurship, public policy, and much more to enhance the future of health and longevity.

9.5 Evolve a Longevity Mindset to Cultivate Health

Peter Diamandis talks about evolving a longevity mindset, made of simple behaviours that will shape your health (Fig. 9.1). Hence, to transform your action today while positively impacting your tomorrow, Diamandis highlights six key areas to focus your energy on [1, 6]:

1. *What you believe*: it is important to think positively about your age. Instead of associating ageing with a "disease", think about how advances in medicine can help you foster your health and longevity.
2. *How you document yourself*: what you choose to read makes the difference about how you approach ageing. Thus, choose carefully how you inform yourself about longevity.

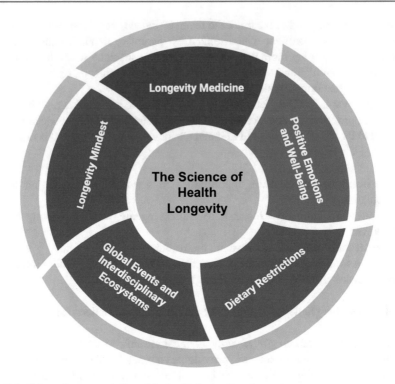

Fig. 9.1 Schematic representation of the health longevity science

3. *The community that surrounds you*: people around you can influence your thoughts and behaviours. Surround yourself with positivity and wise allies. You will not associate your age with death or diseases anymore, but you will be focused on stimulating mental and physical activities to promote the coming healthy years.
4. *Sleep*: it has several functional, physiological, and cognitive benefits to health, especially in ageing and against age-related disease development. Thus, consider making the proper time for sleep.
5. *Nutrition/diet*: among diets and nutritional advice, based on the latest scientific discoveries intermittent fasting seems to have a positive effect on longevity.
6. *Physical exercise*: besides sleep and nutrition, exercise regularly is essential to sustain muscle mass, bone density, and metabolisms.

Moreover, one of the most recent advances in the field of medicine and longevity is gene therapy. Rejuvenate Bio, co-founded by Dr. Church, implements gene therapy to rejuvenate animals to translate this method to reverse age in humans. Indeed, research is focused on specific genes capable of reversing the ageing clock in old animals. This gene-editing method, tested in mice, did show improvement in several age-related diseases such as type 2 diabetes, heart, kidney, and neurodegenerative diseases. In addition, the company is already testing gene therapy in dogs with the hope to begin clinical trials even in humans and FDA approved [1, 6].

Furthermore, another company called Qihan Biotech implements gene-editing and transplantation immunology to create organs that would be transplanted into humans. Currently, genetically-engineered pig organs are created which are resistant to ageing and infections, although more tests are needed before being applied in humans [1, 6]. Thus, controlling gene expression would be the key to reversing the process of ageing, as demonstrated by Dr. Sinclair, professor of genetics at the Harvard Medical School and Co-founder of the journal *Aging*. Three "Yamanaka factors (genes)" were used to reverse ageing in his research. Indeed, these genes were typically expressed in embryos, and so very powerful in driving this process of reprogramming cells. With this method, Dr. Sinclair and his collaborators have already revitalized the eyes of mice affected by glaucoma. However, the team is already planning to continue the research focusing on restoring the liver, muscle, brain, and ears [1].

In conclusion, evolving a longevity mindset means reversing the old concept of ageing, which is characterized by a progression towards "death" and "diseases" into an optimistic opportunity to cultivate health. Indeed, Industry 4.0 and digital transformation are a source of innovation developing what is called "exponential medicine". Therefore, outbreaking methods such as genome sequencing, RNA transcriptomics, vaccines, CRISPR, gene therapies, quantum computing, AI technologies, and many others have already supported the process of ageing, extending health to years of life. Besides medical advances, the new attitude is not considering ageing as a "disease" but a process that can be improved or even reversed with the right mindset and healthy habits [1].

Take-Home Messages
- *Longevity medicine, an emerging field of science, aims to detect biomarkers of age-related diseases with advanced technologies and in the perspective of preventive, predictive, and precision medicine.*
- *Positive emotions, well-being, and life satisfaction are related to a healthier and longer life expectancy.*
- *Dietary restrictions may stimulate geroprotective mechanisms while preventing the development of age-related diseases.*
- *Promoting health longevity through awareness and global events would benefit public health and in response to an increasingly ageing population.*
- *Evolving a longevity mindset through novel habits and proactive behaviour would reverse the concept of ageing from a negative state of being into an opportunity to cultivate health and well-being.*

References

1. Barbazzeni B (2022) Gaining health to your years: science tells you how to live longer and healthier. ExO Insight.. https://insight.openexo.com/gaining-health-to-your-years-science-tells-you-how-to-live-longer-and-healthier/. Accessed 16 February 2022
2. Bischof E, Scheibye-Knudsen M, Show R, Moskalev A (2021) Longevity medicine: ups killing the physicians of tomorrow. Lancet 2(4):E187–E188. https://doi.org/10.1016/S2666-7568(21)00024-6
3. Zimlichman E, Nicklin W, Aggarwal R, Bates DW (2021) HealthCare 2030: the coming transformation. NEJM Catal. https://doi.org/10.1056/CAT.20.0569
4. Diener E, Chan MY (2011) Happy people live longer: subjective well-being contributes to health and longevity. Health Wellbeing 3:1. https://doi.org/10.1111/j.1758-0854.2010.01045.x
5. Green CL, Lamming DW, Fontana L (2022) Molecular mechanisms of dietary restriction promoting health and longevity. Nat Rev Mol Cell Biol 23:56–73. https://doi.org/10.1038/s41580-021-00411-4
6. Diamandis PH (2021) How to impact your longevity today. https://www.diamandis.com/blog/how-to-impact-your-longevity-today. Accessed 3 January 2022

Beatrice Barbazzeni is a proactive scientist and prolific author by *ExO Insight*. With a background in psychobiology, she is currently finishing a Ph.D. in cognitive neuroscience. She likes to challenge herself by joining different research institutes and interdisciplinary teams, with a particular interest in exponential medicine and health longevity. Beatrice is an effective communicator, fluently speaking several languages which allows her to share multicultural experiences and network globally. Last but not least, she has a fantastic passion for fitness, art, and music. The key to her success is discipline and grit; the values that inspire her to become a leader in science.

Space Healthtech: Innovation Base for Longevity

10

Thi Hien Nguyen

Abstract

Research on astronaut health and model organisms have revealed six features of spaceflight biology which guide our current understanding of fundamental molecular changes that occur during space travel. The features include oxidative stress, DNA damage, mitochondrial dis-regulation, epigenetic changes (including gene regulation), telomere length alterations, and microbiome shifts. We review the known hazards of human spaceflight, the effects of spaceflight on living systems through these six fundamental features, and the associated health risks of space exploration. We also discuss the essential health and safety concerns for astronauts involved in future missions, especially in the planned long-duration and Martian missions. While a lot of technologies (downsizing laboratories, training chambers, wearables) were developed for the Space Medicine industry, they also found their applications on Earth. They now allow humans to live both active and healthy lives, and enable their safely entering hazardous environments of deep water as well as deep space. Key genes and polymorphisms that allow us to assess human predispositions to chronic diseases and evaluate astronauts' potential will also be shortly discussed.

Keywords

Space medicine · Longevity · Precision medicine · Biomarkers · Aging · Healthy lifespan

T. H. Nguyen (✉)
spaceconnex, Bad Dürkheim, Germany
e-mail: hien@spaceconnex.com

> **What Will You Learn in This Chapter?**
> - Personalized Medicine benefits from Space Health activities
> - There is a technical overlap between longevity and space health research.
> - Identifying and monitoring relevant biomarkers is a key to precision medicine.

10.1 Introduction

The personalized approach and ability to track the medical conditions in real time would be a huge step toward a healthier lifestyle. Here we can see the obvious synergy among medical approaches that accelerate the space exploration process and are able to mitigate aging-related issues as well. So the question of getting astronauts on other planets is inseparable from prolonging life of Earth population.

10.1.1 Common Biomarkers of Human Longevity and Adaptation to the Spaceflight Environment

Space is not a friendly place for a human to be in. Even if you are aboard a spaceship that seems quite protected from the hazards of the empty void outside, there are still plenty of hazards that you can encounter in every step. Humans are just not supposed to be anywhere but on Earth. The calm and exciting feeling of weightlessness, not to mention magnetic fields and all the sources of radiation, dis-regulates the whole human body. Astronauts enter the orbit in perfect health, but during spaceflight, their bodies undergo physiological changes. We can monitor these changes with specific biomarkers, which stabilize during the stay in space. After the flight, the astronaut's health indicators go back to normal; however, substantial rehabilitation is needed (see Fig. 10.1). Let us examine the reasons for danger and how they impact an astronaut's health.

10.1.2 Radiation

Radiation is definitely a thing to be concerned about when speaking about space missions. The main sources of it are Solar Ultraviolet radiation, Van Allen radiation belt (can be negligible), Galactic cosmic rays, and especially Solar energetic particle events (SPEs). While the average natural radiation on Earth is about 0.0024 Sieverts (Sv) per year, the radiation to hit an astronaut on an interplanetary flight is estimated to be about 0.5 Sv per year. As the lowest annual dosage that increases cancer risk for an acute single radiation dose is about 0.2 Sv and a chronic dose is about 0.4 Sv, the probability that an astronaut flying to Mars develops a malignant tumor is high

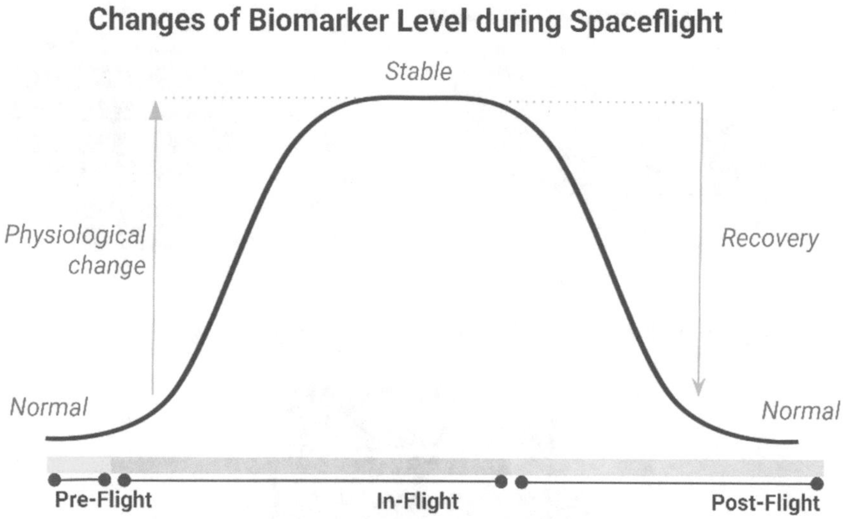

Changes of Biomarker Level during Spaceflight

Fig. 10.1 The physiological changes in astronauts (Source: JAXA)

[1]. There are various types of radiation, but generally, they have the same effect. High-energy protons collide with atoms and molecules and destroy corresponding structures. One of the dangerous effects of radiation exposure is DNA damage. It can lead to cell death, cellular senescence, and tumorigenesis. Luckily, human organisms have various DNA damage repair pathways to ensure genome stability [2]. It is still unclear though if this mechanism works properly in the state of weightlessness. There was an experiment in which fruit flies have died in space from a radiation dose that is not lethal under earth conditions [3].

10.1.3 Weightlessness

The state of weightlessness is probably the most popular reason people want to go to space. However, it is the first thing to harm an astronaut's well-being. Once subjected to the absence of gravity, astronauts experience a formidable fluid shift to the upper body as the gravity does not pull it down. The cardiovascular system of the upper body gets overflowing, inducing cumulative damage to the brain, eyes, and heart. The body triggers some neurohumoral responses to put the fluid back, which leads to negative consequences. Both muscle and bone masses reduce with time (see Fig. 10.2 for the Health Risks of Astronauts on short-term and long-term space flights) while immune cells are less likely to be activated due to the cytoskeleton deformation [4].

The cases of weightlessness also affect the endocrine system. Due to the decreased activity of the sympathetic section of the human brain, which controls adrenal, thyroid, and other endocrine functions, astronauts experience decreased

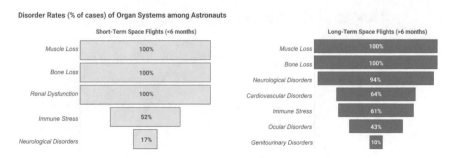

Fig. 10.2 Health Risks for Astronauts: Long-Term versus Short-Term Spaceflight (Source: NCBI)

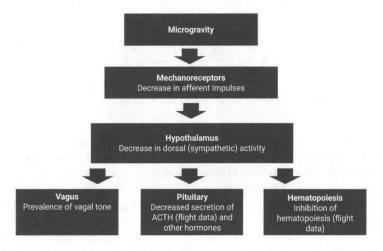

Fig. 10.3 Hypothetical mechanisms for the effects of afferent impulses on autonomic regulation in microgravity (Source: Advances in Space Biology and Medicine)

secretion of hormones, metabolism regulation malfunctions, and increased oxidative reactions levels in the body (see Fig. 10.3 for the effects of Microgravity).

10.1.4 Oxidative Stress

Oxidative stress is a condition in which the balance between **reactive oxygen species (ROS)** manifestation and an organism's biological ability to inhibit oxidative reactions, reduce molecules, and detoxify or reactively repair itself is shaken, resulting in the overall damage to most of the processes in the human body. Under this condition, the ROS production goes into overdrive, which can result in serious biological malfunction, including DNA damage, degradation of protein and lipids, and different neuro-degenerative diseases such as Alzheimer's, Parkinson's, or different forms of sclerosis. With oxidative stress being one of the preceding conditions to more difficult forms of biological damage, it is one of the most important ailments to combat and mitigate through medical research and development [5].

Along with oxidative stress, **DNA damage** is another destructive consequence of the human's presence in space. Caused by space radiation and cases like oxidative stress, the DNA damage can have irreversible negative effects on both the affected persons' and their descendants' genetic and biological structure, resulting in decreased life expectancy and worse overall quality of life due to a damaged immune system. The tests conducted both during and after the spaceflight indicate that genome alterations were potentially related to stem-cell damage and overall cellular instability. While DNA damage is a serious issue to be dealt with, it can serve as a biomarker for human exposure to radiation or other outstanding issues that the human organism is currently affected by [6].

Considering the fact that mitochondria represent the majority of human cell energy supply, **mitochondrial dis-regulation**, which has also been reported as a possible issue in medical research, can also be assumed as a biomarker of human health in space. The main, and the most concerning, issue lying with any dysfunction related to mitochondria lies in their primary function—energy supply. If mitochondria function is disrupted, it leads to dysfunctional metabolism, resulting in cells not getting enough energy to operate properly, henceforth leading them to either become senescent or start decomposing, which causes cell termination. The irregularly functioning metabolism may result in severe metabolic diseases such as obesity, diabetes, and cancer.

One of the most alarming issues that Space Medicine faces in regard to human health is the lasting genetic impact that space exploration may have on human organisms. **Epigenetic changes** may occur due to different factors and effects that space tends to project, like the earlier-mentioned radiation. The exposure to the acute homeostatic disruption and other stress factors may have lasting effects not only on the affected person's genotype but can also affect the descendants of the said person, disrupting their phenotypic properties in a way that would have never appeared in their ancestry line. Therefore, any of such changes, i.e., the Novelty Stress marker—H3 histone phosphoacetylation—or other Social and Material Deprivation markers can act as biomarkers of Human Longevity in regard to human space exploration (Fig. 10.4) [7].

The **telomere length alterations** is another effect that space might have on the human body. Telomeres preserve the stability of the human genome, also acting as possible evidence for other human auto-repair biological systems being relatively intact; hence, there is a lower probability of premature cell senescence. While the telomeres do erode naturally with cell division and aging, they can also be affected by different factors such as space radiation, oxidative stress, different forms of infections, etc. Just like the human body uses the telomere length as the indicator for further "considerations," this parameter and the changes to it may serve as a biomarker of Human Longevity.

The **microbiome shifts** can show the effects that space can induce on the microorganisms and their population size in the human body. With all of the different factors that affect the human organism in space, the entirety of the body

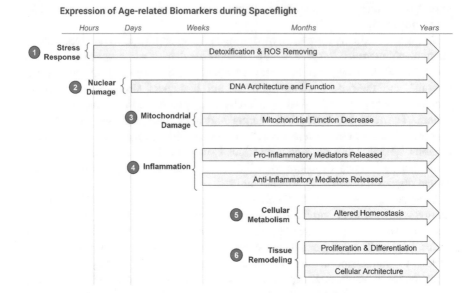

Fig. 10.4 Genomic Biomarkers are Warning Signs for Health in Space (Source: Science Magazine)

is under influence of those inhabiting most of the cavities and surfaces of the human body. As the microbiome is affected, regardless of localization, the body systems may start malfunctioning, therefore leading to significant health effects. Any shifts in the microbial population ratio, when monitored, act as a major biomarker of the state of the human organism. The research into the effects whose shifts in the microbiome are attributed specifically to space is still ongoing as this exact field is not yet fully explored [8].

The human organism per se can act as an indicator of certain factors that might occur in space. It is heavily reactive to the changes in the environment, and considering how hostile the space environment can become, our body tends to react accordingly, showing specific signs and facilitating the necessary changes. That is the primary cause for the implementation of specific biomarkers that would allow monitoring the state of the human ecosystem and properly address the issues that might come up.

These are (see Fig. 10.5):

- Superoxide dismutase (SOD)—oxidative stress-related biomarker
- Bcl-2—human protein that regulates cell death, used as a biomarker for genomic deletions that may cause chronic lymphocytic leukemia
- Lactate dehydrogenase (LDH)—used as a cancer biomarker
- Complex I and Complex II—mitochondrial proteins, could be used as mitochondrial dysfunction biomarkers

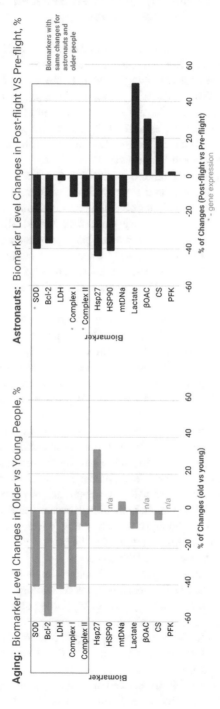

Fig. 10.5 Mitochondrial Biomarkers Changes: Space Flights versus Aging (Source: Mitochondria in Longevity and Space Medicine)

10.1.5 Overview of Technological Overlap Between Longevity and Space Medicine Approaches

As humanity keeps developing new and more advanced technologies for both the Longevity Industry and Space Medicine, it comes as no surprise that, given the close nature of the two, the newest technologies have found usage in both industries, overlapping effectively. While their R & D may be focused primarily on one of the industries, the advancements in one of them, as a rule, find their application in the other. One of the most prolific cases of the technology that was first developed for Space Medicine but later found in Human Longevity is the application of **MDL Algorithms for Remote Health Diagnostics**.

Considering the necessity for remote forms of diagnostics in space exploration, significant attention has been paid to the application of different scientific methods of collecting the astronauts' biodata in space. Such remote monitoring and diagnostics can also benefit the Longevity Industry since they would allow an immediate reaction to any adverse changes in the human body, a better understanding of its current state, and would eliminate the need for constant monitoring of the cases when regular in-person overlook is required. Such technologies have already found usage in certain modern **Next-Gen Wearables** such as **Wearable Health Monitors (WHM)** [9] (Fig. 10.6).

One of the examples of WHM would be a device monitoring the individual's heart activity. With heart issues being one of the most prominent global health problems, such a device would ensure prevention of and quick response to heart dysfunction, therefore lowering the number of heart-related episodes allowing for real-time monitoring designed for everyday usage. Moreover, such a device can be prescribed for people with heart problems since it provides remote monitoring by doctors.

With further development, such a device can also be applied to monitor astronaut's health by adding more functional biomarker tracking [10].

In space, it is not always possible to get the samples for laboratory analysis. That is why the **Downsized Laboratory** development is one of the priorities for both Longevity and Space Industries. Sometimes, urgent analysis is required, and existence of the technology allowing for an immediate laboratory analytical report would be invaluable. With such technology already being developed, more and more laboratorial analytics will become available for the tracking of the biometrics that would otherwise require access to the laboratory.

One of the most promising cases of technology application for both Space Medicine and Longevity is **3D Bioprinting**. This technology breaches the limits of the quantity of biological material required for supporting human life. It allows for the replication of human organs, enabling transplantation of damaged, dysfunctional, or diseased organs to the human body. Despite multiple dangers for humans in space exploration, 3D bioprinting would help in the recovery of human health after exposure to space hazards.

The technological field of **Robotics** and its development provides additional support in the medical field, offering both substitutional functions of nursing and

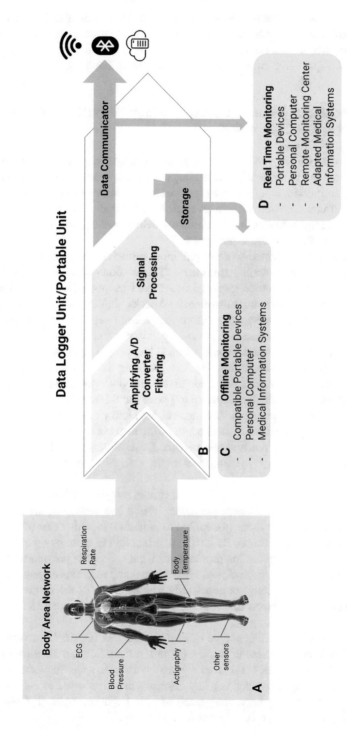

Fig. 10.6 Machine and Deep Learning Algorithms for Wearable Health Monitoring (Source: ResearchGate 2021)

care, and autonomous exploration without the risk to human life. The usage of **Hypoxic Training** is important as it stimulates the presence in space, giving scientists a chance to determine and analyze the biochemical body changes from exposure to the space-like environment while staying on Earth.

10.1.6 Precision Medicine as a Key Strategy Toward Life Expansion and Space Exploration

Recollecting the biblical verse "Your body is a temple," one can say that every temple is unique and requires adjusted tending and maintenance. So is a human body that requires an individual approach to medical treatment due to inherently diverse traits which different humans sometimes exhibit. All of that grows from the phenotypic and genetic prerequisites that are either pre-determined by birth and ancestry or acquired through different life experiences.

As it stands, the manned-space-exploration initiatives require a significant amount of profiling of potential astronauts. This is done to avoid the mission progress being either seriously impaired or delayed by potential issues with the astronauts' health. Through analysis of such setbacks, defining the recurring traits between the astronauts is necessary before the mission sets off. Such traits may include specific versions of genes in personal genotypes and epigenetic markers in phenotypes. This by no means should exclude the people with such traits from participating in the space exploration missions. Defining these traits is important for the necessary precautions during these initiatives, including stockpiling the required medical solutions to provide the necessary response should there be any health issue while the operation is in progress. The medical field known as Precision Medicine focuses on the customization of healthcare, with treatments being tailored to a subgroup of patients or even a single person. Its development will benefit both Space Medicine and Longevity industries as in both cases there is a place for Personalized Medicine.

Polymorphisms represent the capability of certain ranges of DNA in the human genome to exhibit slightly different properties. While one version of the gene can make the "carrier" become more susceptible to radiation, the other one protects its "carrier" somewhat more. This could be an important trait to monitor should such a person enlist in a space exploration program. Still, the concept of polymorphisms is not yet fully explored. The scope of possible outcomes and reactions of gene variants is unpredictable, and they may not necessarily happen. Some of the known chronic disease-related polymorphisms that could affect space exploration include:

- The homozygous 158KK genotype of FMO3 gene that is associated with high risk of chronic heart disease in women but not in men. Solving the issue of this polymorphism would allow for more women involved in space exploration.
- Histidine decarboxylase which is a key determinant of the levels of endogenous histamine that has long been recognized to play important pathophysiological roles during the development of chronic heart failure

- The VEGF gene polymorphism rs699947 that is said to be related to clinical pathology, mortality, and recurrence of hepatocellular carcinoma
- The C→T polymorphism of F10 gene that was significantly (FDR <0.05) associated with Chronic Kidney Disease [11, 12].

10.2 Road Map to Meet Astronaut Medical Requirements

Each individual has its own unique genetics and epigenetic background, as well as an individual health story. Thus, based on the last two parts we are proposing a step-by-step road map to assess the individual genetic predispositions to chronic disorders as well as give basic recommendations to sustain the "astronaut health" (almost ideal health conditions).

As one of the greatest challenges of space travel has been the associated medical issues, we need to get a better understanding of the changes that occur in the human body in the state of weightlessness and in the presence of oxidative stress. The primary medical concerns associated with space include immune and renal dysfunction, loss of bone and muscle mass and liver and heart problems. Various studies have shown variance between men and women in the adaptation to the space environment. This difference has been linked to nearly all organ systems including changes in gene expression, and age-related biomarkers. Health indicators in women have been shown to be different from men's. For instance, women have been shown to have less gravity tolerance compared to men. It has also been shown that in cardiovascular adaptation, females have a reduced ability to maintain venous cardiac output. These together with the varied circadian desynchrony, neurovestibular issues, immunology, behavioral changes and the urogynecology and reproductive issues faced by women call for an individualized form of approach in meeting astronauts' medical requirements.

In the selection of the most fit individuals capable of surviving the extreme conditions of space flight conduction of genetic, biochemical, psychological, physiological, and physical tests are carried out.

Different approaches in medical research have been used to meet astronaut medical requirements including enhancement of radioresistance, increasing tolerance to prolonged weightlessness and human body DNA repair mechanisms using biotechnology. These approaches include the use of preventative medicine and rehabilitation, artificial organs and tissues and research and medical devices. Drugs known as radio protectors are capable of minimizing damage caused by radiation. Some of the techniques developed include 3D-printed organ engineering, artificial medical devices, and Bioregenerative Life Support Systems (BLSS) to provide the astronauts with the required food, water, and air quality in space.

In tackling the loss of bone and muscle mass, the Jackson Laboratory for Genomic Medicine discovered that use of pharmaceuticals to block specific signaling proteins could prevent the loss of bone and muscle mass and also increase their density! Another alternative that is in research is the use of bisphosphonate that has long been used to treat osteoporosis and has been proven to increase bone mass.

Early results show that the space crew can reduce the risk of bone loss and renal stones significantly by combining resistive exercise and a bisphosphonate.

For women in space the gynecological issues like menstrual difficulties can be combated by the use of combined oral contraceptives, IUDs, implants, or injections to suspend menstrual flow as the space environment is not equipped to handle menstrual blood. As more and more women are beginning to explore space, there is a need for continuing studies with female astronauts.

Other protocols include active fitness to prevent loss of bone and muscle tone. In addition, exercise can increase the volume of plasma and blood in the body which can prevent fainting. These exercises in space are facilitated by several equipment such as the cycle Ergo-meter, treadmill, and the resistance exercise device that is a comprehensive exerciser.

After long-term space flights, post-flight injuries are inevitable hence the need for rehabilitation. This would accelerate the recovery and functionality of the space crew to baseline levels.

10.3 Summary

With the growing strain for resources on our planet, exploration of the financial capabilities of the space industry is becoming inevitable. That would mean space exploration, tourism, development, and settlement are slowly becoming a reality. Due to the several challenges associated with the harsh environment of weightlessness, planetary dust, and radiation that can pose serious health issues to humans, space exploration has been limited. Only a few select people who are deemed fit have been able to explore space and even then, are limited in terms of time and radar of coverage.

In order to deeply explore space, the health hazards for astronauts during long-term and short-term space flights have been investigated over the years. There have been more studies on short-term than long-term space flights. Currently, various researches aimed at increasing longevity are underway as NASA is continuously expanding human exploration of space. These include the development of measures to counter health issues and performance. Some of the measures to support human health in space include the use of radio protectors against extreme space radiation, engaging astronauts in space exercises to prevent loss of muscle tone and increase tolerance, use of non-invasive wearable tech to monitor astronauts' physiological data and after space flight rehabilitation, as well as bio-manufacturing in the areas of regenerative medicine, 3D organ printing and drug discovery.

A recent study on a 1-year mission crew has shown a number of mitochondrial-related changes at the genomic and functional levels that shed light on the relationship between the biomarkers after a 1-year flight and aging.

A lot of top companies are now investing in primate research in space. NASA is currently engaged in supporting Commercial life sciences research (CASIS) in partnership with academic researchers and other government organizations to take advantage of the weightless lab. The pharmaceutical community is also not left behind in using the weightless environment to develop and enhance therapies for patients on earth.

A quarter of the development efforts is now dedicated to bioengineering solutions for the space crew against age-related degenerative conditions, bone and eye implants, medical technologies to analyze and support astronaut health. Another quarter is focused on the biotechnology industry while a small percentage is directly dedicated to human longevity in space. The low Earth Orbit (LEO) in space provides a unique environment to investigate new approaches to mitigate disorders related to aging. These include research on drug delivery systems to target cancer, neuro-degenerative disorder therapies like Alzheimer's and regenerative medicine, that is muscle and bone restoration by use of human cell culture. This would in turn reduce the suffering caused by these age-related diseases. Through these studies, scientists have demonstrated that it is indeed possible to extend lifespan. Longevity researchers are now using advanced technologies to discover new interventions and diagnostic tools to extend lifespan by slowing aging [13].

Space exploration has therefore created a potential market for the emerging private sector which will without a doubt result in a drop in the cost of space exploration, advancements in space health, and even space settlements in the coming decades. Furthermore, people will be able to purchase longevity.

Take-Home Messages
- *Developments for space health have a use for healthy lifespan on earth.*
- *Discovering spaceflight conditions reveal the causes of accelerated aging.*
- *Pharma developments such as drug delivery area in AI implementation for aging are very promising.*
- *Low Earth Orbit provides a unique environment for age-related disorders.*
- *Degenerative issues for space travels and their solutions could be very valuable for earth patients.*
- *Biomarkers are indicators for the evaluation of mitochondrial health in Longevity and Space Medicine.*
- *The recent exponential growth of the mitochondria industry is driving the future of space biology and Longevity research.*

Acknowledgments This work was supported by SpaceConneX for Space Medicine and completed in partnership with Space Tech Analytics LLC. The author explicitly acknowledges the contributing work of the analysts Maksym Stoianov and Artem Kompaniiets.

References

1. Hellweg CE, Baumstark-Khan C (2007) Getting ready for the manned mission to Mars: the astronauts' risk from space radiation. Naturwissenschaften 94(7):517–526
2. Scully R, Panday A, Elango R, Willis NA (2019) DNA double-strand break repair-pathway choice in somatic mammalian cells. Nat Rev Mol Cell Biol 20(11):698–714

3. Ohnishi K, Ohnishi T (2004) The biological effects of space radiation during long stays in space. Biol Sci Space
4. Pietsch J, Bauer J, Egli M, Infanger M, Wise P, Ulbrich C, Grimm D (2011) The effects of weightlessness on the human organism and mammalian cells. Curr Mol Med 18(4):201–205
5. Pavlakou P, Dounousi E, Roumeliotis S, Eleftheriadis T, Liakopoulos V (2018) Oxidative stress and the kidney in the space environment. Int J Mol Sci 19:3176
6. Furukawa S, Nagamatsu A, Nenoi M, Fujimori A et al (2020) Space radiation biology for "living in space". Biomed Res Int 2020:4703286
7. Stankiewicz AM, Swiergiel AH (2013) Pawel Lisowski epigenetics of stress adaptations in the brain. Brain Res Bull 98:76–92
8. Afshinnekoo E et al (2020) Fundamental biological features of spaceflight: advancing the field to enable deep-space exploration. Cell 183:1162–1184
9. Fei C, Liu R, Li Z, Wang T (2021) Machine and deep learning algorithms for wearable health monitoring. ResearchGate
10. https://www.qardio.com/qardiocore-wearable-ecg-ekg-monitor-iphone/
11. Yoshida T, Kato K, Yokoi K, Oguri M, Watanabe S, Metoki N, Yoshida H, Satoh K, Aoyagi Y, Nozawa Y, Yamada Y (2009) Association of gene polymorphisms with chronic kidney disease in Japanese individuals. Int J Mol Med 24(4):539–547
12. Ratnasari N, Siti N, Sadewa AH, Hakim M, Yano Y (2017) Difference of polymorphism VEGF-gene rs699947 in Indonesian chronic liver disease population. PLoS One 12(8): e0183503
13. Grigoriev AI, Egorov AD (1992) General mechanisms of the effect of weightlessness on the human body. Adv Space Biol Med 2:1–42

Thi Hien Nguyen is an entrepreneur, futurist, and investor active in the fields of Space Healthtech and Longevity. She is the CEO and founder at Hi Performance Center, a leading performance facility focused on longevity, renowned for its functional medicine approach of nutritional, lifestyle, and training methods for peak performance. Her background in physical therapy, functional medicine, naturopathy, and athlete (multiple marathons and triathlete) has led her into the field of space healthtech. After pursuing a Space Studies Program at the renowned International Space University, Hien founded Space ConneX (SpcX) with a mission to open the space frontier through space education that will foster a better life for all mankind.

Part III

Future Health Value Propositions

Healthcare the Melting Pot of Technology, Humanity, and Confusion

11

Paul Epping and Michael Friebe

Abstract

Will we still recognize healthcare in 10 years from now as the disease care in today's world? This question is currently puzzling a lot of people from all around the world. We still live in the mindset that healthcare will be the biggest burden on the shoulder of our economies. Debates are mainly about the ongoing increase of expenditure due to aging, the use of new expensive technologies, increase in medical knowledge ("we know more about less"), expensive medication, and shortage of clinicians. There are more aspects that can be added to the discussions, but this paper will focus on the above-mentioned five topics and unravel them in the light of how healthcare may look like in 5–10 years from now. And elaborate on the essential question of whether *the current mindsets of health decision-makers are in sync with the exponential dynamics.*

Keywords

Future health · Democratize drug development · Nanobots · Robotic surgery · Microsurgery · Eroon's law

P. Epping
eqxponential, S'Hertogenbosch, The Netherlands
e-mail: paul@eqxponential.com

M. Friebe (✉)
AGH University of Science and Technology, Krakow, Poland

Otto-von-Guericke University, Magdeburg, Germany

IDTM GmbH, Recklinghausen, Germany

FOM University of Applied Science, Center for Innovation and Business Development, Essen, Germany
e-mail: michael.friebe@ovgu.de; info@friebelab.org

What Will You Learn in This Chapter?
- Some exponential technologies and their impact on how we see and perform health
- That these new digital health approaches will lead to a digital transformation (mindset adaptations) with respect to the way we do health business at the moment.
- Interconnected health data systems will learn exponentially faster than any human could ever do.
- How we will democratize Drug Development and Health in general

11.1   Aging

Our life expectancy has increased dramatically over the past century according to our world in data [1, 2]. This counts for all places around the world. Reasons for this improvement can be found in medical and pharmaceutical science. Technology contributed massively to this success over the years. From the use of stethoscopes (which still shows its presence in pockets of clinicians), microscopes to devices helping us to see things inside our body. Big advancements in laboratory testing. All these areas contributed to better treatments for diseases and thus led to the increase in life expectancy.

Fast forward to the current state of these same areas and projection at the end of this decade, we see a massive shift in the focus toward prevention. It is certainly better for the patient (and also a lot cheaper) to prevent a disease from occurring than to cure. Upending the health system from disease (reactive) to prevention (proactive) and lifestyle and well-being.

Back in 1948 the WHO included well-being in her definition of health, shifting away from merely disease orientation [3] It is not the absence of disease. "Immune systems" of organizations are a strong membrane for change as we all know. It took quite some time before healthcare systems were "ready" for this obvious shift to focus on well-being. Insurance systems just now slowly begin to incentivize healthy living [4]. The most important lifestyle activities are sleep, exercise, and healthy food. Although health insurance companies are providing health checks, breast cancer, and colorectal prevention programs, aging will continue to be a (big) burden if the pace of adoption of available solutions is not increasing.

Technology could help all societies around the world to age in good health. The big areas where technology plays a decisive role are screening, diagnostics, treatment, rehabilitation, and telehealth.

What if we change the paradigm from hospital oriented to self-oriented?

Just think about the advancements in sensors, embedded in all sorts of wearables, measuring an increasing number of vital signs [5, 6]. The adoption and recognition of these solutions will change the playfield and is just a matter of time when they will be widely available. The pace of adoption of wearables is growing quickly. However, there are some forces that are responsible for slowing down the pace of recognition of these technologies. This might be rooted in the belief that the data collected by these wearables cannot be trusted if these devices do not have a CE, FDA, FCC, SRRC, IC, KC, MIC, or CCC & CQC certification.

A less anticipated effect on the other hand is that people tend to plan doctor visits when something has been detected and the device or app advises you to see a physician. Many of the "deviations" turn out to be "normal" or harmless but the "findings" are causing a big burden on healthcare systems and thus on the increase of costs. The question is: can that problem be solved?

In our opinion, the answer is "yes"! (Fig. 11.1)

A shift to cheaper, affordable, reliable, and personalized solutions is an ongoing process. We hardly notice that because the improvements are mostly incremental. Over a longer period of time, when we zoom out, we are seeing the difference. The improvements of, e.g., the iWatch, Fitbit, Garmin, Health Mate devices, or iCare (that measure 15 bio signs such as blood pressure, ECG, blood oxygen in a very easy way) have been improved, while becoming cheaper and cheaper (see Fig. 11.1)

We need a different mindset to adopt these new solutions.

Healthcare is a sweet spot for innovative solutions. In the end, nothing is as valuable as our own health. The use of technologies can be observed in healthy longevity (living healthy and living longer), health status tracking (including the information and advises based on the results of the analysis of these data points/sets), early diagnostics (e.g., full body scan, microbiome, liquid biopsy, but also a health checkup every year).

Technology does not seem to be the problem anymore, but awareness, trust, and regulations are as well as behavior. It is known that people are paying more attention to the data captured at the beginning, but after only a few weeks inertia kicks in. Prescribing wearables, self-monitoring solutions (e.g., health trackers or apps) is not yet common practice for physicians.

Kooiman et al. assessed a variety of health trackers and concluded that most of them are reliable [5]. Given the fact that with every update these trackers are getting

Fig. 11.1 iCare self monitoring device measure 15 bio markers very accurately. iCare bio signs tracker (Courtesy: Cloudmed iCare)

better, because the massive amount of data collected from around the world and processed by AI and ML, increases the reliability and predictability dramatically as these algorithms are continuously learning from the acquired and processed data. The more the better!

In the realm of exponential technologies and new versions of a device, we almost always see the repetitive trend that the next version is cheaper, smaller, faster, and more reliable.

Interoperability of the data from these kinds of devices with a personal health record is still a problem because of the semantics and lack of proper standards. We are however sensing a transition from transaction-based systems to intelligent-based systems, which may solve that problem in the very near future. Recording discrete data, using technologies, is not a problem. What is a problem is tapping into the data (allowance to access "propriety" silos and thus the ownership problem), making sense of the data (appropriate algorithms), the visualization of the data (what does it mean?), and capturing contextual data to be added to the mix (the personal layer). Maturing algorithms will be able to do the job by making sense of structured and unstructured (including contextual data) data, sending current standards to obsolesce.

11.2   Expensive Technologies

Modern healthcare technologies are digitized or going to be digitized. The effects of digitization are known. Due to the exponential nature of digital technologies, we are seeing that hardware is getting smaller, faster, and cheaper. Convergence of exponential technologies shows a dematerialization effect on devices.

Despite all the amazing improvements of professional medical technologies, technology does not have a "good name" when it comes down to costs on the healthcare system. Due to the exponential acceleration trend of technologies, also used in modern devices such as sensors, software (e.g., Artificial Intelligence, Machine Learning), new materials, etc., are contributing to the upward trend of its price.

The newest medical devices including the latest technologies are per definition expensive. The main reason is that medical device manufacturers are creating scarcity themselves by adding "must have" features that are of course needed, ahead of competition and therefore more expensive.

It is our belief that this "lock in" is not sustainable and will be broken open this decade, moving toward lower price devices at equal level quality.

Many medical technology-focused startups are entering the market quickly. Usually, they are covering a rather narrow part of the medical world and often do have (super) experts of that particular area on board. Among many overviews that can be found on the Internet, these companies are standing out for their independent investigations CB Insights, Medical Futurist, or Deloitte [7–9].

It is not surprising that in the area of care delivery we see a spike in the use of IoT-related devises whereas in the area of data and platforms AI jumps out. These

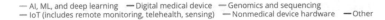

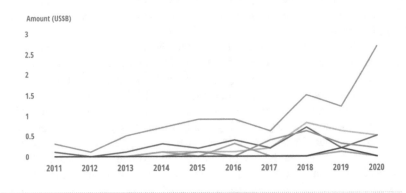

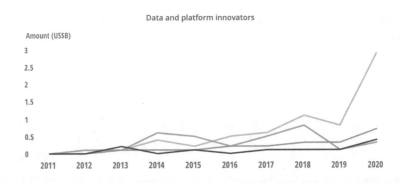

Fig. 11.2 Different innovators are using different technologies (Deloitte) Source: Deloitte Insights

trends will continue to increase, because both IoT and AI are contributing a big part to predictive and personalized health. One of the effects might be that the trend of rising costs can be reversed.

This trend did not escape from the radars of the big health technology incumbents, and they are monitoring the startup market vigorously, trying to acquire or investing in promising companies. Different innovators in the healthcare space are using different technologies (see Fig. 11.2). These technologies usually have been developed by incumbents or research institutions.

11.3   Robotic Surgery. The Disappearance of the "Almighty Surgeon"

A promising area in the medical practice is the development of robotic surgery.

The starting point can be found more than two decades back with the da Vinci surgery platforms.

Millions of patients have had surgery with the help of these hybrid robots. Improvements extended the scope of types of surgeries from abdominal to spine surgeries. Still, the surgeons are being involved but that robots will take over their jobs is just a matter of time. To understand that prediction is not so difficult if we just look at the progress of the area over the past 20 years. The general trend of the appearance of new technologies always shows that it is getting smaller, faster, smarter, and cheaper. Cheaper relative to what it is capable to do. The range of handlings, calculations, etc. is way more extended. To make that clearer, we only need to look at the latest smartphones. Compare today's versions with those from 2 years ago. Calculation speed increased, new applications that are now able to run on your smartphone, better cameras, lidar, etc. All that for practically the same price as the former version. The prices still seem to be steep, but factor in that, at the time of this writing, we are dealing with a shortage of microchips that are driving prices up, perfectly in line with how the market works. The demand is increasing at an exponential rate and that affects the technologies used in surgical robots as well.

> *"If autonomous driving cars are possible, then why can't robots do surgical procedures autonomously? The reason: surgical procedures are less compli-cated, less unpredictable, because in in more of 90% of the cases they can be prepared. Driving is unpredictable, you never know what comes next, specifi-cally in crowded areas like cities."*

11.3.1 Learning Capabilities

Surgical robots are getting smarter due to supervised learning. A real surgeon "teaches" the robot to extend the range of handlings and decisions for certain procedures. Machine Learning algorithms are in place, eagerly awaiting new data. The learning capabilities of robots are close to unlimited and at the speed of light. What does it mean and why is it important to understand that we are at a tipping point of major shifts in this area?

Once robot "A" learns a new handling, all robots, of the same type, can apply that instantly because they are connected and able to use the exact same new features. Robot B just "learned" another new task from a different surgeon somewhere else on the planet and that is new feature is instantly available as well. Theoretically, with hundreds of robots around the world, it is impossible for a surgeon to beat these learning capabilities. On top of that, robots do have immediate access to images (interpret them) and making decisions, adjusting decisions based on an instant overview of vital signs and blood analysis, which they can do themselves while

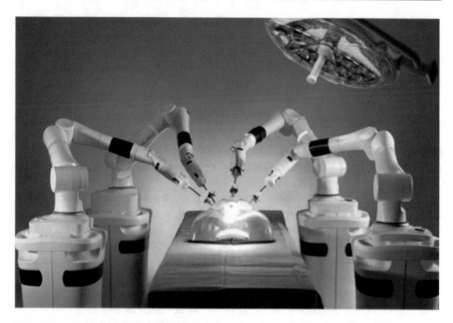

Fig. 11.3 Robotic surgery system. Courtesy: Alcemic

operating. They can also predict certain clinical events based on the current surgical progress in combination with individual real-time clinical data.

We humans only have 6 sensors, robots can technically have thousands and therefore can be more accurate in their physical actions and in the underlying decision-making. In summary, future robots do not need to go to college for a few months to learn a new procedure or visiting conferences to discuss new techniques. Tapping into the network of these robots is probably a smarter and cheaper way to go (see Fig. 11.3). Additionally, new navigation capabilities (a kind of Google maps of your body) are a huge enhancement for the range of (automatic) surgical procedures. And yes, these systems are available 24×7 to help you.

If we think that surgeons are uniquely positioned to do standard surgical procedures, I think that it is time to reconsider that. It goes without saying that rapid advancements in robotic surgery will have huge implications also for our education system (especially for the clinical side of it that has been focussing on diagnosing diseases and then coming up with a therapy), quality controls and respective metrics, insurance policies, medical technology and other industries directly linked to healthcare provision, longevity and demographic shift, waiting lists and regulatory aspects, let alone the resistance to this change from the surgical community itself.

Yes, probably one area that will continue to have surgeons for a while is the domain of traumatology. Although autonomous transportation will reduce traffic accidents dramatically, dangerous work will be executed by (other types of) robots and thus will cause less traumas, there still will be people with traumas that need surgeons (think of weather conditions, earthquakes, etc.).

11.3.2 Robots Are Moving Inside Our Body

An interesting phenomenon that is also on the rise is that of the convergence of nanotechnology, robotics, and artificial intelligence (AI). Current developments of nanorobots are also very promising and will add a new layer of possibilities to the medical area. Upending the paradigm from operating from "out-side-in" to entirely "inside."

"Nanobots," short for nanorobots, are out of the realm of science fiction. Although in vivo and in vitro tests are promising, it still is a long way to actual use in patients. Given the exponential growth of this area, it may not be surprising that the field of nanobots will dominate the medical domain. Remarkable examples are already around and will hit the medical market within this decade. Let us explore what makes us so sure that this will happen.

Energy storage on the microlevel has made major leaps over the past years (Fig. 11.4). Micro-batteries that could power nanobots, are available and are on the downsize curve of even less volume and more power storage [10]. For the time being, these kinds of power sources will be needed until the time that bots will be so smart that they will use the energy that is available around them. Today's nanobots are mainly directed by using external magnetic fields to navigate them.

According to Soto et al., four categories of features and applications can be distinguished: drug delivery, imaging, surgery, and diagnosis [11] (see Fig. 11.5). All these have shown promising results in a variety of tests. Ray Kurzweil thinks that nanobots will play an important role in longevity developments, referring to the aforementioned features [12]. Moreover, he believes and predicts that nanobots will be the new physicians, checking our insights, curing illnesses, and repairing damages. Please note that he described this some 20 years ago and today we see this unfolding rapidly. His updated version of the "Singularity is Near" ("the Singularity is Nearer," expected to be published in spring 2023) will shed light on

Fig. 11.4 Size of Nanobot—tendency to get even smaller in the future. Courtesy: Chemnitz University

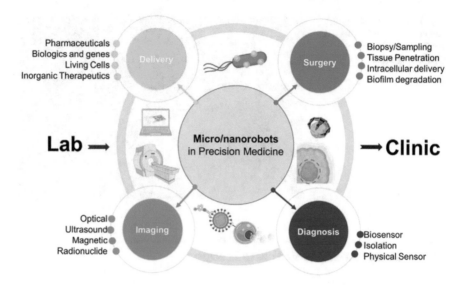

Fig. 11.5 Schematic of the current trends of micro/nanorobotics in precision medicine, including delivery, surgery, diagnosis, and medical imaging applications. (Soto at all, Adanced Science)

the current and future state of nanobots. That, altogether, could become our "upgraded bioartificial" immune systems. This is possible due to the many emerging technologies that are "exploding" at the same time. About fifteen "Gutenberg-moments" (technological development/inventions that lead to transformations) will be hitting the market simultaneously.

11.3.3 Micro Surgery

For the sake of this section, we will carve out the feature of microsurgery. "Nanobots" refers to nanoscale robots, which can accurately build, manipulate, or destroy objects autonomously on a molecular scale. Nanobots can operate in swarms, can be replicated and according to Drexler can be self-replicable [13]. They are bringing a completely new dimension to the field of medicine and particularly to surgery. Imagine small tumors, wherever they may be in your body, can be removed before they can damage tissues and are destroyed unnoticed to the human. Or cleaning arteria and veins from calcification, repairing burst veins, malfunctioning valves in your heart and veins, or destroying viruses before they get a chance to multiply. They are also able to take biopsies from tissues to be examined. Tests have been performed to penetrate tissues in a variety of thicknesses (Fig. 11.6).

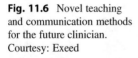

Fig. 11.6 Novel teaching and communication methods for the future clinician. Courtesy: Exeed

11.4   Increase of Medical Knowledge

Information technology is the big engine behind the increase of information in healthcare. The doubling pattern is on an exponential trajectory. Was the doubling pattern of medical knowledge in 1950 about 50 years, and in 1980, about 7 years, in 2018 it is down to 5 months [14]. With that, we easily can calculate and assume that the doubling pattern today is less than 40 days. Not only in volume but what we are also noticing is *that we know more about less*. The latter is the important part because that is where we see the rise of super-specialism. The discovery of nature, including our human body is a continuous path of discovery where "old" insights are not holding forever, because by the time scientists discovered something, they did not know the full picture, which we know a little more at the moment. By knowing I do mean that people have a deep understanding about the reasoning behind a phenomenon, which is in the medical world a prerequisite. In today's world, a medical specialist simply cannot comprehend all that, how good her or his reputation might be.

Every day scientists produce 2500 scientific articles about cancer only. Nobody can keep up with that unless we are embracing AI to help us summarizing the new knowledge for subsequent application and use in treating patients. If you ask medical specialists how many scientific papers they are reading per month then the average number you receive back is about 3, meaning that your knowledge is most likely not based on the latest insights. Even the collective knowledge of medical board meetings, discussing patient cases, pales in the light of what is known that the board did not know that they did not know. This gets even worse regarding the patients of tomorrow.

Instead of putting a lot of effort in trying to read these three papers a month clinicians should work *collectively* on the use of AI that collects the latest information about each specialty, makes summaries, and reads that to physicians while they are commuting. Daily updates in less than 20 min … Once you start with something like that, it is easier to keep up with the latest. We are repeating it one more time: *collectively!*

On a national level, this can be organized, we cannot compete about medical information, after all. Each clinician gets a subscription and during their practice, seamless access to what is relevant (tailored and latest) for her/him. The business case is a no-brainer! Less medical errors, faster decisions, better treatments, shorter hospital stay, better quality of life, etc.

Rosenfeld's insights are representative of what I/we hear very often: single organizations or even individuals are trying to organize the use of AI for these purposes, which, as mentioned above, is probably not a very smart route [15]. It should not be an individual action by a physician to figure out what algorithm is right, what fits best with his/her preferences (which can be different tomorrow), and on top of that, choosing a vendor, whether an incumbent or startup [16]. Make the decisions in a collective manner and do not bother clinicians with algorithms. Even savvy AI specialists do have a hard time to do that properly.

11.5 The End of Expensive Medications?

Recent developments of new COVID-19 vaccines were a precursor of what can be expected for drug discovery in general. The astonishing fast development was mainly possible because of lots of available data and information sharing between scientists all over the world (and willingness to do so), the use of AI and ML, application of gene-editing, high competition, huge investments that were made available globally, and crowd-sourced clinical trials. This unique combination has proven to be successful and broke the tradition of long trajectories of the development of new medications. We know that the average time for the development of a new drug up to now has been somewhere between 10 and 15 years. The huge manual efforts in lab research and testing hypothesis is time-consuming. This will be history soon. The increase in computational power enables researchers to expose Machine Learning algorithms to large drug libraries. This cannot be done with the conventional wet lab work. In silico techniques will help to decrease the time to find a range of potential compounds that could be a match.

Our body is a complicated information processing machinery that we are understanding better and deeper.

The biggest challenges are the biggest opportunities as the saying is going.

This is very true for the quickly emerging field of synthetic biology. The opportunities of the smallest living units of our bodies, at the cellular level, seem to be endless, unprecedented powerful that still only can be understood by our imagination, for good and bad [17].

The science of synthetic biology focuses on, what Michael Levin calls, "cellular decision making." Levin's team developed the first synthetic living cell, the Xenobot, by using the cells from the skin and heart of a frog, the *Xenopus laevis*. With that, they were able to create a new life form that was named after its cell supplier. The team reprogrammed the cells, or better, the decision-making process of the cells. Once we are understanding how decision-making processes work on the cellular level, then we will be able to influence that and create solutions that could solve problems that cannot be solved today. As Hessel already stated for a long time:

"We are not only able to read life, but we are also able to (re)write life."

Rewriting life by manipulating DNA gives rise to writing a new story of life-as-a-whole [18]. The rewriting of life is a big statement. It feels like the rewriting of a complicated combustion engine on paper. How it will work is not clear yet. We do not fully understand how our bodies exactly work. Let alone that we understand the effects of modifications in one part of the body on other parts of the body. Moreover, there is no standard yet or one size fits all.

11.5.1 Democratizing Drug Development

The pharmaceutical industry needs to shift gears as well to keep up with this development and understand that the glorious days of expensive drug discoveries will likely face new times ahead.

The number of new drugs that have been approved is declining and is following the so-called Eroom's law (the number of new drugs approved by the FDA per USD 1 bill R & D and the opposite of Moore's law). As Fig. 11.7 is showing, one can imagine that there will be a theoretical end to the approvals of new drugs by the end of this decade.

In other words, new drugs will become too expensive. Whether this will be the end of the pharmaceutical industry as we know it, is an open question at the moment, but certainly that the roadmap of drug development (thus *not* discovery) will have

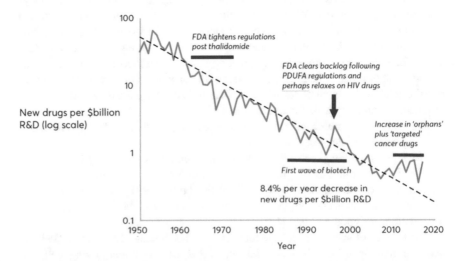

Fig. 11.7 Eroom's law—the number of new drugs approved by the FDA per USD 1 bill R & D. The curve is opposite to Moore's law, which describes the exponential advances in technologies [19]

another direction. Maybe repurposing drugs in common mechanisms of diseases will become such a roadmap [20].

Moreover, this opens the doors for more support of Machine Learning and Deep Learning. This will speed up the process as well as it will decrease costs and may lead to precision medicine, where drugs really will be effective on a personal level.

Modern companies, those who we call Exponential Organizations, do not tend to focus too much on their competition, they rather seek collaboration and see how collaboratively problems can be solved.

The "patent industry" is an example of that showing that it is not a sustainable concept for people who want to make the world a better place to live. That group, Gen-X and Gen-Y and Gen-Z is growing quickly and expressing clearly that they are tired of how we do business. These younger generations are taking over markets including incumbents such as pharmaceutical companies.

Their attitude to collaborate to advance the process of drug discovery does work. For example, young science and research pioneers are collaborating in the Decentralized Trials and Research Alliance (dtra.org).

It is just a matter of time before the readiness has matured to collectively change directions.

11.5.2 Democratizing Health

This is just the continuation of all the efforts in health. A huge societal problem is high cost and with that the in-equal access to high-quality healthcare that we see actually increasing at the moment in the developed world. Future e-patient (e = electronic and engaged or enabled or empowered or by being an expert of their own health [1]) using wearables and artificial intelligence-powered monitoring and health status tools may soon be able to undergo "doctorless" exams. The enabling technology will be very small and even affordable to regions that have big deficits in healthcare provision and delivery. These regions never had sufficient amounts of clinicians available or healthcare services in a close distance or the service that available was too expensive for them. It is foreseeable that the effect and the implementation speed of digital health—and the needed cultural transformation to accept the changes in delivery—might even be faster and more impactful in these areas.

11.6 Summary and Conclusion

Health and healthy longevity are the most important to humans. Exponential technologies will change the way we understand, perform, and manage health in the very near future. Many technologies are already showing the huge impact they could have on providing advanced, personalized, and holistic information that were not available up to now. All these tools and devices will need to be combined however with a mindset and cultural transformation, and with changes in education and health financing. All disruptive changes come with a good and a bad side—the so-called "dual nature." People fear the novel and unknown, especially when their life may depend on it. And existing stakeholders will always try to defend their established position out of fear of losing it. Health innovations need to take these into consideration.

Exponential entrepreneurs are already addressing some of these big challenges.

Take-Home Messages
- We know significantly more about less today—and with that about the interconnecting effects of changes on health and well-being.
- Old medical knowledge (that we still teach) will be replaced by the understanding of a bigger Health picture—this will also need to lead to a change in the way we educate and train clinicians and others in the healthcare sector, including innovators in that space.
- Robots—big ones and small ones—will be making autonomous decisions in the future.
- People and established institutions are always afraid and skeptical of major changes and that is particularly true for the health sector—digital health will lead to a cultural transformation.
- Trust is key to adopting as is a critical and open mindset.

References

1. Mesko B (2022) The digital health course. The medical futurist. www.medicalfuturist.com
2. From https://ourworldindata.org/grapher/life-expectancy, viewed March 17, 2022
3. Health is a state of complete physical, mental, and social **well-being** and not merely the absence of disease or infirmity, WHO
4. Gore A, Hamer P, Pfitzer MW, Jais N (2017) Can insurance companies incentivize their customers to be healthier? HBR
5. Kooiman T et al (2018) Self-tracking of physical activity in people with type 2 diabetes. Comput Inform Nurs
6. Hossain S, et al (2020) Blockchain for the security of internet of things: A smart home use case using Ethereum. International Journal of Recent Technology and Engineering 8(5). ISSN: 2277-3878
7. https://www.cbinsights.com/research/ai-healthcare-startups-market-map-expert-research/
8. Mesko B. https://medicalfuturist.com
9. Deloitte. https://www2.deloitte.com/us/en/insights/industry/health-care/health-tech-private-equity-venture-capital.html
10. Morrison R (2022) World's smallest battery has been designed to power a computer the size of a grain of dust, that could be used as discrete sensors, or to power miniaturised medical implants. Daily Mail
11. Soto F et al (2020) Medical micro/nanorobots in precision medicine. Advanced Science 7(21)
12. Kurzweil R (2005) The singularity is near: When humans transcend biology. Pinguin Group
13. Erik Drexler K (ed) (1992) Nanosystems: molecular machinery, manufacturing, and computation, 1st edn. Wiley

14. Corish B (2018) Medical knowledge doubles every few months; how can clinicians keep up. Elsevier Connect
15. Rosenfeld J (2022) The tech savvy physician: How AI will transform your practice. Medical Economist
16. Stuart S (2019) Human compatible. AI and the problem of control. Viking
17. Wilsdon J (2018) The biomedical bubble: Why UK research and innovation needs a greater diversity of priorities, politics, places and people. Research Gate
18. Webb A, Hessel A (2022) The genesis machine. Our quest to rewrite life in the age of synthetic biology. PubicAffairs
19. Veerappan C PhD, CFA (2020) Breaking Eroom's Law: Biopharma in transition. Diamond Hill
20. Presentation Prof. Dr. H. Schmidt, World Government Forum, Dubai (2019)

Paul Epping is a serial entrepreneur. #65 Thought leader to be followed in 2021. Global keynote speaker. Foresight Mindset.

Background: clinician, psychology, philosophy, and information technology. Co-Founder and Chairman of Xponential, a transformation and foresight company based in Dubai. Chairman Dutch node of the Millennium Project. Double alumnus of the Singularity University and part of an extensive network of innovators and entrepreneurs. Former CIO of several hospitals in the Netherlands. Innovation, exponential technologies including cybersecurity, exponential transformation, systemic and design thinking are red threads throughout his career. International expertise in the selection, implementation, managing innovative information technology solutions, digital solutions in the healthcare sector multi-nationals, smart cities and FinTech. Part-time lecturer innovation, entrepreneurship, and design thinking. OpenExO Head coach exponential transformation. Executive coach.

Michael Friebe received a B.Sc. degree in electrical engineering, an M.Sc. degree in technology management from Golden Gate University, San Francisco, and Ph.D. degree in medical physics in Germany. He spent 5 years in San Francisco, as a Research and Design Engineer with an MRI and ultrasound device manufacturer. He is a German citizen with expertise in diagnostic imaging and image-guided therapies, as a/an Founder/Innovator/CEO/Investor and a Scientist. He is also a Research Fellow with the Technical University of Munich, Munich; an Adjunct Professor with the Queensland University of Technology, Brisbane; and a honorary Professor of image-guided therapies with Otto von Guericke University, Magdeburg, Germany. Since 2022, he has been a Professor of biomedical engineering innovation with the AGH University of Science and Technology, Krakow, Poland. He is a listed inventor of more than 80 patents and has authored over 200 papers, and was part of over 35 Medical Technology Start-Ups. He is a Board Member of four medical technology startup companies and an investment partner of a MedTec-fund. From 2016 to 2018, he was a Distinguished Lecturer of the IEEE EMBC teaching innovation generation and MedTec entrepreneurship. He is also a coach and trainer for OpenExO, and a master Launchpad mentor for the Purpose Alliance.

Democratize Health Delivery

12

Julia Hitzbleck

Abstract

Enabling all individuals to receive the care they need to live healthy lives is the overarching purpose of health innovation. Today we are still far away from reaching this goal and your zip code can tell quite a bit about your health status and care options. But disruption of the health care market is on the horizon and change is accelerating, fuelled by the convergence of several exponential technologies. A strong demonetization of data acquisition and computing power led to rapid advances in the life sciences, followed by dematerialization of complex medical devices into smaller solutions, often powered via standard mobile devices. While data privacy and security concerns are often named as a key risk and regulatory barrier to even more health data acquisition and sharing, the potential for new health care solutions at affordable costs will lead to democratization of health delivery.

Keywords

Health and well-being · Telehealth · Telemedicine · Mobile health · Health literacy · Patient-centered care · Holistic health

J. Hitzbleck (✉)
HI10x GmbH, Mondosano GmbH, Berlin, Germany
e-mail: julia@hi10x.com

What Will You Learn in This chapter?
- Our existing care system will (need to) shift from sick care to health care with new health services emerging.
- Various contestants from other industries are entering the lucrative health market and completely changing current business as well as financial models.
- Technology advances as a key enabler are stepwise providing access to better health and well-being for all.

12.1 Introduction

Ensure healthy lives and promote well-being for all at all ages. This ambitious SDG goal leads the way to a universal right—access to health for all. So how might we follow this purpose and drive innovation toward faster and true democratization of health care delivery?

The UN outlines several specific targets: *achieve universal health coverage, including financial risk protection, access to quality essential health-care services and access to safe, effective, quality, and affordable essential medicines and vaccines for all* [1].

In an ideal world, a (national) health delivery system should provide access to cost-effective health services with consistent quality standards to all citizens when needed. Yet, even in developed countries where such systems exist, it is one of the most unfairly distributed personal rights where location and budget play a crucial role in the quality and services received. We have already come a long way globally, mainly increasing health service coverage in low-income countries and territories by interventions for infectious diseases [2]. But 15% of the global population does not have a financed health system and needs to make out-of-pocket health care payments exceeding in some cases even 25% of their household budgets in competition with other essential spending. The COVID-19 pandemic, while accelerating research collaboration, vaccine and diagnostics production, and regulatory efforts for new therapeutic approaches as well as adoption of telemedicine around the world, also impedes further progress toward this goal by increasing socioeconomic burden and health care personnel burnout. Times of crisis are nevertheless an ideal time for disruptive innovation. This article outlines fast-developing solutions to address the key unsolved problems on the way to democratized health delivery.

12.2   Health Delivery: Status Quo

Health care today is still not equitable. We are now past the challenges of the early twentieth century in preventing and curing illnesses caused by a lack of hygiene. Nevertheless, the overall health care system around intuitive medicine provided by highly trained experts has not managed a shift yet toward embracing the full power of available technology and (big) data-based decision-making. Instead, chronic diseases as the biggest challenge of the twenty-first century are still mainly addressed by finding cures for numerous symptoms instead of clearly focusing on prevention.

With today's lifestyles in the developing and developed world, we still lack health literacy or a clear understanding of what we as individuals need to do for our health, despite access to a wealth of information via the internet [3]. We are used to taking medication as a remedy for our ailments and if symptoms get worse, there is always another therapy or procedure that our expert doctors are keen to perform. For many, this approach seems easier than changing their lifestyle and making sure that several chronic conditions do not occur. Unfortunately, it is not the patient's fault alone. It starts with insufficient health education at school, plenty of (unhealthy) convenient food or lack of access to an affordable, healthy diet, unprecedented stress levels, rising pollution in our environment and health delivery systems incentivized to treat sickness and not keeping people healthy.

With an increasing physician and nurse shortage around the world, new models need to be developed to be able to address demand and provide care for an aging population in most developed countries without posing an even higher financial burden on society. For many developing countries, where no national coverage for health services via sick funds to pay for acute care exists, citizens need to cover expenses via an out-of-pocket system often exceeding their budget, if access to services is available at all.

12.3   Rethinking Health Delivery

To achieve democratization of health care, we need to address all these challenges by further increasing access to knowledge and designing innovative solutions. Not only to train more medical personnel and tech developers for new applications but also for patients and their caregivers. Technology can further empower all stakeholders and help especially patients and healthy individuals to become more responsible for their own care at a level of convenience we know from many other parts of our daily life. To achieve the overarching goal of health and well-being for all, we do not need more of the tech-enabled same approaches. The health care system needs a mindset shift on what health care truly is and how corresponding services are delivered.

12.4   From Sick Care to Health Care Services

Several trends (Fig. 12.1) already support the stepwise mindset shift from reactive sick care with a focus on disease treatments today to proactive health care, where disease prevention and holistic health is the goal. Additional investment in prevention seems like another big check on top for the growing health care cost at first. Keeping not only single individuals but the community healthy long-term addresses the root cause of unhealthy lifestyles and with an aging population the health care financing dilemma. Globally, non-communicable (chronic) diseases, caused by mainly behavioral risk factors, kill 41 million people each year, equivalent to 71% of all deaths, and display the single biggest category in health care expenditures (e.g., 90% of the 3.8 trillion USD spent annually in the USA) [4, 6].

To reduce costs, significant effort is placed for care-intensive groups on managed care systems to provide a continuum of services across different providers while keeping track of treatment results. For individuals with only irregular care needs, usually no such programs exist, and it falls into the hands of the patient to manage care and support coordination between different health institutions. Electronic health records (EHR) have made the portability of diagnoses and test results simpler while reducing service duplication by different providers. At the same time, EHRs enable patients to keep track of their own health data, ideally even providing a translation into lay language or visualizing the results and letting them decide with whom to share their sensitive and valuable health information.

Fig. 12.1 Trends underlining a mindset shift from sick care to health care also supporting the democratization of health care delivery

12.5   Impowering Health Consumers: Convenience and Responsibility

Moving beyond simple internet searches for medical advice, disease information, or exchanging with other patients in online support groups, the empowered patient can not only tap into more knowledge than ever before but also collect and analyze vital signs and indicators via connected health gadgets or just a smartphone. All popular operating systems provide a pre-installed health aggregator application to compile everything from movement data recorded by the device to specific data shared from other apps or sources. Depending on personal interest, data privacy, and security concerns, which are increasingly being addressed by technology providers and legislation, the user turns into the expert of his/her health journey by reviewing data, understanding trends, and setting goals but also being forced to take more responsibility regarding health data and their own care results.

The dematerialization of once arrays of expensive medical devices, operated by trained personnel, into an affordable, everyday device is a key enabler toward democratization of health. Further advances only accelerate from here since the development of software utilizing existing functions of standard mobile devices requires a fraction of the cost previously needed to develop a standalone medical device including regulatory approval. And with more applications emerging by increasing numbers of developers entering this market we see exponential data generation for even deeper insights in previously unavailable depth and width. The tricorder (see Fig. 12.2), a crazy science fiction feature about 50 years ago, is now a reality [8].

Why is this important? Significantly improved compliance and therapy results have been obtained for example with continuous glucose monitoring systems,

Fig. 12.2 Mobile devices—the tricorder in your own pocket [7]

helping patients with diabetes to more accurately dose insulin, track and understand their blood sugar levels 24/7, and changing eating patterns [9]. More recently we see a strong uptake for at-home testing to assess vitamin or hormone levels or even your DNA. Previously such tests could only be ordered by a physician from a certified laboratory. Now they are available online, direct to consumer for at-home testing with user-friendly visualization and explanation of the test results. While cutting out the provider to order and interpret the test, further development of at-home testing solutions are under way, providing even more convenience for the user. Self-testing is part of the new normal since COVID-19 and turning other laboratory tests into simplified test kits, potentially combined with further sample analysis via smartphone camera and app, dematerializes and cuts existing capital and cost-intensive laboratories for many more standard diagnostic tests.

12.6   Redefining Health Care Providers

Simplified test kits for faster diagnostic results, decision-support tools for complex data analysis, like in diagnostic imaging, are now available to support medical professionals in delivering care, which previously required additional expert consultation. Technology advances also help here to turn complex procedures, once only available at specialized clinics, into routine interventions performed by general medical professionals, maybe even as outpatient treatment options. This trend increases the general availability of care options and lowers treatment costs, while at the same time freeing up precious expert time for more challenging cases.

Technology is also enabling how care is provided in communication with the patient. While for the longest time telehealth solutions had slow adoption curves and struggled with regulations, the pandemic removed this hurdle out of pure necessity and added a new channel for patient–doctor interactions. Sharing data from mobile health or telemedicine solutions can replace the physical examination at the clinic in many cases without reducing the quality of data acquisition. In addition to automated, personalized advice provided by notifications, chatbots, or messages in apps, the combination of digital channels with informed patient–doctor conversations, returns the focus on the individual patient and strengthens the human touch of providing care. While offering a strong convenience factor for the patient, without commute and long waiting hours in the clinic, it also increases efficiency by focusing on (virtual) visits that are truly needed.

12.7   Incentivizing Health: New Financing Models or Health Savings Accounts

Mobile health solutions, telehealth, or automated services will be cheaper than the currently available, expert-centered options. In countries with low health services coverage and small health care spending, we have seen fast adoption of mobile and telehealth solutions due to previously physically, and sometimes financially, unavailable services and therefore a huge unmet need, along with fewer regulatory hurdles. Micro-financing solutions for more expensive interventions have developed also for other goods and services in these markets.

In developed countries, we mainly face a financial challenge. Where tax money is used to finance government-owned health care services or allocate sickness funds to private service providers, the aging population with decreasing funding contribution consumes the largest spending due to an increasing prevalence of chronic diseases with age. The younger, proportionally decreasing working population needs to finance health care expenditures in these models [10]. Payers will need to change their operating and financial models soon with the implementation of more cost-effective treatment solutions as well as investments in prevention to keep people healthier.

While this necessary transformation seems obvious on one side, change has been notoriously slow. Health care is a big money pool, and it will be defended by parties with still high profits for now. Till date, technology advances have disrupted every established industry and the increasing cost pressure on the current system along with changing consumer behavior will trigger significant changes toward direct to consumer offers and different payment schemes. The main question is rather whether insurance and other health financing providers are leading the change with new incentives for a culture of health and well-being or defend their business model until the established health care system will be disrupted and the revenue pool shifted to new entrants.

Take-Home Messages
- The democratization of health care is near, with only a few more steps to take. While early adopters still drive the latest tech developments, a price drop in health gadgets like smartwatches, pre-installed health apps, and aggregators on mobile devices are making it mainstream.
- The global pandemic had a positive effect on change in health care—taking global collaboration in the research community to the next level and forcing regulators around the world to lower or remove the bars for telemedicine out of pure necessity.
- There are more (digital) health care tools under development than ever before. By traditional players, big tech companies and anyone who is interested—since access to relevant knowledge is now possible and less upfront invest is needed with the demonetization of technology employed.
- A mindset shift is slowly taking place with a clear trend toward holistic health where health literacy plays a crucial role. While today most offerings are still at an early adopter level, the right opportunities and incentives provided by regulators and payers could significantly accelerate this trend.
- Truly empowered patients not only have more access to their own data, knowledge, the convenience of diagnoses or therapy options at their fingertips but at the same time also increasing responsibility to take actions for their own care.

References

1. United Nations (2022). https://sdgs.un.org/goals/goal3
2. United Nations (2021) Progress towards the Sustainable Development Goals
3. CDC, Centers for Disease Control and Prevention, U.S. (2022), What Is Health Literacy? https://www.cdc.gov/healthliteracy/learn/index.html
4. WHO (2021) Noncommunicable diseases. https://www.who.int/news-room/fact-sheets/detail/noncommunicable-diseases
4. CDC, Centers for Disease Control and Prevention, U.S. (2022) https://www.cdc.gov/chronicdisease/about/costs/index.htm
5. Christensen CM, Grossmann JH, Hwang J (2009) The innovator's prescription – a disruptive solution for health care. McGraw-Hill Education. ISBN 978-1-259-86086-7
7. Source: https://www.economist.com/technology-quarterly/2012/12/01/the-dream-of-the-medical-tricorder; https://www.apple.com/healthcare/products-platform/
6. Empowering Personal Healthcare. https://www.xprize.org/prizes/tricorder#prize-activity
8. https://www.freestyle.abbott/us-en/products/freestyle-libre-app.html + https://www.freestylelibre.de/produkte/freestyle-libre-3-sensor.html
9. WHO (2022). https://www.who.int/health-topics/health-financing#tab=tab_1

Julia Hitzbleck is managing director of the e-health startup Mondosano and the innovation consultancy and venture builder HI10x. Being part of the OpenExO and Purpose Alliance community, she also works as mentor and coach to build purpose-driven, highly scalable organizations with a personal passion for health. Prior experience in the pharmaceutical industry ranges from various R&D functions, corporate development, strategy, culture change and development of Bayer's Innovation Ecosystem. Julia holds a PhD in Chemistry from Syracuse University, New York, and has worked in different research organizations at Monash University, Australia and RWTH Aachen University, Germany.

Health Innovation Process: Definitions and Short Methodology Introductions

13

Michael Friebe

Abstract

Innovation is needed everywhere, but for sure in Health. But what actually is the definition of INNOVATION. Incremental/routine, architectural, disruptive, and frugal, as well as radical will be explained. For the health innovation process, different methodologies and approaches are presented that all are putting the problem definition and evaluation at the beginning. All of them (Design Thinking, Lean Startup, Biodesign, I3-EME, and Purpose Launchpad Health—PLH) are agile and iterative. Design Thinking and Lean Startup are general base principles and the used by the other methodologies dedicated to the health innovation space. The chapter will conclude with some advice on which methods to use and consider.

Keywords

Disruption · Biodesign · Design thinking · Lean startup · Agile innovation · Iterative innovation · Purpose innovation

M. Friebe (✉)
AGH University of Science and Technology, Krakow, Poland

Otto-von-Guericke University, Magdeburg, Germany

IDTM GmbH, Recklinghausen, Germany

FOM University of Applied Science, Center for Innovation and Business Development, Essen, Germany
e-mail: michael.friebe@ovgu.de; info@friebelab.org

What Will You Learn in This Chapter?
- Several agile and iterative Innovation Methodologies
- Basic and more general concepts – Lean Startup + Design Thinking
- And some dedicated to Health Innovations – Biodesign – I3-EME + Purpose Launchpad Health
- Problem understanding and learning through experiments are key to success.
- That PURPOSE is an important starting point.

13.1 Introduction

Let us start with the PAINKILLER statement (Fig. 13.1). What we often do in university-based research is to continue previous research and come up with a new technological or process or workflow improvement. We base these activities on known setups and the main value proposition for these is to make a current awkward process easier, to improve a procedure or workflow, or to add additional accuracy.

But coming from a technical department trying to innovate in the medical space often is creating solutions for a problem that does not even exist or that is not as severe as anticipated. The problem hypothesis was not validated and assumptions lead to the development process. These development processes often are involving the key users or customers (e.g., the patient). You may end up with a PAINKILLER that does not have a PROBLEM to solve. And, you may find that out too late in the development/translation process wasting a lot of time and resources.

And even if it comes up with a worthwhile solution it likely is then an INCRE-MENTAL INNOVATION.

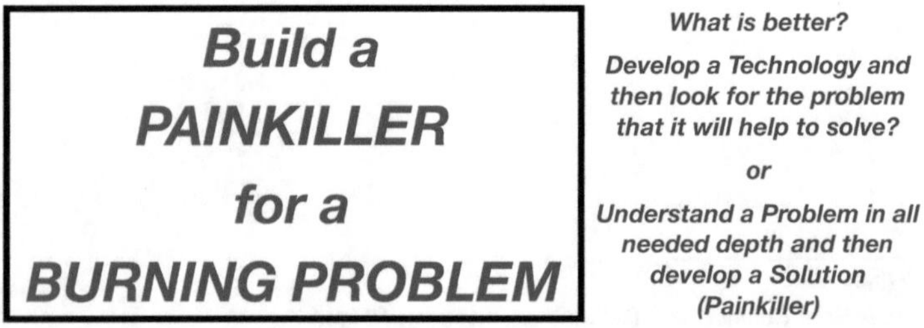

Build a PAINKILLER for a BURNING PROBLEM	*What is better?* *Develop a Technology and then look for the problem that it will help to solve?* *or* *Understand a Problem in all needed depth and then develop a Solution (Painkiller)*

Fig. 13.1 This should be a general goal in an innovation process and certainly is also valid for the HEALTH segment. Often the innovator starts with an idea and an assumption/hypothesis and a problem definition that is unfortunately never really (in)validated before the actual development process starts—especially true for engineers that want to build things. Then we end up with a PAINKILLER that does not have a problem to solve. You need to start understanding the BURNING PROBLEM in depth and then develop—in agile and iterative processes—a PAINKILLER

The term INNOVATION is often misunderstood and contextually misused. Technology or inventions are both pre-steps and only lead to innovation with successful market introduction.

INCREMENTAL refers to improvements to a product/process that already has a market/business model and is based on improving existing technologies. An example is the improvement of Computerized Tomography (CT) systems that continuously add more volume imaging and improved time-resolution capabilities.

Incremental or ROUTINE innovations are best handled by the already successful incumbents in these market segments

DISRUPTIVE INNOVATION in contrast to that is using existing base technologies in new markets/business models ([1–4] including references with direct links to Healthcare Disruptions). New markets means that you create a novel customer group of previous non-users and non-buyers or activate previously neglected users. Taking our previous example from the diagnostic segment a portable and relatively inexpensive Magnetic Resonance Imaging (MRI) system (factor 5–10 cheaper) could now create a new customer (Emergency Physician, Intensive Care Unit) or activate previous neglected customer groups (Urology, Veterinary Medicine, Orthopedic).

The above example provides a FRUGAL INNOVATION. The focus on dramatic (factor of >10) cost reductions—also at the cost of quality, speed, . . .—can be a starting point for a segment disruption of a much larger scale. When these neglected customer groups start to acquire these systems the previous market will likely be changed leading to a MARKET DISRUPTION.

An easy test for a DISRUPTIVE INNOVATION is on whether it is disruptive for ALL established companies and market contenders. In this case, it is also likely that the existing companies will eventually succeed.

Other forms of innovations are ARCHITECTURAL (new technology in a new market/missing business model—e.g., health data platforms, new sensor for audio auscultation and health status assessment) or RADICAL (new technology in an existing market/with a business model—e.g., a novel technology that allows reducing the cost of contrast media injections for diagnostic imaging by a factor of 1.000).

While we embrace in this book the need for DISRUPTIONS for a Future Health. There is a difference though between a DISRUPTIVE INNOVATION and a MARKET DISRUPTION. MARKET DISRUPTION means that new tools, devices, and concepts replace the existing ones complete. HEALTH needs those disruptions to solve the challenges, but it also needs continued incremental innovations. And remember that several incremental steps can also lead to an intentional disruption [2–4].

So have *a disruptive product/process vision and define two or three incremental intermediates that could lead to that.*

The incremental approach has the advantage that a market and business model exists, which also means that you can convince externals to fund the development and market introduction. A DISRUPTION approach requires a lot of initial convincing and goes through a disappointment period with few early innovator customers.

Only if you are able to CROSS THE CHASM—building a bridge from early innovator to early majority use [5]—and make the product attractive to many customers will it actually disrupt—it requires a business model!

13.2 Purpose Oriented Health Innovation Methodology Introduction

What we propose in this book is that you really attempt to understand the problem including future developments and technology support that could help to achieve that. Jumping to a solution to quickly is not always helpful and will likely cause a lot of frustration later on in the development process.

We also suggest that your health innovation activities are purpose driven (e.g., eradicate duplicate and useless medicine), have a *Massive Transformative Purpose* (MTP—the organization's higher aspirational purpose and the change that the team wants to achieve—"Embrace your own Health"), a *Vision* (Longer Term—What does the organization want to become in the future, e.g., 5–10 years?—"Engage and motivate clinical users and stakeholders to use our platforms to avoid duplicate exams!"), and a *Mision* (Shorter term—How do we make our Vision come true?— "Build a user-friendly Data Platform with incentives and gamification!").

With that we can then start to start an initial *Discovery* phase, with the goal to understand the problem space, the customers/users, the market, and collect as much information as possible that we subsequently try to (in)validate in the *Evaluation* phase. In that phase we can create and use initial ideas that we turn into more or less functional demonstrators (so-called Minimal Viable Prototypes) that are used to test and show the *desirability* (is that really what users/customers want), *feasibility* (can it actually be build within a certain period of time and for a certain price), and *viability* (can this create a sustainable operation) among many alternatives.

Several dedicated agile and **ITERATIVE** (continuous improvement by adjusting previous concepts) innovation methodologies are important in that context.

AGILE means that you revisit previous steps to (in)validate new hypotheses, insights, and learnings and adjust your concept based on them. The innovation methods and processes you use must therefore be flexible, interactive, and embrace such an iterative approach.

DESIGN THINKING (IDEO—Empathize/Define /Ideate /Prototype/Test) was not specifically designed for the health innovation space. It is a more general approach that uses in the original version 5 elements (see Fig. 13.2 with ASSESS added as a sixth element). EMPATHIZE (put yourself into the position of user/customer/stakeholder) and DEFINE (the problem accurately) help you to understand the problem. The next two segments, IDEATE and PROTOTYPE are used to explore solutions by the creation of new concept ideas, and through the design of demonstrators. These can then be TEST'ed and evaluated/ASSESS'ed to gain a deeper understanding of how to materialize and scale it ([6] for general information, [7] for Health specific).

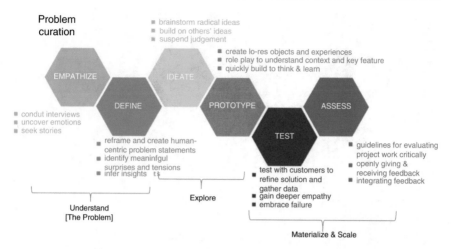

Fig. 13.2 The DESIGN THINKING process and it is 6 segments. Two each for understanding the problem (EMPATHIZE + DEFINE), explore solutions (IDEATE + PROTOTYPE), and materialize and scale the product (TEST + ASSESS) [6, 7]. Figure adapted from Stanford d.school—https://dschool.stanford.edu and from https://www.bmnt.com/post/5-reasons-design-thinking-is-good-for-the-defense-domain

BIODESIGN, a dedicated methodology for the health innovation space. It also uses three main segments for IDENTIFY'ing, INVENT'ing, and subsequently implementing the solutions (please refer to Fig. 13.3 for more details). This is a methodology that dives deep into the market, evaluates competitors, technologies, patents, and also tries to understand the customer dynamics. Regulatory approval is a requirement for health innovations at the moment (possibly even intensifying in the near future) and is also part of that evaluation process. Ideally suited for needs-based health innovation and for routine/incremental innovations also originating from established companies or organizations [8].

LEAN STARTUP/LEAN ENTREPRENEUR (Eric Ries—Build/Measure/Learn and Steve Blank's work) is a very important iteration concept that embraces learnings. An idea/hypothesis is a starting point for which a prototype is built and subsequently a customer experiment designed with the goal to acquire information/data that will be used to validate/verify or invalidate/falsify this idea/hypothesis (Fig. 13.4).

You also use this data to then change/adapt the prototype. LEAN STARTUP is an intrinsic part of all the previously and subsequently listed agile and iterative methodologies providing validated learning and a way to account for the innovation progress [9, 10].

I3-EME—Identify/Ideate/Implement in a multidisciplinary team of Engineers (Scientists), Medical Staff and/or Patients, Economists is a combination of BIODESIGN in a team of users/stakeholders, technologists and health economy/entrepreneurial experts dedicated to a very narrow technology focus. That narrow focus allows to build up an expert domain knowledge in all aspects in the translation

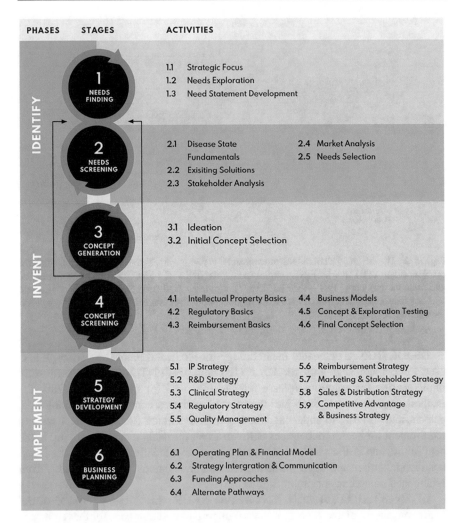

Fig. 13.3 The three main phases of the BIODESIGN process. It starts with IDENTIFY and medical/clinical needs finding and screening. That includes a more in-depth analysis of the existing stakeholders and a market analysis. The next phase deals with IDEATION and INVENTION of solutions that could solve the need. In this phase, you will also check business models and look at existing technologies. A first look on regulatory approval is also part of that step. As this is an agile and iterative process it may be necessary to go back to the IDENTIFY phase to re-check. The next step that follows is to IMPLEMENT the concept to translate it to the market and user/customer. This could be in a Start-Up or through an existing organization. Figure adopted from Biodesign: The Process of Innovating Medical Technologies, Yock P. et al., ISBN13: 9781107087354 [8]

process from bench to bedside originating from the authors lab (see Fig. 13.5 and [11, 12]).

PURPOSE LAUNCHPAD (see Fig. 13.6)—is a general Meta-Methodology with 8 segments (Purpose, People/Customer/Abundance/Viability/Product/Process/

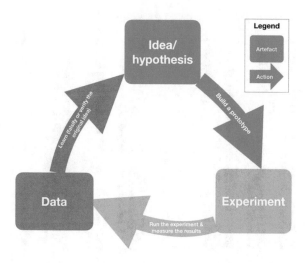

Fig. 13.4 Lean Startup Process—it embraces LEARNINGS and is an integral part of all of the other agile methodologies. For an IDEA/HYPOTHESES a prototype (does not have to be physical—could be a drawing/concept study/cartoon/3D mock-up) is *BUILD*. Then an EXPERIMENT is designed to *MEASURE* results and to collect data. The date is used to *LEARN* and revise/adapt the original idea. Original concept from Eric Ries [9] and Steve Blank [10]. Figure from https://commons.wikimedia.org/wiki/File:nLean-StartUp-Cycle-EN.jpg, Antonsk, CC BY-SA 4.0 <https://creativecommons.org/licenses/by-sa/4.0>

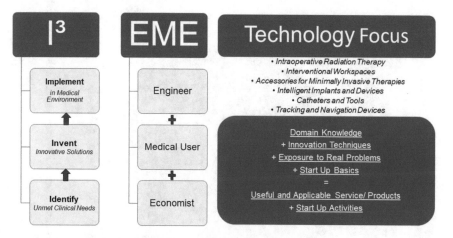

Fig. 13.5 The I3-EME methodology, a combination of Stanford BIODESIGN, Design Thinking and domain expertise originating from a university lab with a narrow technology focus, but with that an expert domain knowledge on all aspects in the clinical translation process in the particular fields [11, 12]

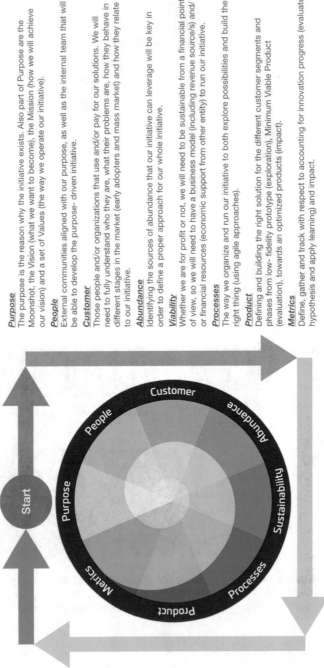

Purpose
The purpose is the reason why the initiative exists. Also part of Purpose are the Moonshot, the Vision (what we want to become), the Mission (how we will achieve our vision) and a set of Values (the way we operate our initiative).

People
External communities aligned with our purpose, as well as the internal team that will be able to develop the purpose- driven initiative.

Customer
Those people and/or organizations that use and/or pay for our solutions. We will need to fully understand who they are, what their problems are, how they behave in different stages in the market (early adopters and mass market) and how they relate to our initiative.

Abundance
Identifying the sources of abundance that our initiative can leverage will be key in order to define a proper approach for our whole initiative.

Viability
Whether we are for profit or not, we will need to be sustainable from a financial point of view, so we will need to have a business model (including revenue source/s) and/ or financial resources (economic support from other entity) to run our initiative.

Processes
The way we organize and run our initiative to both explore possibilities and build the right thing (using agile approaches).

Product
Defining and building the right solution for the different customer segments and phases from low- fidelity prototype (exploration), Minimum Viable Product (evaluation), towards an optimized products (impact).

Metrics
Define, gather and track with respect to accounting for innovation progress (evaluate hypothesis and apply learning) and impact.

Fig. 13.6 The 8 segments of the Purpose Launchpad with the three color coded phases (yellow = Exploration, blue = Evaluation, violet = Impact) [13]

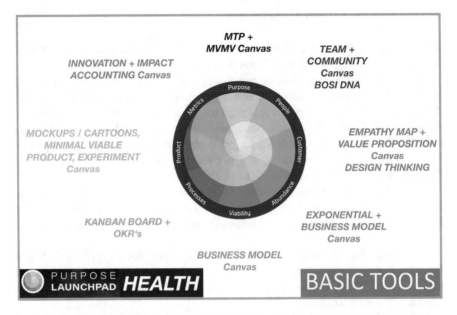

Fig. 13.7 The PURPOSE LAUNCHPAD HEALTH Meta-Methodology based on [H] and subsequently adapted to reflect the special needs and issues when dealing with Health Innovation Issues [t]. For each of the 8 segments for the EXPLORATION (yellow) and EVALUATION (blue) phases. For the IMPACT Generation phase, a sellable product is needed. This typically requires regulatory approval and setting up a startup [14]

Metrics) in three phases (Exploration/Evaluation/Impact Generation) that is based on the core concepts of the previously mentioned innovation approaches. To get insights into each of the 8 segments it suggests the use of specific support tools and canvases. The starting point is the PURPOSE. Why are we actually doing what we are doing? What is the core motivating element of our work? This methodology combines a novel EXPONENTIAL mindset (as part of the ABUNDANCE segment) within a defined, but nevertheless agile circular process, with a visual feedback on the progress and learnings after regular sprints (typically every week a new assessment and To Do backlog will be jointly developed).

We have adapted and optimized the more general PURPOSE LAUNCHPAD for dedicated use in the health innovation process called PURPOSE LAUNCHPAD HEALTH (PLH), which will be presented and used in depth in this book. An overview is shown in Fig. 13.7, which indicates and lists the individual tools that we suggest to use for each of the segments. These tools will be presented and discussed in subsequent chapters.

But it is also clear that the PLH is not always the right tool, especially for incremental and fast innovations of an existing product, or for a problem solution with only a small market and no further scaling potential.

13.3   Summary and Conclusion

There are several relevant methodologies that all have elements that are valuable for generating meaningful healthtech innovations.

Most important is the focus on the unmet clinical need and a deep dive into understanding the problem and not immediately jump to creating a solution. For any new innovation to be successful you need to check the DESIRABILITY, FEASIBILITY, and VIABILITY. The problem understanding toward a valuable product (VIABILITY) can be systematically acquired with the LEAN STARTUP approach in which you validate or falsify a hypothesis through a customer experiment (DESIRABILITY) with a prototype (FEASIBILITY).

The BIODESIGN process typically takes at least 4–6 months and should be started if there is some initial evidence about a medical/clinical need. Not necessarily ideal for a disruptive or architectural innovation qualified through a missing business model.

Acquiring domain and translation knowledge in the complicated health innovation space requires very dedicated expertise. If you focus on a particular segment the I3-EME approach, an extended and more specialized BIODESIGN approach is recommended.

The PLH is the tool of choice for a structured, exponentially and purpose-oriented methodology.

And, Biodesign, I3-EME and PLH take considerable time and commitment to go through several iteration cycles, so not always fitting and appropriate for problem solutions or identified needs that only concern a very small market and cannot be scaled. On the other hand, how do you know that they cannot be scaled and applied for other markets, if you have not invested the time and effort to check?

Take-Home Messages
- Innovation needs the translation of a product/idea to a market. Invention and Technology Developments are just pre-steps!
- Innovation is often misused in that context—often it is attached to something new or different.
- Disruptive Innovation = existing technology in a new/unknown market (Example: new portable, dedicated to certain applications and low-field MRI system that addresses a completely new market at 20% of the cost of a normal general applicable MRI)
- Frugal Innovation is part of Disruptive Innovation = significantly lower cost (Example: handheld Ultrasound system—initially sold as a point of care, but actually universally usable—for 10% of the cost of a normal Ultrasound system).

(continued)

- Radical Innovation is using new technologies in an existing market segment (Example: New technology to lower the capital cost of infusion/injection of pharmaceuticals at only 10% of the cost).
- Architectural Innovation is characterized through new technologies and new market/no business model (Example: Digital Health Platform).
- Agile, iterative, and learning based methodologies should be used.
- LEAN STARTUP is a base methodology that should be used by anyone for any innovation project as part of the other methods.
- BIODESIGN is based on the DESIGN THINKING concept and is used to find and understand an unmet medical need and check, adapt, and provide all the variables to define a product-market fit.
- I3-EME combines the BIODESIGN approach in a more narrow clinical domain with established clinical, technical, and health economics expertise in an innovation lab setup.
- PURPOSE LAUNCHPAD HEALTH is a meta-methodology for exponential approaches to innovation and provides a process with 8 segments and recommended tools to produce learnings and insights.

References and Further Readings

1. Christensen C (2011) The innovator's dilemma: the revolutionary book that will change the way you do business. Harper. ISBN-13: 978-0062060242
2. Christensen C, Bohmer R, Kenagy J (2000) Will disruptive innovations cure health care?. Harv Bus Rev, Sept–Oct 2000 issue. https://hbr.org/2000/09/will-disruptiveinnovations-cure-health-care
3. Christensen C, Waldeck A, Fogg R (2017) How disruptive innovation can finally revolutionize healthcare. Published by INNOSIGHT and the CHRISTENSEN Institute. Available at https://www.innosight.com/wp-content/uploads/2017/05/How-Disruption-Can-Finally-Revolutionize-Healthcare-final.pdf, viewed Feb 28 2022
4. Christensen C, Waldeck A, Fogg R (2017) The innovation health care really needs: help people manage their own health. Harvard Business Review, Oct 30, 2017. https://hbr.org/2017/10/the-innovation-health-care-really-needs-help-people-manage-their-own-health?autocomplete=true
5. Moore C, McKenna R (2006) Crossing the chasm: Marketing and selling high-tech products to mainstream customers. ISBN-13: 978-0062353948
6. d.school (2013) Hasso Plattner Institute of Design at Stanford: an introduction to design thinking process guide. Retrieved March 3rd, 2022 from https://web.stanford.edu/~mshanks/MichaelShanks/files/509554.pdf
7. Abookire S, Plover C, Frasso R, Ku B (2020) Health design thinking: an innovative approach in public health to defining problems and finding solutions. Front Public Health 8:2020
8. Yock P, Zenios S, Makower J (2015) Biodesign: The process of innovating medical technologies 2. Cambridge University Press, Cambridge. ISBN-13: 978-1107087354
9. Ries E (2011) The lean startup: how today's entrepreneurs use continuous innovation to create radically successful businesses. ISBN-13: 978-0307887894
10. Blank S, Dorf B (2020) The startup owner's manual: the step-by-step guide for building a great company. Wiley. ISBN-13: 978-1119690689

11. Fritzsche H, Boese A, Friebe M (2021) From 'bench to bedside and back': Rethinking MedTec innovation and technology transfer through a dedicated Makerlab. J Health Design 6(2): 382–390
12. Fritzsche H, Boese A, Friebe M (2017) INNOLAB - image guided surgery and therapy lab. Curr Dir Biomed Eng 3(2):235–237. https://doi.org/10.1515/cdbme-2017-0049
13. Palao F (2021) Purpose Launchpad Guide V 2.0. https://www.purposelaunchpad.com/methodology - visited and retrieved 10.03.2022. V 1.0 also available at https://www.scribd.com/document/486477523/Purpose-Launchpad - visited and retrieved 10.03.2022
14. Friebe M, Fritzsche H, Morbach O, Heryan K (2022) The PLH - Purpose Launchpad Health - Meta-methodology to explore problems and evaluate solutions for biomedical engineering impact creation. IEEE EMBC Conference Paper, July 2022 - not published yet

Michael Friebe received a B.Sc. degree in electrical engineering, an M.Sc. degree in technology management from Golden Gate University, San Francisco, and Ph.D. degree in medical physics in Germany. He spent 5 years in San Francisco, as a Research and Design Engineer with an MRI and ultrasound device manufacturer. He is a German citizen with expertise in diagnostic imaging and image-guided therapies, as a/an Founder/Innovator/CEO/Investor and a Scientist. He is also a Research Fellow with the Technical University of Munich, Munich; an Adjunct Professor with the Queensland University of Technology, Brisbane; and an honorary Professor of image-guided therapies with Otto von Guericke University, Magdeburg, Germany. Since 2022, he has been a Professor of biomedical engineering innovation with the AGH University of Science and Technology, Krakow, Poland. He is a listed inventor of more than 80 patents and has authored over 200 papers, and was part of over 35 Medical Technology Start-Ups. He is a Board Member of four medical technology startup companies and an investment partner of a MedTec-fund. From 2016 to 2018, he was a Distinguished Lecturer of the IEEE EMBC teaching innovation generation and MedTec entrepreneurship. He is also a coach and trainer for OpenExO, and a master Launchpad mentor for the Purpose Alliance.

Prevention, Prediction, Personalization, and Participation as Key Components in Future Health

14

Beatrice Barbazzeni and Michael Friebe

Abstract

Demographic changes toward the aging population are accompanied by the increasing rate of non-communicable (NCDs) chronic diseases, in which the high demand for continuous patient monitoring, treatment access, prolonged hospitalizations, as well as financial issues represent a global concern affecting the entire healthcare system and management. However, although the onset of symptoms in NDCs might be delayed or reduced, the adoption of preventive measures would be recommended to improve patient health and quality care services. Thus, moving from a reactive (focused on treatment) to a proactive system focused on prevention, the future of healthcare is embraced in the so-called "P4" of medicine: preventive, predictive, proactive, and participatory. Based on this approach, the possibility to predict a disease based on specific biomarkers, personalized treatments, and empowered patients toward their health monitoring while preventing diseases, would become a milestone in the perspective of healthcare innovation.

B. Barbazzeni (✉)
ESF-GS ABINEP International Graduate School, Otto-von-Guericke-University, Magdeburg, Germany
e-mail: beatrice.barbazzeni@med.ovgu.de

M. Friebe
AGH University of Science and Technology, Krakow, Poland

Otto-von-Guericke University, Magdeburg, Germany

IDTM GmbH, Recklinghausen, Germany

FOM University of Applied Science, Center for Innovation and Business Development, Essen, Germany
e-mail: info@friebelab.org

Keywords

Prevention · Prediction · Personalization · Participation · Non-communicable
chronic diseases · Proactive · Healthcare innovation

What Will You Learn in This Chapter?
- The increasing rate of chronic diseases represents a global health concern.
- Increased demand for high-quality care services, system management, and new healthcare approaches to overcome the twenty-first-century challenges
- The need to move from reactive to a proactive healthcare system
- P4 Health is a new concept toward prevention and healthcare innovation
- P4 is a continuous process and exchange between different stakeholders and communities, in which a novel healthcare culture is generated toward health longevity.

In the last decade, the increasing rate of chronic diseases such as cardiovascular and respiratory diseases, cancer, type-2 diabetes, and neurodegenerative diseases became a global concern while impacting the economy of several countries worldwide due to increased healthcare costs [1].

Defined by the World Health Organization (WHO) as non-communicable diseases (NCDs), chronic diseases represent one of the main factors leading to disabilities, decreased quality of life (QoL), and ultimately death due to their high comorbidity with biological aging [1].

Nevertheless, even though the onset of symptoms in chronic diseases might be delayed and reduced by adopting specific interventions (e.g., a healthy lifestyle), the transition from health to disease onset can be initially silent. Therefore, the need to adopt earlier preventive measures would ultimately reduce the impact of consequent interventions and clinical treatments [1].

However, despite the medical progress in improving surgical interventions and pharmacological treatments to address different infectious and other disease types, this clinical approach remains reactive, in which treatments are adopted once the patient has already experienced the onset of negative symptoms. In this regard, the need for a novel method that is preventive, predictive, proactive, and participatory (P4) [1, 2] would be required to generate healthcare innovation while transforming the entire system and moving the needle from sickness to health.

Overall, this P4 approach aims to reduce the increasing rate of NCDs toward a novel perspective in healthcare and in which exponential medicine, the adoption of disruptive technologies, and a patient-centric approach would lead to a proper investigation of environment-biology interplay and a deeper understanding of those mechanisms under disease development. Furthermore, this innovative framework can be seen as a *Continuous P4 Health* (P4H) spectrum.

Different stakeholders and communities come into play, surrounded by an exchange of languages and cultures to promote health longevity and redesign traditional healthcare concepts [1, 3]. Understanding the future of health is about promoting wellness and refers to the possibility of reversing disease stages early in time.

The P4H model can be viewed and understood as characterized by the interconnection of two main components in which the transition from health to disease is structured [1, 4]: *stage of health* and *levels of interventions*. The former refers to the transition from "stage A" (i.e., allostasis) in which individuals are healthy but exposed to a series of environmental factors ("stressors").

The lack of preventive and proactive measures would lead to "stage B" (i.e., allostasis load), which represents a slow and covert progression toward a pre-chronic disease [1]. Furthermore, "stage C" (i.e., progressive allostasis load) is characterized by the manifestation of chronic disease symptoms. Lastly, "stage D" (i.e., allostatic overload) represents the moment in which chronic diseases are diagnosed, and intervention procedures are applied—a typical process of a reactive healthcare model.

Nevertheless, these stages should be located into a potentially reversible continuum process, and in which P4H can be applied at each stage to guarantee the appropriate healthcare procedure while promoting an individual's well-being [5]. Moreover, these stages are considered concerning the levels of interventions.

From *level I*, characterized by global and country-based interventions aimed to improve health over a large population, scaling to *level II* in which healthcare procedures target communities and their environment. In *level III*, health professionals support individuals and relative families, and lastly, *level IV*, where interventions become highly personalized while considering the system-specific physiology of the single individual [6].

Thus, the shift from a reactive to a continuous proactive model would require the participation of multiple stakeholders that interact and collaborate to promote the future of a preventive, predictive, and personalized healthcare system, from a global to individual health [1].

Preventive medicine aims to identify risk factors to either prevent disease development or reduce its impact by adopting early treatments. The investigation of biomarkers results is of primary importance to determine high-risk groups for a certain healthcare issue while matching a physiological and psychological profile.

Prevention can also be implemented in computational modeling analysis combined with advanced imaging techniques to improve diagnosis providing the most suitable treatment to delay disease progression. Regarding prevention, predictive medicine considers genes and genetic variations [7] and individual lifestyles and environmental factors to generate intervention and diagnostic models based on predictive analysis.

Thus, several methods based on single-subject simulation analysis, machine learning, and biomedical imaging techniques would be implemented to extract useful information [8]. From general to individual, personalized medicine aims to consider the complexity of the human system—a product of biological and environmental processes.

Moreover, while replacing a standard and one-size fit approach [1], personalized medicine supports identifying subject-specific target interventions, highlighting the genetic profile and lifestyle [1]. This approach would lead to a system in which

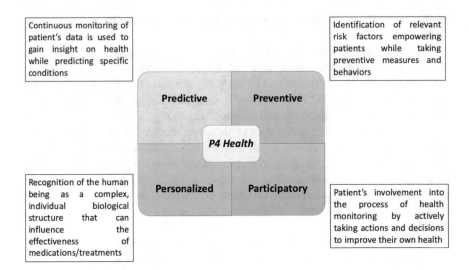

Fig. 14.1 Schematic representation of the P4 Health and related descriptions

patients and individuals are empowered in managing their health, leading to proactive participation.

Hence, participatory medicine represents the last component of this innovative and continuous P4H mode. Here, the individual and their family, who are the central stakeholders involved in healthcare, actively participate in a preventive and predictive process to promote health through education and self-management [7].

Engagement, empowerment, and motivation become the main aspects of locating the patients at the center of this process among clinicians, professionals, and caregivers. Furthermore, the increased availability of health technologies (e.g., wearables, health sensors, virtual reality, clouds) would support the patient's role and collaboration into this new and innovative network centered around the patient's needs, the main focus of participatory medicine [7, 9].

From repairing to regenerating while improving medical procedures and reducing healthcare costs into a novel patient empowerment perspective, the main goal of P4H (see Fig. 14.1) would be to consider the individual as a unique and complex system influenced by biological, environmental, and cross-cultural factors. Into this frame, PH4 promote wellness toward the accurate prediction and prevention of NCDs development [10].

The PH4 approach of FUTURE MEDICINE will be based on advanced technologies, mainly sensors and artificial intelligence used by everyone and affordable enough for everyone (see Fig. 14.2).

Our mainly evidence-based approach that requires complicated and expensive clinical trials and still does not provide personalized outcome information will be replaced or at least substituted with an evaluation and quantification of relevant parameters of personal health baseline. This will subsequently help to predict disease developments or detect diseases in a very early stage and will also help to engage the user in his/her own health.

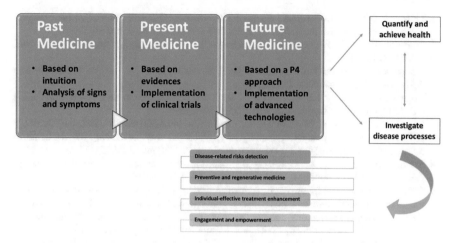

Fig. 14.2 Schematic representation of the evolution of medicine: from past to future

Patient Empowerment through Digital Health tools is the result!

Take-Home Messages
- The shift toward a proactive and preventive healthcare system is necessary to deliver high-quality care services, management, and advanced approaches to overcome the twenty-first-century challenges.
- The P4 of medicine represents a new concept in healthcare based on a continuous exchange between different stakeholders to empower a novel healthcare culture.
- Predictive medicine represents continuous health monitoring to predict the onset of a disease.
- Preventive medicine identifies risk factors related to disease development while encouraging preventive behaviors and self-management.
- Personalized medicine recognizes the individual as a complex system made of specific biological processes and structures to define treatments.
- Participatory medicine aims to empower patients into their health journey through active monitoring, measurement, and decision processes.

References

1. Sagner M, McNeil A, Puska P, Auffray C, Price ND, Hood L, Lavie CJ, Han Z-G, Chen Z, Brahmachari SK, McEwen BS, Soares MB, Balling R, Epel E, Arena R (2017) The P4 Health Spectrum - a predictive, preventive, personalized and participatory continuum for promoting Healthspan. Progress in Preventive Medicine. https://doi.org/10.1097/pp9.0000000000000002
2. Hood L, Auffray C (2013) Participatory medicine: a driving force for revolutionizing healthcare. Genome Med 5:110

3. Flores M, Glusman G, Brogaard K, Price ND, Hood L (2013) P4 medicine: how systems medicine will transform the healthcare sector and society. Pers Med 10:565–576
4. McEwen BS (1998) Protective and damaging effects of stress mediators. New Engl J Med 1998(338):171–179
5. Prainsack B (2014) The powers of participatory medicine. PLoS Biol 12(4):e1001837. https://doi.org/10.1371/journal.pbio.1001837
6. Vogenberg FR, Isaacson Barash C, Pursel M (2010) Personalized medicine: part 1: evolution and development into theranostics. P & T 35(10):560–576
7. Longo UG, Carnevale A, Massaroni C, Lo Presti D, Berton A, Candela V, Schena E, Denaro V (2021) Personalized, predictive, participatory, precision, and preventive (P5) medicine in rotator cuff tears. J Pers Med 11:255. https://doi.org/10.3390/jpm11040255
8. Razzak MI, Imran M, Xu G (2019) Big data analytics for preventive medicine. Neural Comput Appl: 1–35. doi:https://doi.org/10.1007/s00521-019-04095-y
9. Carbonaro N, Lucchesi I, Lorusssi F, Tognetti A (2018) A. Tele-monitoring and tele rehabilitation of the shoulder muscular-skeletal diseases through wearable systems. In Proceedings of the 2018 40th annual international conference of the IEEE engineering in medicine and biology society (EMBC), Honolulu, HI, 18–21 July 2018; Volume 2018, pp 4410–4413
10. European Society of Preventive Medicine (2020) P4 Medicine. Available at: http://www.esprevmed.org/p4-medicine/ (Accessed on August 27, 2021)

Beatrice Barbazzeni is a proactive scientist and prolific author by *ExO Insight*. With a background in psychobiology, she is currently finishing a Ph.D. in cognitive neuroscience. She likes to challenge herself by joining different research institutes and interdisciplinary teams, with a particular interest in exponential medicine and health longevity. Beatrice is an effective communicator fluently speaking several languages which allows her to share multicultural experiences and network globally. Last but not least, she has a fantastic passion for fitness, art, and music. The key to her success is discipline and grit; the values that inspire her to become a leader in science.

Michael Friebe received a B.Sc. degree in electrical engineering, M.Sc. degree in technology management from Golden Gate University, San Francisco, and Ph.D. degree in medical physics in Germany. He spent 5 years in San Francisco, as a Research and Design Engineer with an MRI and ultrasound device manufacturer. He is a German citizen with expertise in diagnostic imaging and image-guided therapies, as a/an Founder/Innovator/CEO/Investor and a Scientist. He is also a Research Fellow with the Technical University of Munich, Munich; an Adjunct Professor with the Queensland University of Technology, Brisbane; and an honorary Professor of image-guided therapies with Otto von Guericke University, Magdeburg, Germany. Since 2022, he has been a Professor of biomedical engineering innovation with the AGH University of Science and Technology, Krakow, Poland. He is a listed inventor of more than 80 patents and has authored over 200 papers, and was part of over 35 Medical Technology Start-Ups. He is a Board Member of four medical technology start-up companies and an investment partner of a MedTec-fund. From 2016 to 2018, he was a Distinguished Lecturer of the IEEE EMBC teaching innovation generation and MedTec entrepreneurship. He is also a coach and trainer for OpenExO, and a master Launchpad mentor for the Purpose Alliance.

Digital Health Business Models: Transformation Is on the Horizon

15

Dominik Böhler

Abstract

Digital Health opens possibilities for novel business models in healthcare. Established and fresh players from outside traditional MedTech and Pharma may change the potential and possibilities of new forms of payment and reimbursement. Hence, we will see new business models arise, which change existing patterns in all areas. The article analyzes the transformational potential by looking at lateral diversification scenarios for both healthcare and non-healthcare players. It supplies ideas for potential future business model configurations.

Keywords

Healthcare business models · Digital health · Health payment · Health reimbursement · Value-based health

What Will You Learn in This Chapter?
- Digital Health comes with new Business Models.
- New players may challenge the existing contenders.
- Different business configurations for digital health offerings

D. Böhler (✉)
Technische Hochschule, Deggendorf, Germany
e-mail: dominik.boehler@th-deg.de

© The Author(s), under exclusive license to Springer Nature Switzerland AG 2022
M. Friebe (ed.), *Novel Innovation Design for the Future of Health*,
https://doi.org/10.1007/978-3-031-08191-0_15

15.1   Introduction

Digital Health offers new ways for integrating insights from distributed patients, connected medical and lab devices, for new forms of personalized treatment, and new payment options. This may stretch the traditional boundaries between Medtech, Biotech, and Digital Health. In addition, it may require traditional healthcare players to also evaluate unregulated markets for user and customer acquisition. Furthermore, non-healthcare players may consider the move into regulated markets for pursuing data monetization strategies - an act of lateral diversification. Real estate, mobility, and energy are among the most obvious examples here.

In this article a brief analysis on the major components of a business model will be provided: Offering, Patient Journey, Value Chain, Financial Model. The potential changes induced by digital health will be analyzed and scenarios for the lateral diversification of both healthcare and non-healthcare companies discussed.

15.1.1 Four Components of Business Models

A Business model is a consistent set of design elements from Innovation, Sales, Production, and Finance activities. It is the logic of how a business earns money. The components of a business model are:

- Offering
- Patient Journey
- Value Chain
- Financial Model

A new business starts with the definition of a problem rooted in a clear purpose. This leads to an offering, which can be marketed in one or many steps of the Patient Journey. Based on the available input factors, a position in the value chain can be chosen and fitting revenue models developed. This perspective focuses on the development of new business and centers around resource-constrained situations, such as in start-ups.

Business Models in Healthcare and Pharma have been strongly influenced by the possibilities of reimbursement strategies (and possibilities) of regulators. This caused a strong focus on the cost-side and hence a focus on producing offerings with unambiguous evidence and configuring cost-efficient value chains. With the advent of digital health new offerings, new steps in the patient journey and potentially new revenue models will be created.

15.1.2 Business Models in Digital Healthcare

Emerging patterns and trends in digital health business models offer insights for both healthcare and non-healthcare companies. In the next sections, I will look at Offering, Patient Journey, Value Chain, and Financial Model.

15.1.3 Offering

While traditional medical devices (including software as a medical device) were mostly marketed in highly regulated product portfolios, Digital Health creates needs to blur these strict distinctions on the level of the product portfolios. It is already visible in market access strategies of start-ups that wellness apps are an important part of customer acquisition. The issues for start-ups and digital solutions are manifold. Two key issues are market access before and after regulations and re-certification of software upgrades in an agile development process.

The potential of new regulatory frameworks (e.g., DiGa in Germany)[1] is to offer easier market access through real-world regulatory processes. While this allows for alternate forms of study design using real-world evidence, it is often mistaken as a market access framework fostering iterative and agile product development. For digital health innovators, the question stays how to market their products to Health Care Providers (HCP). Partnering with a Pharma Company is a practical option, if software components can be frozen and the app marketed as a digital therapeutic solution. A Digital Therapeutic is a digital tool that may be used instead or together with an HCP. It enhances the treatment options and reach, because it may offer medical support independent of place and time, see e.g., Markin [1].

Some players are currently trying to enter with wellness-oriented apps or even games without a clear healthcare value proposition and hence no regulation.

Still, health-relevant data may be produced outside the regulatory framework and used for informing about secondary regulated products. The advantage of the latter strategy is cutting both pharma and HCPs as intermediaries and offering direct-to-patient access.

On the flip side, this means that established consumer-centric companies may be able to re-evaluate their offerings in the light of health-relevant data that could offer market access for regulated apps. This opens either new streams of revenue for these players by integrating backward or partnering with regulated applications.

---

[1]For example, DiGa in Germany: https://www.bfarm.de/EN/Medical-devices/Tasks/Digital-Health-Applications/_node.html

15.1.4 Patient Journey

If regulated products are not the only vehicle to access markets, the nature and initiation of the (digital) patient journey needs to be re-conceptualized. When do we speak of a patient? Once he/she goes and sees a doctor or once someone's data is analyzed with a health focus (e.g., to find relevant biomarkers)? In the analog world, the distinction is clear—I enter the doctor's office, I become a patient. In Digital Health, we need to understand which level of data analysis and what amount of inference is already a true health claim.

This offers interesting entry points into health and healthcare markets for non-traditional healthcare companies. Imagine an energy provider or property developer offering data analytics to digital health applications.

However, if the patient journey starts earlier, e.g., in the consumer sector. What will this mean for market access? In a traditional setting, you would target Key Opinion Leaders (KOL) to influence HCPs. If your customer acquisition happens via yet another digital application your partners need to be knowledgeable about how to reach the exact micro target. The nature of this approach means that the definition of an indication-specific strategy will need an integration across many different data sources.

15.1.5 Value Chain

We discovered that in a digital healthcare system, we can use and integrate personal and public health data from a variety of sources. From the analysis of wastewater over detecting energy usage patterns via smart meters to daily testing of blood or the microbiome, more data will be available that sometimes comes from players far outside the traditional MedTech and Pharma industries.

Understanding these sources of medical information and acting upon them can improve targeting and outcomes of (digital) solutions. However, the variety of partnerships needed to include these real-world sources of evidence is high. Understanding the complexities of data integration along this value chain is key. It is not only about integration formats and standards!

The core question is context control. If data is gathered from a variety of sources over time and space, understanding its origin and context evaluation is key. How might we know about and normalize when an ECG measurement is at sea level and compare it with one from a mountain top.[2]

One solution would be near ubiquitous sensing and identification to gather as much context information as possible. Another would be thinking about new ways to integrate Patients and HCPs in data quality assurance, e.g., with financial incentives for HCPs when generating valid data. In some instances (e.g., Rheumatoid arthritis) pharma companies today already pay small reimbursements to both patients and

---

[2]For example, Guger [2].

HCPs for data sharing and context control. With more diverse data integrations along the value chain, this may need targeted platforms for data generation and exchange. The question of data ownership is another elephant in the room.

15.1.6 Financial Model

Data generation and sharing in Digital Health yields a more value-based approach to healthcare. In most healthcare systems control over reimbursement is exerted via cost-based systems focussing on reducing the number of billable hours and minimizing unnecessary time spent on patients. An HCP can optimize payout most efficiently by seeing many patients for a short amount of time. In a value-based system, a measure of treatment success is the basis for reimbursement. It is generally believed that this may bring a stronger focus on quality and outcome orientation to a health system. See, e.g., Gray [3].

This must mean more options for revenue or equity-based compensation and, as described, indication-independent market segmentation. Two scenarios are on the line:

1. The rise of integrated hierarchical ambulant care or
2. Data commons for independent practitioners

The more ambulant care facilities become integrated hierarchical organizations, the more they can reap the benefits from data-driven strategies. They know and (potentially) own a large and high-quality data set about their patients. This is an immense advantage compared to payers, who do not or are not allowed to own or analyze patient data. This could potentially lead to the health insurers, standing between the patient and the providers, are cut out of the value chain of ambulant care. If ambulant care chains offer flat subscription services, they can effectively distribute the heterogeneity of patient costs themselves and reap the benefits of generating high-quality research data.

On the other end, a data commons strategy might yield the same structures of independent practitioners. Selecting and providing the right set of digital services is a question of strategic IT Management. In a healthcare setting, the added demands on data security yield even more specific resources to be spent on the topic. It is difficult to imagine how single-owned small businesses such as GP offices may be able to keep competitive hybrid service portfolios and effectively understand and manage their patient's data. As an alternative to an integrated hierarchical network, a commons structure may be a viable option to support a decentralized network of independent professionals (Fig. 15.1).

However, through a payer-driven data exchange platform more incentives for HCPs could be generated. Imagine offering data curation or integration challenges or incentives for data-centric cooperation around rare diseases. While currently the patient has the burden of integrating all the ability of dispersed individual practitioners, in a world of well-curated data, competition around the cases could

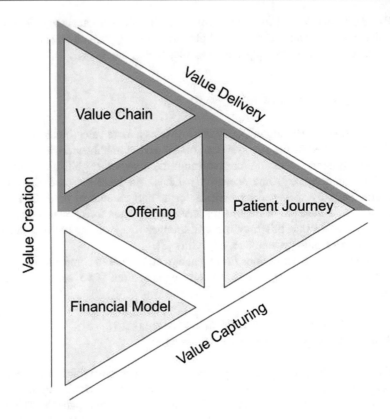

Fig. 15.1 Components and dimensions of business models

appear. Disease management would thus not only focus on reintegration but also orchestrate different practitioners at earlier stages.

15.2   Value Creation, Value Delivery, and Value Capture in Digital Health

Looking at these individual components of business models offers a tentative picture of the future. On a higher level, the dimensions of business modeling Value Creation, Value Delivery, and Value Capturing change. The dimensions Value Creation and Value Delivery were the centric dimensions in a cost-centered healthcare system. Value Capturing played a less dominant role. High regulatory burden left little room for more advanced models of value capturing, which is a unique combination of financial model offering, and patient journey.

However, there are clear limits to what is currently imaginable in healthcare across different cultures and societies. If we look at the business models of scalable digital platforms, these usually also work around auction-based forms of payment.

Amazon, Google, and Facebook center large parts of their revenue models around selling advertising space through competitive bidding. Transferring such a model to healthcare would raise ethical concerns, if primary or emergency care is concerned.

In areas, where there is no acute danger of a lack of access to care, such models may well contribute to improved outcomes and more efficiency. In a world, where new medical and pharmaceutical products are built on the basis of real-world evidence, the need for large amounts of well-curated data delivered in near real-time increases. Providing platforms for competition around a case could yield more dynamic forms of reimbursement for healthcare professionals.

Next to reimbursement from state systems, a redistribution of funds from Clinical Research Organizations to healthcare professionals may yield additional possibilities for a revenue share between them and digital solution providers. In addition, it could create a measurement for the real value of personal health data. If a research system, which is based on real-world evidence yields better and faster outcomes, needs to be evaluated. It may definitely contribute to more distributed and real-time generation and validation of evidence.

Overall, we can see that supporting care with more digital means does not only provide potential for new patient journeys but may question the business models of many stakeholders in the system. If the contributions of both doctors and patients receive more direct valuation through digital platforms, this market for data sharing can improve accessibility and time-to-market for healthcare innovation. Many rigid and bureaucratic elements in generating guidelines, documentation or billing will be replaced by more digital and user-centered solutions. However, for these business models to work, price discrimination and more competition are necessary. In healthcare systems with very individualized treatment, also reimbursement needs to follow individual pricing strategies.

A DRG-system like in Germany that does not follow the speed and collaborative development of a global real-time world may soon be obsolete, because patients are less and less willing to pay for improved digital and personalized care out of their own pockets. While this may sound like a positive effect on cost reduction in healthcare, the tide can turn quickly and render state-funded ambulant care obsolete. In this sense, the trend to ambulantization that the Anglo-Saxon health systems have already seen will also affect Europe. The transformation we witness in healthcare is thus both a transformation from analog to digital and from public to private.

15.3   Summary

The focus of digital health is currently on creating suitable infrastructure for data exchange. Soon the impact of cross-institutional collaboration might change some of the mechanisms currently in place. The blurring of sectoral boundaries might increase and the entry barriers into digital health markets be lowered. This creates the potential for value-based business models that foster patient-centric collaboration. For both payers and pharma, it is time to evaluate all options sensibly as value chains become more diverse and patient journeys may be starting earlier and be governed by new players. The Transformation from an analog to a digital healthcare system will likely go hand in hand with more private and ambulant care.

Take-Home Messages
- Established Consumer Companies will play a key role in marketing digital health solutions.
- Digital Health tools will enter the Health Business much before someone becomes a patient.
- Value-Based Business Models in Health will develop due to new forms of data exchange and reimbursement.
- New forms of organizing (Hierarchical vs. Commons) ambulant care are appearing which help continuously improve digital health service offerings.
- New ways of engaging with patients and medical cases may yield competition for best advice instead of minimizing time and effort.

References

1. Markin S (2019) The emerging world of digital therapeutics. Nature 573:106–109
2. Guger C, Krausert S, Domej W, Edlinger G, Tannheimer M (2008) EEG, ECG and oxygen concentration changes from sea level to a simulated altitude of 4000m and back to sea level. Neurosci Lett 442(2):123–127
3. Gray J (2011) How to get better value healthcare, 2nd edn. Offox, Oxford

Dominik Böhler is Professor for Management in Digital Healthcare. He works on digital business models and care processes in healthcare. He is one of the initiators of the Munich Digital Health Summit and a strategic advisor and business angel. Previously, he led the education department at UnternehmerTUM, where he supported 2000 innovation projects resulting in 150 Startups. He holds a PhD in Information Systems and a degree in business administration from FAU Erlangen-Nürnberg.

Value Propositions for Future Health Developments: Digital, Portable, Connected, Experience-Enhancing, Supportive, Patient-Centric, and Affordable

Beatrice Barbazzeni

Abstract

Over centuries, healthcare evolved into a complex system in which providers, payers, and patients interact at multiple levels. However, a novel proposition in healthcare is needed while supporting preventive, predictive, personalized, and participatory medicine. Hence, to promote innovation and development, moving the healthcare industry forward, a few challenges need to be faced. Increased quality care and services, aging and chronic diseases, reduction of costs, efficient organization, data protection, stress management, and innovative approaches are just a few. Moreover, digital transformation accompanied by advanced technologies brought enormous changes and the need of revising the entire healthcare system. In fact, to foster innovation healthcare should deliver new value propositions, transforming into an ecosystem that is digital, portable, connected, experience-enhancing, supportive, patient-centric, and affordable. Three concepts will be central in this innovative approach: digital clinics, digital patients, and digital devices; developing healthcare toward an empowered and democratized system directly in a pocket.

Keywords

Digital · Portable · Connected · Experience-enhancing · Supportive · Patient-centric · Affordable

B. Barbazzeni (✉)
ESF-GS ABINEP International Graduate School, Otto-von-Guericke-University, Magdeburg, Germany
e-mail: beatrice.barbazzeni@med.ovgu.de

What Will You Learn in This Chapter?
- Healthcare is a complex system made of providers, payers, and patients interacting at multiple levels.
- A novel value proposition is needed to foster innovation in healthcare.
- Digital clinics, digital patients, and digital devices are the three key concepts to generate a novel value.
- Portable and real-time patient monitoring, remote access, data security, telehealth, disruptive and digital technologies, democratization of services, improved quality care are the game-changers supporting a novel value proposition in healthcare.

Across the centuries, the evolution of healthcare has a long history, from the antique use of plant extracts to the discovery of vaccines in the eighteenth century by Edward Jenner, defining the beginning of the modern medicine era [1]. However, through the years the entire healthcare system evolved enormously toward the formation of a hierarchical structure characterized by the interplay of different components such as regulatory procedures, health providers (e.g., doctors, nurses, medical staff), and consumers (e.g., patients and caregivers), insurance companies, medical products providers and suppliers, and diagnostic health services [1] supporting its functioning and background processes. Nevertheless, based on a provider-driven approach, the goal of this modern system was mostly finalized to reduce mortality rates and find innovative treatments to cure diseases, whereas a different scenario can now be faced in the twenty-first century. In fact, the main healthcare challenge focuses on promoting and maintaining wellness in the perspective of a preventive, predictive, personalized, and participatory medicine built around a patient-centric ecosystem and guaranteeing access quality to a variety of healthcare services, managing costs and technological advances [2]. Adopting this novel paradigm would support health longevity, on-demand personalized services, and exponential technologies that would allow longitudinal care toward preventive and outcome-driven healthcare [2].

The understanding of current healthcare is viewed as a mechanical system in which policymakers control procedures, regulations, and processes toward predictable and linear outputs that have only the effect of producing unexpected consequences [3]. Indeed, increased global healthcare costs across countries, reduced and inadequate resources, or unmet clinical needs are just a few of them [1]. Besides, increased longevity results in a higher demand for health services. This fact is also accompanied by the increasing rate of chronic diseases leading to the request for continuous monitoring and access to treatment options, which inevitably brings to hospitalizations, prolonged patient management, and financing issues. In addition, this event, accentuated by socio-demographic changes toward the aging population, generates approximately 78% of healthcare costs due to chronic disease treatments [1, 2], particularly in Germany, Luxembourg, and the Netherlands. Increased deaths also accompany this fact due to medical mistakes in procedures or wrong drug prescriptions.

Hence, supported by disruptive technologies and particularly by the Internet of Things (IoT), the need for innovation would move toward a complex and dynamic system [3] in which different components such as hospitals, clinics, and rehabilitation departments, patients, and caregivers interact non-linearly on different scales; from the patient and family unit to the hospital and government while generating exponential benefits, tangible results, and feasible solutions [3, 4]. Therefore, the importance of valuing nonlinear interactions between parts leads to greater outcomes instead of considering its single contributors. An emergent self-organized system would thus represent the main process of a functional, dynamic healthcare structure in which revised regulatory approaches and policies are redirected toward a common goal: supporting and democratizing wellness with affordable services, global health monitoring, reduction of unnecessary costs around a patient-centric paradigm [3, 4]; a perspective in which patients establish an active collaboration with providers becoming aware and in the power of their care process [2]. In this view of a complex system, to deliver and understand healthcare needs [3, 5], a novel value proposition should be reinvented toward considering and solving healthcare problems as a global concern with related implications [3]. The goal of generating a novel value proposition underlying healthcare services and their complex structure should first consider which stakeholders participate and interact in this process [1].

Primary consumers of health services are patients, represented mainly by the older population [1, 2]. Considering aging as a high-risk factor for developing disabling chronic diseases [2, 6], reducing costs would represent an innovative way of delivering high-quality care, characterized by more effective interaction with clinicians or other professionals, available and unlimited access to personal health information in the perspective of promoting health education, and self-empowerment making informed choices based on value [2, 6]. Furthermore, healthcare providers are physicians, nurses, medical staff, and other professionals that represent the key element to define optimal treatments supporting a patient's needs [2, 7]. In charge of empowering patients with health information, innovative methods to improve communication, exchange insights of health data, reduced administrative costs, and more efficient processes to deliver treatments would add a new value to this complex system [2, 3]. Moreover, considering healthcare as an industry, suppliers of pharmacological drugs and medical services, vendors and providers constitute the background processes to perform and sustain daily healthcare operations. In this regard, revised and optimized procedures to reduce costs [1] and operations would generate innovation while supporting this core system more effectively.

Furthermore, insurance companies play an essential role in this hierarchy, mainly contributing to the global health industry revenue stream [1]. Coping with increased costs due to premiums [1], administration, and disparities across countries, the democratization of health services would represent a valuable achievement in generating a novel value proposition, from reimbursement to outcome-based perspective. In this regard, supported by digitalization, e-Health would represent a feasible solution to promote legible, complete, accurate documentation and billing information [1] even in remote regions. And lastly, data from laboratories delivered to professionals via diagnostic service providers represent detailed and precise information about a patient's health. Again, e-Health and leverage technologies would guarantee high-quality services and information by continuously monitoring individuals toward an accurate understanding of patient needs [1]. Empowering patients to promote self-management through available technologies would allow the management, storage, and acquisition of health data shared in real-time with professionals, guaranteeing constant monitoring into a continuous process of care.

In the context of digital transformation and Industry 4.0, while moving toward a patient-centric and value-based approach to engage patients, providers, and payers, the entire healthcare industry needs to establish innovative care business models to generate not only advantages and optimal service delivery but also guaranteeing solid revenues. The rapid growth of digital technologies would definitely provide telehealth, artificial intelligence-enabled solutions, and blockchain electronic health records for patient care. Accessible at home or in clinics, digital services aim at faster and more efficient delivery of services at any time while reducing the cost of operations and administrative procedures. These interoperability standards of connected digital and wireless devices, cloud-based solutions, and electronic medical data become the building blocks of an effective ecosystem based on digital solutions [6]. Moreover, shifting from a reimbursement value perspective to a service that values the outcome of its intervention, evidence-based medicine policies are the starting point of a game-changer. The mutual interaction between healthcare components would actively educate patients to make conscious decisions on their health [2].

The adoption of a patient-centric design built around the concept of adopting mobile health (mHealth) technologies and services in healthcare resulted from the development of cheaper, available advanced technologies and portable computing devices [6] that changed the way of transmitting health interventions toward a remote approach. A novelty aims to value consumer engagement, cost optimization, and improved treatment outcomes [6]. Nevertheless, to meaningfully handle digital health data, several key questions should be addressed, such as targeting the proper mHealth user, factors enhancing mHealth capabilities, and evidence supporting positive results of implementing such technologies in a clinical environment [6]. Three main components should be considered to gain insight on an outlook: digital devices, digital patients, and digital clinics [6]. Among digital devices, a few are smartphone-connected monitoring devices, implantable and ingestible sensors, wireless and wearables, lab-on-a-chip technologies, and handheld imaging [6]. However, the need to engage patients toward participatory, self-management, self-measurement, and informed decision-making behavior would be necessary to welcome the adoption of digital technologies while understanding which device, service, or process is most effective in mHealth [6].

Moreover, the production of mHealth data requires a clinic that is digital, able to collect and manage effectively and in big real-time data via advanced analytics tools, in the perspective of a personalized and precision medicine approach [6]. The mobile and digital health property characterized by portable, affordable, and user-friendly technologies would represent scalable solutions and cost-effective alternatives reaching new patient populations, particularly in limited and underdeveloped areas [6]. Furthermore, the increasing implementation of medical technologies would require novel regulatory and legislative approvals to guarantee integration, transparency, and interoperability of the amount of digital data into electronic health records (EHRs) [6] while reducing and optimizing clinical workflows.

Therefore, the need to transform the healthcare industry so that different stakeholders actively participate in an integrated care process supported by digitalization and emerging technologies is central to promoting and delivering a new value proposition behind the innovative healthcare business model. Lowering the costs while improving the efficacy and efficiency of health procedures would be facilitated by connected devices, sensors, and wearables [8]. These facilitators respond to various health parameters over different body locations and can continuously assess patients' data remotely, actively toward preventive care, and adapt to current demographic changes and needs [2]. Telemedicine brought an unbeatable advantage in monitoring chronic diseases while giving access to health services reaching rural areas remotely and internationally over those developing nations [2] via digital platforms, mobiles, and devices. The reduction of unnecessary in-office visits [2] resulted in cost reductions and complicated and repetitive administrative procedures. Still, mostly this had a positive impact on patients' behavior and attitude toward care processes. Hence, this increased adherence and responsibilities toward medication and treatments [2]. Moreover, such consistent behavioral patterns would allow health providers and medical industries to test and investigate the optimal intervention and treatment program based on a personalized and participatory framework.

Drivers of precision and personalized medicine toward an evidence-based, preventive, and self-monitoring approach [2] become possible due to the availability of disruptive technologies (e.g., machine learning and artificial intelligence techniques) in data analytics to allow more accurate, complete, and faster diagnosis while optimizing decision-making and prompt actions over treatments [2]. In a connected device system centered around consumer needs, digital technologies proved to ease operations while ameliorating care delivery, patients, and clinician's experience. Indeed, Philipps [2, 9] demonstrated that 57% of patients are interested in monitoring their health with devices, although only one-third of them are willing to share this data with professionals. Thus, translating the process of constantly acquiring, monitoring, and supervising through health mobiles and web platforms into a daily practice would have the effect of providing higher quality care and a better environment to value patients' journey with empathy, offering a direct and more engaging interaction between providers and consumers [2].

Being at the beginning of this twenty-first century, forecasting the future of healthcare toward achieving the vision can be a practical approach to create and evolve toward it. From a reactive, provider-centric, expensive care setting, reimbursement oriented, homogeneous, and unidirectional structure to a complex system characterized by proactive, participatory, predictive, and preventive care centered around empowered patients, valuing outcomes in a cost-effective, digital, and democratized way, this can be the future to foresee in 2030 (see Fig. 16.1) [2]. This approach would be aimed to cope with demographic changes characterized

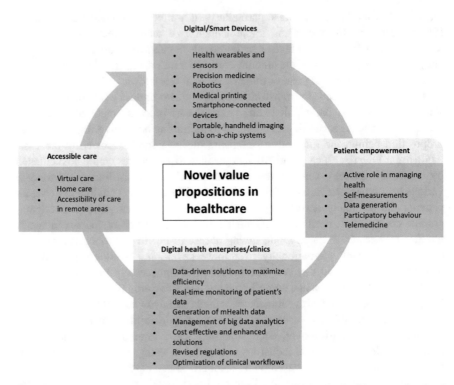

Fig. 16.1 Schematic representation of novel value propositions in healthcare and related description

by an increased rate of chronic diseases and ineffective treatments through personalized interventions, leveraged of quality care over large and diverse populations in the perspective of sustainable health longevity. The digital transformation brought undeniable support into this process, allowing on-demand health services, easiness, management, and insight of big data, effective virtual reality assessments, wearables, and AI-based tracking devices, blockchain, 5G for instantaneous connections while accomplishing the 4PH directly in a pocket [10].

Take-Home Messages
- *Novel value propositions are needed to promote healthcare innovation toward an approach that is digital, patient-centered, personalized, portable, affordable, connected, and experience-enhancing.*
- *Digital clinics, digital patients, and digital devices are the key points around which novel values are generated.*

(continued)

- *Empowering patients through self-measuring and data generation, decision-making and health monitoring is needed when creating the future of health through a participatory behavior.*
- *Revised regulations, management and real-time assessment of big data, as well as, reduction of costs are the solutions to optimize clinical workflow and management of digital health in any clinics.*
- *Virtual care, available at home, hospitals, or in any clinical ambulatory is necessary toward healthcare democratization via remote accessibility.*

References

1. Sneha S, Straub D (2017) E-health: value proposition and technologies enabling collaborative healthcare. In: Proceedings of the 50th Hawaii international conference on system sciences, 2017. http://hdl.handle.net/10125/41260
2. Deilotte UK (2018) A journey towards smart health. The impact of digitalization on patient experience. Available at: https://www2.deloitte.com/content/dam/Deloitte/lu/Documents/life-sciences-health-care/lu_journey-smart-health-digitalisation.pdf. Accessed 31 Aug 2021
3. Burton C, Elliott A, Cochran A et al (2018) Do healthcare services behave as complex systems? Analysis of patterns of attendance and implications for service delivery. BMC Med 16:138. https://doi.org/10.1186/s12916-018-1132-5
4. Lipsitz LA (2012) Understanding health care as a complex system: the foundation for unintended consequences. JAMA 308(3):243–244. https://doi.org/10.1001/jama.2012.7551
5. Hawe P (2015) Lessons from complex interventions to improve health. Annu Rev Public Health 36:307–323
6. Bhavnani SP, Narula J, Sengupta PP (2016) Mobile technology and the digitization of healthcare. Eur Heart J 37:1428–1438. https://doi.org/10.1093/eurheartj/ehv770
7. Aggarwal A, Aeran H, Rathee M (2019) Quality management in healthcare: The pivotal desideratum. J Oral Biol Craniofac Res 9(2):180–182. https://doi.org/10.1016/j.jobcr.2018.06.006
8. Pradhan B, Bhattacharyya S, Pal K (2021) IoT-based applications in healthcare devices. J Healthcare Eng 2021:Article id: 6632599, 18 pages. https://doi.org/10.1155/2021/6632599
9. Philipps (2016) Future health index 2016 report. The capacity to care: Measuring perceptions of accessibility and integration of healthcare systems, and adoption of connected healthcare. See also: www.goo.gl/VDPe5S
10. Reddy M (2021) Digital transformation in healthcare in 2021: 7 key trends. Digital Authorities Partners. Available at: https://www.digitalauthority.me/resources/state-of-digital-transformation-healthcare/. Accessed 31 Aug 2021

Beatrice Barbazzeni is a proactive scientist and prolific author by *ExO Insight*. With a background in psychobiology, she is currently finishing a Ph.D. in cognitive neuroscience. She likes to challenge herself by joining different research institutes and interdisciplinary teams, with a particular interest in exponential medicine and health longevity. Beatrice is an effective communicator fluently speaking several languages which allows her to share multicultural experiences and network globally. Last but not least, she has a fantastic passion for fitness, art, and music. The key to her success is discipline and grit, the values that inspire her to become a leader in science.

(Digital) Patient Journey and Empowerment: Digital Twin

17

Dominik Böhler and Michael Friebe

Abstract

New forms of interaction between doctors and patients are facilitated by a transformation of the patient journey. The advent of better data integration through digital twins allows for completely new forms of interaction, decision-making in unprecedented frequencies. The Digital Twins will also allow a more personalized data analysis leading to predictive information and a more pull-oriented approach in healthcare delivery. This creates real-world challenges for data acquisition, data ownership, and data sharing. We also need to question medical regulation mainly based on evidence and data trust and privacy issues.

Keywords

Patient journey · Digital twin · Data integration · Data ownership · Data sharing

D. Böhler (✉)
Technische Hochschule Deggendorf, Deggendorf, Germany
e-mail: dominik.boehler@th-deg.de

M. Friebe
AGH University of Science and Technology, Krakow, Poland

Otto-von-Guericke University, Magdeburg, Germany

IDTM GmbH, Recklinghausen, Germany

FOM University of Applied Science, Center for Innovation and Business Development, Essen, Germany
e-mail: info@friebelab.org

© The Author(s), under exclusive license to Springer Nature Switzerland AG 2022
M. Friebe (ed.), *Novel Innovation Design for the Future of Health*,
https://doi.org/10.1007/978-3-031-08191-0_17

169

What Will You Learn in This Chapter?
- A framework to analyze future patient journeys
- Digital Twins as technology supporting digital Patient Journeys
- An understanding of Data Gathering, Data Ownership, and Data Sharing

17.1   Introduction

The literature defines digital health as:

> the cultural transformation of how disruptive technologies that provide digital and objective
> data accessible to both caregivers and patients leads to an *equal level doctor-patient
> relationship with shared decision-making and the democratization of care*. [1]

The previous definition of digital health yields several important points about the patient journey and patient empowerment. The interaction and the relationship demands and leads to a certain regularity of interactions, which is in clear opposition to the mostly transactional nature of healthcare today.

The involvement of patients and also the power relationship between patient and doctor is described as shared and democratized in the future. From the perspective of Patient Involvement, both Doctor and Patient will then equally share control over the process.

The third and last element—but possibly the most important and most difficult one—is the cultural transformation that is induced through the upcoming disruptive technologies. This will create an environment where the initiation of a relationship and also the specific nature is not linear anymore, it will be a complex interrelationship between many different nodes of a network. However, the most apparent feature is that the first move to initiating a new interaction between patient and doctor will less often come from the patient, but more often from the doctor or a digital technology that connects patient and doctor. Better relationships and equal distribution of power, decision-making and control lead to a new market which—at least in theory—will have healthcare systems switch from a push logic (patient goes to doctor) to a pull logic (doctor or digital technology initiates health interaction). Or in other words, most "normal" interactions with respect to health could be done in a homecare setting with no or only short digital communication with a physician. While this may soon be the case for a normal individual (healthy) it is different for a patient with known symptoms currently treated or after/during a therapy process or as an after-care or recovery process.

17.2   Analyzing the Patient Journey

To formalize these assertions we would like to discuss the following framework and describe three particular transformations in our future health systems. In particular, we look at the regularity of interactions, the decision-making, and the mode of initiation between an individual/patient and a Health Care Provider (HCP).

> **Dimensions**
> 1. Regularity of Interactions (Once | Multiple Times (Interval) | Continuously)
> 2. Decision-making (HCP is in control | Patient is in control | HCP and Patient share control)
> 3. Mode of Initiation (Reactive | Proactive)

17.2.1 Regularity of Interactions

The regularity of interaction can be differentiated in one-time, several-times or continuous interactions. In traditional health systems, most indications will be treated as one-time interactions, we could also say transactions.

The patient goes to see a doctor, gets a diagnosis, and follows treatment. For the majority of cases where the patient becomes healthy again the interaction ends here. In most cases, the doctor can only assume from a patient that is not showing up anymore that the health problem must have been solved and the patient is healed.

In more complex cases or with chronic diseases multiple interactions over a longer period of time are the norm. A Doctor–Patient Relationship develops from multiple transactions. While this is not unusual as long as a patient is sick, in a digital world we can imagine that this care relationship also continues for the time after the recovery.

Even more so, we can find the potential for a continuous relationship between doctor and patient independent of the patient's health status. While in an analog world this would be prevented by the cost burden of synchronous physical examination, asynchronous remote examination and monitoring in a digital world will allow for such a continuous interaction between doctor and patient. And imagine that there will be artificial intelligence (AI) based systems that constantly learn and that will be able to regularly check the digital data generated by the examination results and contact if anything out of the ordinary is observed.

17.2.2 Decision-Making

Already today we can witness much more informed patients that question the diagnosis and treatment suggestions of their primary care physicians and also of secondary and tertiary care doctors. Extensive virtual exchange and support groups create a significantly better-educated patient. More and more patients use this to their advantage. However, more and more obscure forms of treatment are also popping up at the same time.

With a strict orientation toward evidence for treatment and mostly low digitalization, many patients seek to find alternatives in faster, seemingly modern, and less well-researched forms of treatment. The speed of information exchange on the internet, which is global and real-time, is often in stark contrast with the speed of the regulatory processes and approvals needed for a certified use. Obviously, this context affects how decisions are being made and who bears which part of the risk. And it creates big issues with trust toward the own doctor. It is clear that this evidence-based approach is not a personalized approach. "Evidence" generally means that the statistical results are favorable compared to alternative solutions. But that also means that every one of us is put in a certain category. If a medication works very well for 20% of the patients it is statistically not good, except you are one of the 20%. So would you not approve that process or medication because 80% do not benefit? No, you need to match the individual to the medication. This is a personalized approach that is not following the normal evidence-based clinical validation. So we really need to rethink the regulatory approval process.

In a traditional context, the doctor decides based on the diagnosis for a certain treatment and executes it on or with the patient. In case the doctor is not able to do it then the patient is referred to a specialist or dedicated treatment/therapy provider. While this form of decision-making will certainly prevail for emergency cases also throughout the digital age, the patient will soon influence it with the help of information from other doctors or digital services. This heterogenous and highly dynamic environment can be characterized as a complex adaptive system. This yields new strategies for decision-making. Neither patient or doctor will be able to take these decisions unilaterally and/or alone.

Same as in complex decision-making situations in business, politics or sports, both patients and doctors need to work in a cross-functional team combining multiple sources of data to form a decision under uncertainty. In many cases, only a part of this information can yield the highest possible degree of evidence either due to the recent nature of the data or the inability to process the multitude of signals.

While AI may offer a way out of this dilemma, its limits and thus the limit of Decision-Support-Systems need to be clearly understood. Especially, if the decision is related to a more complex set of indications or differential diagnoses, both data quality and the machine learning models may for a long time not perform as well as simple heuristics and experience.

So it remains important that both patients and doctors are aware of their situation and role and form a joint opinion hearing to a set of different opinions. The bilateral relationship between Patient and Doctor will need to change dramatically, if both patients and doctors want and are allowed to use all available sources of information available. The time needed for that is not available to the clinician at the moment and is also not reimbursed in most healthcare systems. The already discussed shift to more home care will likely free up time for the doctor, but the reimbursement of the doctors time and effort is still unclear.

17.2.3 Mode of Interaction

In the traditional health system, the patient initiates a new interaction. Either by actively going into a hospital or a practice or less actively as in an emergency. Doctors or other health services do not usually start interaction from their end. Mostly because there is just no good reason, if health information is not ubiquitously available.

In a digital environment, we can imagine many situations when a new patient journey may start. In a world where prevention is the focus of healthcare, this may mostly be the norm. We will discuss a set of scenarios for this in the next sections.

1. **The doctor proactively approaches patients based on accessible lab or vital data or new preventive treatment options**.
   Throughout the COVID-19 pandemic, it became obvious that non-pharmaceutical interventions as well as vaccinations were key to tackling an indication that is rapidly spreading globally. In many cases, we have seen very proactive marketing in order to sustain NPIs and to encourage citizens for vaccination. While the final step and decision is still on the patient's side, the amount of influence taken on personal health decisions was staggering. This example is to show that more prevention will lead to more proactive approaches in influencing personal health decisions and thus starting patient journeys.

2. **A digital health service proactively initiates the journey**
   With more and more digital data on patients' health available in real-time, new modes of getting and tracking patients are imaginable. A doctor or another service, e.g., a digital health app could diagnose a certain condition based on available data or suggest a close examination based on digital biomarkers. This systematic and proactive form of health data analysis changes the way the patient journey is initiated and sustained.

3. **A non-traditional digital health service proactively initiates the journey**
   While it is obvious that digital health services initiate a patient journey, in the future we may also see more and more non-traditional healthcare players to offer digital biomarkers and thus also a potential source of data for digital diagnosis or recommendation for an examination. While traditionally the doctor did see very few stages in the value chain between him/her and the patient, this can change drastically in the future. Imagine the following scenario:

A patient get's up in the morning and switches on the light. This is registered in the smart meter and added to the data pool for sleep tracking. Next the person picks up the smart-phone and conducts an internet search, the request and the time is kept by the search engine provider. After getting up and taking a shower, the person opens the fridge and may use the toilet. All these interactions within the house leave a data or energy consumption pattern and over time yield relevant digital biomarkers. Once the person leaves the house and enters a car, the sensor in the seat and the cockpit may offer a variety of biomarkers. All these day-to-day interactions show that far before even thinking about entering a medical environment a person may become a patient.

Undoubtedly, all these data points come with questions regarding data safety and privacy. This certainly needs to be addressed and resolved in combination with adapted ethics guidelines.

17.2.4 Digital Twins Are Fostering the Patient Journey Transformation

From a technological perspective, the transformation of the patient journey is accompanied by several disruptive technologies. With a combination of sensors, and advanced data processing technologies the digital twin effectively enables prevention and accompanies the patient journey

17.3 Digital Twins

A digital twin is an in-silico model of human physiology based on data. It allows for inferences on both the individual and a population. Their use creates a fundamental shift affecting all processes in healthcare. These range from ambulant care to managing a hospital up to clinical care.

This also creates new opportunities for innovation in understanding indications, pharma and medtech. They enable a targeted approach for unique personalized therapy by allowing for simulation and testing on a virtual artifact. This has a direct impact on how innovation in healthcare is being conceptualized and how the go-to-market needs to be organized.

A digital twin can mirror any level of the human organism from cells to organs or complete physiology. This may lead to a more personalized form of medicine in the future (see Fig. 17.1) [2, 3].

Personalized medicine also always means prevention. If I can simulate the effect of treatment, I can also simulate disease progression. This may mean that acute treatment can become better and safer, it may also mean to detect an illness way before symptoms occur. However, without a clear perspective on an integrated patient journey, true digital twins are hard to imagine. This means all patients, healthcare professions, and healthcare institutions need to commit to a systematic use of digital twins along the patient journey.

17.3.1 Digital Twins and the Patient Journey

Digital Twins are essential for the Patient Journey of the future. Through this systematic form of data gathering and processing new conclusions can be drawn earlier and more frequently. A digital twin however is not a static concept. It is frequently updated with the latest personal data. The patient journey plays an important role in this process. However, with the perspective on the patient journey shown in the last section, practical issues of data integration need to be considered,

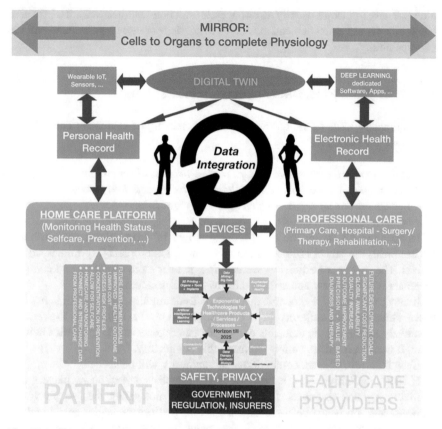

Fig. 17.1 Illustration adapted from Friebe [2]—a Digital Twin that mirrors everything known from the human organism—from cells to organs or complete physiology. This personalized data put together from data points collected at home (IoT, Wearables, other Sensors) and the information obtained in the professional care setting. Exponential technologies will combine these points, but safety and privacy issues need to be addressed, just as changed regulations. The government has a huge steering and legislative role

especially if the data needs to be acquired across public and private institutions from within and from outside the healthcare sector. Data integration, ownership, and sharing are the crucial issues in this regard.

While most current research is focused on formal frameworks for technical data integration and sharing, a question rarely asked is around the incentives and practical obstacles of data integration.

It is plausible that legally enforced data integration for all players in regulated markets may be successful. However, with relevant biomarkers from data outside the regulated first healthcare market it is questionable how a complete dataset of a patient leading to a digital twin may be created.

Next to clearly medical data being governed by medical institutions it is especially interesting to integrate data from other industries like energy, mobility, and fitness. These industries usually do not operate with non-healthcare claims. Effective integration would either demand incentives to annotate or more regulation to enforce interoperability.

17.3.2 Digital Twins, Data Ownership and Sharing

This ultimately leads to the question of data ownership of healthcare data. Here a clear trade-off is visible:

- *A state-governed regime where data resides on central servers or*
- *A private personal regime, which allows for distributed data storage or*
- *A private corporate regime, which offers a hybrid solution.*

Independent of where the data is being stored in a centralized or distributed environment, the process of governing the semantic interoperability frameworks is crucial. The process of efficiently exchanging on ever more changing and complex schemata will need new and effective forms of decision-making.

If they are based on democratic discourse in a community, competition in market-based structures, or authority in a bureaucracy is to be defined. However, for effectively answering the demands of a highly heterogeneous and dynamic environment, it needs a requisite variety to match this complexity.

Usually, a mix of trust, competition, and authority will be needed.

If interoperability councils just focus on established healthcare players and publicly regulated resources, they will most likely not gather the full picture of relevant data. If private players focus too much on platform strategies enabling data-centric business models, they will not maintain trust with both patients and healthcare professionals.

Finally, if individuals try to focus too much on autonomy and refrain from sharing, effective machine learning on the data is impossible.

Transformation of Healthcare and new Patient Journeys thus also require an exchange with new players and thinking beyond the current boundaries of the first healthcare market. The future of organizing healthcare will be borderless.

17.4 Conclusion

Concluding this article, we have created a language that allows us to characterize the level of transformation in a patient–doctor relationship along a digital patient journey. We offered three scenarios and promising examples for how this transformation may materialize in the near future. In addition we briefly looked at technologies supporting this transformation, which can be subsumed under the concept of the digital twin.

For the future healthcare system, this creates new challenges of Data Integration, Data Ownership, and Data Sharing across an increasingly complex Patient Journey with public and private stakeholders from within and outside the healthcare system.

This creates a complex adaptive system and needs new forms of decision-making on the semantic schemata to enable data integration, sharing, and ownership.

We need to extend collaborative efforts beyond the first healthcare market.

Take-Home Messages
- The Patient–Doctor relation will change from an exclusive PUSH to a PULL.
- Regularity of Interactions + Decision-making + Mode of Initiation will change with Digital Health.
- The clinician will be enabled to continuously be in touch with the patient journey.
- Individualized digital data from home care and professional care will merge and create a personalized profile.
- Health Data Ownership is still unclear.
- The Future of Healthcare will be without any boundaries.
- Many new challenges with respect to Data Integration, Data Ownership, and Data Sharing need to be dealt with - huge opportunities!

References

1. Meskó B, Drobni Z, Bényei É, Gergely B, Győrffy Z (2017) Digital health is a cultural transformation of traditional healthcare. mHealth 3:38. https://doi.org/10.21037/mhealth.2017.08.07
2. Friebe M (2020) Healthcare in need of innovation: exponential technology and biomedical entrepreneurship as solution providers (Keynote Paper). In: Proc. SPIE 11315, medical imaging 2020: image-guided procedures, robotic interventions, and modeling, 113150T (16 March 2020). https://doi.org/10.1117/12.2556776
3. Voigt I, Inojosa H, Dillenseger A, Haase R, Akgün K, Ziemssen T (2021) Digital twins for multiple sclerosis. Front Immunol. https://www.frontiersin.org/article/10.3389/fimmu.2021.669811. https://doi.org/10.3389/fimmu.2021.669811

Dominik Böhler is Professor of Management in Digital Healthcare. He works on digital business models and care processes in healthcare. He is one of the initiators of the Munich Digital Health Summit and a strategic advisor and business angel. Previously, he lead the education department at UnternehmerTUM, where he supported 2000 innovation projects resulting in 150 Startups. He holds a PhD in Information Systems and a degree in business administration from FAU Erlangen-Nürnberg.

Michael Friebe received a B.Sc. degree in electrical engineering, M.Sc. degree in technology management from Golden Gate University, San Francisco, and Ph.D. degree in medical physics in Germany. He spent 5 years in San Francisco, as a Research and Design Engineer with an MRI and ultrasound device manufacturer. He is a German citizen with expertise in diagnostic imaging and image-guided therapies, as a/an Founder/Innovator/CEO/Investor and a Scientist. He is also a Research Fellow with the Technical University of Munich, Munich; an Adjunct Professor with the Queensland University of Technology, Brisbane; and an honorary Professor of image-guided therapies with Otto von Guericke University, Magdeburg, Germany. Since 2022, he has been a Professor of biomedical engineering innovation with the AGH University of Science and Technology, Krakow, Poland. He is a listed inventor of more than 80 patents and has authored over 200 papers and was part of over 35 Medical Technology Start-Ups. He is a Board Member of four medical technology start-up companies and an investment partner of a MedTec-fund. From 2016 to 2018, he was a Distinguished Lecturer of the IEEE EMBC teaching innovation generation and MedTec entrepreneurship. He is also a coach and trainer for OpenExO, and a master Launchpad mentor for the Purpose Alliance.

Part IV

Innovation Methodology Basics

Stanford Biodesign as Base: *Empathy and Patient Centricity as the Main Driver*

18

Holger Fritzsche

Abstract

Today's healthcare challenges with increasing unmet clinical needs require faster innovation generation. To address this challenge, in 2001, Stanford created a fellowship called Biodesign to teach on-demand innovation in healthcare. For more than 20 years now, research and development institutions worldwide have adopted and further established the innovation approach based on Biodesign. Successive health innovations are primarily based on a structured innovation teaching program. It teaches the step-by-step approach of problem identification (Identify), the creative development of new solutions (Invent), and the necessary considerations for economic implementation (Implement). In this case, what is needed above all is access to resources, experts, and the respective stakeholders are within reach. A shared vision, empathetic understanding of medical users, and a patient-oriented focus play a primary role in functioning as a productive unit and increasing the success of innovations and their adaptation. The Biodesign-based health innovation program wants to generate more innovations in health care and implement them quickly to generate benefits for patients and medical users.

Keywords

Stanford biodesign · Innovation generation · Student training · Innovation · Invention · Needs finding

H. Fritzsche (✉)
INKA Application Driven Research, Medical Faculty, Otto-von-Guericke-University Magdeburg, Magdeburg, Germany
e-mail: holger.fritzsche@ovgu.de

What Will You Learn in This Chapter?
- The core elements of the BIODESIGN methodology
- Innovation generation for Health requires domain experts
- Patient orientation and Empathy are key

18.1 Introduction

In healthcare, innovation is defined as adopting best practices proven successful. In other words, innovation is synonymous with invention and commercialization. This includes implementing such practices to improve patient treatment, diagnosis, health education, and disease prevention, with the long-term goal of improving quality, safety, outcomes, efficiency, and costs. In addition, the value of innovation in healthcare is constantly moving toward proving that innovation can deliver better or equivalent outcomes at a lower cost. For the patient's innovation to be reached, it is necessary to be familiar with the market, the regulations, and the needs of the medical practitioners and patients that enable the successful adaptation and dissemination of innovations. Accordingly, contact and access to experts are needed to enable innovation to be reaching the patients. Reasons for the failure of innovations in healthcare are the lack of a conscious innovation process and the assumption that invention equals innovation and a lack of (empathic) understanding of the pains of medical users and patient needs. The innovation itself must be based on a well-defined unmet clinical need to increase their success. Therefore, the identified unmet clinical need needs to be well-monitored, investigated, validated, and critiqued in all respects. A deep dive into characterization involves input and observation in unattended moments from physicians, patients, nurses, technicians, and others involved in patient care.

The process of identifying an unmet clinical need involves understanding the disease state and the market surrounding care. And the condition for turning an idea into innovation is finding suitable market needs with regulation and reimbursement channels. Therefore, based on an unmet clinical need, the invention must find the right market fit. For this purpose, regulation and reimbursement approaches should be considered early enough in the idea generation and validation process. Next, a business model must be created, including financing, marketing, sales, and distribution. It is a nonlinear, iterative process that requires frequent reviews. For this, a health innovator must master an agile and iterative innovation approach.

18.2   The Biodesign Process

Stanford Biodesign introduced the Innovation Fellowship in 2001 as the first postgraduate educational experience for teaching biomedical technologies and their implementation in patients. It is a 10.5-month, full-time training experience in which aspiring biomedical technology innovators learn a demand-driven process for developing new medical devices and preparing them for use in patient care. The needs-driven process includes three key phases: (1) identifying and screening critical unmet healthcare needs; (2) inventing and testing new technologies to meet them; and (3) developing detailed implementation plans to bring the products to market (see Fig. 18.1).

18.2.1 The Identify Phase

The process starts with the objective to discover and find the right strategic focus for the innovator. Background research, clinical observations, and clinical stakeholders' interviews are conducted to discover unmet clinical needs. Different types of problems that may result in significant opportunities will be explored. Then these research outcomes and problems will be translated into a clinical need statement that is accurate, descriptive, and solution independent. After having formulated various need statements, the role of the disease state around the need will be researched. Understanding each stakeholder's perception and reaction regarding the medical need will be anticipated. Crucial stakeholders who are involved in decision-making will be recognized. Moreover, the market around the need and disease statement will be analyzed to understand and recognize key considerations in choosing a target market. After doing all the research, the need with the highest potential will be selected.

18.2.2 The Invention Phase

Various ideas will be generated based on the previously selected need, and ideation techniques like brainstorming and idea clustering will be used. Within the ideation process, the generated solution concepts will be compared against the criteria from the need specification to determine which concepts should be pursued. After that, the most promising concepts will be screened through intellectual property risks, regulatory risks, reimbursement risks, and business model risks for concept prioritizing. Having done that, the concept will be explored by turning the idea into a real prototype. With that, design requirements will be created, and high-level technical specifications will be generated for product feasibility. The prototypes will be thoroughly tested within simulated use cases. Finally, a final concept will be selected.

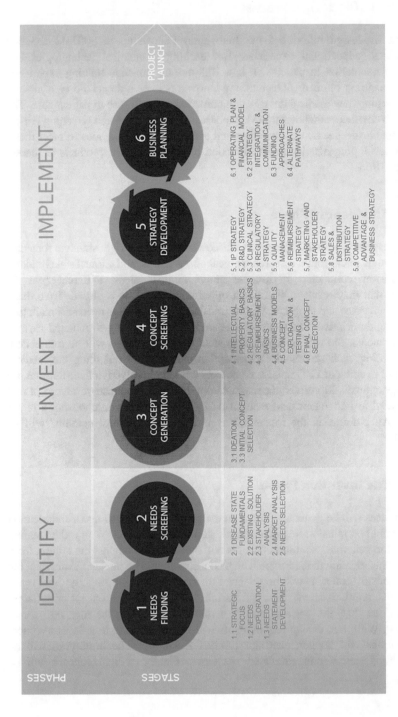

Fig. 18.1 Biodesign [Yock, 2015], the process of innovating medical technologies split into the phases identify, invent, implement and the associated stages and activities

18.2.3 The Implementation Phase

After the idea concepts have been extensively tested, the invention will be implemented. Indeed, successful implementation distinguishes innovation from invention. Final IP strategies, research and development strategies, clinical strategies, regulatory and reimbursement strategies, as well as sales and distribution, financial and business strategies, are determined to create a viable business model.

As seen in the description, many different people are involved in the process of innovation generation and implementation. Since introducing the Biodesign-based innovation approach, many Biodesign-based innovation teaching programs have been initiated and established around the world.

18.3   Empathy Toward the User, Patient Centricity as Driver

What does it mean to be empathetic? Empathy is part of our emotional intelligence. It is the ability to sense what others are feeling without saying it. Patients must be at the center of everything we do because their health, well-being, and healthcare are at stake. With empathy playing an increasingly important role in medical technology, there is probably no other industry where empathy is as important as healthcare. Empathy shows us how we can understand medical users and patients. Being empathetic puts the patient or user at the center, creating a community where people trust each other and a caring environment that supports a patient-centric culture. The steps used in Biodesign to observe, understand and evaluate the clinical environment, to get in touch with patients and medical users to understand their needs, fears, and desires can flow directly into the initial conception and development of medical technologies. This increases the success of a Medtech development and improves the customer satisfaction of the medical user and the patient experience.

18.4   Limitations

As described above, biodesign uses the methodology to solve biomedical issues in interdisciplinary teams. Biodesign offers the opportunity to develop and integrate new, better, and more suitable technologies for the biomedical sectors. It is essential to emphasize the need to develop an interdisciplinary approach between the different sectors, their points of contact, and their interest in the human being as a unique entity.

18.4.1 Interdisciplinary Barriers

The creation of interdisciplinary is essential for developing the biomedical sector, the study of problems related to the highly complex and multifaceted areas of biomedical engineering:

- Engineering mainly focuses on technological aspects, realization, and feasibility.
- Design deals with handling and interaction of user and device, ergonomics and quality of use are essential features.
- Economists calculate opportunities and strategies for markets and materials.
- Doctors look after the well-being of the patient and everyday clinical practice.

The biodesign approach attempts to establish compatibility throughout the project process. Nevertheless, approaches, experiences, understanding and goals of the individual parties are often opposed and leads to fear, poor communication, and short-term thinking. However, a shared vision, good communication, empathetic understanding and structured management and strong leadership can bridge these hurdles.

18.4.2 Disruptive Innovation

Disruptive innovations are new products or services that radically change existing structures and entire markets. However, radical changes are rarely possible in today's highly organized and regulated medicine. On the one hand, these regulations and restrictions are essential to ensure patient safety and clinical workflow. However, on the other hand, they hinder the integration of new exponential technologies into the clinical routine for prevention, diagnosis, treatment, and care. This overregulation applies mainly to new digital applications such as AI, deep learning, telemedicine, virtual realities, mobile apps, and blockchain technologies—because the regulations here are often far behind the state of the art.

18.4.3 Costs and Financing

Current challenges like long and cost-intensive development cycles and complex cost structures for reimbursement in the healthcare sector, ethical approvals, new certification rules, and the impact of the new medical device regulatory guidelines can hinder innovation and pose significant challenges for R&D and the transfer from bench to bedside. The qualitative and valid implementation of innovation projects in application-oriented and translational research is associated with additional financial and time burdens for all stakeholders. A lack of funding for staff, study design, approval, and patenting can lead to a rapid abandonment of an innovative approach. These hurdles can only be overcome with an established research structure, strong industrial partners, continuous financing models, and a shared vision.

18.5   Conclusion

Modern medicine is a rapidly changing, innovation-centric discipline with deep ties to medical technology, ideally suited to applying a structured innovation process. The Stanford Biodesign Process provides a step-by-step approach to developing new biomedical technologies that begin with a deep understanding of unmet clinical needs and emphasizes to the medical users and patients. It provides a standardized process to put this needs assessment into practice. By leveraging this process, needs eventually lead to solutions that advance the field and directly benefit patients. This agile and disruptive innovation methodology, with the active role of the clinician at every stage, gives better insights into the unmet clinical needs and engages all stakeholders to generate growth through innovation. It is based on high-frequency, low-cost testing and iterative improvements that encourage flexibility and quick responses to validated feedback from all stakeholders involved. Medical device development and use are subject to regulations and approval activities for market entry. The early integration of consideration will increase the success of innovative technical developments.

> **Take-Home Messages**
> - Stanford Biodesign is an innovation training program focused on need-based innovation.
> - This process can be applied to any field of medicine.
> - Identify, Invent, Implement describes a systematic evaluation of healthcare needs, invention, and concept development.
> - Create an innovation mindset and culture within your group that supports transformative thinking.
> - Skills to create that mindset include risk tolerance, patience, encouragement of creativity, management of conflicts, and networking activities.
> - Understand and emphasize the user's goals, tasks, and pains (e.g., Surgeon, Nurse).
> - Patient-oriented thinking; as if you were the patient to motivate for the best option.

References and Further Reading

1. Wall J, Wynne E, Krummel T (2015) Biodesign process and culture to enable pediatric medical technology innovation. Semin Pediatr Surg 24(3):102–106. https://doi.org/10.1053/j.sempedsurg.2015.02.005. Epub 2015 Mar 2
2. Yock P, Zenios S, Makower J (2015) Biodesign: the process of innovating medical technologies, 1st edn. Cambridge University Press, Cambridge

3. Augustin DA, Denend L, Wall J, Krummel T, Azagury DE (2019) The biodesign model: training physician innovators and entrepreneurs. In: Cohen M, Kao L (eds) Success in academic surgery: innovation and entrepreneurship. Success in academic surgery. Springer, Cham. https://doi.org/10.1007/978-3-030-18613-5_7
4. Yock PG, Brinton TJ, Zenios SA (2011) Teaching biomedical technology innovation as a discipline. Sci Transl Med 3(92)
5. Dutta N, Dhar D (2021) From industrial design to healthcare innovation—a comparative study on the role of user-centered design and stanford biodesign process. In: Chakrabarti A, Poovaiah R, Bokil P, Kant V (eds) Design for tomorrow—volume 3. Smart innovation, systems and technologies, vol 223. Springer, Singapore. https://doi.org/10.1007/978-981-16-0084-5_55
6. Fuerch JH, Wang P, Van Wert R, Denend L (2021) Turning practicing surgeons into health technology innovators: outcomes from the stanford biodesign faculty fellowship. Surg Innov 28(1):134–143. https://doi.org/10.1177/1553350620984338

Holger Fritzsche is currently working at the Otto-von-Guericke-University Magdeburg (OVGU) where he is part of the INKA research group. He takes care of the design and organization of the Innovation Laboratory for Image-Guided Therapies, which is funded by the EFRE (Europäischer Fonds für regionale Entwicklung) and coordinates the Graduate School T^2I-^{2}-Technology Innovation in Therapy and Imaging.

Purpose Launchpad Methodology: Introduction

19

Oliver Morbach and Michael Friebe

Abstract

To evolve initial ideas into viable healthcare innovations and scalable businesses, the Purpose Launchpad framework dedicated to Health Innovations is introduced. Both authors are certified Purpose Launchpad Mentors (Master Level) and contributed to the development of the Purpose Launchpad Guide and Innovation Meta-Methodology. For each of the three phases in the Innovation Process (Exploration/Discovery, Evaluation/Validation, Implementation/Impact Generation) eight segments (Purpose, People, Customer, Abundance, Viability, Process, Product, Metrics) are used for a structured learning process. We will present recommended tools for each of the segments covering the first two phases. The Implementation/Impact Generation phase requires an existing team and a working prototype in a start-up or in a dedicated product marketing team as part of an intrapreneurial process. We will present the visual feedback of the process that can be used to determine the next action items to evolve the project and that is also a valuable tool for discussions with potential investors, as it focuses on the problem, the initiative's purpose, and on validation questions that prove the need to address the problem.

O. Morbach (✉)
pro.q.it Management Consulting GmbH, Roschbach, Germany
e-mail: oliver.morbach@proqitconsult.com

M. Friebe
AGH University of Science and Technology, Krakow, Poland

Otto-von-Guericke University, Magdeburg, Germany

IDTM GmbH, Recklinghausen, Germany

FOM University of Applied Science, Center for Innovation and Business Development, Essen, Germany
e-mail: info@friebelab.org

Keywords

Purpose launchpad · Purpose development · Problem discovery · Problem evaluation · Validation · Impact generation

What Will You Learn in This Chapter?
- Purpose Launchpad (PL) Innovation Methodology
- The three phases of the PL—Exploration, Evaluation, Impact
- The eight segments (Purpose, People, Customer, Abundance, Viability, Product, Process, and Metrics)
- The PL tools, targets, and the repository

19.1 Purpose Launchpad

Innovation starts with numerous questions and shall result in a viable offering and ideally a scalable business or organization. What is the problem to be solved, who are the people having this problem and what is their situation, what are potential solutions and what goal do they serve, who are the people working on the solutions, what is driving and which values are important to them, how do they collaborate, what technologies and processes shall be used, how will the solutions be implemented, what skills and competencies are needed and when? And many more!

Different mindsets are needed along the evolution of an idea into a solution that fits the market:

- *Exploring the problem(s), addressable customer segments and potential solutions*
- *Evaluating the assumptions and hypotheses that are initially formulated around the above*
- *Building and optimizing a viable and scalable organization after the hypotheses are validated*

Tools and checklists can be helpful along this evolution as long as they are suitable to solve the problem at hand and are used with the appropriate mindset. Otherwise, tools may determine a solution and limit the thinking—known as Maslow's Hammer [1] that tends to treat all problems from a hammer's perspective as nails. Tools and checklists do not guarantee success but rather guide the mind and actions to find multiple ways—one of which may be successful in the end. Here it is important to not look for the ideal and only solution but to check many potential solutions and quickly dismiss the ones that are not appropriate or do not work in the context.

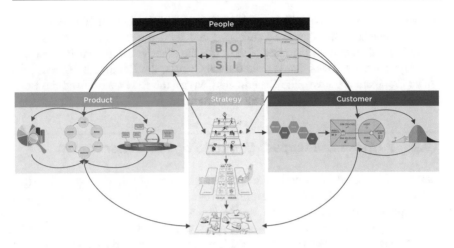

Fig. 19.1 The Purpose Launchpad methodology helps the involved People to understand the Customer needs and desires, to formulate and execute a Strategy, and define a Product that fits the market [3]. Originally published under CC-BY-SA with permission

Especially in the early phases keeping an explorer mindset is key—looking beyond the obviously visible, listening to what is not said or hearable, expecting that what you do not expect may contain the most valuable data or information.

It should be clearly stated here that INNOVATION is an often-used word that is typically associated with creating something new or different. What it should contain and what Innovation Labs should do is to make sure that there is a very deep understanding of the problem space, some foresight of future developments that may affect the problem (e.g., how will technologies influence the current way of addressing the need and problem) and to ensure that the solution that is developed satisfies a large enough need. The inventions need to be connected to some form of commercial need otherwise they should not be considered a true INNOVATION [2].

Purpose Launchpad was developed by Francisco Palao based on his personal experience in evolving ideas into purpose-driven initiatives. The Purpose Launchpad Guide provides the relevant information on Purpose Launchpad. Its first release was published in July 2020, the version we use as a basis is dated September 2021 with the input from more than 150 contributors around the globe. It can be downloaded for free under www.purposelaunchpad.com [3]. Note that the latest version of Purpose Launchpad was published in July 2022, incorporating learnings and improvements from applying it in many more practical cases.

The Purpose Launchpad combines TEAM, CUSTOMER, STRATEGY and PRODUCT in one framework and uses different tools to analyze the deficiencies and provide solutions and insights (see Fig. 19.1).

Later on in this book, we will present you the Purpose Launchpad Health (PLH), an adapted version dedicated to the health field that will help to create products that fit their market, which requires understanding the customer or user journey and to create and subsequently validate Storyboards, Cartoons, Prototypes and Minimum

Viable Products—which will be addressed in the following section. For that it uses—in line with the Purpose Launchpad and its iterative approach focused on execution and learning—a number of tools like the Massive Transformative Purpose (MTP) canvas, the Team Canvas, the Ethics Canvas, the Community Canvas, the Value Proposition Canvas, the Experiment Canvas, the ExO Canvas and the Business Model Canvas. All of which will be introduced.

19.2   Purpose Launchpad: Methodology and Assessment

Purpose Launchpad is an agile framework to help people operate with the right mindset to successfully generate new initiatives and/or evolve existing organizations (start-ups or corporates) to make a massive impact.

The Purpose Launchpad mindset is expressed as a set of principles (see Fig. 19.2). The Purpose Launchpad Evolution of ideas/initiatives comprises eight segments integrating proven innovation tools and techniques in an iterative way with time-bound sprints and an assessment after each sprint to measure progress along the dimensions.

19.2.1 Purpose Launchpad Principles

At the core of Purpose Launchpad is the mindset expressed as a set of principles. The framework linked to this mindset is developing eight segments—Purpose, People, Customer, Abundance, Viability, Processes, Product, Metrics—in the order stated by using different tools.

---

1. Purpose over problem, and problem over solution.
2. Exploration outcomes over optimization.
3. Customers' insights over market research.
4. Abundance over scarcity.
5. Meaningful incomes over investment.
6. Mindset over processes and tools.
7. Validated learnings before building.
8. Qualitative metrics over quantitative metrics.
9. Purpose-driven synergies over competition.
10. Long-term impact over short-term profit.

Fig. 19.2 The Purpose Launchpad principles. Source: Purpose Launchpad Guide, September 2021 [3]

19.2.2 Ways to Implement Purpose Launchpad

Purpose Launchpad can be implemented using different approaches: either as a mindset, as a guided toolkit or as an agile framework. Depending on how Purpose Launchpad is implemented, different elements will be required.

1. Implementing Purpose Launchpad as a mindset will mainly require considering the ten principles that are mentioned in Fig. 19.2, always taking into account the Purpose Launchpad Phase the initiative is at. This alone can be very powerful to properly evolve initiatives and people's mindset.
2. Implementing Purpose Launchpad as a guided toolkit will require to keep implementing the principles and, also, to leverage a set of tools and suggested actions that can be found in the Purpose Launchpad Repository. Doing so will leverage the power of many validated innovation methodologies and tools that Purpose Launchpad integrates.
3. Implementing Purpose Launchpad as an agile framework will require to keep implementing the principles, using the knowledge base and additionally the Purpose Launchpad Sprint process to provide the team and the initiative with extra rigor when it comes to execution. Doing this will evolve the initiative in an iterative way taking into account validated learnings to minimize risks and maximize impact.

The agile framework approach as per (3) above will be used as basis for the following sections.

19.2.3 Purpose Launchpad Phases

The evolution of initiatives applying Purpose Launchpad follows typically three stages—exploration, evaluation, and impact (see Fig. 19.3). Depending on the evolution stage of an initiative, a different mindset is needed. The tools and checklists proposed are applied corresponding to the stage and in an iterative way.

With its focus on learning, Purpose Launchpad helps avoiding waste efforts and finding the way to solutions that fit the problem and to develop products that fit the market.

When starting to evolve new ideas, little is known about the problem(s) and potential solutions.

Exploring both what is known and being open-minded about what is not known helps discovering all different possible paths to develop the initiative centered around the purpose. Running initial rapid and soft experiments (e.g., customer interviews, product mockups) without building anything at this point will allow us to partially evaluate hypotheses in order to pick the most promising path to further develop our initiative.

The **Evaluation** phase is about running real tests beyond testing just the idea. This is done in the eight different segments in order to evaluate whether our partially

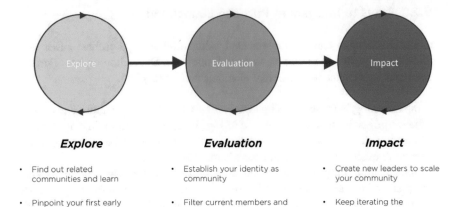

Explore   **Evaluation**   **Impact**

- Find out related   - Establish your identity as   - Create new leaders to scale
  communities and learn   community   your community

- Pinpoint your first early - Filter current members and   - Keep iterating the
  adopters and learn   attract new ones   community

Fig. 19.3 The Purpose Launchpad stages. Source: Purpose Launchpad Guide, September 2021 [3]. Originally published under CC-BY-SA with permission

validated hypotheses are true or not. The key during the evaluation phase is maximizing the validated learnings while minimizing the investment in terms of time and other resources. The focus is on iterating the initiative based on the validated learnings in order to prepare it to scale.

The **Impact** phase is about optimizing and scaling the initiative to make an impact and create a viable business model. This should only be done after having properly implemented the exploration and evaluation phases. Then we will be ready to make a massive impact—and to create a sustainable business model. In this phase, it is time to think about how to reach the mass market, how to improve the user experience of our solutions, how to optimize and even automate operations.

People involved will need to be focused on execution in order to take the initiative to the next level.

While shown as a sequence in Fig. 19.3, the Purpose Launchpad phases follow a cumulative approach meaning that when in evaluation or impact phase, the mindset and approach of the previous phases need to be kept. Validated learning in the Evaluate or Impact stage may cause initiatives to pivot and revert to previous stages.

The fact that Purpose Launchpad phases are not a set of steps that follow a linear process but a set of modes to be applied in a cumulative way is the main reason why the Purpose Launchpad icon shows an iterative cyclical process where the different Purpose Launchpad Phases' colors are shown as layers, meaning that the previous ones are kept active, too.

19.2.4 Purpose Launchpad Axes

The Purpose Launchpad Axes (Purpose, People, Customer, Abundance, Viability, Processes, Product, and Metrics) are the key areas that need to be developed in order

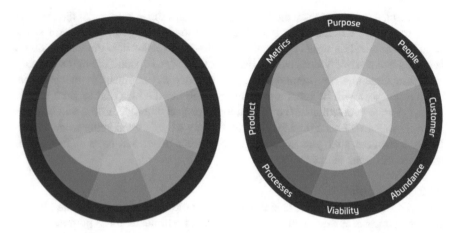

Fig. 19.4 The Purpose Launchpad icon and axes. Source: Purpose Launchpad Guide, September 2021 [3]. Originally published under CC-BY-SA with permission

to successfully define and/or evolve purpose-driven innovations in healthcare (see Fig. 19.4).

Though many innovations start with a problem first, the first dimension in Purpose Launchpad is the Purpose. It formulates what will be the impact for customers when one or many solutions corresponding to problems that are connected with this purpose are successfully implemented. This follows the first principle "Purpose over problem, and problem over solution." The Purpose is essential to evolving from an Exploration Phase to subsequently create Impact, but it is often quite abstract to start with the purpose without even knowing who the customer is and what the product or service could look like. So we advise to have an initial idea about the purpose of the operation, but to not get stuck here at an early stage. All the segments including the "Purpose" can be revisited at any time.

In order to find the right communities and the right team to achieve the purpose, the next dimension to be addressed is People. And, after Purpose and People, the process will follow with the rest of the axes: Customer, Abundance, Viability, Processes, Product, and Metrics.

All the axes are related, i.e., to develop one axis there needs to be a minimum progress on the other ones. Hence, Purpose Launchpad develops all the axes at the same time following an iterative and agile approach. Short feedback and learning cycles are pursued before doing another iteration of the full loop taking into account the previous progress in all axes.

19.2.4.1 Purpose

The purpose is the reason why healthcare innovation is developed. It guides the evolution of the innovation in order to eventually make a positive impact in the world, while being sustainable for the environment in a responsible way.

Connected to the purpose, the Moonshot describes the way to make the Purpose come true using an ambitious, quantified and time-bound goal, the Vision states what the organization wants to become, the Mission explains how the vision will be achieved and a set of Values are the basis for the way to operate the organization.

Please remember the comment for the Purpose development in the previous chapter. Purpose will be most important for evolving the idea and to get a buy in by all the great people that are connected to the project, but the Purpose will become more obvious to all once the idea is being developed. It is often better to revisit the Purpose segment after you understand the process better and have already some results in the other segments.

19.2.4.2 People

This axis comprises both those external communities that relate to and are aligned with the purpose, as well as the internal team that will develop the purpose-driven innovation.

As for external communities, it will be key to identify communities of people and/or organizations that are aligned to our purpose or that can help in some way to make it true. Eventually, an own community may be built up.

As for the team, it will be key to find the right people at the different Purpose Launchpad phases (exploration, evaluation, and impact), since each phase will need different types of mindset and skill sets. Here it is important to analyze the current team and to find out what expertise is needed for the next steps. It is quite unlikely that the team is perfect at the beginning and that the role that everyone plays or wants to play is actually existing. It should also be clear that the abilities of the individuals in a team context are probably the most important success factors for a future implementation.

19.2.4.3 Customer

Those people and/or organizations that use and/or pay for the solutions developed. Fully understanding their problems is key: these are challenges related to the underlying purpose, and how customers in different stages in the market (early adopters and mass market) behave and how they relate to the innovation initiative.

At the very beginning, the focus will be on exploring who is the customer (and who is not). In order to do so, a Customer Development process (or something similar) may be run. At some point, Early Adopters have to be found to start testing the value proposition and, finally, the mass market has to be reached.

19.2.4.4 Abundance

In order to develop a scalable business model, identifying the sources of abundance that the innovation initiative can leverage is key in order to define a proper approach. Defining and testing how to connect with those sources of abundance and how to properly manage them in order to eventually scale and make a massive impact will need to be done.

19.2.4.5 Viability
Whether for profit or not, sustainability from a financial point of view is essential—i.e., a business model including revenue or other financial source/s and/or needs to be in place to develop the innovation initiative.

19.2.4.6 Processes
The way to organize the innovation initiative to both explore possibilities and build the right thing and also run the daily operations in an efficient way.

19.2.4.7 Product
Defining and building the right solution for the different customer segments. Different approaches are used for the different Purpose Launchpad Phases (low-fidelity prototype in exploration phase, Minimum Viable Product in evaluation phase and optimized products in the impact phase).

19.2.4.8 Metrics
There are two types of metrics to be defined, gathered and tracked: innovation accounting and impact accounting.

Innovation accounting comprises the key qualitative and quantitative data required to evaluate hypotheses, to learn and to properly evolve the innovative initiative.

- Qualitative data is key for the exploration phase, in order to discover those things that are even not known to be not known.
- Quantitative data focused on measuring value will be key for the evaluation phase, in order to understand the real value delivered to customers.
- Quantitative data focused on growth will be key for the impact phase, in order to understand how to scale the business/impact.

Impact accounting is a set of metrics that help to track whether an impact is made or not, and the size of it. Among these metrics are:

- Moonshot: describes how the Purpose is being made come true using an ambitious, quantified, and time-bound goal.
- Moonpath metrics: These metrics are defined and measured as intermediate goals before starting to make progress on the Moonshot.
- Moon-landing metrics: measured once real progress toward the Moonshot is made. The goal is to measure the key and positive implications that are happening just because the Moonshot is being achieved. These metrics may be defined by taking into account the SDGs (Sustainable Development Goals) targets.

19.2.5 Purpose Launchpad Assessment and Radar

The Purpose Launchpad Assessment evaluates the status of each dimension of Purpose Launchapd through a questionnaire. For each dimension of the Purpose Launchpad axes, three questions are asked with three answers each—one for each evolution phase (explore, evaluate, impact), yielding a score from 0 to 3. See Fig. 19.5 for the visual progress radar that is calculated based on the answers of the online questionnaire for each of the eight segments—shown in Fig. 19.6 the part for the PURPOSE segment.

The scores for all dimensions are combined in a spider graph called the Purpose Launchpad Radar. It gives a very simple visual overview in which phase the different dimensions are—e.g., customer may be in evaluation phase while the product is still in exploration. The inner (yellow) area corresponds to the exploration phase, the middle light blue one to evaluation and the outer purple one to impact.

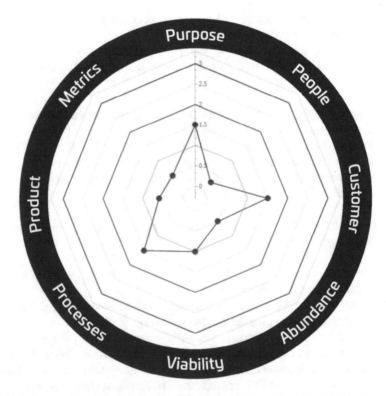

Fig. 19.5 Purpose Launchpad Radar example. Source: Purpose Launchpad Guide, September 2021 [3]

Purpose

Do you have a Purpose or Massive Transformative Purpose (MTP)?

○ No yet
○ Yes, but we have not defined our Moonshot, Vision, Mission and Values
○ Yes, and we have also defined our Moonshot, Vision, Mission and Values

Have you done any real progress in relation to your Moonshot?

○ Not yet, and we don't even know how to do it
○ Not yet, but we believe we know how to start getting real progress
○ Yes, we have started to get some real progress towards our Moonshot

Have you already generated a (massive) impact in relation to your Purpose?

○ Not at all
○ Yes, we are on track and moving with high speed towards achieving our Moonshot
○ Yes, we have already achieved our Moonshot and we may define the next one

Back Next

Key Definitions

- Purpose and MTP
- Moonshot
- Vision
- Mission
- Values

Fig. 19.6 The Purpose Launchpad Assessment, sample questions for Purpose dimension. Every segment has such a questionnaire page that is subsequently used to create the visual progress radar (see Fig. 19.5). Source: https://members.purposealliance.org/assessment/new/

Analyzing the Purpose Launchpad Radar shows what phase the innovation initiative is in. The dimension with the minimum score determines what phase the initiative is in. In Fig. 19.5, Purpose, Customer, Viability and Processes dimensions are in the evaluation phase while all others are in exploration, hence the initiative is in the exploration phase.

19.2.6 Purpose Launchpad Phase Targets

Depending on the Purpose Launchpad Phase the innovation initiative is in, different types of targets need to be achieved for each of the Purpose Launchpad Axes. Figure 19.7 shows them for all the three phases, not talking about tool and methodologies but about the essence to be achieved. As previously mentioned we want to concentrate in the chapter on the first two phases of the process.

TARGETS	Explore	Evaluation	Impact
Purpose	Define your identity: why the initiative exists (purpose), what you want to be and how you will do it	Check your identity and iterate it as you learn with real experience	Promote your purpose externally and expand your identity and culture internally
People	Connect with Purpose-oriented communities to find out personas, problems, and solutions. Also, connect with possible team members, and focus on exploration profiles.	Keep engaging with communities to find early adopters and others. Also, keep evaluating the team members with real action and make changes in the team (if needed).	Keep engaging with communities to find the early majority, in order to build a strategy to scale your reach. Also, extend the team (if needed) to focus on execution.
Customer	Start discovering what could be your customer segments.	Start selling to early adopters.	Jump from early adopters to early majority.
Abundance	Define how to connect and manage abundance.	Start implementing and validating the way we connect with abundance.	Start implementing and validating the way we manage abundance.
Viability	Define your business model to be financially sustainable.	Evaluate whether your business model can generate revenue and iterate it (if needed).	Optimize your business and operational model to scale your initiative.
Processes	Simplify processes to focus on learning.	Start defining and executing operations to deliver value to early adopters, while learning from them and iterating our solution based on it.	Improve processes to optimize operations and define sales processes to scale, while keep the learning approach.
Product	Define Value Proposition and learn about it.	Build/iterate MVP to satisfy early adopters.	Evolve from MVP to a product focused on delivering great user experience.
Metric	Focus on qualitative insights.	Measure value.	Measure growth & Impact.

Fig. 19.7 Purpose Launchpad targets per phase [3, 4]

19.2.7 Purpose Launchpad Sprints

The heart of Purpose Launchpad—when implemented as an agile framework—is a Sprint, a time-box of 1, 2, or more weeks (depending on the level of evolution of the initiative) during which the team makes real progress evolving the Purpose Launchpad Axes. A new Sprint starts immediately after the conclusion of the previous Sprint.

The roles, events, and artifacts of a Purpose Launchpad Sprint are described in detail in the Purpose Launchpad Guide.

19.2.8 Purpose Launchpad Resources

There is a set of resources that facilitate the proper implementation of Purpose Launchpad, the ones for the PURPOSE LAUNCHPAD HEALTH are explained in detail in this chapter.

The different available Purpose Launchpad Resources useful to implement Purpose Launchpad as different levels:

- The Purpose Launchpad Assessment—useful to implement Purpose Launchpad for all levels (mindset, guided toolkit, and agile framework).
- The Purpose Launchpad Repository—useful to implement Purpose Launchpad as guided toolkit and agile framework.
- The Purpose Launchpad Board—useful to implement Purpose Launchpad as agile framework.

19.2.8.1 Purpose Launchpad Repository

The Purpose Launchpad Repository (see Fig. 19.8 for details) suggests a series of things to do, depending on the evolution phase the innovation initiative is in, in order to make progress toward the targets set in each of the different Purpose Launchpad Axes. Also, the Purpose Launchpad Repository shows the key challenges that initiatives face during the different phases for each axis, and specific advice on how to overcome them. We propose to use a limited set of tools for the segments in Exploration and Evaluation phase as summarized in the Purpose Launchpad Health approach shown in Part 6, Chap. 26.

19.2.8.2 Purpose Launchpad Board

The Purpose Launchpad Board (see Fig. 19.9) is a tool that helps teams to implement Purpose Launchpad as an agile framework by managing the Purpose Launchpad Sprint's elements, such as the backlog. The Purpose Launchpad Board also shows the evolution of the initiative by tracking key elements, such as the Purpose Launchpad Phase, Purpose Launchpad Mood, and the Moonshot.

Further resources can be found under https://purposealliance.org/join/, like Purpose Launchpad Implementation Lessons Learned.

When implementing Purpose Launchpad as an agile framework, a set of methodologies and tools can be used to leverage the power of validated techniques that will help to evolve the innovation initiative in an easier way and share a common language about key concepts among the team.

Purpose Launchpad recommends a set of proven tools and methodologies for each Purpose Launchpad Axis and Phase (see Fig. 19.10). Others can be used that support making progress on the individual axes.

More important than using any specific tool or methodology is to use and develop the corresponding mindset, remembering the Purpose Launchpad principle: "Mindset over processes and tools."

AXES	Purpose	People	Customer	Abundance	Viability	Processes	Product	Metrics
EXPLORATION TARGETS	Define your identity with the initiative (yourself, what you want to be and how you will do it).	Connect with Purpose-oriented communities to find out potential users, members, and positive teammates and focus at exploration phase.	Start discovering who would be your customer segments.	Define how to connect and manage abundance.	Define your business model to be financially sustainable.	Simplify processes to focus on learning.	Define Value Proposition and learn about it.	Define the metrics that really matter (focus on qualitative insight).
EXPLORATION SUGGESTED ACTIONS	• Define the Purpose (using the MTP Canvas) • Define moonshot, vision, mission and values (using the MVMV Canvas)	• Connect with MTP-related communities • Run Design Thinking processes with the communities you connect in order to find out with different customer segments. • Fill the Empathy Map Canvas for different type of personas. • Evaluate entrepreneurial capabilities and/or bring entrepreneurs and leadership to the team • Bring mentors/advisors with great experience with startups to the project • Team alignment with MTP, Vision, Mission & Values • Fill the Team Canvas	• Define customer segments • Define customer segments' pains using the VPC • Evaluate customer segments' pains • Pick customer segments (and their pains) to focus on (more than one in platform-based business models and only one in others)	• Think about the sources of abundance (using the Abundance Canvas). • Define ExO Attributes using the ExO Canvas, to connect and manage abundance.	• Define customer segments, value proposition & revenue model • Take into account the ExO Canvas to build the BMC • Check financial resources for next milestone of achieving first sales • Build Cashflow Projection based on a business logic • Find resources (money, time, community help, etc.) to achieve first sales and/or financial support to run the initiative.	• Implement Purpose Launchpad Sprints in order to properly evolve the initiative. • Remove processes and roles to low-fidelity minimum • CEO and/or Founding Team directly connected to customer as product owner	• Define how a product/service could look like in alignment with the Value Proposition (by doing a low-fidelity prototype). • Test product approach (showing or explaining the low-fidelity prototype to users)	• Be open to insights and qualitative information beyond a current hypotheses • Define an innovation accounting system with the metrics, focused on hypotheses validation (percentage of validation of customer segments' hypotheses) • Define an impact accounting system witht he metrics focused on impact. • Track metrics
EVALUATION TARGETS	Check your identity and iterate it as you learn with real experience	Keep engaging with communities to find early adopters and others. Also, keep evaluating the team members with real action and make changes in the team (if needed).	Start selling to early adopters.	Start implementing and validating the way we connect with abundance.	Evaluate whether your business model can generate revenue and iterate it (if needed).	Start defining and executing operations to deliver value to early adopters, while learning from them and iterating our solution based on it.	Build/iterate MVP to satisfy early adopters.	Measure value.
EVALUATION SUGGESTED ACTIONS	• Evaluate MTP with community level to validate hypotheses • Iterate MTP, vision, mission and values • Define impact of your MTP (to the external world)	• Engage community to the next level to validate hypotheses • Connect with community to find early adopters • Establish team members' responsabilities, focused on learning/searching	• Define personas for the customer segment/s • Test Personas • Look for early adopters • Test first sales with early adopters • Iterate customer segments • Design / Iterate Customer Journey Canvas • Market Opportunity Analysis (TAM, SAM, SOM)	• Validate SCALE ExO Attributes • Iterate ExO Canvas	• Evaluate Business Model • Pivot, if needed • Iterate Build Cashflow Projections • Check financial resources for next milestone of achieving product/market fit • Find resources (investment, time, community help, etc.) to achieve product/market fit	• Define processes for customer support, focused on learning (integrated with experiments and metrics) based on team setup • Iterate the Culture Design Canvas to have the right way to operate at this level of development	• Build MVP • Test MVP with real customers (early adopters) • Iterate low-fidelity MVP - To keep combining Agile and Lean to build the right thing in the right way • Be open to design A/B (and other types of) experiments	• Re-define key metrics, focused on product/market fit (NPS, Stickiness, etc.) • Track metrics
IMPACT TARGETS	Promote your purpose externally and expand your identity and culture internally.	Keep engaging with communities to find the early majority, in order to build a strategy to scale your reach. Also, extend the team (if needed) to focus on execution.	Jump from early adopters to early majority.	Start implementing and validating the way we manage abundance.	Optimize your business and operational model to scale your initiative.	Improve processes to optimize operations and define sales processes to scale, while keep the learning approach.	Evolve from MVP to a product focused on delivering great user experience.	Measure growth & Impact.
IMPACT SUGGESTED ACTIONS	• Track impact link to MTP • Massive promotion of MTP	• Engage community to the next level to connect with abundance • Establish team members' responsabilities, focused on scale • Extend team / roles with execution capabilities in different areas (admin, marketing, sales, production, services, etc.)	• Use early adopters help to get new customers (mass market) • Extend the Value Proposition Canvas with Mass Market • Define sales pitch & materials	• Validate IDEAS ExO Attributes • Iterate ExO Canvas	• Evaluate Business Model • Iterate, if needed • Check financial resources for next milestones	• Define sales processes, focused on learning (integrated with experiments and metrics) • Iterate the Culture Design Canvas to have the right way to operate at this level of development	• Iterate high fidelity product and adapt for early majority	• Re-define key metrics, focused on growth (Viral, LTV, AARRR, etc.) • Track metrics

Fig. 19.8 Purpose Launchpad Repository per phase [3, 4]

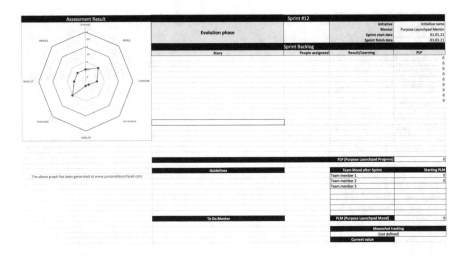

Fig. 19.9 Purpose Launchpad Board [3, 4]. This board is used to suggest next steps, find people that can do these tasks and subsequently evaluate the results and the associated learnings. The PLP = Learning Points are used to visualize the learning progress. The PLM = Mood of the Individual team members is tracked to see on whether there are issues around the process or the team dynamics. An example is shown in the subsequent paragraphs

Take-Home Messages
- We need to invest a lot of time initially to understand the problem and define a purpose for our actions and ambitions.
- Only when we understand the depth of the problem for the community of users we can define a solution that solves the clinical issues.
- We often do not validate our hypotheses and too quickly jump to create solutions that will subsequently likely fail because the clinical and user need was not properly checked.
- PLH provides a set of tools for the eight segments of the Purpose Launchpad for a first definition of the clinical needs, the health stakeholders, solution ideas based on the first Health-related innovations in an Exploration phase.
- Following the method and using the tools will significantly decrease the risk of failure and help to gain innovator, health stakeholder, and investor confidence.
- The visual radar obtained after a questionnaire-based assessment helps to define the next steps for the innovation progress.

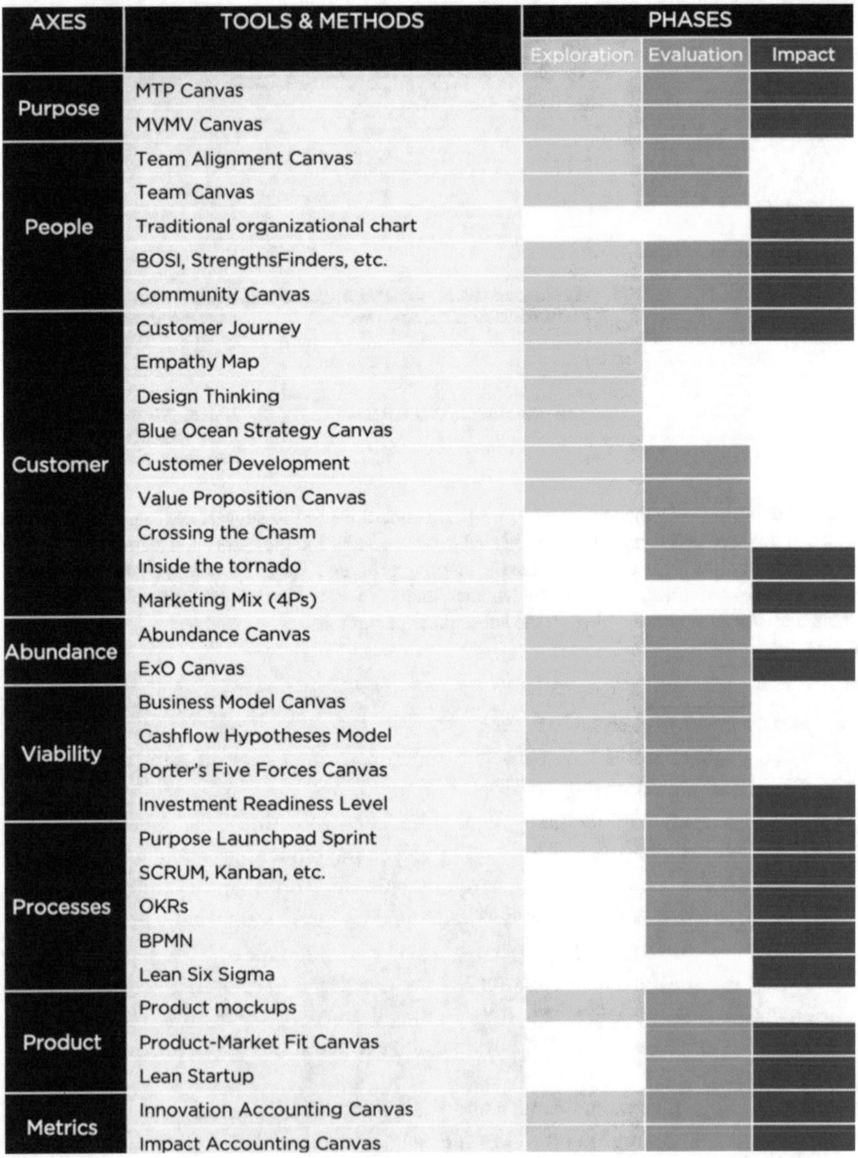

AXES	TOOLS & METHODS	PHASES		
		Exploration	Evaluation	Impact
Purpose	MTP Canvas	■	■	■
	MVMV Canvas	■		■
People	Team Alignment Canvas		■	
	Team Canvas	■	■	
	Traditional organizational chart			■
	BOSI, StrengthsFinders, etc.	■		■
	Community Canvas	■		■
Customer	Customer Journey	■		■
	Empathy Map	■		
	Design Thinking	■		
	Blue Ocean Strategy Canvas	■		
	Customer Development		■	
	Value Proposition Canvas	■	■	
	Crossing the Chasm		■	
	Inside the tornado		■	■
	Marketing Mix (4Ps)			■
Abundance	Abundance Canvas	■	■	
	ExO Canvas	■		■
Viability	Business Model Canvas	■	■	
	Cashflow Hypotheses Model		■	
	Porter's Five Forces Canvas		■	
	Investment Readiness Level			■
Processes	Purpose Launchpad Sprint	■	■	
	SCRUM, Kanban, etc.		■	■
	OKRs		■	■
	BPMN			■
	Lean Six Sigma			■
Product	Product mockups	■	■	
	Product-Market Fit Canvas		■	■
	Lean Startup		■	■
Metrics	Innovation Accounting Canvas	■	■	
	Impact Accounting Canvas		■	■

Fig. 19.10 Purpose Launchpad Recommended Tools and Methodologies per axis [3, 4]. We actually focus in this section on the first two phases and obviously on Health-related innovation generation. For that, we propose to use a reduced number of tools that we summarized under the name Purpose Launchpad Health (PLH) ([3] and Fig. 19.9)

References

1. Maslow's Hammer, also known as "Law of the Instrument" refers to the observation that people being familiar with a tool or an approach tend to use them also in cases other tools or approaches would be more appropriate. Abraham H. Maslow, The psychology of science: a reconnaissance, p 15
2. Friebe M (2020) Healthcare in need of innovation: exponential technology and biomedical entrepreneurship as solution providers. In: Proc. SPIE 11315, medical imaging 2020: image-guided procedures, robotic interventions, and modeling, 113150T (16 March 2020); https://doi.org/10.1117/12.2556776
3. Purpose Launchpad Guide, The manual on the agile framework and the mindset, Developed and sustained by Francisco Palao with the input of 150+ contributors around the world, September 2021, Offered for license under the Attribution Share-Alike license of Creative Commons, accessible at http://creativecommons.org/licenses/by-sa/4.0/legalcode and described in summary form at http://creativecommons.org/licenses/by-sa/4.0/
4. Friebe M, Fritzsche H, Morbach O, Heryan K (2022) The PLH—purpose launchpad health—meta-methodology to explore problems and evaluate solutions for biomedical engineering impact creation. In: IEEE EMBC conference 2022, Glasgow, conference paper

Oliver Morbach is Managing Director of pro.q.it Management Consulting GmbH. He is a coach, consultant, trainer, experience designer, and mentor working to enable co-creative processes and environments that unleash people's innovative potential in line with their purpose, by leveraging technology, lean-agile methodology, and exponential growth mindset. Over the last 25+ years, he held various senior and executive management roles in customer service and information technology organizations in the European ICT sector. Prior to that, he worked in defence software engineering. He holds a computer science degree (Diplom-Informatiker) from the University of the German Armed Forces Munich.

Michael Friebe received a B.Sc. degree in electrical engineering, M.Sc. degree in technology management from Golden Gate University, San Francisco, and Ph.D. degree in medical physics in Germany. He spent 5 years in San Francisco, as a Research and Design Engineer with an MRI and ultrasound device manufacturer. He is a German citizen with expertise in diagnostic imaging and image-guided therapies, as a/an Founder/Innovator/CEO/Investor and a Scientist. He is also a Research Fellow with the Technical University of Munich, Munich; an Adjunct Professor with the Queensland University of Technology, Brisbane; and an honorary Professor of image-guided therapies with Otto-von-Guericke-University, Magdeburg, Germany. Since 2022, he has been a Professor of biomedical engineering innovation with the AGH University of Science and Technology, Krakow, Poland. He is a listed inventor of more than 80 patents and has authored over 200 papers and was part of over 35 Medical Technology Start-Ups. He is a Board Member of four medical technology start-up companies and an investment partner of a MedTec-fund. From 2016 to 2018, he was a Distinguished Lecturer of the IEEE EMBC teaching innovation generation and MedTec entrepreneurship. He is also a coach and trainer for OpenExO, and a master Launchpad mentor for the Purpose Alliance.

Design Thinking for Innovations in Healthcare

20

Dietmar Georg Wiedemann and Michael Friebe

Abstract

Design Thinking—following a path from Empathize/Problem DEFINITION/ Solution IDEATION/PROTOTYPE creation/and TESTING—leads to more successful and sustainable innovations compared to other conventional problem-solving methods in health care and other public health adjacent fields. Applying Design Thinking to health care could improve patient-, provider-, and community- satisfaction, efficiency, and collaboration. The goal of this paper is to describe how design thinking works and how it could be applied to address the many challenges of future healthcare.

Keywords

Design thinking · Prototyping · Healthcare innovations · Experimentation · Validation

D. G. Wiedemann
PROVENTA AG, Frankfurt/Main, Germany
e-mail: d.wiedemann@proventa.de

M. Friebe (✉)
AGH University of Science and Technology, Krakow, Poland

Otto-von-Guericke University, Magdeburg, Germany

IDTM GmbH, Recklinghausen, Germany

FOM University of Applied Science, Center for Innovation and Business Development, Essen, Germany
e-mail: info@friebelab.org; michael.friebe@ovgu.de

207

What Will You Learn in This Chapter?
- *The DESIGN THINKING methodology and the five steps in that process with HEALTH examples*
- *That the initial three phases are essential for a product design definition and for subsequent customer testing*
- *Quick and cheap prototypes help to understand, fine-tune and re-define the problem statement and the solution idea*

20.1   Introduction

Internet of Things, smart technologies, and digitalization are core elements driving the fourth industrial revolution and have already significantly impacted our society. As Barbazzeni and Friebe [1] stated: "*In a futuristic scenario, we can imagine that scientists will be supported by technologies, carrying out numerous experiments, managing big datasets, producing accurate results, increasing communication, openness and collaboration among the worldwide scientific community, where ethics, regulations and social norms will always be observed.*"

According to Altman et al. [2], health care systems require permanent innovation while meeting the needs of patients and providers. However, products often remain unused because they do not account for human context, need, or fallibility. Thus, scientists have to address the needs of patients and providers when designing new health-related interventions or system processes.

Design Thinking is a relatively new way to incorporate user needs and feedback throughout the development process. "*Design thinking is a human-centred approach to innovation that draws from the designer's toolkit to integrate the needs of people, the possibilities of technology, and the requirements for business success.*" [3].

Design Thinking is an innovative problem-solving approach leveraging insights from the end-users of new products, services, and experiences. Rapidly prototyped and iteratively refined solution options result in best-fit solutions.

Abookire et al. [4] argued that Design Thinking leads to more successful and sustainable innovations than conventional problem-solving methods in health care and other public health adjacent fields. Scientists have optimized patient, provider, and community satisfaction, efficiency, and collaboration using Design Thinking. Fritzsche et al. [5] consider Design thinking and entrepreneurship education as significant drivers of successful innovation. Thus, Design Thinking knowledge should be considered an essential tool for future clinicians, biomedical engineers, health IT system engineers, given its promising nature as an effective problem-solving method. Integrating Design Thinking may also equip innovators in health care with crucial skills to understand and more effectively approach complex challenges.

This article introduces the Design Thinking framework for a healthcare audience and shows how the framework can foster new approaches to complex and persistent healthcare problems through human-centered research, collective and diverse teamwork, and rapid prototyping.

The following section provides an overview of the Design Thinking process based on d.School Stanford's five phases; Sect. 20.3 then describes a design thinking case study. The paper closes with eight practical tips for use and integration of Design Thinking in biomedical and health-related innovation projects.

20.2   How Design Thinking works

Brown [3] describes three stages in the design thinking cycle:

- The inspiration stage embodies the initial problem or opportunity.
- The ideation stage encompasses the development and refinement of ideas.
- The implementation stage involves the introduction and application of the derived solution.

d.School Stanford's five hexagons (as shown in Fig. 20.1) is one of the most often used visual presentation of the Design Thinking process:

1. EMPATHIZE, where the goal is to understand the audience for who you are designing
2. DEFINE stage involves describing the point of view and needs of the individual
3. IDEATE includes brainstorming to produce as many creative solutions as possible
4. PROTOTYPE phase in which potential solutions are crafted to be able to manipulate and identify flaws, and finally
5. TEST, which includes sharing the prototype with the target users to obtain feedback and lead to modifications

The following section summarizes the d.school's Design Thinking Process Guide [6]. For simplicity, the process is articulated here as a linear progression. However,

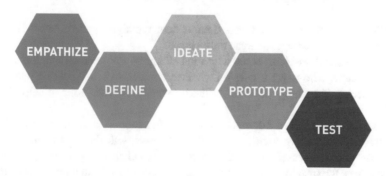

Fig. 20.1 Phases of the design thinking process. Source: d.school, Hasso Plattner Institute of Design at Stanford [6]

iteration by cycling through the process multiple times and iterating within a step is critical, e.g., creating various prototypes or trying variations of a brainstorming topic with various groups. The scope narrows as the team takes multiple cycles through the design process. The team moves from working on the broad concept to the fine details, but the process still supports this development.

The overall goal of the design process is to evaluate different alternatives with respect to their desirability, feasibility, and viability. This is quite difficult in the health context, as every country has a different starting point with respect to services, there is a multitude of stakeholders involved (the one that needs something are often not the ones that order or pay for the use), and it is a rather complicated and time-consuming process to get new products and processes integrated.

20.2.1 Empathize

Empathy is the focus of a human-centered design process. The goal of the Empathize phase is to understand the people within the context of the design challenge. It is all about understanding how and why they do things, their physical and emotional needs. To empathize, a designer observes and engages. Designers have to view users and their behavior in the context of their lives. Engaging means having a conversation with the users based on semi-structured interviews. Always asking "Why?" is essential to reveal a deeper meaning. Designers should visually get all the impressions and information about the user by using pictures of users, quotes on sticky notes, or user journey maps. This synthesis process leads to the define phase.

For Health Innovation Design it is good to start with the PATIENT in that phase. It should be clearly stated that with PATIENT we do not only refer to a sick person, but also to someone, who is interested in their health or who could become a patient at some point in the future. If there is a genetic problem in the family, and historic family problems in that area, one would for example be very interested to monitor early indicators that could lead to early interventions.

If you put yourself into the position of that patient you will of course need to check the reality. All this is part of the Design Thinking process. Let us start with giving you some examples of very early EMPATHY phase questions:

- What issues, feelings, problems, dreams does a dialysis patient have? How can you ease, improve, maybe even solve these?
- What fears and problems come with food allergies? Put yourself into the position of someone, who never knows on whether certain foods will cause a potentially life-threatening allergic reaction. What for example can be done to help identify these ingredients?
- A patient that received a coronary artery stent (a support structure placed after an intervention opening narrowed or closed arteries with the goal to prevent fast reclosure of the vessel) knows that there is a high chance of a restenosis, a repeat closure of the treated artery. How would you feel knowing that this could happen with very little early indications? What can be done to allow easy monitoring at home?

20.2.2 Define

The goal of the Define phase is to craft a meaningful and actionable problem statement, i.e., a point-of-view that provides focus and frames the problem. A good point-of-view is inspiring, includes criteria for evaluating competing ideas, and empowers the team to make decisions. The point-of-view should be discrete, not broad.

A recommended transition step into the Ideation phase is to create a list of "How-Might-We" (HMW) questions, i.e., brainstorming topics that flow from the problem statement. These topics are subsets of the entire problem, focusing on different aspects of the design challenge. When moving into the Ideation phase, designers can select various topics and try out a few to find the sweet spot where the team can churn out many compelling ideas.

Figure 20.2 shows one HMW question for one of the case studies used in this book—EASYJECTOR, a device with the ultimate goal to reduce the complexity and cost of contrast media injection during diagnostic imaging procedures. The HMW question and the possible answer via SO THAT is accompanied in this case with a first conceptual IDEA (the next phase of the DESIGN THINKING process).

20.2.3 Ideate

The Ideate phase includes brainstorming to produce as many creative solutions as possible for building prototypes. Besides brainstorming, other ideation techniques are bodystorming, mind mapping, and sketching. It is essential to separate the generation of ideas from the evaluation of ideas to defer judgment.

Please note the "as many" as possible creative solutions. It is not the goal to come up with THE idea, but to be able to select among alternatives. It is also a mindset

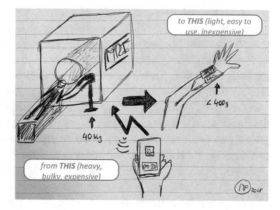

Fig. 20.2 A good way for the Define phase is to use How Might We questions to frame a potential problem/issue and the stated goal with a So That answer. On the right, you see a very early drawing/sketch of a possible solution—IDEA phase—which could already be used for initial validation (Test) experiments

problem that innovators work too quickly on one particular idea, fall in love with it, and have a hard time to later on admit that this was the wrong concept. It is better and easier in this phase to follow a process of eliminating (or parking for a later re-evaluation) ideas early on.

d.school's Design Thinking Process Guide [6] suggests a considered selection process. The team should set three voting criteria, like "the most likely to delight," "the rational choice," and "the most unexpected." The team should consider the two or three ideas that receive the most votes for prototyping. With this selection process, the team preserves the innovation potential by carrying multiple ideas forward.

And remember to park any idea, even if it was not part of the first selection process!

It is not important to immediately and early on seek the right path to achieve the goal, but it is important to eliminate the wrong ones.

20.2.4 Prototype

The Prototype phase generates prototypes iteratively that are intended to answer questions that get the team closer to the final solution. In the early stages of a project, that question may be broad. Therefore, the team should create cheap prototypes still producing valuable user feedback. The prototype and the question may get refined in later stages of the project.

According to Dosi et al. [7], prototypes trigger interaction and enrich conversations, help in generating joint ideas and identifying opportunities, and enable the testing and evaluation of assumptions. Prototypes limit the common cognitive biases that hinder innovation processes and help teams communicate across disciplines, functions, hierarchies, and organizations and stakeholders. Prototypes include a broad portfolio of artifacts and visual representations, such as functional representations of a product, early mock-ups made of different materials, presentations, or live executions of service, product, or business model. In the context of Design Thinking, a prototype can be anything that a user can interact with, like a wall of post-it notes, a gadget, a role-playing activity, or even a storyboard.

Prototyping does not have to be complex and expensive, as Brown [3] describes the design of a new device for sinus surgery. As the surgeons told the ideal physical characteristics of the instrument, a designer grabbed a whiteboard marker, a film canister, and a clothespin and taped them together. Or see the sketched prototype of the EASYJECTOR idea in Fig. 20.2 that were followed by more tangible 3D printed prototypes—still non-functional concept studies—and accompanies by a software visualization for subsequent customer experiments (see Fig. 20.3).

Another prototyping alternative in an early phase could be a user story based. Figure 20.4 shows such a short story that integrates the findings from the EMPA-THIZE phase for patients and clinical support staff for the DIALYSIS patients in Egypt. The problem as described does not exist in many countries, but was considered an issue during a workshop at a university in Cairo [8].

This story was then used to confirm the setup, the hypotheses with respect to the waiting and disease infection problems.

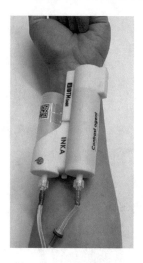

Fig. 20.3 Some more advanced prototypes—3D printed model without actual functionality, just meant to be shown as part of a process and workflow optimization experiment and a "fake" smartphone app that allows potential users to play around and give feedback. Courtesy www.easyjector.com

20.2.5 Test

d.school's Design Thinking Process Guide [6] highlights that prototyping and testing are sometimes entirely intertwined (see also Fig. 20.3). What the team is trying to test and how the team will test that aspect are critically important to consider before creating a prototype. Therefore, planning and executing a successful testing scenario is often a considerable additional step after creating a prototype.

The Test phase includes sharing the prototype with the target users to obtain feedback to refine prototypes and solutions in the subsequent iterations. The team learns more about the user by building empathy through observation and engagement. Sometimes testing reveals that the team did not get the solution right and failed to frame the problem correctly.

Going back to the example as illustrated in Fig. 20.3, the 3D printed demonstrator was used to measure and illustrate the improved handling and the time it takes to prepare a patient and with that to indicate time savings of the entire process. This was a very important experiment to validate one of the value propositions (and HOW MIGHT WE/SO THAT assumptions as shown in Fig. 20.2). It took a mere 1 week to design this prototype with cost of less than 500 €.

The software prototype was used to show a user the process management tasks, including warehouse management, ordering, billing of the service, connection to the patient and hospital management system. Most importantly was it used to collect insights and get improvement feedback from the user. Such a non-functional smartphone app can be designed with commercially available apps that are free or very affordable. Even an inexperienced user took just 2 days to create this valuable

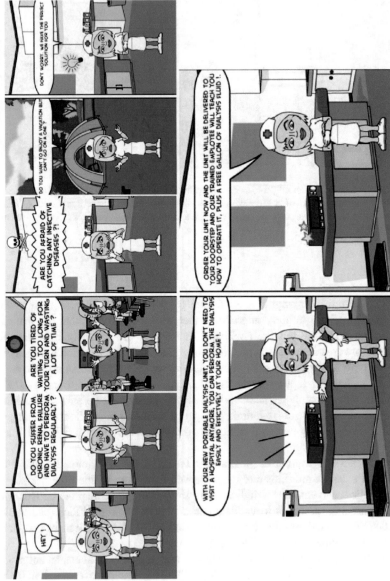

Fig. 20.4 Another form of a very early prototype developed after the Empathize and Define phase as a way of checking the identified problem hypotheses [8]

validation tool. The actual work is in designing the experiment and defining the metrics. Very well spend time!

20.3   Design Thinking in Action

Hou et al. [9] used a five-step design thinking approach to collect information on the requirements and expectations of Taiwanese women with breast cancer concerning a mobile app (Table 20.1).

Table 20.1 Description of the five-step design thinking process

Phase	Goal	Methodologies
Empathy	To understand the way women with breast cancer do things and why, their physical and emotional needs, and what is meaningful to them	• Review of the literature • Review of existing breast cancer apps • Interviews with the Taiwan Breast Cancer Foundation staff • Focus group discussion on subjects' experiences at different stages of breast cancer treatment
Define	To bring clarity and focus to the design. The goal is to craft a meaningful and actionable problem statement	• Brainstorming on "How may we use the mHealth app to support you through your cancer fighting journey?" • Categorization notes with similar information needs • Ranking the importance of information needs
Ideate	To generate ideas and get innovative solutions for women with breast cancer	• Sketching an mHealth app interface on smartphone cardboard mockups (done in 3 subgroups) • Demonstration of mockups for disease self-management • Voting the preferred mockups as a reference in the prototype phase
Prototype	To show a solution that can talk to women with breast cancer without investing a lot of time and money	• Design of mHealth app simulation by nursing informatics graduate student investigators (using a Prototyper tool)
Test	To get feedback on the prototype from women with breast cancer and then find the right level of optimization of the prototype and solution	• Individual interviews to evaluate the simulated breast cancer mHealth app through the open-ended question, "Tell me your recommendations for each function in this simulated app" • Transcription of the recorded interviews according to a guideline for transcription to avoid inconsistency in transcript styles • Confirmation of the transcripts by each participant to confirm content accuracy

In this, they discussed eight topics including treatment, physical activity, emotions, diet, health records, social resources, experience sharing, and expert consulting.

The design thinking approach contributed to the app development and reduced the gap between end-users and developers. The interactive prototype and individual interviews helped the team identify content and begin prototyping before designing the actual app. Hou et al. [9] highlighted the strengths of the design thinking approach, i.e., user-centered design and cultural sensitivity.

20.4 Conclusion

Design Thinking is a very powerful methodology in biomedical and health innovation projects moving from patient (or other health stakeholders, like hospital, clinician, health insurance) empathy toward a process or product that meets the clinical and health needs.

It is essential to spend a lot of time on the first three phases (EMPATHY, DESIGn, IDEATE) and iteratively revisit the previous phases once new insights have been obtained. The EXPERIMENT and TEST phases can be done with inexpensive prototypes to check and validate workflow or process improvements and to receive customer feedback that is then used to either discard or improve a current idea. Investing time in the experiment design, metrics, and determining success or failure criteria is very well spent. Experiment should however be designed so that they do not confirm things that are already known or ones that have no effect on the outcome or overall idea if confirmed or invalidated.

The concluding Take-Home Messages section shows eight practical tips [3, 10] for the use of Design Thinking.

Take-Home Messages
1. *Value diversity in your innovation team. Design Thinking and successful problem-solving require cognitive characteristics such as open-mindedness, suspension of judgment, and a bias toward action. Diversity and collaboration across disciplines, viewpoints, and backgrounds are highly recommended.*
2. *Begin at the beginning. Use Design Thinking at the start of the innovation process as it will help explore more ideas more quickly than a team could otherwise.*
3. *Take a human-centered approach. Innovation should consider human behavior, needs, and preferences along with business and technology considerations. Human-centered design thinking will capture unexpected insights and produce innovation that mirrors consumers' wants.*
4. *Try early and often. Encourage teams to create prototypes early as possible. Metrics, e.g., the average time to the very first prototype or the number of users exposed to prototypes, help measure progress.*

(continued)

5. *Seek outside help. Expand your innovation team and search for opportunities to co-create with providers and patients. Therefore, use social networks to grow the adequate size of the innovation team.*
6. *Mix big and small innovation projects. Manage an innovation portfolio including shorter-term incremental ideas and longer-term revolutionary ones. Business units should drive and fund incremental innovation. Top management should initiate revolutionary innovation.*
7. *Budget to the pace of innovation. Design thinking delivers fast results, yet the route to a marketable product or service can be volatile. Do not constrain the pace by relying on unsuitable budgeting cycles. Be prepared to rethink your funding approach as teams learn more about opportunities throughout the project.*
8. *Find talent any way you can. Look to hire from interdisciplinary study courses that teach Design Thinking and entrepreneurship in the context of health care (for an overview see Fritzsche et al. 2021). You may even be able to train clinicians, biomedical engineers, and health IT professionals with the right attributes to excel in design-thinking roles.*

References

1. Barbazzeni B, Friebe M (2021) Digital scientist 2035—an outlook on innovation and education. Front Comput Sci 3
2. Altman M, Huang TTK, Breland JY (2018) Design thinking in health care. Prev Chronic Dis 15
3. Brown T (2009) Change by design: how design thinking transforms organizations and inspires innovation
4. Abookire S, Plover C, Frasso R, Ku B (2020) Health design thinking: an innovative approach in public health to defining problems and finding solutions. Front Public Health 8
5. Fritzsche H, Barbazzeni B, Mahmeen M, Haider S, Friebe M (2021) A structured pathway toward disruption: a novel HealthTec innovation design curriculum with entrepreneurship in mind. Front Public Health 9
6. d.school, Hasso Plattner Institute of Design at Stanford: an introduction to design thinking process guide (2013) Retrieved 3 Jan 2022 from https://web.stanford.edu/~mshanks/MichaelShanks/files/509554.pdf
7. Dosi C, Mattarelli E, Vignoli M (2020) Prototypes as identity markers: The double-edged role of prototypes in multidisciplinary innovation teams. Creat Innovat Manag 29(4)
8. Friebe M (2017) Healthcare translation and entrepreneurial training in and for Egypt—case study and potential impact analysis. Open J Bus Manag 5:51–62. https://doi.org/10.4236/ojbm.2017.51005
9. Hou I, Lan M, Shen S, Tsai P, Chang K, Tai H, Tsai A, Chang P, Wang T, Sheu S, Dykes P (2020) The development of a mobile health app for breast cancer self-management support in Taiwan: design thinking approach. JMIR Mhealth Uhealth 8(4)
10. McLaughlin JE, Wolcott M, Hubbard D, Umstead K, Rider TR (2019) A qualitative review of the design thinking framework in health professions education. BMC Med Educ 19:98. https://doi.org/10.1186/s12909-019-1528-8

Dr. Dietmar Georg Wiedemann is Member of the Management Board at Proventa AG, a German IT consultancy providing management, innovation, and technology consulting services. He helps organizations to plan and execute their digital transformation journey using agile frameworks and best practices. As an Agile Coach, he stands for continuous improvement of business value and is driver for necessary changes and adaptations in the organization. He earned his Ph.D. in Mobile Commerce at the Department of Information Systems at the University of Augsburg.

Michael Friebe received a B.Sc. degree in electrical engineering, M.Sc. degree in technology management from Golden Gate University, San Francisco, and Ph.D. degree in medical physics in Germany. He spent 5 years in San Francisco, as a Research and Design Engineer with an MRI and ultrasound device manufacturer. He is a German citizen with expertise in diagnostic imaging and image-guided therapies, as a/an Founder/Innovator/CEO/Investor and a Scientist. He is also a Research Fellow with the Technical University of Munich, Munich; an Adjunct Professor with the Queensland University of Technology, Brisbane; and an honorary Professor of image-guided therapies with Otto von Guericke University, Magdeburg, Germany. Since 2022, he has been a Professor of biomedical engineering innovation with the AGH University of Science and Technology, Krakow, Poland. He is a listed inventor of more than 80 patents and has authored over 200 papers and was part of over 35 Medical Technology Start-Ups. He is a Board Member of four medical technology start-up companies and an investment partner of a MedTec-fund. From 2016 to 2018, he was a Distinguished Lecturer of the IEEE EMBC teaching innovation generation and MedTec entrepreneurship. He is also a coach and trainer for OpenExO, and a master Launchpad mentor for the Purpose Alliance.

VPC to BMC to Exponential Canvas: Canvas Interconnectivity for Exponential Scaling

21

Katarzyna Heryan

Abstract

In this chapter, an outline of useful innovative tools (Value Proposition—Business Model Canvas—Exponential Canvas) is given with a focus on the analysis of their interconnectivity. They are especially powerful when used together, synergizing efforts toward the facilitation of innovation generation. Although these business-oriented and well-established tools are dedicated to designing a product-market fit, they are also particularly useful for the healthcare innovation generation process. By providing a broader perspective, all persona engaged in the development process, as well as the community around them, can be understood and their needs included so that the proposition fits their expectations, and value-based trust and Engagement can be established. The use of future-oriented tools (Exponential Canvas) allows you to reliably assess the current state of the organization (ExO survey) and plan a strategic development, and scaling-up transformation. In the paper, we describe how to make use of each exponential attribute to adapt a current business model for abundance.

Keywords

Value Proposition Canvas · Empathy map · Persona canvas · Business Model Canvas · ExO Canvas · Scaling-up · Experiments and assumptions testing · Canvas interconnectivity

K. Heryan (✉)
Department of Measurement and Electronics, AGH University of Science and Technology, Krakow, Poland
e-mail: heryan@agh.edu.pl

What Will You Learn in This Chapter?
- Tools for business model planning and creating innovation with respect to a particular stakeholder's needs and expectations:
  - Value Proposition Canvas
  - Empathy Map
  - Persona Canvas
  - Business Model Canvas
- Future-oriented tool Exponential Canvas that provides scaling-up attributes and enables to define organization's mission and vision
- How to use exponential canvas and its attributes to change/adapt the business model for transformation

21.1   Methods Overview

The **Value Proposition Canvas (VPC)** provides tools for extensive analysis of customer segments (gains, pains, jobs) to understand the customers' needs. It aims to mutually create the customer's profile (choosing the right customer, developing an understanding of hers/his needs) and tailor the value proposition for her/him in particular.

To deeper immerse into the client's perspective both empathy map and *persona* canvas are useful. **Empathy map** helps to put yourself in a position of a client, stakeholder, or patient. Empathize with hers/his needs, thoughts, fears, understand actions taken, grasping that person's perspective so that the innovation itself, as well as the way of presenting it to a specific person, is in line with empathy map findings.

Persona canvas provides similar to the empathy map, but broadened, insights into the community concerned by the innovation. It reveals *persona* needs, fears, headaches, opportunities, and jobs, together with environmental determinants (negative and positive).

All three are shown in Fig. 21.1.

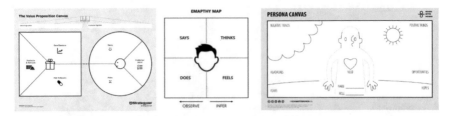

Fig. 21.1 Helpful tools related to designing value proposition and understanding customer and community needs. Left: Value Proposition Canvas used for matching product/solution with customer profile/needs [1], Middle: Empathy Map, Right: Persona canvas [2]

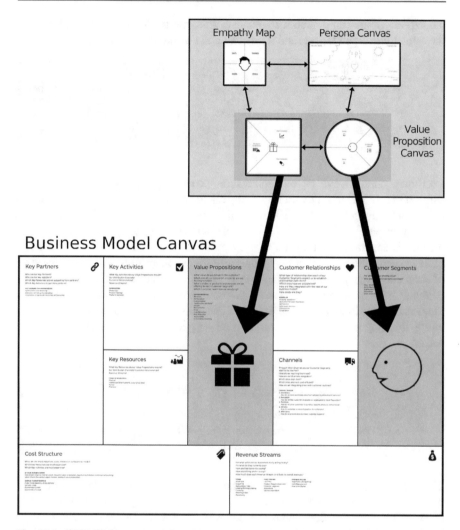

Fig. 21.2 BMC [3] interconnectivity with VPC, Empathy Map, and Persona Canvas—VPC and BMC courtesy www.strategyzer.com

It is advisable to iteratively work together on a persona canvas, empathy map, and VPC to harmonize all components with each other and work out the best representation of the customer, her/his values, pain, joy, potential benefits, and losses, obligations, expectations, requirements, but also other external conditions that may affect hers/his perception of our proposition. To be exact, not just any proposition, but a **Value Proposition** (VP)—see Fig. 21.2. Thanks to the understanding and empathizing with the position of an innovation beneficiary (and a client), we can offer a solution that is valuable and engaging. Through building trust and

Engagement one can pursue sustainable profits, not the other way around, and, at the same time, avoid failure, and rejection due to misunderstanding of expectations.

> ☼ **Analysis performed over** Value Proposition Canvas, Empathy Map, Persona Canvas **directly defines Business Model Canvas** (BMC): Value Proposition, Customer Segment, and Customer Relationships **Fig. 21.2.**

Customer, client, recipient, and innovation beneficiary are used interchangeably since in the context of innovation, and especially healthcare innovation, the commitment, and **Engagement** of all individual stakeholders, including the patient and clinician in particular (even if, for example, she or he is not going to buy a new diagnostic apparatus) and creating a solution that meets their needs is crucial. Lack of their perspective inclusion may jeopardize the whole project. While, for the sake of simplicity, it may seem tempting to only prepare one model, it should in principle be separate for each target group and each product variant. Otherwise, generalization may result in the loss of essential details. If this is too much to start with—maybe it is worth focusing on one client's profile and a specific product tailored to her/his needs?

> ☼ **The perspective of all parties, including individual stakeholders, patients, and clinicians should be reflected in the business model (BM). For that, creating a dedicated VPC (and persona canvas, empathy map) and BMC might be helpful.**

The briefly aforementioned **BMC**, proposed by Alexander Osterwalder, is a strategic management and startup template for business model development [4]. It is a visual chart of nine blocks that comprehensively and clearly describe a company or product's value, infrastructure, customers, and finances (Fig. 21.2). This enables analysis of the relationships between them and shaping the business strategy. In comparison to other business model templates, it is powerful through its simplicity and clarity. The **BMC** describes the following ten key fields [4] (Table 21.1).

With the presented tools we can decide on the **Value Proposition**, tailor it to specific clients (**Customer Segments, Customer Relationships, Channels**), analyze company assets (**Key Partners, Activities, Resources**), and socio-economic dependencies, check whether the overall business model (including **Costs** and **Revenue Streams**) makes sense in general, and find out what is needed to achieve it.

Table 21.1 Key fields of the business model canvas

Customer Segment	Differentiated by specific customer profile (e.g., teenager vs. middle-class representative in his thirties) or function in an organization (e.g., drivers in Uber)
Value Proposition	Products/services that provide value for a specific Customer Segment
Channel	How an organization communicates with and reaches each Customer Segment to deliver a VP
Customer Relationship	The types of relationships a company establishes with specific Customer Segment
Revenue Sources	The way that an organization generates income from each Customer Segment
Key Resources	The most important assets required to make a BM work
Key Activities	The most important things an organization must do to make its BM work
Key Partners	The network of suppliers and partners that make the BM work
Cost Structure	All costs incurred to operate a BM

 What is missing here is the broader future-oriented perspective—what is the organization's mission and vision, where we see ourselves in 5, 10,... years. And most importantly, how to objectively assess our situation and find out how we can strive for the intended changes and exponential development.

For that, we need a future-oriented tool—**Exponential Canvas (ExO Canvas)**, together with an evaluation ExO survey [5] or/and Purpose Launchpad [6].

The **Exponential Canvas** is a tool dedicated to designing highly scalable organizations by leveraging new organizational techniques and accelerating technologies [7, 8]. The **ExO Canvas** describes the organization in terms of **key attributes** required for exponential development. Exo surveys allow us to quantitatively measure the implementation level of each attribute, allowing us to find out which areas should be particularly worked on, and to define both short-term and long-term goals. With subsequent survey evaluations, we can reliably track the implemented changes and analyze their influence on organization development.

The **ExO Canvas** also defines the **Massive Transformation Purpose** (MTP)—the higher, aspirational purpose for the organization's existence, and the driving force that helps us, in moments of crisis and doubts, make key decisions through the prism of the awareness of the mission behind the organization's operation. It makes sure that the team thinks holistically about the business and prevents them from getting stuck on details.

There are **10 exponential attributes** presented in Fig. 21.3 and briefly described in Table 21.2.

Fig. 21.3 The ExO model refers to attributes associations with particular hemispheres [7], the ExO Canvas template—originally published under CC-BY-SA with permission www.openexo.com [8]

Table 21.2 Exponential attributes

MTP	Massive Transformation Purpose	Reflects an organization's aspiration, and purpose of existence. It describes the change in the world that an organization wants to achieve
SCALE = external mechanisms	Staff on Demand	Using external resources for core business: refers to people who work without being regular employees, leveraging the talent of people
	Community and Crowd	Group of individuals who are drawn to—and aligned with your MTP and that you interact with regularly (community) or not (crowd)
	Algorithms	Are used to analyze the abundance of data, automate processes
	Leveraged Assets	Reflects the essential shift to scale-up: from owning things to leveraging them
	Engagement	Using techniques to gain and grow the loyalty and engagement of your team, customers, and community
IDEAS = internal mechanisms	Interfaces	Bridge the external and internal world of an organization, takes data created by the SCALE attributes, and analyze, manage it internally
	Dashboards	Provide real-time information on your business (culture of transparency and trust internally and externally), tracking key metrics
	Experimentation	Validating your assumptions before making significant investments, rely on creativity and logical analysis of results
	Autonomy	The self-organized organization with teams operating with a decentralized authority (self-government)
	Social Technologies	Tools that allow improving the communication and collaboration within the company, with staff-on-demand, and inside the organization's community

The left side of the canvas (violet in Table 21.2) refers to the organization's flexibility in business model scaling and external mechanisms applied to build and engage the **Community and Crowd**, to use **Staff on Demand** and **Leveraged Assets**, and to leverage **Algorithms**. The acronym SCALE is derived from the first letter of attribute names that can be associated with the right hemisphere functions (creativity, growth, uncertainty—Fig. 21.6).

The five on the right side (green in Table 21.2) refer to the organization's internal mechanism: **Interfaces, Dashboards, Experimentation, Autonomy, Social Technologies** (acronym: IDEAS) that are by contrast typically identified with the left hemisphere functions (order, control, stability—Fig. 21.3).

Scale attributes are focused on reaching abundance and **ideas** on managing to achieve exponential growth.

21.2 Analysis of Canvas Interconnections: ExO Canvas and BMC

In this chapter, we analyze whether and how exponential attributes determined with **ExO Canvas** can be implemented in the **BMC**.

> The value of Exponential Canvas can only be appreciated once actually applied on and in combination with a BMC.

Therefore, here, we take the perspective of subsequent ExO attributes and analyze which **BMC** field they influence directly, and how to adapt the **BMC**.

21.2.1 Staff on Demand SoD

Enables a company to adapt to external uncertainties by the capacity to reduce fixed **Costs** (limiting the number of full-time employees).

- Provides access to exceptional talents (**Key Partners**) without the necessity of regular employment (fixed **Costs** reduction—**Cost Structure**). This indirectly improves the **VP** through the **Engagement** of highly skilled people and keeps the company's skillset up-to-date. It gives more flexibility to optimize the allocation of **resources** and **activities**.
- Higher quality of the employees (best available external talents) will lead to a more precise and better **VP** and a better **Cost Structure**.
- Without access to highly specialized employees through **SoD** (too expensive to hire them on a regular employment contract), some **Key Activities** and services could not be delivered, and through this a better **VP**.

Depending on what is the subject of the analysis, **SoD** may also constitute **Customer Segments** for which a separate **BMC** and dedicated **Value Proposition** can be prepared (e.g., uber drivers perform **Key Activities**, but are also a client and for that reason should not be accounted as partners, but customers). In essence, it can be discussed whether staff-on-demand should be accounted for as **Key Partners** since employment temporality is crucial [9]. See Fig. 21.4 for the affected **BMC** segments once employing this attribute.

Community and crowd CaC is composed of customers, **Key Partners**, fans who can be:

- Employed as **Staff on Demand** and add value to **Key Partners** sector (available, engaged, already familiar with the topic)

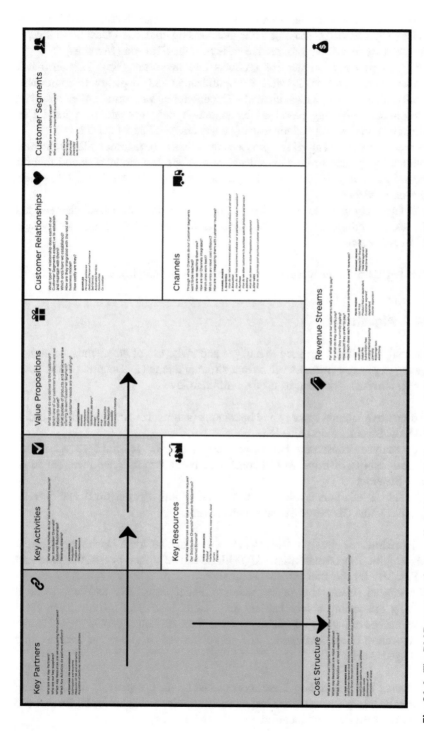

Fig. 21.4 The BMC segments that are affected by the Staff on Demand attribute of the ExO Canvas — BMC courtesy www.strateyzer.com

- Leveraged for fundraising, idea generation, design, distribution, marketing, and sales (**Key Activities**) shaping **VP** without adding up to fixed **Cost Structure,** on the contrary, providing prospective sales revenue (**Revenue Streams**)
- Engaged as early adopters of products and reviewers (**Key Partners**) hold important value toward iterative **VP** optimization and adaptation to **community** needs (that also includes clients)—**Experimentation, data collection (Key Resources)**. Working based on **Engagement** does not add to a fixed **Cost Structure** that would include testers, reviewers, and market research
- Advocating and engaging prospective clients (**Customer Relationship**), influencing people to be engaged/addicted to specific product—brand faithfulness, can be a force that makes a product accepted or abandoned (**Channels, Revenue Streams**).
- Helping a company to be more agile (quicker reaction to market changes, and trends, collecting a large dataset of feedback to improve **VP** without adding up to **Cost Structure**)

See Fig. 21.5 for the affected **BMC** segments once employing this attribute.

21.2.2 Algorithms

In general, algorithms are used to analyze and make use of the abundance of data, allowing to generate business value from different internal and external data sources (**Key Resources**). Then can be used on different levels:

- To improve internal strategy and operations (automatize **Key Activities**) leading to **Cost Structure** optimization
- To analyze consumer behaviors and decisions to improve **Customer Relationships,** distribution and communication **Channels,** and therefore **Revenue Streams**
- To deliver additional value to customers (**Value Proposition**) and increase product value (**Revenue Streams** enhancement)

Algorithms are closely related to means of data collection: **Dashboards, Social Technologies, Experimentation. Algorithms** can be written by **Staff on Demand.** See Fig. 21.6 for the affected **BMC** segments.

Leveraged Assets change the business model completely and are crucial to removing the scaling ability barriers (e.g., cloud instead of infrastructure, SaaS, renting instead of owning). Radically decreases the amount of capital needed to start a new venture (**Cost Structure**). The cost-effectiveness allows for faster scaling. **Leveraged assets**:

- Influence changes in **Key Resources** owned by the company, that are delivered either by a **Key Partners** or **Customer Segments** (additional **BMC** for that one, e.g., Uber drivers) and therefore reduce the fixed **Costs** structure

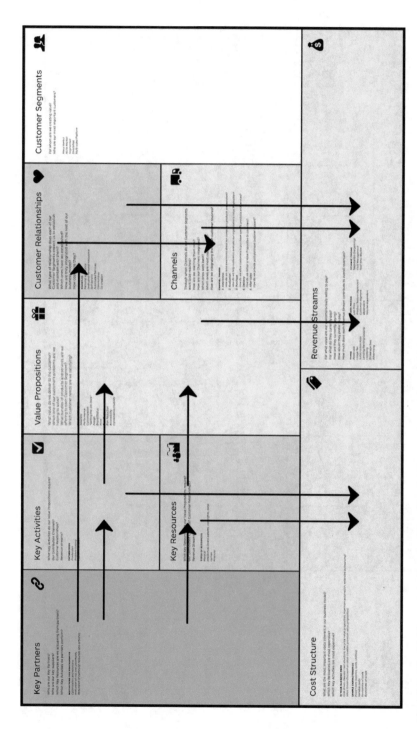

Fig. 21.5 The BMC segments that are affected by the Community and Crowd attribute of the ExO Canvas

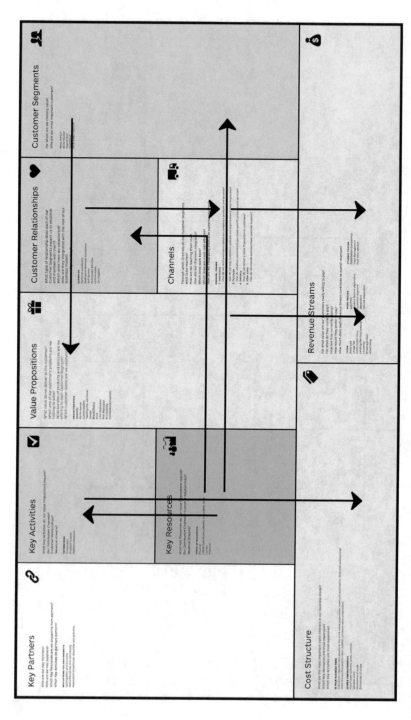

Fig. 21.6 The BMC segments affected by the Algorithm attribute of the ExO Canvas

- Change the **VP** (new services or products) and can open up new **Revenue Streams**

See Fig. 21.7 for the effect of **Leveraged Assets** on the **BMC**.

Engagement is about building a relationship with customers, team members, and the **Community and Crowd** around to add value to the business model in the following ways:

- Provide valuable insights about the customers, community members, and market to empathize with the customer need and to better tailor the **Value Proposition** and through that increase the **Revenue Streams**
- **Engagement** techniques and tools can also increase the sense of belonging, and employees' contribution so they work with more dedication toward the **VP**.

For **Engagement** related to **CaC**, please refer to this attribute directly (compare). In short:

- **Engagement** of **CaC** provides benefits such as an abundance of data delivered (**Key Resources**) from hypothesis testing and product validation (**Experimentation**) to improve the **Value Proposition**, analyze distribution **Channels**, and through that provide higher income (**Revenue Streams**) while the **Costs** are reduced by resources and data provided by them at no or little cost (compare **CaC**)
- **Engagement** of **CaC** influences and brings prospective clients (**Customer Relationship**) to improve product recognition and desirability (**Revenue Streams**)

See Fig. 21.8 on how **Engagement** can be used to change/adapt the **BMC** segments.

Interfaces are a bidirectional bridge between the internal and external world of the organization to make use of the abundance of data created by the scale attributes. **Interfaces** are the means required to enable effective scaling-up, so their influence on **BMC** is indirect through other exponential attributes (please refer to them). However, the goals are clear here, improve/modify/change the **VP**, and balance **cost/revenue** effectiveness by any resource. **Interfaces** might be used in the following ways:

- To manage **Leveraged Assets** and **Staff on Demand** with the use of **Algorithms**
- To gather data through **Experiments** and process it with the use of **Algorithms**
- To reinforce customer **Engagement** and better experience the product or technology
- To manage social relationships of employees (**Engagement**, dedication, motivation, internal interactions) and facilitate communication. Both with the aid of **Social Technologies**

See Fig. 21.9 for the **Interfaces** attribute on the **BMC** segments.

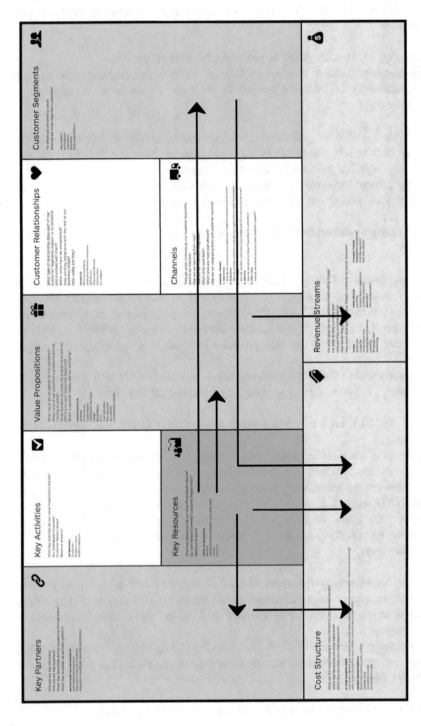

Fig. 21.7 The BMC segments affected by the Leveraged Assets attribute of the ExO Canvas

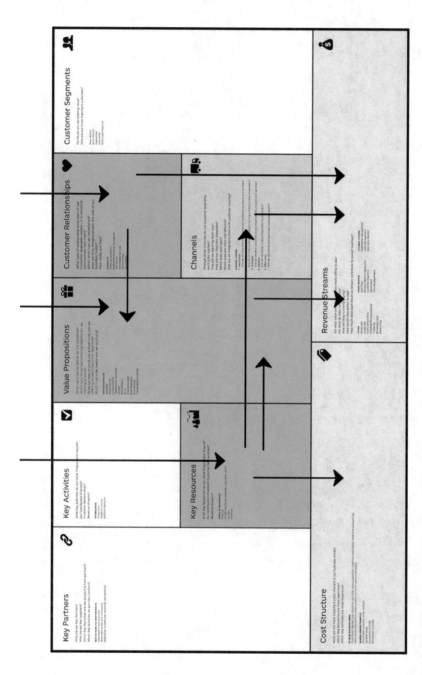

Fig. 21.8 The BMC segments affected by the Engagement attribute of the ExO Canvas

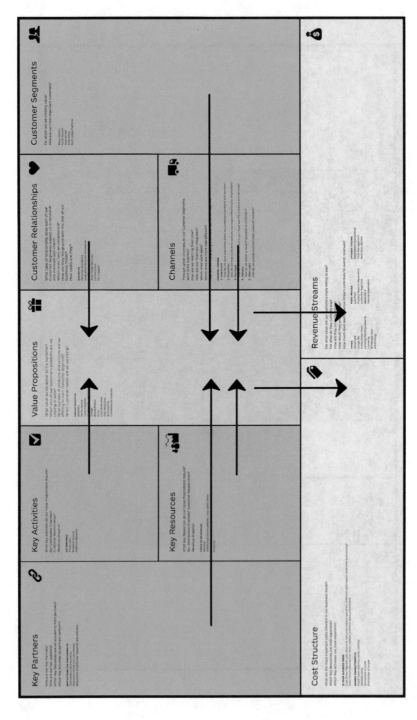

Fig. 21.9 The BMC segments affected by the Interfaces attribute of the ExO Canvas

Dashboards provide the possibility of one-way interaction with data to measure and observe processes in real time to manage them through **Interfaces**. They enable efficient tracking of key metrics (visualizations) to provide a culture of data transparency for internal and external parties (company **Key Resources**).

- To observe staff and **Staff on Demand** and community-driven activities (e.g., spatial distribution of Uber drives can be analyzed through **Dashboards** with the **Algorithms** aid, but modified through **Interfaces**)
- To observe and analyze customers' activity
- To analyze and assess the results of data evaluation gathered through **Experiments** and processes with the use of **Algorithms**
- To provide insights into the organization's data for all parties (KPI tracking)

See Fig. 21.10 for the effect of the **Dashboards** attribute on the **BMC** segments.

Autonomy is self-organized with teams operating with a decentralized authority. It gives the team authority to make decisions in the best interest of all: organization, customers, community, and in line with MTP. In healthcare innovation, best practices include lean and agile frameworks to enable team interdisciplinary synergy. Taking all parties in place (designers, engineers, patients, clinicians, stakeholders, investors) Team of Team approach is best suited (Fig. 21.11). Since an employee's motivation comes mainly from **Autonomy**, sense of purpose (MTP), and satisfaction (**Dashboards, Engagement**, KPI, data transparency, trust) such a flat hierarchy increases employee empowerment and enables the development of a unique culture that keeps the team and the organization together. In **BMC** it is reflected by directly changing the **Key Activities** that implicitly contribute to improved **VP, Customer Relationships, and Key Partners relationship** (and choice). See Fig. 21.11 for the effect of the **Autonomy** attribute on the **BMC** segments.

Social Technologies are the means to create **Engagement both** within the organization and with the **Community and Crowd**:

- Blogs, social networks, and facilitate internal company communication and collaboration (**Engagement,** management), and help to overcome organizational barriers (**Key Activities** performed)
- **Social Technologies** are the basis for the creation of a **Customer Relationships** (**Engagement**) essential for analyzing customer needs and requirements (pains and gains), validating MVP test hypotheses, performing **Experiments**, also used directly as a communication **Channels** reinforcing product **Engagement** and sales (**Revenue Streams**).

See Fig. 21.12 for the effect of the **Social Technologies** attribute on the **BMC** segments.

Experiments are essential to validate assumptions before making significant investments.

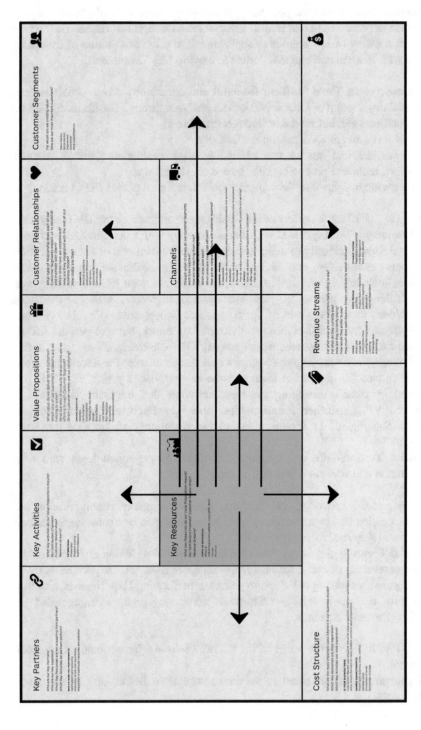

Fig. 21.10 The BMC segments affected by the Dashboards attribute of the ExO Canvas

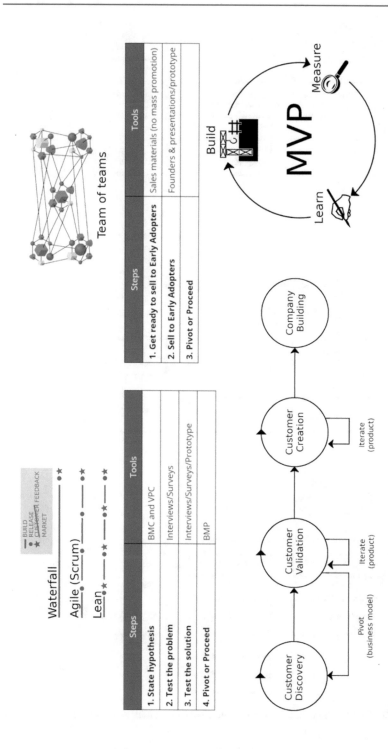

Fig. 21.11 Comparison of Waterfall, Agile, and Lean methodologies, the Team-of-Team. Running Experiments through building an MVP, testing it (hypothesis testing, measuring), learning out, and applying new insights to either BM (pivot) or MVP

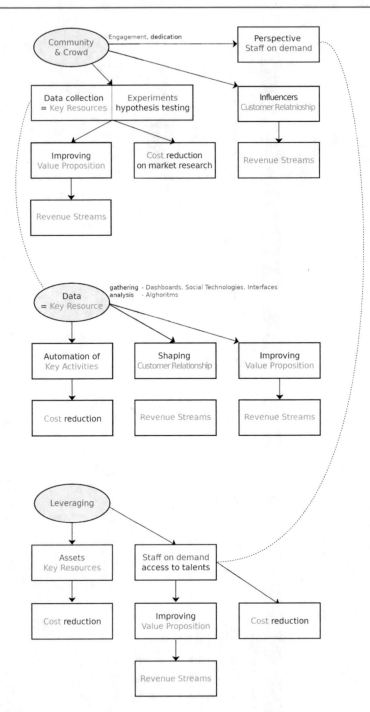

Fig. 21.12 Strategies to define BMC based on exponential attributes analysis (ExO Canvas and Survey)

> *"There's No Transformation Without Experimentation"*
>
> Brian Bell, CEO—Split Software, Forbes Councils Member, Jun 21, 2021

The **Lean start-up methodology** (Fig. 21.11) followed up by a Stanford **BioDesign** approach [10] is a good strategy to develop meaningful solutions that meet clinical needs, are tailored to the socio-economic background, and include all stakeholders' perspectives (patient, clinician, customer). In comparison to the Waterfall methodology, Lean is particularly suited to healthcare innovation generation. Due to the high **Costs** of implementation and certification, it is of great value to verify the **Minimum Valuable Prototype** (MVP) as soon as possible, and thus to have the option of changing the strategy (pivot), or withdrawing without investing a lot of money and time. The ability to iteratively validate the MVP and gain market feedback (Lean) is crucial to creating meaningful and clinically relevant products accepted by the community.

Hypothesis testing, and verification, together with MVP validation and iteration are crucial since the lack of a customer profile/market test was declared as the main reason for failure in a startup/business/research/industry project dealing with healthcare innovation. However, it is important to know **what makes good Experiments** and what does not (please see the dedicated chapters in the book on that). According to Astro Teller (Chief of Moonshots, Google X), a bad **Experiment** is either when you know the answer from the very beginning or you are not able to make use of revealed insights. A good one is, therefore, the opposite, and best if both input and output can be measured quantitatively.

Experiments define/change/the entire Business Model and therefore **BMC**.

21.3 Creating an Exponential BMC

Technology is enabling organizations to reach entirely new markets in massive and viral ways—to create a solution for many in many places (even if locally the number of people is not significant). Exponential organizations can close the gap between our growing population and the resources they need [11]. By design, an Exponential Organization must adequately address all relevant limits to its growth and must continue to make sure it does over subsequent stages of growth.

The key strategies to transform insights of **ExO attributes** to **BMC** to create an exponential scalable and flexible business model, are summarized in Fig. 21.12, and include the following:

- Reduce ownership toward outsourcing (people, and resources), and as a result reduce fixed **COST (LEVERAGING ASSETS** instead of owning **KEY RESOURCES, OUTSOURCING Staff on Demand,** and **Community and Crowd** as **KEY PARTNERS**).

- Invest in an **engaging Community and Crowd** that advocates the product (**Channels, Customer Relationships**), provides user-empowered customization, and **Experimentation** environment, and therefore is the source of data (**Key Resources**), and supports MVP validation and product launch (**Key Activities**). For that, a dedicated communication platform including social media **Channels** is needed.
- Analyze and make use of data abundance: invest in **Dashboards**. interface, and **Algorithms** to do so (**DATA** as a **KEY RESOURCE**).
- **DIGITIZE** the existing **Value Proposition** or add digital features or **Channels**.
- Create such **Channels** to enhance the product **value** (perception, **Engagement**) by using the multi-modal and **social communication** platform.
- Define **Key Activities** to provide scalable processes, and enable automatization through technology (activities analysis with **Dashboards**, **Interfaces**, **Algorithms**) to disrupt traditional manufacturing or delivery methods.
- Use **Key Activities** to perform rapid cycles of **Experimentation** and learning. Working in a Lean environment, people are more comfortable taking responsible risks, and through this provide data that can be studied and learned from.
- Apply **Algorithms** to drive nearly every part of the business: **Key Activities, Resources, and Partners**.
- Make **Autonomy** (self-managed teams) a **Key Resource.** Employees share real-time insights and experiences, which speeds up decision-making, and opens up the possibility of creating an exponential business model. For that, they need a supportive and open culture decentralized environment.

These actions aim at creating a **Value Proposition** that with the help of VPC, persona and empathy map are tailored to specific **Customer Segments**, and to define new **Revenue Streams**.

21.4 Conclusion

Understanding the mechanics of novel business models is one of the most important skills that all innovative leaders must develop. To design exponential business models, an exponential imagination is required. Canvas provides indispensable tools to support business leaders with this process. When used all together, the greatest profit can be obtained by making use of their interdependencies to shape and strengthen the new business model.

With canvas, we can put ourselves in the shoes of the client, customer, stakeholder, and investor. It forces us to constantly change our perspective and analyze whether it is beneficial for each party, what are the gains, what is missing, did we forget about something? These tools are met to make sure we did not.

Canvas provides the outline of the initiative/start-up/company's current situation—a comprehensive overview that differentiates the things you need to work on that otherwise might go unnoticed. With our subjective minds sometimes it is difficult to notice some things and see the broad picture by ourselves. This is what

we assume is not necessarily true and should be validated. Then the conducted Experiments help us to question our assumptions. With this, we can reformulate the canvas and tailor the business model to specific needs that are aligned with economical factors, and further drive the development process.

Once we are on the right track, there are additional tools to evaluate the process that regularly performed facilitates the path to innovation. Again, it is advisable to update BMC, personals, VP, and ExO Canvas regularly so that to understand the company ecosystem, keep track of changes, and focus on the right burning problems at the right time.

Some may be reluctant and claim it is all under control without all these tools. But why not make sure we did what we could to proceed right? And once not right—notice it at the right moment—and adopt! Be flexible—not sentimental. It is also a matter of trust and respect for others engaged in the process. A form of flat hierarchy engaging all employees in a process, where tools advise us, support, and justify the decisions made. Leader authority comes from understanding the motivations (data transparency policy), ability to make hard decisions, and experiences that so far proved the right actions were taken.

A lot of effort can be put into innovation generation. But if the process is not well documented, some things could be forgotten, or difficult to follow from a long-term perspective (comprehension of cause-effect relationships) which may have a negative impact on the investors' perception. Regular assessments and canvas, together with other KPIs, are indeed what they expect to analyze the company's condition, using the language and methodology that is understandable and commonly accepted.

One can say that *instead of actually working on development we only keep on talking designing, and evaluating.* Oh, yes, especially at the beginning—when everything needs to be planned from scratch and once we are not yet familiar with the tools—yes, it takes time. But it is the best investment you can make.

Preparation of the comprehensive strategy for development and its assessment, indeed reformatting, reshaping ideas, tailoring them to actual needs, not assumptions, having consciousness on a mission, vision, and tools for motivating co-workers, engaging them with the process (a sense of purpose and a feeling of belonging), organizing work, delegating the tasks, noticing issues that need to be taken care at the right time (technical and person), planning required actions both short- and long-term. It is easy to get lost in all of this without canvas comprehensive support. And what is important is having the answers ready for all questions that might appear (you will see that). This is the gain.

Take-Home Messages
- The power of the methodology comes from the canvas synergy: a coherent business vision is created through the interaction of individual elements of different canvases requiring mutual iterative adjustment.
- Assumptions testing and hypotheses validation through experiments and early adopters' feedback shape the value proposition by iterative adaptation of canvas individual elements (canvas interconnectivity).

(continued)

- The canvas should be regularly updated to include validation results, as well as market dynamics, and internal changes.
- Above all, it is a mindset, not the processes or the tools, that is the key factor to successfully developing a new exponential business model.

References

1. https://www.strategyzer.com/canvas/value-proposition-canvas—viewed and downloaded April 5, 2022
2. https://www.designabetterbusiness.tools/tools/persona-canvas—viewed and downloaded April 5, 2022
3. https://www.strategyzer.com/canvas/business-model-canvas—viewed and downloaded April 5, 2022
4. Osterwalder A, Pigneur Y (2010) Business model generation: a handbook for visionaries, game-changers, and challengers, vol 1. Wiley, Hoboken, NJ
5. https://exqsurvey.com/—viewed April 5, 2022
6. https://www.purposealliance.org/post/we-all-can-make-an-impact-purpose-launchpad—viewed April 5, 2022
7. Ismail S (2014) Exponential Organizations: Why new organizations are ten times better, faster, and cheaper than yours (and what to do about it). Diversion Books, New York
8. https://openexo.com/exo-canvas/—viewed and downloaded April 5, 2022
9. https://blog.openexo.com/five-steps-to-designing-an-exponential-business-model—viewed and downloaded April 5, 2022
10. https://biodesign.stanford.edu/about-us/process.html—viewed April 5, 2022
11. https://www.businessmodelsinc.com/how-to-design-exponential-business-models/—viewed April 5, 2022

Katarzyna Heryan is a passionate biomedical engineer (graduated with distinction from Computer Science and Medical Electronics in 2014), employed at Department of Measurement and Electronics, AGH-UST Krakow, Poland. Following her desire to create meaningful innovations in healthcare, she took the training at the University of Lund, Sweden on soft skills and entrepreneurship, design thinking, business model canvas, coaching, and teaching in higher education—Train the Trainers Transformation.doc. She also studied biomedical signal processing and medical image analysis, ML/DL at ETH Zurich (School on Biomedical Imaging) and Masaryk University (Advanced Methods in Biomedical Image Analysis). She is the author of more than 25 original research papers and four patents.

Innovation Methodology I³ EME: Awareness for Biomedical Engineers

22

Holger Fritzsche

Abstract

With the innovation approach, I³ EME: Identify, Invent, Implement, together with Engineers, Medical users, and Economists, possible interdisciplinary working methods of an engineer in this segment are to be presented. Undergraduate and graduate students of Biomedical engineering cannot imagine much of the work of a medical technology engineer and are challenged in the first semesters with mathematics, engineering, and scientific fundamentals. With the help of this approach, students should be shown a perspective very early on and be able to compensate for the initial problems mentioned with the basics. During their studies, these students have little to no contact with expectations, particularly with the unique features of medical technology. For many, medical technology is only known in its use/application, without knowing why specific innovations are created in this area and how they can be kept up. In addition to the points already mentioned, the innovation methodology should also help students classify medical technology as a melting pot of all possible technologies and thus help to support the integration of new approaches. A fundamental requirement is, of course, to know the doctor's work environment and to learn empathy toward the problems and the patient. In addition, it must be made clear to future developers at a very early stage that new technologies and processes and transfer projects must also be financed. The Innovation concept I³ EME is intended to teach interdisciplinary innovation generation for new products and services.

H. Fritzsche (✉)
INKA Application Driven Research; Medical Faculty, Otto-von-Guericke-University Magdeburg,
Magdeburg, Germany
e-mail: holger.fritzsche@ovgu.de

Keywords

Innovation generation · Innovation tools · Interdisciplinary research · UNMET clinical need · Technical translation

What Will You Learn in This Chapter?
- I^3 EME stands for identify–ideate–implement with engineers + medical staff + economics, a combination of Stanford biodesign and design thinking.
- Essential for biomedical engineering student training as it focuses on the patient
- Innovation in the health segment is to leave your laboratory building and experience to empathize.

22.1 Motivation

Biomedical engineering students should learn to improve exciting or create usable new technologies and systems used by doctors for more accurate, faster, and cheaper diagnosis and therapy and provide patients with better care and treatment. Scientific basics, imaging technologies, tools used, how they work, how to use them are an integral part of university education. Actual interaction with clinicians, active participation in medical procedures, understanding of health economics, and a structured technical approach to creating innovation in the field is rarely part of the curriculum. However, especially for medical technology, development, research, and transfer should be based on medical needs due to many peculiarities for applicability like regulations and ethical considerations. Observing the clinical application and the environment in close cooperation with the actual users is not as disruptive but easier to recognize and implement, provides relatively quick feedback, and can be used as a teaching tool for interdisciplinary innovation generation. Although it seems evident that medical technology innovations should be carried out in close collaboration between clinical users, engineers, and economists, the reality is most of the university curricula are different. New devices, procedures, and systems are only innovative if the actual invention is perceived as valuable by the clinical user and is converted into a commercial product. Against this background, basic knowledge of health economics seems indispensable for future medical technology innovators.

22.2   Innovation Methodology I³ EME

With the conception of the innovation approach, essential core points of innovation methodologies like Stanford Biodesign [1] and Design Thinking [2] were fused and taught under the innovation concept I³ EME [3] within the student training. The combination of the Biodesign process (iterative interaction between Identify an Unmet Clinical Need—Invent a potential solution—Implement a verified solution in a product = I³) and the interdisciplinary exchange of competencies and the cooperation between the Engineers, the Medical Users and the Economist. In addition to the basics of medical technologies and innovative development techniques, there are also the basics of health economics such as value proposition canvas, business canvas, and lean startup.

22.2.1 Identify an Unmet Clinical Need

Small teams of 3–4 students should be sent to live interventions to see, experience, observe and analyze an actual surgical procedure to identify unmet clinical needs and work toward a solution.

The live observation creates an initial empathetic understanding of clinical staff such as doctors and nurses and the perception of the patient as the central point of happenings. In addition, the observation of structure, management, and technical use away from the daily routine reveals a multitude of improvement possibilities.

- *The objective is to discover and find the proper strategic focus. The sequence that is followed:*
- *Background research, observations within the clinical setting, and interviews with clinical stakeholders are conducted to discover unmet clinical needs.*
- *Different types of problems that may result in significant opportunities will be explored.*
- *The research outcomes and issues are then translated from the student group into a clinical need statement that is accurate, descriptive, and solution-independent.*
- *After having formulated various need statements, the role of the disease state around the need will be researched.*
- *The understanding of each stakeholder's perception and reaction regarding the medical need will be anticipated.*
- *Crucial stakeholders, which are involved in decision-making, will be recognized.*
- *Moreover, the market around the need and disease statement will be analyzed to understand and recognize key considerations in choosing a target market.*
- *After having done all the research, the need with the highest potential will be selected.*

22.2.2 Invent a Potential Solution

The interdisciplinary group evaluates and discusses the identified needs in the technical, economic, and medical context and checks for feasible solutions. Innovation tools and creative processes help adopt new processes, technical possibilities, and regulatory requirements. So, these solutions are reviewed for ease of implementation, potential gain for stakeholders, and resolution of related issues. Proven analysis tools such as pain and gain analysis and value proposition canvas help here.

With the overall picture of the unmet clinical needs, the need for innovation, and what is technically/regulatory possible, initial ideas and approaches can be implemented as prototypes and discussed and tested with the clinical staff.

Based on the previously selected need, various ideas will be generated. Ideation techniques like brainstorming and idea clustering will be used. Within the ideation process, the generated solution concepts will be compared against the criteria from the need specification to determine which concepts should be pursued. After that, the most promising concepts will be screened with respect to existing intellectual property risks, regulatory risks, reimbursement risks, and business model risks for concept prioritizing.

Next, the concept will be explored by turning the idea into an actual prototype. Design requirements will be created, and high-level technical specifications will be generated for product feasibility. The prototypes will be thoroughly tested within simulated use cases.

Finally, a final concept will be selected.

22.2.3 Implement a Verified Solution

If an idea/prototype is convincing, a specific project to be worked on can be developed. Market exploration, patent research, regulatory, and certification are the following steps that must be incorporated into the development.

No process step is completed here—all findings and experiences flow into the overall concept and are re-evaluated with every progress.

After the idea concepts have been extensively tested, the invention will be implemented or discarded or discontinued. Reasons for the latter two are missing team, missing enthusiasm, no real problem, no financing, no stakeholder support, and many more.

Options are the transfer of the technology to an existing operation or via a Start-Up. Successful implementation distinguishes innovation from invention. Finally, final IP strategies, research and development strategies, clinical strategies, regulatory and reimbursement strategies, sales and distribution, financial and business strategies are determined to create a viable business model. For a graphical summary on the I^3EME concept see Fig. 22.1.

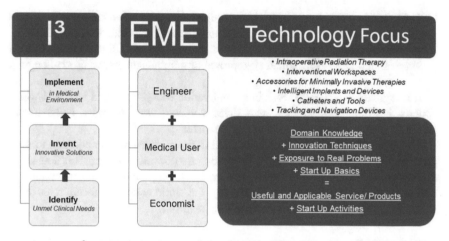

Fig. 22.1 The I³-EME methodology, a combination of Stanford BIODESIGN and Design Thinking and here a focus on patient empathy and patient benefits

22.3   Key Themes and Twenty-First Century Soft Skills that Promote Success

Students work with the I³EME concept within an established innovation ecosystem with clinical and commercial partners. In addition to technical expertise, essential skills of the twenty-first century will be imparted, especially in interdisciplinary exchange.

According to Blevins et al., three fundamental principles are essential to increasing the probability of success. These principles proved to consistently contribute to successful innovations within the Silicon Valley ecosystem and leading universities worldwide [4]:

- The first theme of a good innovation ecosystem is a "culture of both collaboration and innovation." If any key healthcare stakeholders are not present or invested in the common goal of producing innovation, then tension, frustration, and failure will result. Therefore, each stakeholder must share a common goal to collaborate appropriately.
- The second characteristic of a good innovation ecosystem is the "systematic commitment to technology transfer." Along with a culture of support, there must be a commitment to technology transfer to move from discovery or invention to a commercial reality. Therefore, the stakeholders mentioned as prerequisites should strongly support the process. However, so should other medical and non-medical industries because, as mentioned earlier, they could be an excellent distribution, acquisition, or licensing partner. Also, a symbiotic relationship between faculty research teams, university technology transfer, and industry partners is essential to achieve a common goal of improving medical care.

- The third characteristic of a good innovation ecosystem is "an adjacent ecosystem of real-world experts." Bringing any new technology from the "bench to the bedside and back" requires knowledge from real experts. In particular, difficult innovation phases like regulatory approval or reimbursement can be challenging.

It is a great advantage to work in an innovation ecosystem filled with real-world experts and other stakeholders who share a vision, work within a collaborative culture, and are committed to technology transfer.

These principles should be demonstrated to the students and, in addition to their professional skills, they encourage the students to develop further their soft skills such as critical thinking/problem solving, creativity, communication, and collaboration through an interdisciplinary exchange—the world economic forum rates these as the essential soft skills of the twenty-first century [5].

22.4   Outcome Measures

The development of medical technology and the generation of innovations is a complex and multidisciplinary task. Added to this are regulatory requirements for safety and use. Biomedical engineers with a naïve view are lost in this process, and various innovative methodologies are available to steer through this process. Experiences and insights from established innovation methodologies such as design thinking and Stanford Biodesign have been incorporated. The presented innovation concept I^3 EME intends to introduce the exciting work mode of a medical technology engineer, who identifies problems in work with the clinicians. And then—also considering economic issues—in prototypes and then possibly after discussion with the users implemented as medical devices.

On the one hand, the students are brought into direct contact with the future environment, learn direct contact with the user, observe how medical technology users work, and learn to recognize problem approaches to propose subsequent improvements.

The students are offered a collaborative and communicative innovation environment to work creatively. Students can apply their professional skills in a structured innovation process and improve their soft skills through interdisciplinary group work.

Take-Home Messages

The combination of different technical expertise, creative solutions, and teamwork leads to the following learning objectives:

- Innovation can be learned and implemented in a structured manner and is necessary, especially in biomedical engineering.

(continued)

- Interdisciplinary teams improve the innovation process and are the basis for successful developments in medical technology.
- Medical technology is a combination of different technologies.
- Medical technology is a technology for physicians and patients; one has to understand the working methods and the problems of the "customers."
- Teamwork and interdisciplinary approaches promote the development of communication skills, critical thinking, and creativity, as well as other soft skills.
- Having fun with medical technology and the course of study, plus providing perspectives

References

1. Yock P, Zenios S, Makower J (2015) Biodesign: the process of innovating medical technologies, 1st edn. Cambridge University Press, Cambridge
2. Brown T (2008) Design thinking. Harv Bus Rev 86(6):84–92, 141
3. Friebe M, Boese A (2016) I3 EME — Innovation Awareness für Medizintechniker. Magdeburger BEITRÄGE ZUR HOCHSCHULENTWICKLUNG 4:24–27
4. Blevins KS, Azagury DE, Wall JK, Chandra V, Wynne EK, Krummel TM (2018) How good ideas die: understanding common pitfalls of medtech innovation. Med Innov Concept to Commer:117–127. https://doi.org/10.1016/B978-0-12-814926-3.00012-7
5. Soffel J (n.d.) Ten 21st-century skills every student needs. World Economic Forum [Online]. Retrieved from: https://www.weforum.org/agenda/2016/03/21st-century-skills-future-jobs-students/

Holger Fritzsche is currently working at the Otto-von-Guericke-University Magdeburg (OVGU) where he is part of the INKA research group. He takes care of the design and organization of the Innovation Laboratory for Image-Guided Therapies, which is funded by the EFRE (Europäischer Fonds für regionale Entwicklung) and coordinates the Graduate School T^2I^2- Technology Innovation in Therapy and Imaging.

Part V

Ethics + Health Innovation

Integrating Ethical Considerations into Innovation Design

23

Sabrina Breyer and Christian Herzog

Abstract

Ethical considerations should be integral to the development process right from the start. We elaborate on the underlying rationale, especially pertaining to the difference between regulatory compliance and ethical reflection. We further present responsible research and innovation in health as an appropriate framework for integrating ethical considerations during innovation processes. We briefly discuss how responsible research and innovation practice may pay off and provide concrete tools in the form of a list of possible ethical issues as well as introductions into methodologies that can aid in managing responsible research and innovation practice as a collaborative team effort.

Keywords

Ethical considerations · Responsible research and innovation · Responsible innovation in health · Methods · Ethics of technology

---

What Will You Learn in This Chapter?
- Start with Ethical considerations very early in the Health Innovation Design process
- A Framework based on responsible research and innovation in Health
- That ethical design considerations can have a positive impact on innovations' success

---

S. Breyer · C. Herzog (✉)
Universität zu Lübeck, Ethical Innovation Hub (EIH), Lübeck, Schleswig-Holstein, Germany
e-mail: Sabrina.breyer@uni-luebeck.de; christian.herzog@uni-luebeck.de

Why Does Ethics Deserve Its Own Chapter?
A comment from the Editor, Michael Friebe

Several years ago, I taught a class with Christian Herzog titled ETHICAL DESIGN CONSIDERATIONS FOR MEDICAL TECHNOLOGY DEVELOPMENTS. This was a very exciting combination of philosophical base knowledge with innovation design and entrepreneurship. I distinctly remember the discussion between us on whether ETHICS and ETHICAL CONSIDERATIONS as part of a novel medical technology product development should be examined and dealt with very early on in the process (Christian Herzog) or rather later after some validation of the desirability, viability, and feasibility confirmation (Michael Friebe). In this 2019 lecture, we combined the famous VALUE PROPOSITION and BUSINESS CANVAS from Alexander Osterwalder and several other tools with an ETHICS CANVAS [1].

It is clear that these new developments, technologies, changed business models, focus on the individual's health data, democratization of health will need an intensive discussion and consideration of ethical design considerations during the innovation, development, and translation process.

Thank you for providing this comprehensive chapter because:

- Democratization is needed, so health and healthy longevity is NOT an issue of personal wealth or the quality of a country's healthcare system.
- We are moving from a human-controlled to a semi-autonomous and soon to an autonomous medicine with many ethical and trust issues that need to be addressed.
- ETHICS is and needs to be in flux, adjusting to the circumstances and developments. Cultural, life circumstances, and political systems often influence ethical understanding.
- In a global world with global data exchange, we need to have a common ethical base.
- Disruption—we talked about the need for that and the issues that come with it—will come with new and currently unclear Business Models.
- The effects of Exponential Developments cannot be foreseen—refer to the 6Ds of exponential developments with a "chaos" and many unknown after the disruption.
- Big Data, complex unsupervised AI models, and Federated Learning methods will be very helpful and possibly will cause more disruption than all other developments together, but the results could also be biased and are often not explainable, just as
- Health data protection and privacy is essential to build trust but also comes with issues—are we protecting a minority and do not allow a majority to benefit? Or is that essential to ensure that the system intelligence will not pass us?

23.1   Introduction

During typical innovation processes and their technical implementation, economic effort and legal conformity tend to dominate. Ethical issues, in turn, are often not addressed directly. What is their measurable benefit? Will their consideration result in a better product? Can a business afford the time it takes to consider ethics systematically?

However, it is safe to assume that most protagonists in innovation are intrinsically motivated to innovate responsibly. Especially in the healthcare sector, responsible innovation is often comprehended as being satisfactorily considered, e.g., by focusing solely on enhancing the patient's well-being. This, however, disregards potential further and relevant ethical and societal impacts of the innovation as well as the interests and requirements of the other actors in the socio-technical ecosystem.

But even if there is an awareness of potential socio-ethical impacts and their importance is acknowledged, the challenge of identifying "what," "when," and "how" to proceed largely remains unclear.

This chapter aims to clarify *why* ethical design considerations are relevant, *when* and *how* to address them, *which* socio-ethical issues should be taken into account, and *how* to start innovating in a responsible way that addresses the needs and challenges of the entire health care ecosystem. This is the outline by which this chapter traverses the basics of responsible innovation and ethical considerations in health care.

23.2   Why Socio-ethical Considerations Should Be Integral to the Development Process

Before proceeding with details on potential socio-ethical issues and how to address them, perhaps the question of why and when to address them requires some more elaboration. In his 2021 Harvard Business Review article "Thinking Through the Ethics of New Tech...Before There's a Problem," Beena Ammanath frames it like this:

> *Our ability to ethically manage and increase trust in our tech tools is expected to only gain in importance in coming years as technology evolves, accelerates, and reaches more deeply into our lives. This will likely challenge every company and business, and it may have bruising lessons for the organizations that fail to keep pace.*

Simply put, the consideration of ethical and societal implications is necessary to avoid negative socio-ethical impacts:

- That may thwart the positive vision and idea of the innovation in question
- Which can lead to loss of trust in both product and brand
- And which will ultimately hamper economic success

In the (economic) worst case, the innovation does not get the market approval, or it can fail on the market because of justified and unanticipated public controversy. According to analysts at CBInsights,[1] 42% of all startups fail because they do not try to solve a relevant societal challenge. Furthermore, Deloitte has assessed that companies being successful in the digital technologies sector are already seriously considering ethical aspects [2]. Being compliant with regulatory requirements does not necessarily guarantee positive or beneficial societal effects. Rather, truly considering socio-ethical matters aims at accomplishing a positive societal impact by aligning the innovation and its development process with a specific set of values consistent with societal needs. This indicates a necessity to endeavor beyond the regulatory requirements and standards. In essence, this requires focusing on what should and should not be done, which implies developing a positive (but ethically informed) vision. This approach should be rather familiar to startups or those initiating new ventures. It is, hence, a misconception that ethics impose yet another regulation-like layer limiting the things that may be done. Rather, it is a guiding compass for trying to guarantee positive impact and success.

In his TechCrunch column "MVP versus EVP: Is it time to introduce ethics into the agile startup model?"[2] Anand Rao puts this perfectly:

> *Startups typically have one shot at success, and it would be a shame if an otherwise high-performing product is killed because some ethical concerns weren't uncovered until after it hits the market. Startups need to integrate ethics into the development process from the very beginning, develop an EVP based on RAI and continue to ensure AI governance post-launch.*

However, strong voices deny that some of the socio-ethical effects–good or bad–are part of their responsibility as technology developers or providers. It is often stated that technology is but a tool, and those who put it to use should bear the blame if something goes awry. This view is misguided as it neglects the inherent value-ladenness of technological innovation. Prominent voices from the World Economic Forum give an idea of how technology and society interrelate:

> *Technologies have a clear moral dimension—that is to say, a fundamental aspect that relates to values, ethics, and norms. Technologies reflect the interests, behaviors, and desires of their creators, and shape how the people using them can realize their potential, identities, relationships, and goals [3].*

This highlights that technology is not value-neutral. The recognition of this matter of fact by the World Economic Forum follows a long-lasting debate about the interrelation between technology and society: Do societal needs and values shape the technology (social constructivism), or does technology evolve nearly autonomously, thereby determining new societal conditions (technological determinism)? There are

---

[1]CBInsights (2019). The Top 20 Reasons Startups Fail. https://www.cbinsights.com/research/startup-failure-reasons-top (accessed January 28, 2022).

[2]https://tcrn.ch/32Q0qDD (accessed January 18, 2022).

significant grounds to assume that, on average, a mixed form—called soft techno-logical determinism or weak societal constructivism [4]—is prevalent.

Accordingly, a perspective on societal norms and opinions should be factored into the innovation process in addition to normative ethics. For instance, surveys of the European Union have shown that robotics and AI, in principle, have a good reputation. In contrast, the wide-scale implementation of these technologies for digitization and automation is met with skepticism [5]. For instance, while the implementation of AI is favored in settings that aim at preventing immediate harm, in settings involving particularly vulnerable groups, e.g., older people, children, or patients, the implementation of robotics and AI is mostly rejected.

However, if the innovation should be developed in a responsible and societally desired way by anticipating the effects of its deployment, the so-called Collingridge Dilemma arises [6]. The Collingridge Dilemma covers two equally undesirable alternatives: at the beginning of the technology development, there exists much uncertainty about the ethical and societal impact of the targeted innovation. At the same time, many parameters of the innovation could be adjusted. This relation reverses when the innovation has been deployed.

> **Box 23.1 Take-Aways: Why Socio-ethical Considerations Should Be Integral to the Development Process**
> Technology is not value-neutral.
> Addressing ethical concerns is key to success.
> Addressing ethics means first and foremost to form a positive vision of an innovation.

On the other hand, the Collingridge Dilemma demonstrates the challenge that the system of values can change dynamically after introducing the innovation onto the market [7]. For instance, new types of medical imaging or diagnostic procedures offer new options to evaluate someone's state of health and act accordingly. Ultrasound imaging, for example, offers a range of very helpful diagnostics but has also sparked a moral debate about abortion. Google classes may be extremely useful in the professional domain but have been widely rejected by the public.

Hence, taking ethical considerations seriously also means anticipating value conflicts, engaging with them, and finding the right path to the market that avoids risks and leads to economic success and societal benefit. Therefore, it is paramount that socio-ethical considerations are recognized and implemented as an integral part of the development process. Even though the Collingridge Dilemma may never be entirely solved, and projections of the future are doomed to be inaccurate, engaging in ethical discourse is a means of staying in control and, perhaps, being able to avert disaster. What would that mean? Have a look at Box 23.2 for a primer of what a dystopian or a utopian future of AI in health could look like.

Box 23.2 A Dystopian and Utopian Vision of (Un-)Ethical AI in Health
Dystopian Vision
- Algorithms trained on biased data amplify and perpetuate discriminating effects, as well as social inequality.
- Technology developed based on hypes and not with a clear orientation toward improving clinical outcomes may render the health system more expensive.
- Medical personnel may act inadequately if not well-trained on and onboard with new technology. Good healthcare will become more difficult.
- Automation is mainly motivated to cut costs, the quality of care, however, decreases.

Utopian Vision
- Data-based Algorithms increase clinical objectivity and reduce biases, as well as discriminating effects.
- The application of technology is directed toward improving clinical outcomes and societal utility.
- Medical personnel is involved in technology development and deployment and is trained in its responsible utilization.
- AI systems aim at enhancing human potential in medical practice.

23.3   The Difference between Regulations and Ethics

To clarify what the integration of socio-ethical evaluation into the development process entails, it needs to be clearly distinguished from mere compliance with norms and regulations. Adherence to laws and regulations cannot, in itself, amount to doing ethics proper justice. Likewise, it would be naive to assume that ethical use of technology can be achieved without regulations sanctioning improper conduct. As we will see, ethics is rather about pushing for the best possible conditions and consequences, while regulations are predominantly targeted at prevention. Ethics is about what should be done rather than what may be done. Think, for instance, about novel AI-powered medical equipment that significantly lowers the expertise or practice required to perform diagnostics. Ethical issues in self-diagnosis (issues of care, anxiety, coping, and dealing with other emotional responses) or the socio-ethical impact such technology may have on the physician–patient relationship transcend regulation. Similarly, issues such as ensuring responsible human–machine interaction, power relations, as well as modes of communication and cooperation largely elude regulation. This is because:

- The technology is potentially so innovative that it spawns a completely novel set of concerns to be addressed. Even if you could conceive of a set of regulations that need to be enacted, it would probably take years until this would ensue. In that regard, ethics precedes regulation, and you are traversing uncharted territory.

- There is considerable uncertainty concerning the occurrence of potentially adverse socio-ethical effects. Accordingly, a range of rather minor innovation and human-interface design tweaks as well as non-technical means, such as raising awareness, integrated training, and fostering a respectful approach to an innovation's usage, etc., may be more likely to reap the innovation's benefits while circumventing the potential issues.

- Legal means simply cannot provide the proper incentives. Neglecting to design to prevent system responses or modes of interaction that endanger psychological well-being, provide unfair advantages, or do not work toward respectful cooperation may almost always elude potential sanctioning. However, your approach to considering the ethics of your innovative idea may well determine its economic success.

However, extensive public discourse about the ethical issues surrounding innovative technology—and AI in particular—has led to the emergence of high-profile ethical guidelines, such as those by the EU, OECD, and WHO, that lay out ethical principles to follow. Aiming at further operationalizing ethics, standards such as the IEEE P7000 suite are being developed, which have an almost soft law-like appeal to it: you either comply with the standards or risk going out of business. Hence, technology developers—especially in healthcare—are frequently bemoaning the high regulatory burdens imposed on them—even more so, as ethics is supposedly joining the list of things to consider.

While we do not want to belittle the importance of these standards, let alone the aforementioned ethical guidelines, both standards and lists of principles do not constitute ethics. Authors such as Rességuier and Rodrigues [8] are justified in pointing out that the formulation of ethical principles is but the documentation or proposition of a potential consensus on moral norms. Accordingly, rather than considering this consensus as a fixed signpost that guaranteedly leads to an ethical innovation, it is far more important to continuously reflect on and question an innovation's underlying premises, values, as well as potential socio-ethical effects.

> **Box 23.3 Major Ethical Principles Proposed by the European Commission, OECD, and WHO**
>
> *Protect human autonomy:* Technological solutions should respect a human's self-determination, provide sufficient means to reject system-proposed decisions, refrain from manipulation, deception, or enacting illegitimate pressure or even force. The EC's guidelines further introduce the principle of preventing harm (non-maleficence) to denote that a human's dignity should remain inviolable and technological solutions should put special

(continued)

Box 23.3 (continued)

considerations on the needs of vulnerable persons and not exploit power asymmetries.

Promote beneficial solutions: Depending on the institution that proposed this criterion, what constitutes beneficial is elaborated on and emphasized in terms of sustainable economic growth (OECD), human well-being and health (WHO). This principle is notably absent from the EC's set of principles, which is otherwise closely adapting Childress' and Beauchamp's Principles of Biomedical Ethics that specifically distinguish between beneficence and non-maleficence.

Ensure explainable solutions: A typical AI-related new addition to guidelines as compared to previous sets of ethical principles (Childress' and Beauchamp's Principles of Biomedical Ethics in particular), principles of explainability, transparency, and accountability are all featured in either of the EC's, OECD's or WHO's pamphlets. All of these aim at ensuring that someone human remains accountable for a system's effects and there is means to incentivize and ensure due diligence.

Foster fairness: The principle of fairness or justice demands that individuals or groups are not subject to discrimination, exploitation, unfair biases, or stigmatization. The principle also demands that humans may contest and seek proper redress against the effects of technological systems.

For small-scale enterprises and startups, this may even be good news. Rather than being an additional load of paperwork to file, ethics can become much more closely aligned with the agile and vision-driven process dominant in successful innovation processes. As Anand Rao writes in his TechCrunch column "MVP versus EVP: Is it time to introduce ethics into the agile startup model?"[3] ethical principles can be used as a starting point for considerations. But after that, the team should constantly check alignment with the company's values and socio-ethical goals. This is not entirely different from pursuing an innovation oriented toward customers' needs but, of course, requires broadening the scope. In that regard, ethics is open-ended and transcends checklists or any form of clear-cut criteria that must be fulfilled. As previously stated, this must remain so, largely because your innovation hopefully boldly goes where no ethicist before has gone. However, ethicists have thought carefully about potential abstract ethical issues in anticipation of innovative ideas that may be challenging. Therefore, ethics guidelines will help you get started with reflecting potential issues *in the context of your specific innovation*.

These principles have, in fact, converged to some degree. The European Commission's "Ethics Guidelines for Trustworthy AI" (2019), the OECD's Recommendations of the Council of AI (2019), and the World Health Organization's

---

[3] https://tcrn.ch/32Q0qDD (accessed January 18, 2022).

guidance report on the "Ethics and Governance of Artificial Intelligence for Health" (2021) all mention the principles of *autonomy, beneficence, explainability* and *fairness* in some form, cf. Box 23.3.

Box 23.4 Take-Aways: The Difference between Regulation and Ethics

Ethics is forward-looking and pre-emptively helps to anticipate and account for value conflicts. Regulation is backward-looking and establishes specific rules trying to prevent negative impacts.

Ethics is open-ended and aims at being or achieving the best. Regulation is codified ethics combined with mechanisms for sanctions.

Ethics is about what should or should not be done. Regulation defines what can or cannot be done.

Ethics can consider "softer" issues that elude regulations, such as some forms of unfair power asymmetries in informal settings. Regulation mostly considers issues that can be objectively proven.

Adhering to these principles and letting them guide one's ethical evaluation further elevates another distinction between regulations and ethics. Rather than limiting what *can be done*, these principles highlight what *should be done* with technology. Again, this change of perspective does not amount to another regulatory checklist but should rather fit into the creative mindset of innovation that keeps positive socio-ethical effects firmly in sight.

23.4 Ethical Considerations from the Beginning of the Development Process: Principles of Responsible Research and Innovation

To advance toward a responsible innovation design, there is a need to shift the mindset toward ethical considerations throughout the whole development process. For this, the academic community has developed an approach called Responsible Research and Innovation (RRI), which provides a set of conceptual pillars for reflecting upon socio-ethical aspects.

In general, RRI is defined as

> [...] *a transparent, interactive process by which societal actors and innovators become mutually responsible to each other with a view to the (ethical) acceptability, sustainability and societal desirability of the innovation process and its marketable products (in order to allow a proper embedding of scientific and technological advances in our society)* (von Schomberg [9]: 19).

Accordingly, RRI addresses the interplay between diverse actors with respect to their mutual responsibility within the innovation's socio-technical ecosystem to uncover how the innovation can be designed beneficially for society. Two specific but intertwined dimensions can be considered [9: 20]:

- *Product dimension:* this is where the specific ethical and societal implications of the (technological) innovation are analyzed and evaluated from a normative point of view.
- *Process dimension:* this is where methodological approaches and organizational configurations are defined, enabling ethical evaluation and reflection in a participative manner, i.e., together with all those affected.

Burget et al. [10]: 9) highlight this aspect of inclusion and define RRI "as an attempt to govern the process of research and innovation with the aim of democratically including, early on, all parties concerned (...)." Thereby, integrating the stakeholders' interests and concerns at an early stage becomes another important aspect of RRI. In short, the RRI approach demands a continuous analysis of the ethical and societal implications throughout the entire development process and product lifecycle.

Furthermore, the conceptual dimension of RRI provides essentially four principles that should be integrated into the process: anticipation, reflexivity, inclusion, and responsiveness ([10]: 9–12, [11]: 189). In the following, these principles are briefly illustrated in the context of health technology development, referring to the analysis of Demers-Payette et al. [11]. Their analysis is based on "three mixed focus groups on the needs and challenges of health care systems that technological innovations should address" ([11]: 189). Focus groups comprised users, developers, and managers related to health and technology.

23.4.1 Anticipation

Technology push strategies risk misalignment between health needs and technological solutions in the healthcare sector. For instance, Demers-Payette et al. [11] stress that a focus on the financial impact on hospitals of the medical innovation falls short of taking seriously the impacts on the processes professionals or patients are subjected to. Consequently, intended use and hence, the benefit may not be realized.

In addition, engaging in anticipatory activities, such as scenario building, vision assessments, etc., enables identifying new opportunities for innovation as there is a deeper understanding of the future impact on the health care system concerning clinical, human, technological, and financial resources ([11]: 194). Nevertheless, the challenging interplay of changing societal environments and needs makes a solution-oriented innovation design complex and difficult to predict. The following provides a brief primer on what dimensions may be anticipated and could be subject to a more in-depth analysis:

- Anticipation of implementation challenges (adoption, rejection, etc.)
- Anticipation of changes in workflow

- Anticipation of feedback effects (changes that may affect even the assumptions that were underlying the innovation process itself)
- Anticipation of value conflicts and the enabling of ethically relevant options for action

In conclusion, the anticipation of technology impacts aims at considering the potential risks and opportunities of the technology and its effect on society and the environment [10]. As a result, unforeseen and undesirable consequences can be prevented.

23.4.2 Reflexivity

The principle of reflexivity focuses on the values that are—or should be—inherent to the development process and the innovation. The development team should reflect upon and critically question their explicitly or potentially implicitly self-imposed normative ethical framework that guides the activities, assumptions, and decisions during the innovation process. For instance, patient-centric innovation could be misconstrued by entirely focusing on safety issues. At the same time, principles like shared decision-making or the actual needs and wishes of the patients are left aside, resulting in a disregard of the patients' dignity and autonomy ([11]: 195). More globally, efforts focusing on increasing life expectations may neglect making sensible trade-offs between a medical intervention and quality of life possible. Tensions between lucrative technical solutions in clinical niches, population needs, and an adequate reflection of the normative context for the intended use of the medical innovation arise, concerning which innovators need to strike a balance. This entails the requirement of considering the "continuum of care, lifecycle, clinical ethos" ([11]: 193), which also contains the balancing act of different power dynamics, e.g., between health professionals, funding of health establishments, and regulatory requirements, to clearly point out the relation between the different expert's assessment and the resulting allocation of the responsibility.

For example, in referring to the ethical principles listed above, the following value conflicts and ethical issues may be relevant:

- Autonomy: designs may be paternalistic, disempower shared decision-making, or incur emotional responses limiting self-determined action.
- Beneficence: Designs may trade population-level (economic) benefits versus individual patients' health benefits.
- Fairness: alternative or more conservative healthcare practices may be disrupted, or benefits may not be fairly allocated.
- Explainability: system designs may undermine the medical personnel's responsibility and potential to engage in shared decision-making.

At this point, besides the interrelation of reflexivity and anticipation, the dimension of inclusion plays an important role.

23.4.3 Inclusion

Inclusion plays a significant part in responsible design considerations because the integration of stakeholder interests and external, societal stakeholders can influence decision-making during innovation processes. In the entire socio-technical healthcare system, different actors with various expertise and needs (e.g., producers of medical innovations, health professionals, patients) move in different worlds and may lack knowledge of others' work routines, environments, or interests. That is why inclusion aims at bridging the distance between the actors to create a space where the different points of view can be shared, and participants can learn from each other ([11]: 198). Certainly, the involvement of a wide range of stakeholders is not without concern. For instance, particularly for the most vulnerable–the patients–participation involves opportunity costs. The benefit of getting your voice heard involves the cost of sparing the time and energy to participate. On behalf of the developing system, considerable efforts are required to open up the development process, facilitate accessibility, and acquire and motivate others to participate. Further, the benefits of valuable feedback incur the risk of disseminating intellectual property. Even if a willingness to participate exists, efforts are needed to reduce the "lack of a common language" ([11]: 199). After surveying the different interests, how can these be balanced and evaluated in the context of the innovation design? While inclusion per se does not ensure responsible innovation, in conjunction with the principles of anticipation and reflexivity, it can provide a social legitimacy ([11]: 198) and–with this–economic success. Consequently, including interests from the socio-technical ecosystem is highly relevant to share responsibilities and advance toward a more democratically legitimized technology development. But to include also means to respond, which is described in the following.

23.4.4 Responsiveness

Medical innovations are nourished by the interrelation of clinical practice, scientific knowledge, and technological developments leading to an innovation process as a complex exercise structured by institutional rules, policies, funding, licensing, etc. ([11]: 200). For instance, an upcoming change of law or emerging demands by society can adjust the innovation design necessary. In addition, stakeholder participation, ideally, is not a one-off exercise, and repeated consultations can reveal novel input, concerns, and reflections—particularly from non-professional stakeholders. Responsiveness entails adapting to, acting upon, and taking seriously the input obtained from activities in anticipation, reflection, and inclusion. In this context, the challenge of standardization and, at the same time, the different methods of operation in various clinical environments show that the utilization phase also needs to be considered. Thus, responsiveness goes beyond reactive behavior and seeks to make the process and the innovation flexible and adjustable or even resistant concerning potential objections and changes in the requirements. In conclusion, responsiveness means responding to the different circumstances of clinical environments and diverse contexts of use to develop a responsible and effective medical innovation.

So far, our discussion mainly tackled the high-level concept of RRI. In the following, we will dig deeper into how this translates to issues in health more specifically.

> **Box 23.5 Take-Aways: Principles of RRI**
> Anticipation: establish processes that help to anticipate the socio-ethical effects of product deployment.
> Reflexivity: reflect upon the values embedded in both organizational processes as well as the product.
> Inclusion: include diverse stakeholders.
> Responsiveness: adopt strategies to be able to react to stakeholder input and adjust development processes and goals accordingly.

23.5 Responsible Innovation in Health

With a focus on the healthcare sector, Pacifico Silva et al. [12]: 4ff.) have developed a specific definition for Responsible Innovation in Health (RIH):

> RIH consists of a collaborative endeavour wherein stakeholders are committed to clarify and meet a set of ethical, economic, social and environmental principles, values and requirements when they design, finance, produce, distribute, use and discard socio-technical solutions to address the need and challenges of health systems in a sustainable way ([12]: 5).

RIH adopts the principles of RRI and especially highlights the aspect of collaboration among the different stakeholder groups. Herein, Pacifico Silva et al. [12]: 5) do not deny the challenges and tensions such an extensive collaboration comprises, but rather emphasize the need and opportunities in integrating the complementary expertise across different aspects of health innovations (e.g., design and usability, regulation, financing, production, etc.), in addition to the normative dimension of considering the interests and needs of the stakeholders. The term "socio-technical solutions" is construed as technology that always requires interaction with individuals regardless of the complexity or simplicity of the technology ([12]: 5). That is why health technologies should address the diverse needs of the health systems in the context of the well-being of individuals throughout their lives.

To make a significant contribution in solving healthcare challenges, there are different aspects of responsibility and sustainability to be considered before and while starting a business or developing an innovation. The RIH framework by Pacifico Silva et al. [12] (see Fig. 23.1) provides an entry point into necessary and useful thoughts about starting a responsible startup or creating a responsible innovation in health. The RIH framework identifies and structures five so-called

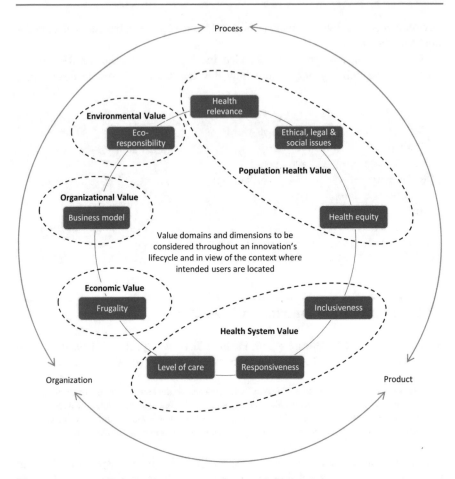

Fig. 23.1 Responsible innovation in health framework ([12]: 9)

value domains comprising nine dimensions with ethical considerations in the inter-related sphere of process, product, and organization. The value domains—population health, health system, economic, organizational, and environmental—will be described next.

23.5.1 Population Health Value

The first value domain focuses on population health, which is about promoting the collective—in contrast to the individual—health needs for solving health inequalities. Even though health innovations that benefit individuals should not be devalued, emphasis should also be put on innovation that aims at tackling population-wide needs and inequalities. This emphasis is justified as there is a need to overcome reluctance with respect to investments into research and development that seek to provide remedies for diseases in smaller and less affluent markets ([12]: 6). The challenge of health equity is a supplementary dimension. It stresses the differing perceptions of "good health" among social groups and the complex interaction of the individual settings (e.g., socioeconomic status, vulnerability factors, social position) relevant to healthcare disparities. These disparities stretch from issues in accessibility and affordability of adequate care to the comprehensiveness of medical therapy options ([12]: 6). From another perspective, the innovation itself and the induced impact on its context of use can potentially both cause or increase inequity and disparities. For instance, built-in biases can lead to the exclusion or even discrimination of specific social groups ([12]: 7). Furthermore, emerging technologies—especially artificial intelligence-based technologies—may produce outputs that cannot be easily explained. Educational status may be a factor that limits how well therapy decisions can be made in a shared, informed, and substantiated way, potentially exacerbating discrepancies in the quality of healthcare concerning social status. Hence, RIH aims at a perception of responsible innovation that provides a benefit that does not vary according to social factors and promotes health relevance and equity.

23.5.2 Health System Value

The health system value domain evaluates the degree of the innovation's contribution to tackling contemporary health system challenges, such as challenges due to changes in demography, shortages in qualified personnel, and other supply gaps. This domain comprises two links to RRI principles: *inclusiveness* and *responsiveness*. Besides highlighting how important it is that the outcome of innovations and its development processes are shaped inclusively by opening up for stakeholders and societal actors to diversify the input and to have a democratic development process, Pacifico Silva et al. [12] underline the usefulness of early identification of who and how to include. A "well-justified set of stakeholders and the ways in which their inputs will or will not be integrated into the innovation" ([12]: 7) is paramount. This helps prevent power asymmetries in the choice of the innovation design, albeit not every aspect of inclusion can be foreseen. For innovations in health, inclusiveness requires special consideration, as development teams are typically dealing with highly vulnerable stakeholders whose consultation and engagement pose additional challenges. In terms of RIH, responsiveness is substantiated as creating flexible and adaptable innovations to emerging system-level challenges in health service delivery, human resources, and governance. All of these are predominant challenges in the international peer-reviewed literature on health systems ([12]: 7), identifying a need for products that reduce the effort to better handle the growing health demand due to population growth and an aging population. Self-care products can also support solving this challenge but require taking into account the ethical implications of enabling patients to self-diagnose or engage in self-therapy. This makes a high level of usability and comprehensibility necessary, which demands an innovation design closely aligned with the user's needs. A further aim can be to achieve far-reaching subsidiarity in the health system. Thus, the local healthcare needs can be mitigated more directly, helping to increase availability, especially in rural areas ([12]: 7).

23.5.3 Economic Value

There is a need for high-performance but lower-cost innovations to increase affordability, e.g., in rural areas, to increase equity in health and hence increase the reach of innovation to more people while using fewer resources. Pacifico Silva et al. [12]: 8) suggest some possibilities to accomplish frugality, such as (i) simplifying already-existing techniques and technologies, (ii) acquiring new technologies tackling enduring problems, (iii) expanding solutions in diversifying their use for different purposes, and (iv) developing low-tech solutions to meet specific local needs. This means frugality can be achieved by reducing the use of resources and in the sense of focusing on core functions to make the production, use, service, etc., easier, more durable, and leaner. This also provides the opportunity to have a low-cost development process, leading to rather affordable products for low- and middle-income countries.

23.5.4 Organizational Value

The organizational domain examines the *business model* and its strategies concerning ethical value creation and promoting social entrepreneurship. Developing purpose-oriented innovations can reduce tensions between creating economical and societal value. These aspects must not be mutually exclusive if the organization or the project is geared toward economic profit and considers the qualitative, ethical added value. However, these qualitative added values are difficult to quantify, especially in monetary terms. A possible way to overcome the obstacle of economic effectiveness while still pursuing a philanthropic agenda is to start a hybrid organization. This can be a social business, joint venture, or non-profit organization that aims at benefitting society as a whole. Hence, the organizational value domain promotes social entrepreneurship as an effective business model for contributing to the resolution of health system challenges and for innovating responsibly in the context of the approach of RIH.

23.5.5 Environmental Value

The dimension of *eco-responsibility* considers the environmental impact throughout the whole product lifecycle. In the healthcare sector, this mainly refers to the amount of energy and raw materials needed and the special factor of hazardous materials as well as single-use supplies. The analysis of the product lifecycle to reduce negative environmental impacts includes, among others, the conscious choice of used resources (e.g., recycling, avoiding chemicals, heavy metals, etc.) and its procurement (e.g., requirements for suppliers), the carbon footprint, the utilization (e.g., energy consumption), and the end of life (e.g., remanufacturing, biological degradation) ([12]: 8). Consequently, eco-responsibility is ethically important to protect the planet and prevent far-reaching negative impacts on people's well-being caused by pollution and excessive use of natural resources.

Box 23.6 Take-Aways: Responsible Innovation in Health

Population Value: increase attention toward collective needs and health inequity.

Health System Value: respond to challenges of the health system itself.

Economic Value: deliver high performance and affordability.

Organizational Value: engage in responsible business strategies to provide value to users and society.

Environmental Value: reduce environmental impacts throughout an innovation's lifecycle.

Having looked at what it takes to engage in responsible innovation in health, we will now come back to the question of whether and how it pays off to do so.

23.6   Entrepreneurial Motives and Advantages of Responsible Innovation

To quickly showcase how responsible innovation can accommodate both business-oriented and more altruistic motives, we adopt the taxonomy by Garst et al. [13], comprising instrumental, moral, and relational motives, cf. Fig. 23.2.

Instrumental motives focus on the economic success and benefit of the startup or company. These motives can be distinguished into short-term and long-term outcomes. How does responsible innovation contribute to these? Ethical and sustainable activities can directly increase the short-term profit by cost reductions due to resource efficiency and improving capabilities. However, they can also incur high additional costs, such as investments in personnel, infrastructure, and time. The long-term outcome seeks to sustain "pragmatic legislation" ([13]: 3) by deepening business relations, for instance, in cooperating with suppliers to improve the performance regarding sustainable and ethical aspects, by managing the reputation to enhance the attractivity for investors as well as employees, and by fulfilling consumer demands. A competitive advantage can be being prepared and staying ahead of legislation by anticipating risks due to emerging demands ([14]: 37). Companies and economic associations may prevent state regulation by proactively acting on ethical and societal challenges, substituting them with effective self-regulation.

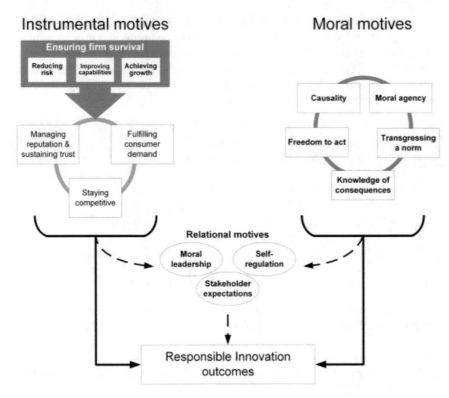

Fig. 23.2 Overview of the motives for acting upon social responsibilities through innovation ([13]: 11)

However, these instrumental motives are mostly self-serving and aligned with external incentives ([13]: 3). RRI will pay off in the medium to long term if done right. Rarely, there will be benefits to reap already in the short term.

In the sphere of *moral motives*, an economic entity identifies itself as a moral agent as part of society and consequently complies with and respects the relevant socio-ethical and cultural requirements. Moral motives are intrinsic, meaning motivation is not based on profit but rather stems from philanthropy. By pursuing socially desired and ethically accepted outcomes, a company could be said to hone its part of a social contract, thus being granted a social license to operate without extensive regulation. Moral motives seek to serve society and thereby support well-being and societal welfare.

The *relational motives* to innovate responsibly integrate instrumental and moral motives to create synergetic value. The aim is to serve the relational interests of the company and its stakeholders. Stakeholders are not only comprised of shareholders and customers but also include external societal stakeholders. This contributes to acquiring "cognitive legitimacy" ([13]: 4) by proactively developing an innovation that aligns with economic and societal needs and sustains the social license to operate. This can also be achieved by complying with industry standards or norms or implementing institutional self-regulation. Transparently engaging in public deliberative processes can also increase social acceptance and trust in technological innovations, further strengthening the social license to operate ([14]: 38). Incorporating stakeholder expectations comprises another competitive advantage, as it effectively translates into a market pull strategy that substantiates a guaranteed product alignment with target group interests. Additionally, close relations with stakeholders can support recognizing new market opportunities by identifying new trends and evolving needs ([14]: 39).

It should now be clear *why*, *when*, and *how* to address ethical considerations in general. The remainder of the chapter will be dedicated to specifics: *Which* socio-ethical issues should be taken into account, and *what* tools can be used to get you started innovating in a responsible way.

23.7   Ethics Is Not a Checklist Issue: But This Can Get You Started

The previous sections have elaborated extensively on a general theme: the consideration of ethics is highly context-dependent, may involve a vast range of stakeholders, and pertains to an open-ended endeavor, rather than something that can—or should be—done as on a one-off basis. Therefore, ethics cannot be easily operationalized into a checklist. The outcomes of ethical evaluations need to be taken as preliminary ones and need constant revision. While many activities related to RRI could be formulated as a kind of (iterative) checklist, such a representation alone could create the impression of a single necessary ethical audit as a gatekeeper that, once passed, justifies, or even ennobles, all resulting outcomes.

However, ethical considerations rather need a continuous, critical, and attentive process that seeks to continuously question and improve current designs. As cumbersome as this might sound, agile development teams, especially startups, do just that. They just need to add the "ethical dimension" to their purpose-driven endeavor.

However, even though ethics is not a checklist issue, this does not mean that there are no tools or guidelines that can get you thinking in helpful directions. The following provides an overview of general issues of health-related AI as an inspiration to what ethical considerations could take into account.

23.7.1 An Overview of Ethical Issues in (AI-based) Health Technology

For providing an idea of the potential range of ethical issues, in the following, a list of several topics occurring in the context of (AI-based) health devices is presented. This overview is derived from abstracted examples and is not intended to be exhaustive. Every innovation has its individual context of use and its specific ethical implications. No ordering with respect to the relevance of the ethical issues is imposed. There is an emphasis on AI-based health technology, but most items translate well to health technology in general.

23.7.1.1 Autonomy

Innovative technologies which involve at least partial automation entail a shift from human to automatic control. This often also incurs (the risk of) reduced human autonomy. Specific ethical issues can be identified as/from:

- **Technological opacity**: opaque medical technology may limit means of shared decision-making and informed consent, potentially limiting a patient's autonomy. However, technological opacity may also result in an implicit shift of authority from physician to technology provider. Responsibility issues (a gap) may ensue.
- **Paternalism**: technology designs that healthcare practice grows dependent on have the potential to limit a patient's autonomy, e.g., by being inflexible and non-adaptable to the patient's needs.
- **Self-care**: homecare devices for self-diagnostic or self-therapy may indeed enhance the patient's autonomy. However, negative psychological effects, such as anxieties or compulsory disorders, may effectively reduce a patient's capability to act autonomously.

23.7.1.2 Accountability

When technologies adopt medical expertise promising to provide more profound diagnoses, responsibility and accountability need to be potentially re-allocated:

- **Asymmetries in informational power**: datafication of health may render medical personnel incapable of taking responsibility, shifting significant power toward medical technology suppliers.
- **Black-box technology:** technologically opaque solutions may epistemically limit the information conveyed by medical personnel.

23.7.1.3 Bias

Biases ensue when algorithm outputs are systematically skewed, producing potentially unfair or discriminatory outcomes. The increasing technological complexity and automation challenge the identification and prevention of bias in different dimensions:

- **Existing bias**: the development team's unconscious or conscious biases may enter the technology design.
- **Technical bias**: the availability or cost-effectiveness of particular sensors, actuators, or other technological items may skew potential use-cases or applicability.
- **Emerging and feedback bias**: socio-cultural conditions may change (potentially as a feedback effect of putting a particular technology into use), but technological solutions may remain unadjusted to them.

- **Data bias**: over or underrepresentation and the absence of characteristics/parameters in the training dataset may cause discriminatory output.
- **Algorithmic bias**: wrong classifications of parameters and misdirected inferences made by algorithms may be particularly crucial because of their detection and prevention difficulty.
- **Automation bias**: preference for suggested decisions from automated systems while disregarding contradictory information from the human perspective may increase the challenge of overreliance.

23.7.1.4 Privacy and IT-Security

Datafication and the need for a large amount of data to run systems reliably may jeopardize the assurance of privacy in (at least) three contexts:

- **Contextual dimensions of privacy:** medical analyses, e.g., regarding the genome, can reveal information not only about the patient in question but may also extend toward, e.g., relatives.
- **Risk of inference:** Data fusion may allow deanonymisation procedures.
- **IT-security:** insufficient means to ward off attacks on medical software can amount to privacy breaches, data poisoning, system disruptions, or model stealing.

23.7.1.5 Data Quality, Context, and Provenance

Monitoring the data quality, context, and provenance is important to ensure a validated transferability of the trained system to new types of applications. There are diverse aspects to be considered:

- **Inherent uncertainty and ambiguity of medical phenomena:** formalized processes in data acquisition may generate a false sense of certainty and unambiguity of medical phenomena, leading to potentially wrong conclusions and, hence, patient harm.
- **Disregard of interobserver variability:** manually annotated data which can be falsely treated as indisputable ground truth, may lead to wrong assessments of probabilities in decision support systems.
- **Disregard of interpretive frameworks in data acquisition and processing:** a disregard of interpretive frameworks employed to make sense of medical data severely erodes the potential to draw objective conclusions from it. Unjustified belief in the objectivity of data can lead to wrong conclusions and inferences, potentially yielding severe patient harm.
- **Disregard of contexts in data acquisition:** contexts such as informal processes within the medical workflow may contribute to potentially harmful misinterpretation and miscorrelation of events.
- **Phenomenological limits of data acquisition:** rich qualitative medical data may often not be amenable to formalized documentation and data acquisition practices. Loss of relevant data and insights might ensue.

23.7.1.6 Socio-technical Challenges

Emerging technologies influence their use of contexts leading to change in the societal environment and may cause several socio-technical challenges:

- **Deskilling:** extensive use of medical technology may cause significantly increased harms and damages once devalued and lost manual skills are required.
- **Incompatibilities of data acquisition and demands in medical workflow and documentation**: loss of important data may ensue if documentation and data acquisition processes are designed to be overly formalized, inflexible, and non-accommodating of specific forms of informal documentation.
- **The inflexibility of data-based systems to accommodate rich and possibly-evolving ontological systems:** medical taxonomies may be revised and, hence, evolving away from fixed ontologies toward more fluent and difficult to quantify terminologies, such as spectrum disorders, for instance. Without revising the technical systems, medical workflow and assessments may be impaired to the detriment of the patient.
- **Disregarding the nonlinear nature of the medical workflow and its inherent tacit knowledge**: forcing medical personnel or patients into rigid technologically guided processes, documentation practices, and means of assessment may incur stress, medical errors, and patient harm.

23.7.1.7 Safety and Reliability

The reliable operation of technologies, especially in the health care sector, is paramount to guarantee patient safety:

- **Technological opacity:** unintelligible algorithms may hamper medical software maintenance and debug, yielding potentially wrong outputs.
- **Unwarranted ontological assumptions:** specifically, unchecked or misinterpreted ontological assumptions strongly influence data acquisition, pre, and post-processing and may result in entirely wrong inference systems that cannot account for particular symptoms, diseases, internal and undocumented medical workflow, or other forms of tacit knowledge.
- **Reproducibility:** a lack of accurate and trustworthy assessments of the potential for improving clinical outcomes and an overemphasis on technological accuracy metrics risk negative health impacts.
- **Functional safety:** unpredictable responses to failure, a lack of backup routines, or redundant system designs may prove to be hazardous.

23.7.1.8 Erroneous Definition of Innovation Goals

Relying on a biased, limited, or incomprehensive definition of overall innovation goals or specialized technical objectives (optimization criteria, safety and accuracy measures, etc.) may contribute to missing the actual expectations put forward toward health innovation:

- **Disregard of clinical utility and improved health outcomes over technical performance metrics:** the limits of translating technical performance limits, such as the area under the curve (AUC) or the receiver operating characteristic (ROC), into improved health outcomes may yield economic losses or even patient harm.
- **Disregard of qualitative performance indicators over quantitative metrics:** simplistic quantitative health care metrics, such as life expectancy, may fall short of capturing deeper qualitative goals. Accordingly, e.g., for particularly vulnerable people, well-being and quality of life may reduce.

23.7.1.9 Malicious Intent or Misconduct
Occasionally, tensions between economic pressures and an innovator's social responsibility may lead to misconduct, perhaps even in deceit for personal gain:

- **Fabricated data:** algorithms may be designed based on fabricated data to the extent that it violates the rules of good scientific practice.
- **Untransparent communication:** the active obfuscation of critical voices or technological disadvantages or limitations may ultimately lead to patient harm.

The above list is not exhaustive but may give an impression of which aspects, causal relations, or even criteria to consider when reflecting on the ethics of innovative technology. Next, we will briefly present some tools that help to organize the process of ethical reflection and planning for responsible innovation

23.7.2 Tools for Enabling Ethical Reflection

In the following, we will briefly introduce and outline methods for engaging in responsible innovation. The methods typically promote a collaborative, iterative process and will yield an outcome documenting all ethical, legal, and societal considerations that have emerged during discussions. Documenting is also vital when engaging in ethical reflection, as it provides a means to monitor progress, improve compliance and organize quality assurance.

The methods will progress from simple to more complex. However, all feature low entry barriers and are mostly designed to be conducted by the developing team rather than by professional ethicists or similar.

23.7.2.1 Consequence Scanning
Consequence scanning[4] is a method that focuses on the potential impacts an innovation might have on people or the environment, ways to mitigate them, and, hence, keeping in line with a company's set of values.

The method requires little to no preparation and no ethical expertise. It is most easily adopted in agile development frameworks during initial conceptional phases.

---

[4] https://doteveryone.org.uk/project/consequence-scanning/ (accessed February 2, 2022).

However, its results should be re-evaluated using the same method every time the innovation enters a new phase of concretization: From vision to a roadmap to release and even every time a new feature is introduced. The method is focused on three main questions:

- What are the intended and unintended consequences of this product or feature?
- What are the positive consequences we want to focus on?
- What are the consequences we want to mitigate?

The method should be conducted collaboratively, involving everyone from the core development team to business specialists, including external stakeholders, such as users. The method should be set up as a 45 to 1-hour workshop, structured into two phases: ideation and action.

Phase 1: Ideation

- The item/feature of interest with its intended consequences is introduced.
- All participants quietly think about the consequences regarding the first question.
- Similar consequences are clustered.
- Additional consequences can be added.

Phase 2: Action

- Consequences are sorted into action categories (act, influence, or monitor).
- Consequences are voted on as either positive or in need of mitigation.
- Consequences are discussed, and responsibilities are assigned.
- Express ideas about the next steps.

The method is freely available and comes with materials for organizing ideas or prompting them by cards that capture examples.

Keeping a record of each consequence scanning event is important. The tool's authors propose to keep a log of each consequence, mark it as intended/unintended, positive/negative, note down its action category, potential means for mitigation, priority, underlying hypotheses, action plan, measure to assess its success, and the time scale on which to act.

23.7.2.2 The Ethics Canvas

Another tool, the so-called "Ethics Canvas,"[5] cf. Fig. 23.3 is designed to appeal to developers familiar with the Business Canvas. The Ethics Canvas allows to systematically identify relevant groups of people, technology impacts, and potential value conflicts to design innovative solutions. Its conception is a perfect fit to the rationale presented in this chapter, as it purports that considering ethics goes beyond

---

[5] https://www.ethicscanvas.org (accessed February 2, 2022)—originally published under CC-BY-SA license with permission.

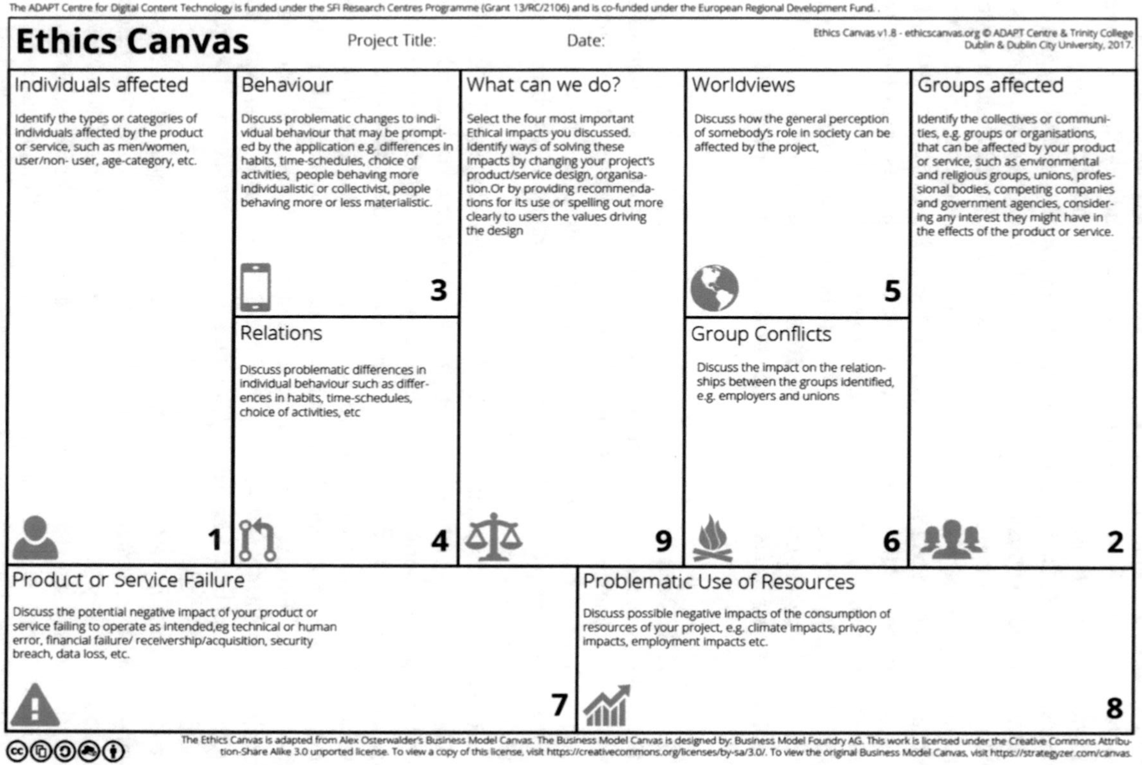

The ADAPT Centre for Digital Content Technology is funded under the SFI Research Centres Programme (Grant 13/RC/2106) and is co-funded under the European Regional Development Fund.
Ethics Canvas
Project Title:
Date:
Ethics Canvas v1.8 - ethicscanvas.org © ADAPT Centre & Trinity College Dublin & Dublin City University, 2017.
Individuals affected
Identify the types or categories of individuals affected by the product or service, such as men/women, user/non- user, age-category, etc.
1
Behaviour
Discuss problematic changes to individual behaviour that may be prompted by the application e.g. differences in habits, time-schedules, choice of activities, people behaving more individualistic or collectivist, people behaving more or less materialistic.
3
Relations
Discuss problematic differences in individual behaviour such as differences in habits, time-schedules, choice of activities, etc
4
What can we do?
Select the four most important Ethical impacts you discussed. Identify ways of solving these impacts by changing your project's product/service design, organisation.Or by providing recommendations for its use or spelling out more clearly to users the values driving the design
9
Worldviews
Discuss how the general perception of somebody's role in society can be affected by the project,
5
Group Conflicts
Discuss the impact on the relationships between the groups identified, e.g. employers and unions
6
Groups affected
Identify the collectives or communities, e.g. groups or organisations, that can be affected by your product or service, such as environmental and religious groups, unions, professional bodies, competing companies and government agencies, considering any interest they might have in the effects of the product or service.
2
Product or Service Failure
Discuss the potential negative impact of your product or service failing to operate as intended,eg technical or human error, financial failure/ receivership/acquisition, security breach, data loss, etc.
7
Problematic Use of Resources
Discuss possible negative impacts of the consumption of resources of your project, e.g. climate impacts, privacy impacts, employment impacts etc.
8
The Ethics Canvas is adapted from Alex Osterwalder's Business Model Canvas. The Business Model Canvas is designed by: Business Model Foundry AG. This work is licensed under the Creative Commons Attribution-Share Alike 3.0 unported license. To view a copy of this license, visit https://creativecommons.org/licenses/by-sa/3.0/. To view the original Business Model Canvas, visit https://strategyzer.com/canvas.

Fig. 23.3 The Ethics Canvas (https://www.ethicscanvas.org/ (accessed February 23, 2022))—originally published under CC-BY-SA license with permission

regulatory compliance, ethical considerations need to be articulated right at early development stages, and ethical reflection is a task for everyone in the development team, which should be a collaborative effort [1].

Based on this rationale, the Ethics Canvas is modeled after the Business Model Canvas by Osterwalder and Pigneur [15]. The canvas is structured as alternating between the micro and the macro level, considering first individuals, then groups, first behaviors, associated issues, then worldviews, and potential conflicts.

The canvas is intended to be traversed in order with the first phase concerning the stakeholders affected comprising fields one and two. The second phase analyses all potential impacts regarding behavior, worldviews, product failures, and resources, spanning fields three to eight. Phase three is about solutions and concerns field nine.

The canvas is quite verbose about what the respective question's intents are. However, despite the wording 'problematic' occurring often, the canvas lends itself to document and discuss positive ethical aspects as well.

Hence, once again, this tool can act as a way to quickly and intuitively map and document ethical considerations. It can be refined iteratively and used in early conceptual phases, but it can also be broken down and used on a single product feature level.

23.7.2.3 The RRI Roadmap

While the previous methods have focused on the operative, the approach of the so-called "RRI Roadmap" [16], cf., Fig. 23.4 concerns issues of responsible innovation, ethics, and potential means of mitigation from a strategic perspective. Hence, the RRI Roadmap explicitly references the time frames in which challenges may emerge, and actions should be taken.

The RRI Roadmap is the result of a large European research project, PRISMA,[6] aimed at bringing the principles of RRI into industrial practice. The methodology is comprised of six steps outlined in Fig. 23.5. Again, these highlight the strategic potential and, hence, as a first step, the need to obtain the endorsement of an organization's C-level personnel toward RRI values. Accordingly, this method aims at realizing the principles of RRI not only on the product level but also on the process level. Further steps such as context analysis and materiality considerations also encompass establishing organizational processes, e.g., those utilizing the tools mentioned above, beyond the actual product-specific stakeholder analyses, risks, or barriers.

After experimentation with RRI methods in specific contexts, the process of determining an RRI Roadmap also incorporates the definition of operative management tools, such as key performance indicators. The roadmap itself is a means to document the journey and steps planned, mandating a long-term RRI strategy.

This discussion can only provide a cursory introduction to the strategic aspects of RRI implementation. However, we encourage the reader to browse through the

---

[6]www.rri-prisma.eu (accessed February 2, 2022).

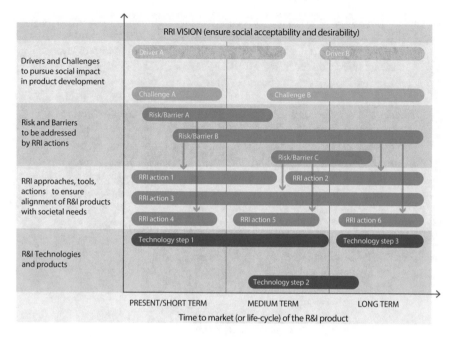

Fig. 23.4 Visualization of the RRI Roadmap (https://www.rri-prisma.eu/wp-content/uploads/201 9/08/PRISMA_RRIRoadmap_template-1.jpg (accessed February 23, 2022))

application examples on the project website to get an idea of how to apply the method.

23.8   Summary

This chapter clarified *why* ethical design considerations are relevant, *when* and *how* to address them, *which* socio-ethical issues should be taken into account, and *how* to get started with tools that help to innovate responsibly. When insightful, we elaborated on the meanings and specific issues arising in the domain of health technology.

Take-Home Messages
- Socio-ethical considerations should be integral to the development process because technology, especially in health, is inherently value-laden. Furthermore, ethical considerations contribute to an innovation's success.
- Ethics is different from regulatory compliance in that ethics are open-ended and translate into what should be done rather than what can be done.

(continued)

- Ethical considerations should be integrated right from the beginning of the development process because it is a means to uncover how innovation can be designed beneficially for all stakeholders. Responsible Research and Innovation provides the appropriate framework for this.
- Responsible innovation in health incurs a range of specific and important issues. We outlined a few of them to provide a less abstract definition of the framework in context.
- Entrepreneurs may follow responsible innovation practice for motives resulting from self-interest and altruism. Benefits are mostly medium and long-term, but generally, RRI can be assumed even to pay off.
- Ethics cannot be done by ticking boxes on a checklist. However, it is good to get some experience about potential issues. We have listed a non-exhaustive set of ethical issues to lower the entry barrier.
- Entry barriers can also be lowered by engaging in collaborative, workshop-type methods to identify and discuss potential ethical issues. Remedies can be planned and evaluated.

Fig. 23.5 Steps toward an RRI Roadmap (https://www.rri-prisma.eu/road-map-rri-for-companies/ (accessed February 23, 2022))

References

1. Reijers W, Koidl K, Lewis D, Pandit H, Gordijn B (2018) Discussing ethical impacts in research and innovation: the ethics canvas. Proc IFIP International Conference on Human Choice and Computers 537(September):299–313. https://doi.org/10.1007/978-3-319-99605-9
2. Bannister C, Sniderman B, Buckley N (2020) Ethical tech: making ethics a priority in today's digital organization. Deloitte Review 26:54–65
3. Philbeck T, Davis N, Larsen AME (2018) Values, ethics and innovation rethinking technological development in the fourth industrial revolution (Issue August). World Economic Forum
4. Grunwald A (2011) Responsible innovation: bringing together technology assessment, applied ethics, and STS research. Enterprise and Work Innovation Studies 7:9–31
5. Bird E, Fox-Skelly J, Jenner N, Larbey R, Weitkamp E, Winfield A (2020) The ethics of artificial intelligence: issues and initiatives (Issue March). European Parliament; European Parliamentary Research Service; Scientific Foresight Unit. https://doi.org/10.2861/644
6. Collingridge D (1980) The social control of technology. St. Martin's Press; Pinter
7. Kudina O, Verbeek PP (2019) Ethics from within: Google glass, the Collingridge dilemma, and the mediated value of privacy. Science Technology and Human Values 44(2):291–314. https://doi.org/10.1177/0162243918793711
8. Rességuier A, Rodrigues R (2020) AI ethics should not remain toothless! A call to bring back the teeth of ethics. Big Data Soc 7(2):205395172094254. https://doi.org/10.1177/2053951720942541
9. Von Schomberg R (2013) A vision of responsible research and innovation. In: Owen R, Heintz M, Bessant J (eds) Responsible Innovation. Wiley, London
10. Burget M, Bardone E, Pedaste M (2017) Definitions and conceptual dimensions of responsible research and innovation: a literature review. Sci Eng Ethics 23:1–19. https://doi.org/10.1007/s11948-016-9782-1
11. Demers-Payette O, Lehoux P, Daudelin G (2016) Responsible research and innovation: a productive model for the future of medical innovation. Journal of Responsible Innovation 3(3):188–208. https://doi.org/10.1080/23299460.2016.1256659
12. Pacifico Silva H, Lehoux P, Miller FA et al (2018) Introducing responsible innovation in health: a policy-oriented framework. Health Res Policy Sys 16:90. https://doi.org/10.1186/s12961-018-0362-5
13. Garst J, Blok V, Jansen L, Omta OSWF (2017) Responsibility versus profit: the motives of food firms for healthy product innovation. Sustainability 9:2286. https://doi.org/10.3390/su9122286
14. Blok V, Inigo E, et al. (2020) Recommendations for the development of a competitive advantage based on RRI. Report Project RRING, Part of the European Union's Horizon 2020 Research and Innovation Programme Under Grant Agreement No 788503, https://rring.eu/wp-content/uploads/2020/08/D-5.1.-RECOMMENDATIONS-FOR-THE-DEVELOPMENT-OF-A-COMPETITIVE-ADVANTAGE-BASED-ON-RRI.pdf
15. Osterwalder A, Pigneur Y, Bernarda G, Smith A, Papadakos T (2014) Value proposition design: how to create products and services customers want. Wiley
16. Porcari A, Pimponi D, Borsella E, Mantovani E (2019) PRISMA RRI-CSR Roadmap 710059. PRISMA RRI-CSR

Sabrina Breyer is a practical philosopher and ethicist in the interdisciplinary fields of RRI and CSR. In 2017, she received a B.A. in Combined Studies of Economics and Ethics/Social Sciences from the University of Vechta. Her bachelor thesis comprises a concept for respecting human rights in business processes and throughout the supply chain. In 2020, she received an M.A. in Practical Philosophy of the Economy and the Environment from Kiel University. In her master thesis, she analyzes the Sustainable Development Goals in an industrial context. Currently, she focuses on the research of ethical and societal aspects of AI.

Christian Herzog is a transdisciplinary researcher originally trained in engineering at Hamburg University of Technology, Germany, where he received his B.Sc, M.Sc., and Ph.D. degrees in mechatronics and control theory between 2005 and 2015. Since 2015, he is with the Institute for Electrical Engineering in Medicine at the University of Lübeck. Having engaged in teaching about ethics and sustainability issues of technology since 2011, from 2016 he has shifted his research focus toward the ethics of engineering, innovative technologies, and artificial intelligence in particular, having received an M.A. in Applied and Professional Ethics from the University of Leeds in 2020.

Part VI

Health Innovation Design

Case Studies Used Throughout the Book: Innovation Categories Explained

Michael Friebe

Abstract

Three actual Health Innovations from the editors' lab are presented and used in the subsequent chapters as case studies to illustrate some of the process steps of the presented innovation methodologies. They represent frugal, disruptive, radical, and traditional/incremental innovations. All are still active (April 2022) and public information as well as scientific references are available describing the innovation idea and initial results. All of them went through a process of exploring and defining the customer and their problem, ideating problem solutions, check and validate the solution, and develop initial minimal viable prototypes. The agile and iterative processes and methodologies for that (Design Thinking, Lean Start-up, Biodesign, I3-EME, Purpose Launchpad Health) are explained in subsequent chapters. The ultimate goal was and is to create dedicated start-ups that search—and hopefully succeed in finding—a sustainable and scalable business model.

Keywords

Disruptive innovation · Incremental innovation · Health business model

M. Friebe (✉)
AGH University of Science and Technology, Krakow, Poland

Otto-von-Guericke University, Magdeburg, Germany

IDTM GmbH, Recklinghausen, Germany

FOM University of Applied Science, Center for Innovation and Business Development, Essen, Germany
e-mail: michael.friebe@ovgu.de; info@friebelab.org

M. Friebe (ed.), *Novel Innovation Design for the Future of Health*,
https://doi.org/10.1007/978-3-031-08191-0_24

What Will you Learn in This Chapter?
- Different Innovation categories using case study examples
- An approach for developing radical/frugal innovation based on an exclusive focus on cost
- That there are likely plenty of untouched radical innovation opportunities for the Healthcare segment

24.1 Introduction

For demonstration purposes, we often refer to and use three actual case studies from the Innovation Lab of the Editor.

BODYTUNE (BT), EASYJECTOR (EJ), and ULTRACLEAR (UC) were chosen among more than 30 active Innovation Projects and over 100 identified UNMET CLINICAL NEEDS, because they address different innovation segments and can be categorized as a sustaining/traditional/radical or disruptive Innovation. Please see the details for these projects in Table 24.1.

All of the case studies presented have gone—at least to a large extent—through the innovation generation process described in this book.

They address an initially validated UNMET CLINICAL NEED (RP and BT), or provide significant improvements (Cost, Business Model, Handling, Quality, etc.) to an existing problem/process (EJ), or will provide novel monitoring and health data information that can be used for advanced diagnosis in certain clinical settings or for monitoring your own health status at home (BT).

For all of them, intellectual property (patents, algorithms) was secured, the start-up process initiated, and initial results have been published. You can get additional information by visiting the web pages or by checking the additional literature information. They are meant and used as living examples.

It should be mentioned that these are 3 projects out of 30—from a University-based Health Innovation Lab—that actually got to that point. This also means that 27 were not pursued beyond an initial check/thesis/prototype. Main reasons were that these were not "worth" to be continued because the problem that they addressed was not big enough, or in some cases turned out to not even be a problem.

Or, they were improvements of existing products or processes that are better handled by the companies, institutions that are already active in that field. We always attempted to secure the technologies to be able to transfer them subsequently to external interested parties. But it is not easy to do that from within a university and it is recommended to rethink the actual innovation generation and transfer policies of university-based research setups. That discussion of this topic would fill a book on its own.

24.2   Innovation Categories Covered with these Case Studies

Lets shortly define some of the key categories that will be explained and dealt with in more depth in subsequent chapters:

Radical Innovation = An innovation that has the potential to destroy existing business models. Radical innovations are rare but could come with a large reward for the innovator. Often radical and **disruptive innovations** are used in the same context. The effects are essentially the same, but the starting points are different. Radical innovations typically start with a known business model, while disruptive innovations lead to new business models.

Modular/Incremental /Traditional/Sustaining Innovation refers to changing a component or module in a design, process, or business model and generates significant improvements. However, the customer base is well defined and the business model in general is clearly established.

Frugal Innovation is a term not often used, but it can easily be explained through the use of frugal. Advanced and High-Tech Healthcare provision uses very expensive and complicated equipment that is not really needed or does not really provide a benefit for a large majority of cases. A technology that effectively could be used effectively for this large majority at a fraction of the cost ($10\times$ or $100\times$ less) would open completely new markets and make technology available also for areas and applications that cannot afford these systems.

Health Innovation can start from within a research environment with dedicated clinical and/or technical expertise. The perceived and analyzed problems (and the subsequent solution ideas) are therefore typically related to a known segment (EJ, UC). The problems typically deal with improvements to existing approaches and technology/clinical solutions. Not surprising that these will therefore typically produce incremental innovations.

EJ is a little different, as it is a continuation of a development that started in 2006. At the time a novel injector technology was developed with a clear goal to fit into the existing workflow and business model.

For the EJ development, a different problem analysis approach was used based on following base questions derived from the Blue Ocean Innovation methodology of identifying a Value Innovation. Such a value innovation is at the intersection of identifying and raising the most important or most needed features while eliminating all the ones that are not really needed and with that to attempt to significantly reduce the cost.

For that case we started by asking the following questions:

- *Which features for Contrast Media Power Injectors are absolutely needed? Which are just optional? Which are not needed? ... and what is needed from the clinical side?*
- *How might we be able to just focus on these essential ones so that we can reduce the cost of the expensive device (currently € 20,000) significantly (by a reduction factor of more than 10x − < €2000) while maintaining the clinical requirements?*

Table 24.1 The three example and case study projects: Bodytune, Easyjector, and Radclear plus some links with further explanations, scientific works

	INNOVATION Category and Keywords	CASE STUDY DETAILS		Additional Resources
BODYTUNE	**Disruptive** Homecare Biometric Health Status Base Monitoring Auscultation Vascular Flow Personalized Medicine	This is a novel project and product idea that was developed in a joint effort between a commercial profit oriented small company and an innovation lab at a University Hospital. The coarse idea was on whether it is possible to develop a device that would be able to assess a vascular stenosis just by holding an external device to the artery (carotid with device held to the upper neck), "listen" to the signal and use advanced signal processing and machine learning to diagnose and characterize a progress or regress of arteriosclerosis. This could then be used to monitor at home and inform the patient and the clinical staff of any critical situation. The team did not really know about the depth of the problem and the difficulties of implementing such a device as part of healthcare delivery. On the other hand, it also did not understand the possibilities of such a technology for a multitude of other applications. The project was actively mentored for several month and a start-up is being created with a new understanding of the problem and newly identified opportunities that now include audio based biometrics. With this approach audio was now identified as a powerful tool to create a personal baseline (picture on the right shows the wavelet transformation of 6 different users for left and right carotid artery) of many different body emitted sounds (auscultation), including the vascular flow, some selected heart sounds, coughing and swallowing. This		https://doi.org/10. 3390/s21196656 https://www. bodytune-online.de

	allows a personalized biometric baseline and with that can detect any deviation from that baseline. With big data and federated learning it should be possible to use this to predict disease development.			
EASYJECTOR	***Radical Frugal Disruptive*** Contrast Media Injection Infusion MRI CT US	This is another product idea that was developed in a joint effort between a commercial profit-oriented small company (www.idtmt.de) and an innovation lab at a University Hospital (www.inka-md.de). Magnetic Resonance Imaging (MRI) requires for enhancing certain diagnostic image procedures the vascular injection of contrast media. This is a process that can be done (for higher accuracy and process compliance) with power injectors. To operate these devices in the MRI environment (very strong magnetic and electric fields) they need to be purpose designed. The cost for the systems is currently around € 20,000 with additional operational cost of € 2000 per annum for the system and € 20 per patient for the consumables. No real innovation was introduced in that segment for over a decade. We analyzed the clinical process in depth with the goal to provide a disruptive solution with respect to cost (> factor 100 cheaper) and handling improvements plus associated time savings. After first sketches, several very crude mock-ups were build and tested on whether they solve the identified customer needs. After validation 3D models were created and a mock-up software app build.		https://www. easyjector.com https://doi.org/10. 2147/MDER.S106338

(continued)

Table 24.1 (continued)

	INNOVATION Category and Keywords	CASE STUDY DETAILS		Additional Resources
RADCLEAR	*Incremental Sustaining Traditional* Hybrid imaging Ultrasound Nuclear Handheld Imaging Point of Care Image-Guided Therapies Sentinel Lymph Node	Handheld gamma cameras have started to come into play in surgical scenarios to provide surgeons with real-time nuclear image inside the operating rooms. These cameras are designed to imagine small and superficial structures and come in handy in sentinel lymph node studies. Although their small size makes them suitable for radio-guided surgery, the lack of structural information in their acquired images is a significant drawback of these scanners. To address this unmet clinical need, we introduce a handheld gamma-ultrasound (US) scanner that makes it possible for the surgeon to see both anatomy and function of their target. The scanner's compact size can ideally suit surgical settings and is expected to give the user flexibility and maneuverability required for radio-guided surgery. This would solve several problems/create opportunities: • US and Gamma Imaging Planes are the same • Only one hand is needed for the imaging, which leaves one hand for the biopsy • Accurate determination and selection of the sentinel lymph node • Combination with tracking to create 3D volume images Several issues were not initially thought through though: 1. Availability and permission to use radionuclide is difficult in many countries—handling often requires a nuclear imaging specialist 2. Reimbursement (payment) for this service is not established in many locations		https://doi.org/10.1515/cdbme-2021-2036

So it starts with domain and process knowledge and really has only one major goal—keep the needed features and performance with a dramatic cost reduction. When you think in terms of 10× or greater improvements/reductions then you always need to rethink the entire setup. That is what we did with the EJ and "surprisingly" anticipate now a cost reduction of >100× with added benefits in reduction of procedure errors, and procedure time while improving the handling. These were not anticipated.

> *A good advice for possible frugal or radical health innovations is to look at expensive and complicated setups and processes with the only goal to reduce cost dramatically while maintaining the key performance parameters in place.*

You would be surprised how many health segments and technologies have not experienced major innovations in the last decades because—despite high cost that was eventually accepted due to the lack of alternatives—users though that it works actually quite well and the providers of the technologies were not interested to change their lucrative business.

There is a clear business model for the EJ case already in place.

BT is different in that respect. It proposes to use novel audio-based informations to create a biometric health baseline. Once you have a baseline you can determine deviations/changes and possibly will be able to associate positive or negative health events with that or to even be able to predict or diagnose certain sicknesses. All quite unclear on how that will be integrated into the health system, who will pay for that device/technology in a homecare setup. So an innovation that could lead to doctor-less diagnosis or valuable monitoring and health status assessments, but also something that has no clear business model. Very promising initial results, a great potential vision, but without a clearly identified short-term business model. A clear disruptive Innovation.

While you can hypothesize about the future use cases and applications, these innovations require a lot of convincing toward potential finance providers that often do not have a long-term vision and the patience you need for these types of innovations. On the other hand, these are the innovations that could help to transform health and provide solutions that help tackle some of the big health challenges (in-equal access, demographic changes, cost issues). These small powerful innovation ideas are also significantly more sustainable compared to the big and scarce systems or complicated processes that they would be able to replace.

24.3 Summary

The presented three case studies are all not implemented yet, but were deeply explored and partially customer validated. They are examples that are used throughout the book to illustrate the innovation methods and to illustrate the terminology. There are plenty of starting points for health innovation generation. If you are a domain expert clinical or technological you will likely come up with ideas on how to improve the current setup/process. This will then typically produce traditional/sustaining/incremental innovations. Nothing wrong with that, they are needed and there are business models in place already. To address the big health challenges or cause big changes you need to think more disruptive and radical with respect to innovation design and also question existing approaches, workflows, processes, and setups on whether they need to be that complicated and expensive.

Take-Home Messages
- There are plenty of innovation opportunities in the health domain.
- Look at expensive setup that have not seen a significant innovation for a long time.
- Look at expensive setups where user says that there is no innovation needed, because it works quite well at the moment.
- Apply a $10\times$ thinking process as often as you can and as often as it is feasible, as this forces you to rethink current solution setups and approaches.
- Disruptive and radical innovations come with a high risk of failure but also with a high reward for the innovator and society.

Michael Friebe received a B.Sc. degree in electrical engineering, M.Sc. degree in technology management from Golden Gate University, San Francisco, and Ph.D. degree in medical physics in Germany. He spent 5 years in San Francisco, as a Research and Design Engineer with an MRI and ultrasound device manufacturer. He is a German citizen with expertise in diagnostic imaging and image-guided therapies, as a/an Founder/Innovator/CEO/Investor and a Scientist. He is also a Research Fellow with the Technical University of Munich, Munich; an Adjunct Professor with the Queensland University of Technology, Brisbane; and an honorary Professor of image-guided therapies with Otto von Guericke University, Magdeburg, Germany. Since 2022, he has been a Professor of biomedical engineering innovation with the AGH University of Science and Technology, Krakow, Poland. He is a listed inventor of more than 80 patents and has authored over 200 papers, and was part of over 35 Medical Technology Start-Ups. He is a Board Member of four medical technology start-up companies and an investment partner of a MedTec-fund. From 2016 to 2018, he was a Distinguished Lecturer of the IEEE EMBC teaching innovation generation and MedTec entrepreneurship. He is also a coach and trainer for OpenExO, and a master Launchpad mentor for the Purpose Alliance.

Why Is Healthcare Different with Respect to Innovation and Entrepreneurial Activities?

25

Sys Zoffmann Glud

Abstract

This paper seeks to highlight why innovation for and innovation within the Healthcare sector is quite different than for other industries. Even within the life science arena, there is a difference between the development of a new drug and the development of a new medical device.

Likewise having success with a health-tech/medical device startup requires overcoming particular pitfalls that startups for other markets are not experiencing. Selected pitfalls are highlighted and explained.

The paper also explains three overarching topics that any health innovator should be concerned with.

Keywords

Health innovation design · Healthcare innovation needs · Medical device development · Patient-centric medicine · Value design

What Will You Learn in This Chapter?
- That it is hard to get the problem definition right in the healthcare scene
- It is very difficult to determine the value (or how valuable) of the solution to a problem in healthcare is.

(continued)

S. Z. Glud (✉)
BioMedical Design, Institute of Clinical Medicine, Faculty of Health, Aarhus University, Aarhus, Denmark
e-mail: sys@clin.au.dk

- While healthcare deals with solving patient problems the patient is rarely the customer of a solution
- Three overarching themes that make it different to innovate for healthcare.

25.1 Getting the Problem Definition Right Is Difficult if You Have No Access to the Clinical Setting

Einstein's famous quote "If I had an hour to solve a problem I'd spend 55 minutes thinking about the problem and five minutes thinking about solutions."

To have success with innovative solutions in any type of industry requires getting the problem definition right.

And you have to get the problem definition right before you have burnt off all your resources on building the wrong solution to an ill-defined problem.

What does it mean to get the problem definition right? Apple successfully invented the iPod and there was no obvious problem beforehand, right? People already had portable music players so what is new?

Or maybe the truth is that there actually was a problem to be solved with the iPod. Most people were just not able to express what their problem was as they could not imagine how things could work differently from the existing solutions.

The truth is that there was a latent need for integrating hardware and software, i.e., allowing people to smoothly and effortlessly synchronize music from the library with the portable music player. And Apple spent plenty of time and hours of observation to define that problem, you can be sure of that.

Whereas Apple was already an established company with the resources and competences to investigate and research unmet user needs the same is not the case for young startups not to mention the sole innovator with a good idea.

Further, it gets even more complicated when wanting to invent new healthcare solutions for either patients or healthcare professionals. The doors to the healthcare sector are generally closed for the outside world to do exploration. And few patients openly share the details of their health and the treatments they are undergoing.

Likewise, healthcare professionals tend to not communicate about their work outside their field of expertise. This limits the access to critical information that is not voiced, but rather is belonging to what is called tacit knowledge, i.e., the domain-specific inherited knowledge which is not written down and hence not accessible for outsiders like for instance the genius engineer with a "great" idea.

Hence, unless you are employed in the healthcare sector it is quite difficult to acquire in-depth knowledge about what is actually needed of new solutions to diagnose, treat and care for patients (also see Fig. 25.1 on that issue).

When it comes to healthcare product innovation, if you do not have a proper problem definition you have a high risk of failing to properly articulate the value proposition and you are very likely to suffer from not picking the right clinical endpoint to underpin your case.

Fig. 25.1 In the current Health innovation landscape many academic researchers, companies, and startups are pursuing ideas founded often on a limited understanding of the whole picture of a clinical need

The truth is that it is simply really difficult to get the problem definition right when wanting to invent solutions that make life better for patients and healthcare professionals.

25.1.1 Many Obvious Problems but Only a Few Are High Priority

Even if you are lucky enough to be either an insider to a healthcare system and seemingly have got the problem understanding right, you can go wrong about how high a priority the problem has. Healthcare systems will notoriously have an endless list of relevant clinical needs to be solved, but they cannot afford or practically cope with changing too much at a time.

Any innovator needs to be curious about how valuable a solution to a clinical need is likely to be at the point in time in the future where the solution could be ready for adoption. Assessing pain points, and willingness to pay together with interpreting political agendas and long-term (societal) health trends is beneficial.

25.2 Understanding and Navigating the Complexity of the Healthcare System Is a Key Barrier to Innovation Adoption

Healthcare systems and healthcare delivery is highly complex in terms of stakeholder interests, business models, traditions, cultures, and inter-human variations in physiology which complicates clinical outcome evidence!

25.2.1 The User, the Champion (of a New Solution) and the Paying Customer Is Rarely the Same Person

Think of a procedure like self-catheterization which people with urinary incontinence or other urinary diseases need to perform. In this situation, the patient will need to establish healthy and sanitary routines at home where they place a urine catheter themselves in order to drain the bladder.

The product they will be using is likely to be the same product as the healthcare professional is introducing them to during demonstration sessions at the clinic, hence the healthcare professional is championing a solution while the actual user is the patient and the payer made be a mix of either insurance, public payment, or with a co-out of pocket payment from the user.

Patients vary greatly in how they live and how well they are able to comply to health procedures hence the interface between the catheter product, the catheterization procedure and the use-instructions and use-surroundings will likewise vary greatly and can cause a multitude of risks which the product needs to be able to mitigate.

The inventor of a new healthcare solution needs to understand such variation in workflows and use-situations, deeply. Not to mention decision pathways of product selection. Who decides which brand of catheters that the healthcare professional introduces to the patient?

The inventor needs to know this in order to know who to influence.

25.2.2 Selling Healthcare Products Is Not as Straight Forward as Selling a Car

Diversity in business models of health systems means diversity in purchase incentives and in buyer center decisions.

From country to country healthcare is organized differently and this means that there is a new healthcare business model to adapt to in all countries, in some countries like Germany there can even be big differences from region to region. Any healthcare inventor needs to be able to answer the question of "who is paying for your solution and do they have the money for it?"

If the care provider cannot get the cost for the new technology covered through reimbursement or other benefit streams they will likely not buy.

25.2.3 Siloed Organizations Prevents Sound Uptake of Solutions that Addresses Preventive Health Measures

Another complication added to adoption of healthcare technology innovations is the fact that healthcare budgets are compartmentalized. Meaning that it is difficult for a technology to be adapted, if the associated health and financial value propositions gains are acquired elsewhere than the department that carries the cost for the equipment.

An example of this:

> *Imagine a technology to be used during cesarean sections in order to minimize longterm health side-effects like the child developing type I diabetes. In such situation the birth department is only getting an extra cost and while the society will in the long run experience fewer type I diabetes cases caused by c-sections. But "society" is not helping the birth department with financial support to cover the cost since short-term financial decisions are governing.*

25.3 Being Bold and Fast as an Entrepreneur and Operate as a Century-Old Professional Medical Device Company

The third challenge put on healthcare innovators embarking on the entrepreneurial journey of putting a new valuable solution on the market is regulation and the demands for documentation. From day one in the development process, all steps and all changes and updates to a solution needs to be documented in a quality management system. Further, the immense paperwork required to obtain regulatory approval, i.e., documentation of safety and effect also needs to be started early on.

Quality management and regulatory approval are work tasks, which are very poorly fitted to the iterative nature of good innovation processes and it is a heavy administrative task, which often does not appeal to the fast lived entrepreneur. However, it is a compulsory task to manage and at a very professional manner regardless of whether the exit strategy is IP selling to incumbents or full-go-to-market.

Three overarching elements have been mentioned here, but truth is that there are many more.

Yet, these difficulties should not scare people away, rather use it as a fuel to drive you forward. Since innovating for healthcare is so challenging, the more important difference can be made for patients and healthcare professionals [1].

25.3.1 Adoption of New Health Technology Depends on Expert Recommendations Based on Evidence

Healthcare innovation is also different from other areas of innovation because there is a requirement for evidence of a positive clinical outcome. And that is not trait forward to establish. It takes money, it takes a long time and many things can go wrong. Furthermore, often experts disagree about the solidity of clinical research results and if the experts of a medical society are not agreeing on the evidence then approval and adoption of the technology will take longer and require further evidence. It is critical that the innovator early on identifies which clinical outcomes the decision-makers want to see.

25.4 Reflection on Action Questions

- Are you sure that you are solving a true clinical problem?
- Do the relevant stakeholders agree that it is a high-priority problem worth solving and that they would pay for?
- With how many patients or healthcare professionals have you validated the problem you are trying to solve?
- Countries are different—does the same problem exist in other countries as well?

Take-Home Messages
- *Implementation of new technology starts at day one in the innovation process.*
- *From early on in the process the innovator has to imagine how a solution should and would fit into workflows or use situation of patients.*
- *The large diversity of stakeholders and the complex organization of healthcare require that the innovator dedicate time and effort to get a 360-degree view on everything from clinical problem understanding to implementation pathways and barriers.*
- *The innovator needs to understand the problem to be solved in details, and*
- *needs to know the workflow surrounding the problem and the key stakeholders who influence the adoption of solutions.*

Reference

1. Herzlinger R (2006) Why Innovation in Health Care is so Hard, Harvard Business Review 2006. Free access via https://hbr.org/2006/05/why-innovation-in-health-care-is-so-hard

Sys Zoffmann Glud , Managing Director of BioMedical Design Novo Nordisk Foundation Fellowship Programme, Sys Zoffmann Glud has graduated from Aarhus University with an MSc in Molecular Biology and PhD in Nanomedicine. Leaving the research career behind, Sys has since 2012 worked in the field of healthcare innovation and entrepreneurship applied through university courses and continuing education training programs. Through her work she has trained +300 students, healthcare professionals, and industry experts in needs-driven healthcare innovation and commercialization, resulting in more than 10 startups and attractive innovation talents who keep applying the methodology they have been trained in for years.

Purpose Launchpad Health (PLH) Methodology Introduction

26

Michael Friebe, Julia Hitzbleck, Dietmar Wiedemann, and Oliver Morbach

Abstract

Healthcare Innovation ideas originating from biomedical engineering departments are rarely based on a deep understanding of a problem, but are often based on coming up with an engineering solution that does not meet an Unmet Clinical Need, is too complicated, bulky, costly, and does not consider global developments. For an impactful innovation design, it is essential however to properly understand the clinical issues, forward project the effect of exponential technologies and other global developments. Health and healthcare are in need of disruptive ideas for preventive, predictive, personalized solutions that engage the individuals to pave the way toward real healthcare. We have adapted a novel meta-methodology for dedicated use for health-related innovation generation. This novel health dedicated meta-methodology based on the PURPOSE

M. Friebe (✉)
AGH University of Science and Technology, Krakow, Poland

Otto-von-Guericke University, Magdeburg, Germany

IDTM GmbH, Recklinghausen, Germany

FOM University of Applied Science, Center for Innovation and Business Development, Essen, Germany
e-mail: michael.friebe@ovgu.de; info@friebelab.org

J. Hitzbleck
HI10x GmbH, Mondosano GmbH, Berlin, Germany
e-mail: julia@hi10x.com

D. Wiedemann
PROVENTA AG, Frankfurt/Main, Germany
e-mail: d.wiedemann@proventa.de

O. Morbach
pro.q.it Management Consulting GmbH, Roschbach, Germany
e-mail: oliver.morbach@proqitconsult.com

LAUNCHPAD is dependent on interdisciplinary team and innovation work and heavily relies on a good understanding of the current clinical processes and needs as well as on a future projection of global health delivery developments. The clinical perspective is essential and meaningful and impactful innovation can only be developed by validating desirability feasibility, and viability, which needs clinical, engineering/technical, as well as economic expertise.

Keywords

Purpose Launchpad Health · Exploration · Evaluation

What Will You Learn in This Chapter?
- Purpose Launchpad Health (PLH) Innovation Toolset
- For Health Innovations typically only the first two PLH phases are initially relevant.

26.1   Introduction

Health-related innovation ideas are often initiated in a technical department without a clear understanding of the actual clinical and patient needs. The apparent problems that are addressed by the new idea are rarely validated and also are often just incremental improvements of currently implemented and used systems.

Future healthcare should be more predictive and lead to personalized prevention rather than to only focus on fixing the actual health problems [1]. Development activities should attempt to identify product and process ideas that will lead to a shift from the current sick-care to an actual focus on maintaining personal health and with that avoid or greatly reduce the getting sick part (Fig. 26.1).

Students in technical departments, like biomedical engineering, learn in depth, but often lack the ability to understand and solve problems with an empathetic,

Fig. 26.1 Future Health should be predictable, lead to the prevention of diseases, is personalized, and will allow the individual to actively participate. The developments should obviously lead to better outcomes, improve the experience of patient and clinician, but should also address the cost issue to ensure a more democratic health delivery. If cost is not significantly reduced it will not lead to disruption and to sustainable human health on a global level

economic, and global point of view. While it is clear that inventions will only become innovations when *desirability* (does the market need it?), *feasibility* (can it be built?), and *viability* (is anyone willing to spend money for it?) are positively evaluated, the needed questions and validations are not initially asked and applied to gain development and usability insights. It is much more common that we base our development activities on assumptions and then quickly come upon with a perceived solution idea that we then start to build and test.

We do anticipate a significant disruption for future health developments with the convergence of different exponentially developing technologies. Robotic surgery for example is currently just a telemanipulation system controlled by a surgeon, but when this is combined in the future with machine and federated learning, advanced sensors and intraoperative imaging a fast move toward semi-autonomous and auton-omous control and operation is imaginable. There are several other areas where sensors, machine learning, big data, genetic information, 3D printing could lead to completely new ways of analyzing health developments that could disrupt healthcare delivery and with the related health business models.

Depending on the country you live in healthcare is very different from an offering, quality, and cost perspective point of view and is embedded in different health business models. Some may adapt to these upcoming new technological abilities much faster than others, because of a lack of alternatives and cost issues. This will likely lead to fast changes in health-related delivery offerings and business models in certain regions.

With that, we believe that a dedicated innovation meta-methodology is needed that explores the actual clinical needs and problems, analysis the environment including future technologies, validates assumptions and hypotheses, and subse-quently evaluates solutions via rudimentary prototype ideas to find a product/market fit that justifies an impact initiative [2–4].

A good recipe would be to find a purpose–product fit in a problem EXPLORA-TION phase and validate some of the initial hypothesis with many customer and problem-oriented experiments. Once that has been achieved—most likely with many iterations and revisions of the initial problem understanding—it is time to build MVPs, minimal viable prototypes (e.g., rudimentary models, sketches) that help to evaluate the product idea toward a product/market fit.

This validation of desirability, viability, and feasibility is often the core initial work of start-ups to define and validate an impactful product and business model. Start-Ups are probably the ones that will focus much more likely on disruptions (making current products and processes obsolete) as existing stakeholders often have little interest in such a change.

The three phases of a development process from an idea toward implementation are shown in Fig. 26.2.

The innovation model that we propose is based on the Purpose Launchpad (PL) methodology and adapted to the Health Innovation environment (PLH—Purpose Launchpad Health) defining the interaction of the PEOPLE that are involved in this innovation activity so they understand the needs and desires of the CUSTOMER, are able to define a proper future-oriented STRATEGY and a PROD-UCT that is attractive and fits the market needs [2, 3].

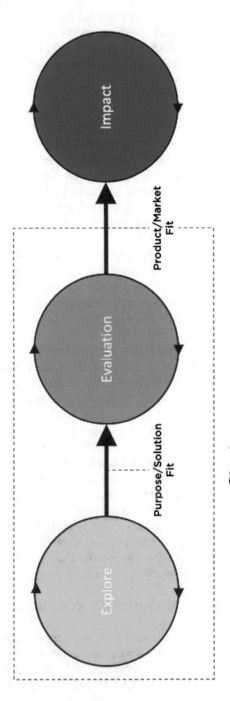

Fig. 26.2 Innovation starts with an exploration of the problem space and a deep understanding of the needs and pains of the potential users/customers. To check that we need to work with lots of hypotheses and related experiments to validate in an EXPLORATION phase a Purpose/Solution fit that justifies to advance to an EVALUATION phase (see [1])

26.2   PLH Methodology

26.2.1 The Eight Segments of PL for the Different Phases

The individual segments have different goals for the different phases. For the PURPOSE, for example, the exploration phase should be used to define the reason for the initiative and your personal role. With more information and insights, you will certainly re-iterate and re-define them in an EVALUATION phase and be more outspoken about it in the IMPACT phase.

Figure 26.3 shows the different goals for the different segments and phases.

Especially for Start-Ups and ideas created in a lab environment, it is not always good to start with the PURPOSE as many things first need to fall in place. But a start-up needs to have a clear understanding among all involved about the WHY of the operation.

26.2.2 The PLH Innovation Tools for Exploration/Evaluation

The Health field is quite special and it is essential to check and analyze the needs of the stakeholders as well as the individual health business models that vary from country to country.

In one country the main customer is the hospital, in another the individual doctor, and in a third several clinical services are offered in a pharmacy. This is why we defined a small set of tools that help to understand the problem and to develop a purpose– and product–market fit for health-related innovations. All the mentioned tools (see Fig. 26.4) are publicly available, attached to this paper (Addendum), and you can also find explanations on their use in the internet and in the references.

TARGETS	Explore	Evaluation	TARGETS	Explore	Evaluation
Purpose	Define your identity: why the initiative exists (purpose), what you want to be and how you will do it	Check your identity and iterate it as you learn with real experience	Viability	Define your business model to be financially sustainable.	Evaluate whether your business model can generate revenue and iterate it (if needed).
People	Connect with Purpose-oriented communities to find out personas, problems, and solutions. Also, connect with possible team members, and focus on exploration profiles.	Keep engaging with communities to find early adopters and others. Also, keep evaluating the team members with real action and make changes in the team (if needed).	Processes	Simplify processes to focus on learning.	Start defining and executing operations to deliver value to early adopters, while learning from them and iterating our solution based on it.
Customer	Start discovering what could be your customer segments.	Start selling to early adopters.	Product	Define Value Proposition and learn about it.	Build/iterate MVP to satisfy early adopters.
Abundance	Define how to connect and manage abundance.	Start implementing and validating the way we connect with abundance.	Metric	Focus on qualitative insights.	Measure value.

Fig. 26.3 PL segments and goals of each segment for the different phases—limited to EXPLO-RATION and EVALUATION, Source: Purpose Launchpad Guide [1]

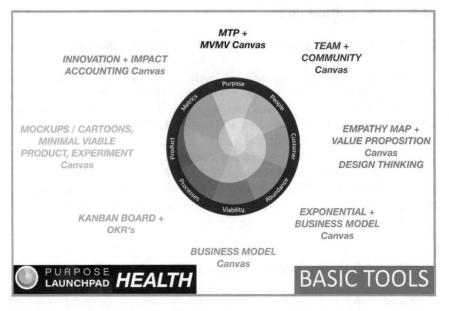

Fig. 26.4 For each of the segments in the EXPLORATION and EVALUATION phase, specific innovation tools are proposed. For the PURPOSE for example the MVMV (*M*assive Transformative Purpose, *V*ision, *M*ission, and *V*alues) canvas. Or PROCESSES the definition of Objective and Key Results (OKR). See [1] for Purpose Launchpad Logo

Presented are the MVMV (*M*TP, *V*ision, *M*ission, *V*alues) canvas used for the PURPOSE segment. The MTP, Massive Transformative Purpose, is an organization's higher, aspirational purpose and describes the change in the world that a team wants to achieve. The initiatives vision is what we want to become in the future, the mission describes the way on how you will make your vision true, while the values are indicators of the way that the organization is operated.

The TEAM canvas is centered around the purpose and highlights the capabilities and ambitions of the members. Through an empathetic view on the CUSTOMER (persona canvas) and by evaluating the customers' pain and gain, it is possible to analyze the problem and define the value propositions of a potential PRODUCT and get closer to a purpose– and product–market fit. For that, you use the VALUE PROPOSITION CANVAS (VPC), one of the most important tools of PLH. The VPC is also used together with the BUSINESS MODEL CANVAS (BMC—VPC and BMC are both from Strategyzer, Alexander Osterwalder) to analyze exponential scaling possibilities by connecting to ABUNDANCE with the help of the Exponential Canvas (by OpenExo). The use of these three tools in relation to each other is explained in a separate chapter [4–9].

For the Process Segment we obviously recommend to follow a structured, but nevertheless agile and iterative process like the PLH, and add other tools to it like a KANBAN board or dedicated Objective and Key Result (OKR) measures. These can also be used for innovation METRICS in combination with other Innovation Accounting tools (e.g., number and quality of experiments). And for the PRODUCT segment, it is advised to build as many as possible easy demonstrators. That could be an easy drawing/sketch, a story that shows how the proposed product/service/ process is used, or a cartoon image/movie. There are plenty of very easy to use— most are free as well—software applications available on the market. A product could also include simulations or 3D printed "mock-ups." Of course, you can also build real working prototypes, but especially for the EXPLORATION phase, this is not recommended as it takes too much time and typically does not have the same effect than a well-prepared drawing/movie.

26.2.3   The PLH Innovation Process

An experienced PL mentor should guide the team through the process that is in a lot of ways organized like other agile methodologies (e.g., SCRUM).

An independent person or the university professor would be ideal for that role as they are not involved in the day-to-day process.

A PLH innovation SPRINT (= regular repeating process) can be weekly or biweekly. The team itself should meet daily to discuss the open issues and next things they want to do to gain insights or validate hypotheses. In a weekly planning meeting with the mentor, the progress will be presented and a new activities list defined, called the project BACKLOG. The mentor will also discuss the learning and the new insights with the team and inquire about the mood of the team. This is quite important as it may show problems with the team and process early on.

26.2.4   Innovation Accounting: Learning as Essential Measure

What has the team actually learned during the PLH sprint process is the essential measure of progress. And it is also essential to embrace failure and invalidated hypothesis with learning. The process is iterative and follows the lean start-up principle of BUILD something, MEASURE the outcome, and LEARN from it. The previously created backlog is used as a base for the discussion with the team. In(validated) results are valuable learnings and new hypotheses are as well (which then need to be subsequently tested).

26.3   Results

The PL provides learning points (1 point for a new insight, 2 points for a validation/ invalidation of a hypothesis/assumption), measures the weekly team mood on a scale from 1 (poor) to 5 (excellent), and provides a visual feedback presenting the progress from one sprint to the next based on a standard and publicly available questionnaire [5].

Figure 26.5 shows a PLH innovation initiative from the authors' lab over many months and the associated visual progress. It also shows the learning curve and total

Sprint	Start	1	2	3	4	5	6	7	8	9	10	11	12	13
Progress	3,3	5	6	6	8	9	8	7	8	10	9	3	8	4
Total Progress	-	5	11	17	25	34	42	49	57	67	76	79	87	91
Mood	3,3	3,0	2,8	4,3	4,5	4,3	4,3	4,2	4,0	5,0	3,8	4,0	4,1	
Evolution Phase	-	Explore	Explore	Explore	Explore	Evaluate	Evaluate	Evaluate	Evaluate	Evaluate	Evaluate	Evaluate	Evaluate	Evaluate

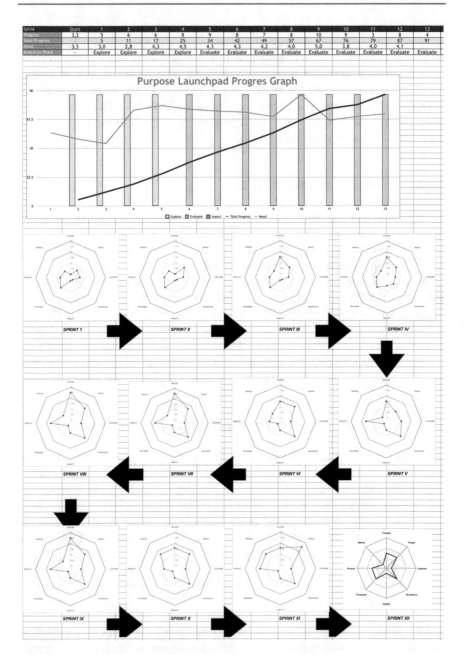

Fig. 26.5 The learning progress (points—blue line), mood graph of the team (green line), and the progress radar of a PLH initiative over many sprints/months, for Purpose Launchpad Assessment, see [1]

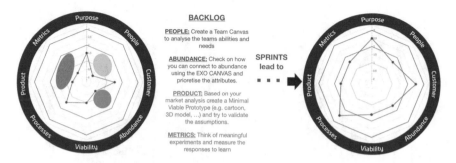

Fig. 26.6 A progress assessment produces a visual progress radar that can be used to define a backlog of items for the next sprint toward an initiative (right) that has progressed toward a product–market fit

learning points (top graph (dark blue line) and the mood variations (green line). There was a mood dip from sprint 1 to 3 that came from frustration and unclear expectations. Valuable information that was quickly corrected through additional training.

Figure 26.6 shows such a progress radar that is used to formulate a backlog of items that should be accomplished in the upcoming sprint. The example shows problem areas (low values) for an evolution from DISCOVERY (yellow) to EVAL-UATION (blue) phase for the PEOPLE, ABUNDANCE and PRODUCT/METRICS segments and the actions that should be taken to gain progress.

The radar is calculated based on a questionnaire (see Fig. 26.7) with three questions and three options for every question for each of the segments. As a general rule of thumb (not completely true, as there are interrelations between the segments), if you check the first option you are in the EXPLORATION, second in the EVAL-UATION, and third in the IMPACT GENERATION phase.

The PLH was also used in a 5 ECTS lecture/seminar (10 × 5 h on-site lecture, 3 × 3 h online mentor sprint sessions), where undergraduate and graduate biomedical engineering students ($n = 26$, 9 teams) were asked to define, improve, and validate an unmet clinical need (the problem understanding) all the way toward designing initial MVPs [2].

26.4   Conclusion

It is important to first understand the problem space that you are trying to address to find a purpose/solution fit for biomedical innovations.

In an exploration phase, you gain insights that help you define prototypes that can be evaluated to gain valuable insights for a future product–market fit that you test with MVPs.

Purpose

Do you have a Purpose or Massive Transformative Pur... (MTP)?
- Not yet
- Yes, but we have not defined our Moonshot, Vision, Mi... and Values
- Yes, and we have also defined our Moonshot, Vision, M... and Values

Have you done any real progress in relation to your Moonshot?
- Not yet, and we don't even know how to do it
- Not yet, but we believe we know how to start getting re... progress
- Yes, we have started to get some real progress towards... Moonshot

Have you already generated a (massive) impact in rela... to your Purpose?
- Not at all
- Yes, we are on track and moving with high speed towar... achieving our Moonshot
- Yes, we have already achieved our Moonshot and we m... define the next one

People

Does your team have doers with truly explorer mindset?
- Not yet or we don't know yet (still knowing the team in deep)
- Yes, but with limited entrepreneurial experience and/or the team performance is not being the right one
- Yes, and either the core team and/or our external mentors have significant experience with innovative initiatives

Are you connected with communities related to your Purpose in order to learn?
- Not yet
- Yes, we have started initial interaction with communities related to our Purpose to learn from them (not necessarily to sell to them) anything yet
- Yes, we are deeply connected and interacting with communities related to our Purpose both to learn from them, and to scale sales

Does your team have people with execution-oriented mindset?
- Not yet
- Yes, we have team members focused on execution, but we don't know yet all the positions that we will need to execute all the operations in the future
- Yes, we have a high-performance team, and we are fully focused on scaling our operations

Abundance

Have you identified sources of Abundance that may be relevant for your initiative?
- Not yet
- Yes, but still not connected to either our value proposition to clients or the way we can scale.
- Yes, and those sources of abundance could help us to provide a better value proposition and to scale our operations.

Have you validated the way you are connecting with Abundance?
- Not yet
- Yes, we have started to implement mechanisms to connect with Abundance, but we haven't finished it yet all and/or all the hypotheses related to this are not validated yet.
- Yes, we have connected with Abundance, and have validated we can do it!

Have you validated the way you are managing Abundance?
- Not yet
- We have started to implement mechanisms to manage Abundance, but we haven't finished it yet and/or all the hypotheses related to this are not validated yet.
- Yes, we are managing the Abundance, and have validated we can do it!

Viability

Have you defined the key elements of your business model?
- Not yet
- Yes, we have defined Customer Segment(s), Value Proposition(s) and Revenue Model(s), but don't have resources to get first sales
- Yes, we have defined Customer Segment(s), Value Proposition(s) and Revenue Model(s), and we think we have resources to get first sales

Do you have revenue from your current business model?
- No yet
- Yes, we have revenue but no enough resources to make changes in order to achieve happy Early Adopters
- Yes, we have revenue and either enough resources to happy Early Adopters or we have already achieved it

Is your initiative sustainable from a financial point of view?
- No, and we don't have resources to evolve the initiative to become sustainable
- No, but we have resources to evolve the initiative towards financial sustainability
- Yes, we are sustainable from a financial point of view

Processes

What processes do you have to define and develop your product/service?
- The founders and/or core team are deeply involved in customer discovery and product definition activity
- We have a dedicated Product team reporting to the founders and/or executives, defining the Product
- We have different people for UX, UI and/or Product Management

Do you have processes for running operations to delivery value?
- No, we don't have users / paying customers, nor operational processes yet
- Yes, we have users / paying customers and have developed processes to delivery value
- Yes, we have users / paying customers and have developed processes to both delivery value and learn

Do you have systems and processes to run sales and/or scale usage?
- No, we don't have users / paying customers, nor sales/promotion processes yet
- Yes, we have users / paying customers, and standard sales/promotion processes to grow our customers/users base
- Yes, we have users / paying customers, and our sales processes are focused on learning, with experimentation strategies that allow us to scale!

Customer

Have you identified your customer segments (and their pains)?
- No, we haven't defined the customer segments (and their pains) yet
- Yes, we have defined customer segments (and their pains), but we have not validated our key Hypothesis yet
- Yes, we believe we have a set of customer segments (and their pains) validated after running some experiment/s

Have you validated your customer segments with first sales and/or usage of your solution?
- No, we haven't even identified potential Early Adopters yet
- No, we have identified Early Adopters, but they haven't paid for our product/service(s) and/or used it yet
- Yes, we have Early Adopters (or clients) paying for our service/product and/or using it, and they are happy about it

Do you have a growing number of customers/users who are (very) satisfied with your Product/Service?
- Not yet
- No, but we are deeply understanding differences between Early Adopters and Mass Market, and we have defined a strategy to go to the mass market
- Yes, we are growing in a Mass Market (beyond Early adopters)

Product

Do you have a product definition (using a low-fidelity prototype)?
- Not yet
- No, but we will have already defined our Value Proposition
- Yes, we believe we have validated our Value Proposition by running experiments including a low-fidelity prototype

Have you defined/built your Minimum Viable Product (MVP)?
- No yet
- Yes, but Early Adopters are not yet satisfied with it. We are iterating.
- Yes, and Early Adopter/customers are (very) satisfied with it

Have you built a Product that has been validated through growing customer/user adoption?
- Not yet. We don't have a Product OR customer adoption is not growing
- We have a Product and customer adoption is growing, but we are still stuck with Early Adopters
- Yes, our product adoption is growing continuously in the Mass Market

Metrics

Are you capturing insights and qualitative metrics related to problem-solution fit?
- Not yet
- We are running experiments to evaluate our Hypotheses but not capturing the data yet
- Yes, we are running experiments and tracking qualitative metrics and information related them

Are you tracking key metrics related to product-market fit?
- Not yet
- We have defined value metrics related to the product, but not tracking them yet
- Yes, we are tracking metrics related to the value of the product (e.g. NPS, Stickiness, etc.)

Are you tracking key metrics related to grow?
- Not yet
- We have defined growth metrics, but not tracking them yet
- Yes, we are tracking growth metrics and we are scaling our initiative based on them

Fig. 26.7 The Questionnaire that needs to be filled out to create the visual radar and with that to determine the project status and items that still need to be done. For every segment there are 3 questions and for each of those 3 options. For Purpose Launchpad Assessment, see [1]

A validated market with well-done customer experiments is also very important when talking to investors in case of a start-up, probably more essential than having a cool technical solution that has not yet found a problem to solve.

Take-Home Messages
- We need to invest a lot of time initially to understand the problem and define a purpose for our actions and ambitions.
- Only when we understand the depth of the problem for the community of users we can define a solution that solves the clinical issues.
- We often do not validate our hypotheses and too quickly jump to create solutions that will subsequently likely fail because the clinical and user need was not properly checked.
- PLH provides a set of tools for the 8 segments of the Purpose Launchpad for a first definition of the clinical needs, the health stakeholders, solution ideas based on the first Health-related innovations in an EXPLORATION phase.
- Following the method and using the tools will significantly decrease the risk of failure and help to gain innovator, health stakeholder, and investor confidence.
- The visual radar obtained after a questionnaire-based assessment helps to define the next steps for the innovation progress.

Acknowledgments The PL and PLH artifacts in Figs. 26.3, 26.4, 26.5, 26.6 and 26.7 are provided by Purpose Alliance under Creative Commons Attribution-ShareAlike 4.0 license.

Dr. Francisco Palao is recognized as the developer of Purpose Launchpad and we are thankful for his contributions, enthusiasm, and support.

All shown canvases and innovation tools (see Addendum) are available under Creative Commons Attribution-ShareAlike 3.0 or 4.0 license.

References

1. Purpose Launchpad Guide, The manual on the agile framework and the mindset, Developed and sustained by Francisco Palao with the input of 150+ contributors around the world, September 2021, Offered for license under the Attribution Share-Alike license of Creative Commons, accessible at http://creativecommons.org/licenses/by-sa/4.0/legalcode and described in summary form at http://creativecommons.org/licenses/by-sa/4.0/., and Assessment Information, www.purposelaunchpad.com, viewed 30. Jan. 2022
2. Friebe M, Fritzsche H, Heryan K (2022) The PLH - purpose Launchpad health - meta-methodology to explore problems and evaluate solutions for biomedical engineering impact creation. IEEE EMBC 2022, Glasgow, Conference Paper
3. Fritzsche H, Barbazzeni B, Mahmeen M, Haider S, Friebe M (2021) A structured pathway toward disruption: a novel HealthTec innovation design curriculum with entrepreneurship in mind. Front Public Health 9:715768. https://doi.org/10.3389/fpubh.2021.715768

4. Blank S, Christensen C, Godin S, Pigneur Y, Osterwalder A. Value proposition Canvas https://www.strategyzer.com/canvas/value-proposition-canvas
5. Test your value proposition. https://www.strategyzer.com/blog/posts/2015/2/17/roadmap-to-test-your-value-proposition
6. Osterwalder A, Pigneur Y. Business model generation: a handbook for visionaries, game changers, and challengers business model canvas. Canvas can be downloaded at https://www.strategyzer.com/canvas/business-model-canvas also available via https://en.wikipedia.org/wiki/Business_Model_Canvas
7. Exponential Canvas. OpenExo. Canvas can be downloaded at https://openexo.com/exo-canvas/
8. Pfortmüller F, Luchsinger N, Mombartz S Community Canvas. https://community-canvas.org
9. Ivanov A, Voloshchuk M. Team Canvas. Canvas can be downloaded at http://theteamcanvas.com

Michael Friebe received a B.Sc. degree in electrical engineering, M.Sc. degree in technology management from Golden Gate University, San Francisco, and Ph.D. degree in medical physics in Germany. He spent 5 years in San Francisco, as a Research and Design Engineer with an MRI and ultrasound device manufacturer. He is a German citizen with expertise in diagnostic imaging and image-guided therapies, as a/an Founder/Innovator/CEO/Investor and a Scientist. He is also a Research Fellow with the Technical University of Munich, Munich; an Adjunct Professor with the Queensland University of Technology, Brisbane; and an honorary Professor of image-guided therapies with Otto von Guericke University, Magdeburg, Germany. Since 2022, he has been a Professor of biomedical engineering innovation with the AGH University of Science and Technology, Krakow, Poland. He is a listed inventor of more than 80 patents and has authored over 200 papers and was part of over 35 Medical Technology Start-Ups. He is a Board Member of four medical technology start-up companies and an investment partner of a MedTec-fund. From 2016 to 2018, he was a Distinguished Lecturer of the IEEE EMBC teaching innovation generation and MedTec entrepreneurship. He is also a coach and trainer for OpenExO, and a master Launchpad mentor for the Purpose Alliance.

Julia Hitzbleck is managing director of the e-health startup Mondosano and the innovation consultancy and venture builder HI10x. Being part of the OpenExO and Purpose Alliance community, she also works as a mentor and coach to build purpose-driven, highly scalable organizations with a personal passion for health. Prior experience in the pharmaceutical industry ranges from various R&D functions, corporate development, strategy, culture change and development of Bayer's Innovation Ecosystem. Julia holds a PhD in Chemistry from Syracuse University, New York, and has worked in different research organizations at Monash University, Australia and RWTH Aachen University, Germany.

Dr. Dietmar Wiedemann is a Member of the Management Board at Proventa AG, a German IT consultancy providing management, innovation, and technology consulting services. He helps organizations to plan and execute their digital transformation journey using agile frameworks and best practices. As an Agile Coach, he stands for continuous improvement of business value and is driver for necessary changes and adaptations in the organization. He earned his Ph.D. in Mobile Commerce at the Department of Information Systems at the University of Augsburg.

Oliver Morbach is Managing Director of pro.q.it Management Consulting GmbH. He is a coach, consultant, trainer, experience designer, and mentor working to enable co-creative processes and environments that unleash people's innovative potential in line with their purpose, by leveraging technology, lean-agile methodology, and exponential growth mindset. Over the last 25+ years, he held various senior and executive management roles in customer service and information technology organizations in the European ICT sector. Prior to that, he worked in defence software engineering. He holds a computer science degree (Diplom-Informatiker) from the University of the German Armed Forces Munich.

Part VII

Purpose Launchpad Health

Purpose Launchpad Health: Exploration and Evaluation Phases—Actual Case Studies

27

Michael Friebe and Oliver Morbach

Abstract

We will explain the Purpose Launchpad Health (PLH), a dedicated version of the Purpose Launchpad that was already presented. For each of the 8 segments, specific tools are recommended and their use is explained using two medical technology case studies (one more disruptive—Bodytune, the other fitting into the category of radical innovation—EASYJECTOR). The goal is to evolve the project idea from a DISCOVERY phase to an EVALUATION phase. For health-related products, programs, and processes, this typically leads to a start-up, transfer to industry through licensing or sale, or is discarded because the EVALUATION phase did not validate a need or the team is not willing to continue. The third phase of the PLH is the IMPACT generation, one that requires a sellable/usable product and revenues. This can only be achieved with ethical review, clinical studies depending on the risk category of the product, regulatory approval, and excited customer that are willing to pay for the offering. This phase will be dealt with in a separate chapter.

M. Friebe (✉)
AGH University of Science and Technology, Krakow, Poland

Otto-von-Guericke University, Magdeburg, Germany

IDTM GmbH, Recklinghausen, Germany

FOM University of Applied Science, Center for Innovation and Business Development, Essen, Germany
e-mail: michael.friebe@ovgu.de; info@friebelab.org

O. Morbach
pro.q.it Management Consulting GmbH, Roschbach, Germany
e-mail: oliver.morbach@proqitconsult.com

Keywords

Purpose Launchpad Health · PLH · Discovery · Unmet clinical need · Problem evaluation · Validation

What Will You Learn in This Chapter?

- Purpose Launchpad Health and its recommended tools
- Examples of how to use the tools using two case studies
- How to use this agile and iterative innovation methodology around a purpose
- How to create a backlog and next validation steps leading to learnings using a visual feedback

Case Studies
Case Study BODYTUNE (see www.BODYTUNE.online):

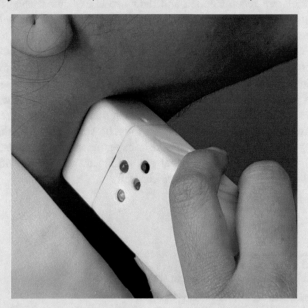

This is a novel project and product idea that was developed in a joint effort between a commercial profit-oriented small company (www.idtmt.de) and an innovation lab at a University Hospital (www.inka-md.de). The coarse idea was on whether it is possible to develop a device that would be able to assess a vascular stenosis just by holding an external device to the artery (carotid with device held to the upper neck), "listen" to the signal and use advanced signal processing and machine learning to diagnose and characterize a progress or

(continued)

regress of arteriosclerosis. This could then be used to monitor at home and inform the patient and the clinical staff of any critical situation. The team did not really know about the depth of the problem and the difficulties of implementing such a device as part of healthcare delivery. On the other hand, it also did not understand the possibilities of such a technology for a multitude of other applications. The project was actively mentored for several months and a start-up is being created with a new understanding of the problem and newly identified opportunities that now include audio-based biometrics. With this approach audio was now identified as a powerful tool to create a personal baseline (picture on the right shows the wavelet transformation of 6 different users for left and right carotid artery) of many different body-emitted sounds (auscultation), including the vascular flow, some selected heart sounds, coughing, and swallowing. This allows a personalized biometric baseline and with that can detect any deviation from that baseline. With big data and federated learning, it should be possible to use this to predict a disease development. For technical details see a publication—https://doi.org/10.3390/s21196656

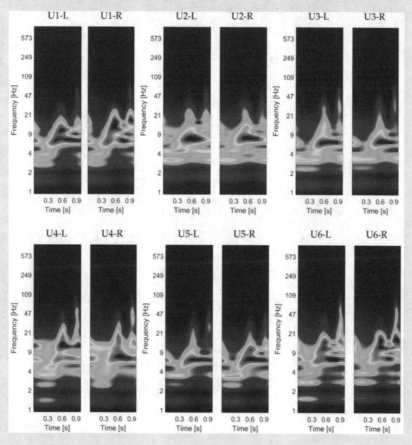

(continued)

Case Study EASYJECTOR (see www.easyjector.com):

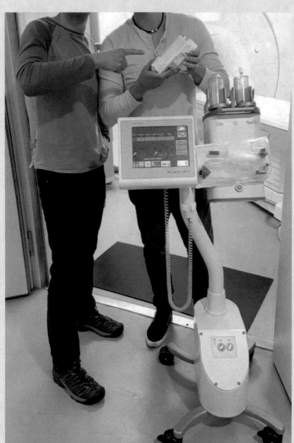

This is another product idea that was developed in a joint effort between a commercial profit-oriented small company (www.idtmt.de) and an innovation lab at a University Hospital (www.inka-md.de).

Magnetic Resonance Imaging (MRI) requires for enhancing certain diagnostic image procedures of the vascular injection of contrast media. This is a process that can be done (for higher accuracy and process compliance) with power injectors. To operate these devices in the MRI environment (very strong magnetic and electric fields) they need to be purpose-designed. The cost for the systems is currently around € 20.000 with an additional operational cost of € 2.000 per annum for the system and € 20 per patient for the consumables. No real innovation was introduced in that segment for over a decade. We analyzed

(continued)

the clinical process in depth with the goal to provide a disruptive solution with respect to cost (> factor 100 cheaper) and handling improvements plus associated time savings. After the first sketches, several very crude mock-ups were built and tested on whether they solve the identified customer needs. After validation, 3D models were created and a mock-up software app build. For some base details of the medical and technical needs see a publication—

 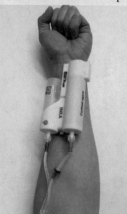

https://doi.org/10.2147/MDER.S106338

27.1 Tools for Innovations in Healthcare along the Purpose Launchpad Dimensions

Our experience working with many Health initiatives has led to develop and suggest some basic tools that we dubbed Purpose Launchpad Health (see Fig. 27.1 and [1]—PLH) and we will be explaining them using the BODYTUNE and EASYJECTOR Casestudy examples.

The PLH is based on a methodology developed and introduced by Francisco Palao and the Purpose Alliance community [2, 3].

What we show in these real case studies is the actual knowledge of the team at a certain time. A typical innovation journey starts with defining some initial hypotheses. The team then designs prototypes or interview questions to get more insights and possibly even a validation of the hypotheses or to formulate another new hypothesis.

This very important process of collecting data and insights is then used to define the customer needs (right side of the Value Proposition Canvas—explained later) and derive value propositions of a potential idea/solution (left side of the VPC).

This is an iterative process that gets into more and more detail (see Purpose Launchpad chapter for the iteration process). All of the tools that we present in this chapter are supposed to be used in this iterative manner—following Eric Ries's

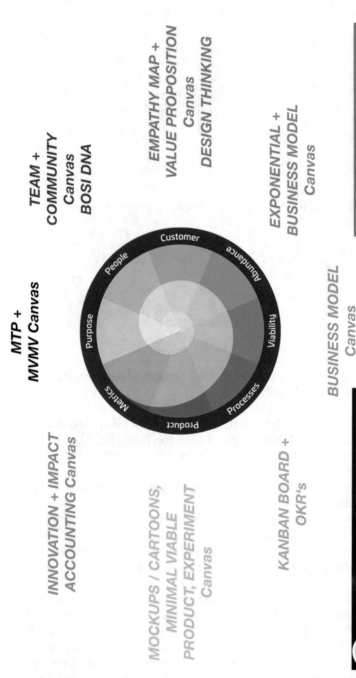

Fig. 27.1 The recommended basic tools to follow the Purpose Launchpad Methodology for Health-related innovation generation for the first two phases. For the subsequent Impact Generation/Implementation phase, other Health-related questions and issues need to be checked and resolved—e.g., regulatory approval, ethics and privacy issues. Courtesy www.purposealliance.org and Francisco Palao

build, measure, learn approach [4] and illustrated in Fig. 27.2 using the easyjector example (more details later).

A good starting point for evaluating and initiating the process (also for intrapreneurial activities/projects/ideas in larger operations) is to use a template as shown in Fig. 27.3. This can be used all the way at the beginning of the process or as a summary after the tools that we present have been used and the first validated insights have been established.

We start the tool presentations for the PLH as indicated in Fig. 27.4 with the PURPOSE segment. And then follow the tool presentation for the other segments in a clockwise manner.

The PLH is a flexible and agile methodology that allows you to skip one segment as long as you revisit it at a later point in time. It also makes sometimes only sense to spend time on a segment once insights in other segments have been generated. That is quite alright and encouraged. It is highly recommended to involve an external mentor that helps the process (equivalent to a SCRUM Master), that is not part of the core team, to ask questions and suggest the next steps to be taken in the process.

27.2 Purpose Tools

27.2.1 Massive Transformative Purpose (MTP) Canvas

The MTP Canvas is a tool that can help both people and organizations to discover the purpose and also to define other related elements (such as the Moonshot, which is called "Footprint" in the MTP Canvas).

The MTP Canvas includes nine different blocks divided into three different areas:

- The Hero (individual, team, or organization) with his/her values, longings, and superpowers
- The World (environment to be impacted) described by the kingdom (scope), inhabitants (for whom a positive impact shall be created) and the challenge to be tackled
- The Journey with the footprint, the path and the Massive Transformative Purpose (MTP)

In the context of Purpose Launchpad, the MTP Canvas can be used from the very beginning and evolved during the rest of the Purpose Launchpad Phases at the same time that we evolve the initiative.

BODYTUNE's MTP for example is defined as "Disrupting Disease Diagnosis and Prediction" (see Fig. 27.5) and the MTP canvas was used to develop this 5-word statement after many iterations.

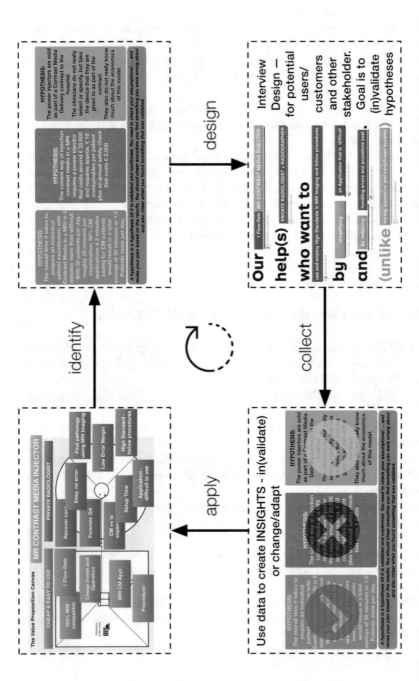

Fig. 27.2 All the tools that we present in this chapter are based on iterative processes that typically start with hypotheses that need to be validated or invalidated. For that, customer interviews (initially without any own idea/potential solution) are designed and the collected data evaluated and subsequently the results used to formulate/change the customer needs (right side of the VPC) and to formulate value propositions (left side of the VPC). The VPC (top left) and the elevator pitch (bottom right) originate from www.strategyzer.com

Initiative Submission

Name of Initiative:

Your Idea

Team and Support
Lead: lead@novelidea.com
Member: member@novelidea.com
Project Sponsor : sponsor@novelidea.com

Elevator Pitch   "My team [product/service name] will develop [a defined offering] to help [target customer] to [solve a problem with primary benefits] unlike [existing alternatives] with [secret source]."

Opportunity

Background

Briefly describe the context of your opportunity. What are the triggers? *(external market changes like technology, regulations, ecosystem, internal organizational changes)* Help the evaluator to better understand the idea.

Customer

Who are your customers?
Where does your solution come into play?
Which pain point are you trying to solve?
Which benefit are you trying to offer?

Solution

What are potential solutions on the market?
What are their current limitations?
What is your right to win?
Why is now the right time?
What is your main objective for a solution?

Technology

Key Technologies

Which exponentially developing technologies will be used?
Which technical resources will be used?
Which partners do you envision?

Readiness Level

How mature are the technologies you are using? *(research, market intro, mainstream)*
What is the TRL (technology readiness level) of your solution?

Roadmap

What is your development timeline?
Which technologies are converging?
Where do you see tech platform potential?

Impact

Value for Customers

Estimate new/additional value for customers

Revenue/Savings

Estimate new/additional revenue/savings

Scaling Potential

Cross-country / -function / - division / -BU

Further Benefits

Intangible assets like *reputation, partnership attractiveness, customer engagement, talent attraction, IP/licensing fees*

Fig. 27.3 A good starting point for the PLH process, or a summary after some PLH sprints, is this template that covers the identified (or assumed) Opportunity, the Technology that is proposed, and the Impact that this will create for customers and the organization

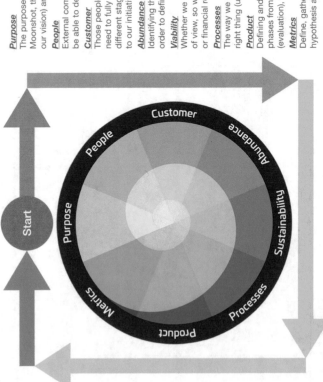

Purpose
The purpose is the reason why the initiative exists. Also part of Purpose are the Moonshot, the Vision (what we want to become); the Mission (how we will achieve our vision) and a set of Values (the way we operate our initiative).

People
External communities aligned with our purpose, as well as the internal team that will be able to develop the purpose- driven initiative.

Customer
Those people and/or organizations that use and/or pay for our solutions. We will need to fully understand who they are, what their problems are, how they behave in different stages in the market (early adopters and mass market) and how they relate to our initiative.

Abundance
Identifying the sources of abundance that our initiative can leverage will be key in order to define a proper approach for our whole initiative.

Viability
Whether we are for profit or not, we will need to be sustainable from a financial point of view, so we will need to have a business model (including revenue source/s) and/ or financial resources (economic support from other entity) to run our initiative.

Processes
The way we organize and run our initiative to both explore possibilities and build the right thing (using agile approaches).

Product
Defining and building the right solution for the different customer segments and phases from low- fidelity prototype (exploration), Minimum Viable Product (evaluation), towards an optimized products (impact).

Metrics
Define, gather and track with respect to accounting for innovation progress (evaluate hypothesis and apply learning) and impact.

Fig. 27.4 The different segments of the PLH in a short summary, also indicating the starting point and the direction of addressing the different segments. This is the suggested approach, but this is a flexible and agile methodology that allows you to jump over segments and revisit them later. Also, the purpose is sometimes difficult to define with a new team … so pause and get back to it once you have gained more insights in other areas. Courtesy www.purposealliance.org and Francisco Palao

27.2.2 Moonshot–Vision–Mission–Values (MVMV) Canvas

The MVMV Canvas is a tool to define the key strategic elements of the initiative and is very powerful to help us to see the high-level picture of it [5]

The elements (blocks) included in the MVMV Canvas are Purpose as defined in the MTP (see Fig. 27.5 left), Moonshot (a very ambitious goal toward the purpose for the next decade), Vision (What do we want to become in the future?), Mission (How will we make our vision true?), and Values (What is the way we operate as an organization/team). BODYTUNE's Moonshot as stated in Fig. 27.5 is "A personal diagnosis tool owned by all by 2030."

27.3   People Tools

The People axis is both about the external communities that relate to and are aligned with the purpose, as well as the internal team that will develop the purpose-driven innovation.

27.3.1 BOSI Assessment (Internal Team)

The BOSI assessment (https://bosidna.com/assessment) helps people to understand its entrepreneurial DNA profile, based on four types of Entrepreneurial DNA: Builder, Opportunist, Specialist, and Innovator.

In the context of Purpose Launchpad, the BOSI test can be very helpful when forming a new team or exploring for new team members, in order to find the complementary mindsets and styles.

See Fig. 27.6 for the different profiles and a sample report for one of the authors of this chapter.

27.3.2 Team Canvas (Internal Team)

The Team Canvas is an open-access free tool to organize team alignment around a purpose and brings members on the same page, resolve conflicts, and build productive culture, fast. Figure 27.7 shows the Team Canvas for BODYTUNE.

In the context of Purpose Launchpad, the Team Canvas can be very useful at the early stages of initiatives to see the high-level picture of the team and identify possible gaps that may need to be fulfilled.

27.3.3 Community Canvas

The Community Canvas is a framework that helps people and organizations build stronger communities. It provides a template for anyone who brings people together.

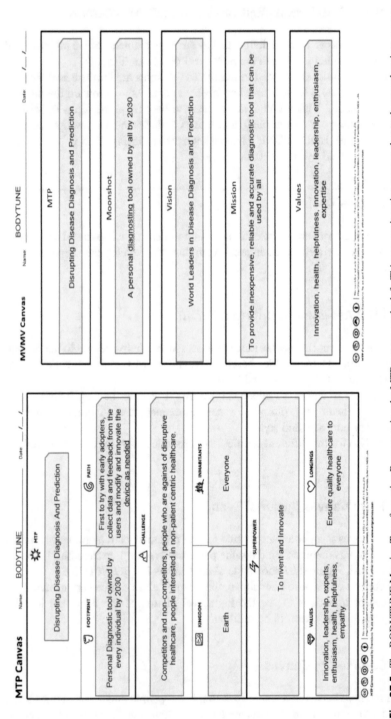

Fig. 27.5 The BODYTUNE Massive Transformative Purpose using the MTP canvas on the left. This was an iterative process that took several sprints among the team. So do not get frustrated if it does not fall into place immediately. It may take some time and the completion of other segments first. On the right, the MVMV canvas for BODYTUNE is shown. Canvas originally published under CC-BY-SA license with permission from [5]

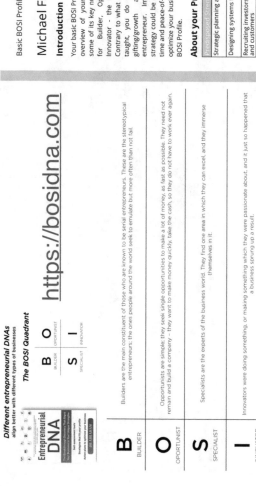

Fig. 27.6 The BOSI (Builder, Opportunist, Specialist, Innovator) profile helps to form heterogeneous teams that are made up of individuals with all the needed DNA. Of course, this is not meant as a selection, but more as an awareness tool. On the right, you see an example profile for one of the authors

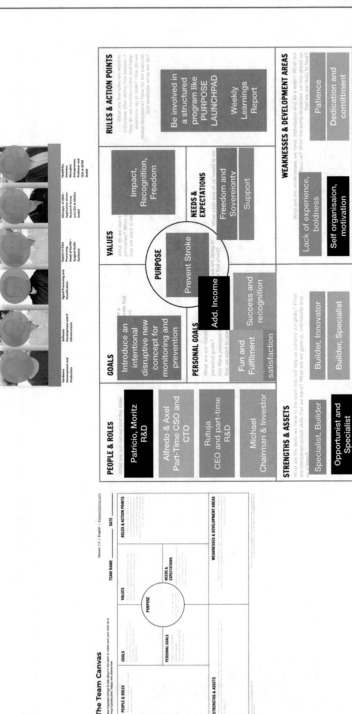

Fig. 27.7 The TEAM canvas for the BODYTUNE project helps to align the individual goals, strength, and weaknesses with the roles in the team. It helps to formulate project/team goals and values and also helps to define the purpose. Team canvas originally published under CC-BY-SA license with permission Dave Gray and www.xplane.com

Figure 27.8 shows an empty Community Canvas, as well as the three different sections. In the center the IDENTITY, the top EXPERIENCE is used to explore the community members' perspectives, while the bottom STRUCTURE deals with the operational elements needed to run this community.

In the context of Purpose Launchpad, the Community Canvas can be very powerful at the exploration phase in order to define potential communities to engage and build in the future. Also, in upcoming phases to evaluate the community approach, and to keep iterating it based on learnings.

Figure 27.9 shows the first Community Canvas for BODYTUNE. The team was surprised at the time to see—after using this tool—not only what important insights for the core project could be extracted, but how big a community around the team/project/purpose could really be.

27.4   Customer Tools

Healthcare innovations need to be centered around the key customers involved—those who are treated, those who treat or take care of others, those who offer services and products, and also those who are responsible for setting the rules, regulations, and processes. The Purpose Launchpad Customer axis focuses on customer needs and corresponding offerings.

27.4.1 Value Proposition Canvas

The Value Proposition Canvas is as a tool to ensure that there is a fit between the product and market. The Value Proposition Canvas can help ensure that a product or service is positioned around what the customer values and needs.

In the context of Purpose Launchpad, the Value Proposition Canvas can be used in different moments where the team working on the innovation is trying to better understand the customer(s) and/or design a value proposition for them.

The Value Proposition Canvas includes two building blocks:

- RIGHT SIDE—Customer profile for each customer segment to identify your customer's major Jobs-to-be-done, the pains they face when trying to accomplish their jobs-to-be-done, and the gains they perceive by getting their jobs done.
- LEFT SIDE—Value proposition to define the most important components of your offering, how you relieve pain and create gains for your customers.

A product–market fit is achieved when the products offered as part of the value proposition address the most significant pains and gains from the customer profile.

Completed Value Proposition Model provides necessary input for the Business Model Canvas.

It is highly recommended to start with the customer side, validate the assumptions/hypotheses there, and then dedicate time to formulate the value

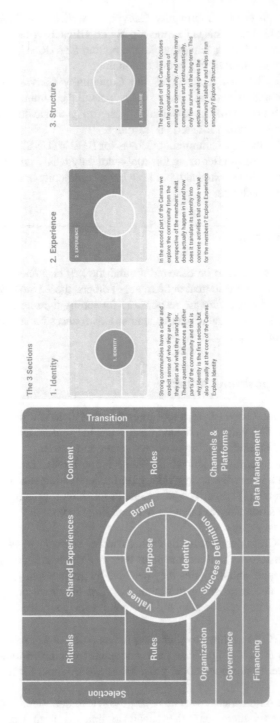

Fig. 27.8 The COMMUNITY CANVAS with its three sections and section elements. Canvas courtesy Fabian Pfortmüller and Sascha Mombartz and community-canvas.org

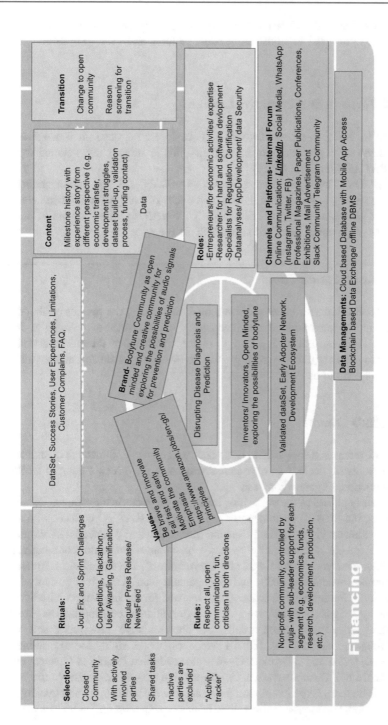

Fig. 27.9 The BODYTUNE community canvas around the purpose (center). This was a very stimulating experience with surprising results for the team

propositions (see Fig. 27.10). Of course, this process changes when you have a product and you need to adjust the product features and value propositions to the customer.

27.4.2 Persona and Empathy Canvas

In order to describe representatives of a customer segment in a tangible and personalized way, personas can be defined—this is another less detailed form of an EMPATHY Map (Fig. 27.11).

For BODYTUNE, we used a PERSONA CANVAS for understanding the headaches, pains, opportunities, hopes, and needs of our potential customers. Figure 27.12 shows the empty PERSONA CANVAS followed by three persona definitions of potential users of the BODYTUNE device.

27.5 Abundance Tools

In order to develop a scalable business model, identifying the sources of abundance that the innovation initiative can leverage is key in order to defining a proper approach. Defining and testing how to connect with those sources of abundance and how to properly manage them in order to eventually scale and make a massive impact will need to be done.

27.5.1 ExO Canvas

The ExO Canvas is a tool to help organizations connect and manage abundance in order to eventually scale in an exponential way using the most common 11 attributes that Exponential Organizations use.

In the early phases of the innovation initiative, the focus is often on the left side of the canvas, the SCALE attributes that are about the HOW TO connect with abundance: whether and how Staff on Demand (use staff when you need it), Community & Crowd (create a community to early on support you and your activities), Algorithms (can you acquire data and learn from it), Leveraged Assets (can you outsource non-essential assets), and Engagement (can you actively involve your community in the innovation process) can be used for the innovation initiative.

Being hypotheses first, they are tested and (in)validated or prioritized as shown in Fig. 27.13 with (red) green marks. As the initiative progresses and value propositions are defined, the IDEAS attributes on the right side are defined as hypotheses, tested and (in)validated—whether and what Interfaces (connections to the external side), Dashboards (what is important to measure and observe and how to communicate that), Experimentation (constant internal validation and novel hypothesis definitions), Autonomy (how free is the organization/individual in decision

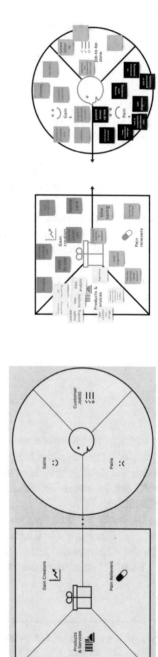

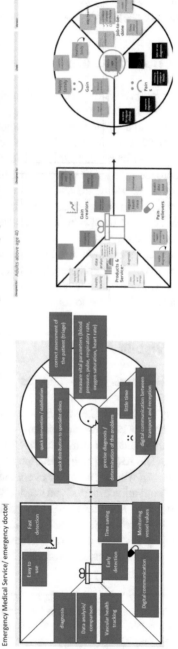

Understand the „Customer" (right side of the VPC), design a „Solution" (left side of the VPC), validate the assumptions (!!) and find commonalities

Fig. 27.10 Top left shows an empty VPC followed by three different potential VPCs for three different potential BODYTUNE customers. This is one of the most powerful and important tools in the PLH innovation generation process. You need to understand who your customer is—and often the first identified potential customer is

making), and Social Technologies (involvement and connection of the team members) can be used to internally manage the abundance.

In the context of Purpose Launchpad, the ExO Canvas helps to define, evaluate, and iterate the approach to scale in an exponential way.

27.5.2 Test and Learning Cards

Testing hypotheses along the evolution of an innovation initiative and also as part of the Experimentation attribute in the ExO Canvas is crucial to avoid waste efforts and eventually find a viable and scalable solution path.

The TEST and LEARNING cards as shown in Fig. 27.14 help to get to an objective experiment setup. It is again seen in the context of the Lean Entrepreneurship approach of Eric Ries, Build—Measure—Learn.

It focuses on testing the riskiest assumption for the innovation at hand. Falsifiable (!) hypotheses along with this assumption are formulated and the experiment setup is defined. The right hand side documents the results from the experiment, the conclusion if any and next steps.

The experiment can be considered to be successful in any case—if there is a clear confirmation of the hypotheses, the team can carry on with the assumption at hand. If not, this greatly helps to avoid waste of effort and money. And if the experiment is inconclusive—it is time to improve the experiment.

27.6 Viability Tools

27.6.1 Business Model Canvas

The Business Model Canvas (BMC) is a strategic management template used for developing new business models and documenting existing ones. It is the next step after defining and validating the customer segment and the value propositions from the VPC. The respective informations from the VPC can be entered in the BMC segments as a whole (so not for every customer individually).

In an Innovation process, you need to validate the Desirability (VPC, Persona and Empathy Map, Interviews, minimal viable prototypes), Feasibility (Is it actually possible to make? Is it realistic? Can it be done with the resources and team?), and the Viability (Is there a viable business model?). The BMC combines now all of these points in a one-page "business-plan."

Fig. 27.10 (continued) not the one that you want to and need to address—and what their problems are and gains could be to define valuable Value Propositions. Do not start with your value proposition and then analyze the customer, but do it the other way around. Especially, if you want to create a new product/service. And as always: all these points need to be validated! VPC Canvas courtesy www.strategyzer.com

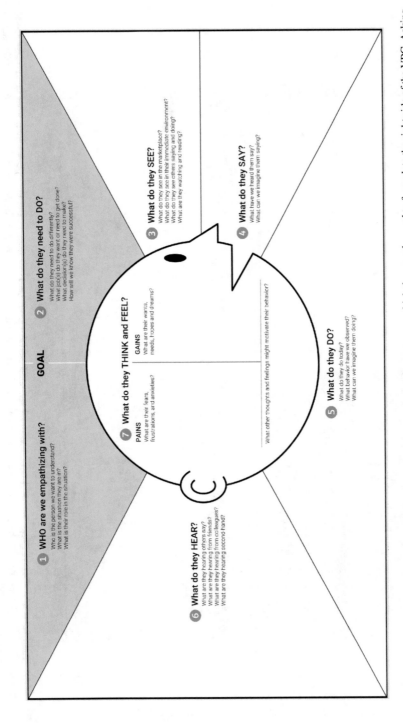

Fig. 27.11 An EMPATHY map can be used for a comprehensive analysis of a customer, which then can be used to formulate the right side of the VPC. Asking a potential customer does often not yield a correct understanding of the customer. Updated Empathy Map Canvas (https://gamestorming.com/update-to-the-empathy-map/) courtesy Dave Gray and www.xplane.com

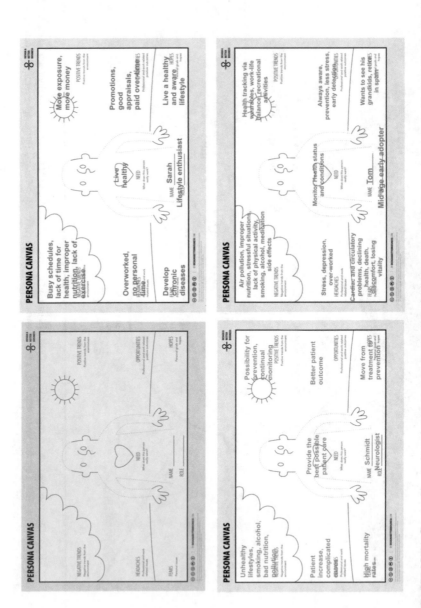

Fig. 27.12 The PERSONA Canvas (Top Left) and three identified potential BODYTUNE Users/Customers and their positive/negative trends, headaches, opportunities, hopes, and fears. This can help to understand the problem and shape a solution idea. Canvas downloaded from www.designabetterbusiness.com originally published under CC-BY-SA license

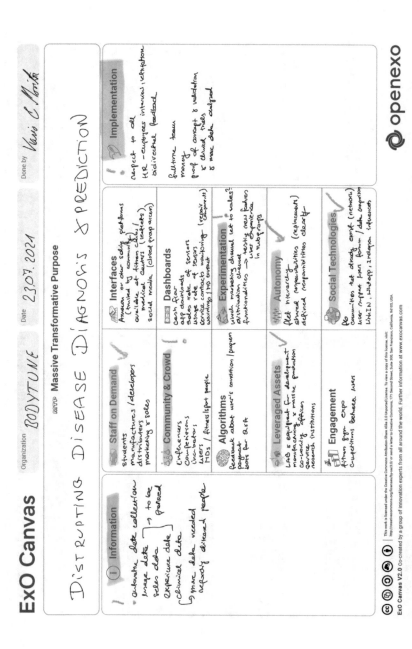

Fig. 27.13 The EXPONENTIAL Canvas for BODYTUNE with the 10 attributes plus the MTP. The left attributes (abbreviated SCALE) are dealing with external interactions, while the right ones (abbreviated IDEAS) are defining the internal operation. It is not necessary to cover all these attributes but identify the four or five that are probably the most impactful—again something that needs to be validated. For BODYTUNE the green highlighted ones were assumed to be the ones that the team wanted to focus on initially. Exponential Canvas originally published under CC-BY-SA license with permission www.openexo.com

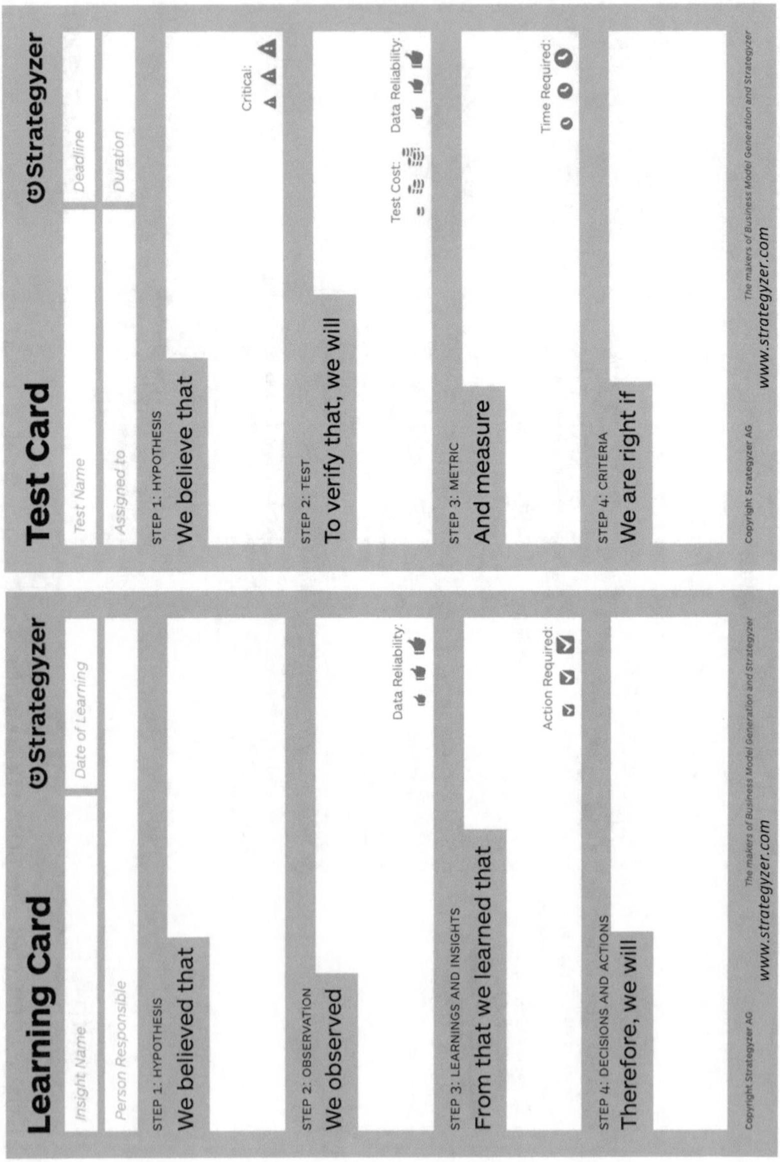

Fig. 27.14 The EXPERIMENT CANVAS is a great tool to evaluate the riskiest assumptions and helps to define the setup for an experiment to (in)validate the hypotheses. The experiment should not be designed to prove an already validated point or produce a result that would not change anything. Courtesy www.strategyzer.com

The center (VALUE PROPOSITION) lists the reason that differentiates the idea from anything that is on the market. The right side of the BMC lists the customer and how one can actually reach and interact with the customer, while the left side indicates the actual product/service implementation requirements. What resources are needed, what partnerships would be good to have, and what activities need to be accomplished.

Cost and Revenue segments help to identify the needed capital resources for developing and for the subsequent operation and present the revenue ideas.

The BMC allows to present all the needed points on one page and they also allow to quickly adapt the assumptions and hypotheses once new information has been collected or the results of validation experiments have been obtained.

Figure 27.15 shows one of the first BMC versions of BODYTUNE. The original BMC is available for download from www.strategyzer.com

It is key to think about different ways of business models to eventually find the "right" one.

27.6.2 Ethics Canvas

Healthcare innovations in particular are subject to ethical questions that relate both to the people being treated and the resources being used for treatment. In the BODYTUNE case, the ethical questions are most likely uncritical from a user perspective—a device being usable (and affordable) by anyone to self-apply and detect/prevent potential diseases. On the other hand, if the individual data is used in big data analysis and/or federated learning the privacy rights need to be ensured and the machine learning algorithms and the data used for the clustering and decision making needs to be free of any gender or demographic bias. And of course, the user needs to be informed about what anonymized data is used for generating the results. Trust in the algorithm development, data usage, and the result output are essential.

The situation is however quite different in genetic treatment, e.g., when detecting genetically caused diseases in unborn children. Figure 27.16 shows an empty Ethics Canvas that helps structure the thinking and answering the relevant ethical questions of an innovation. For more details on ethics, please refer to the dedicated chapter in this book.

27.7 Processes, Product, and Metrics Tools

Tools for the processes and product axes of Purpose Launchpad are introduced in separate sections below, as they address both how to evolve within the Purpose Launchpad approach and how to manage processes and product development and sales in the different phases of an innovation initiative.

We will still continue BODYTUNE as a case study, but introduce a second one here mainly used for the demonstration of the Product Tools. This second one is EASYJECTOR, a novel disruptive product idea for the injection of contrast media

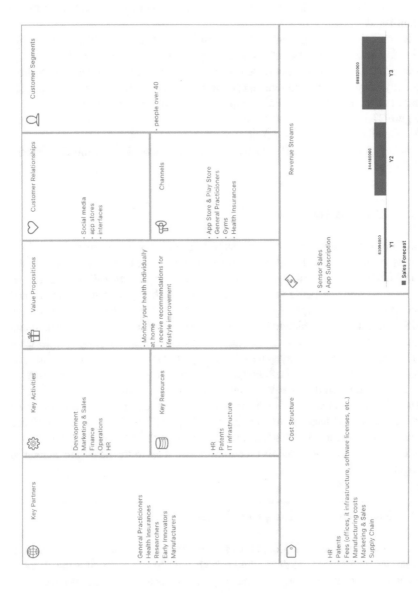

Fig. 27.15 An early version of a BMC for BODYTUNE, that only lists one customer and only two value propositions. But it already lists ideas for all the other segments that were subsequently evaluated and tested in greater detail. The advantage of the BMC is the fact that it is a living document and that it can easily be adjusted to new learnings and insights

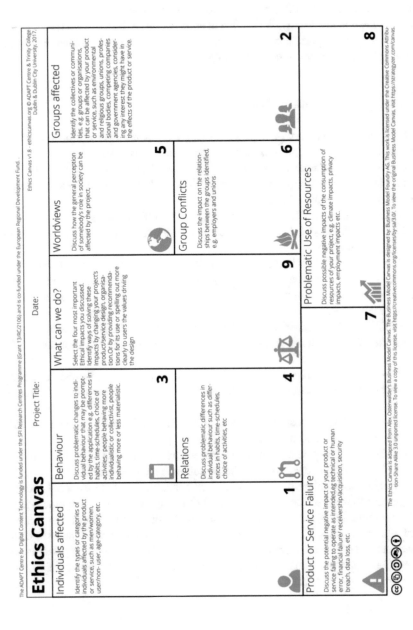

Fig. 27.16 The Ethics Canvas is especially helpful and needed for Health products that use health data for machine learning and that share this data. Bias, privacy, and other issues need to be checked. Please also refer to the Ethics chapter in this book. Originally published under CC-BY-SA license with permission Wessel Reijers, Business Model Foundry AG, https://www.ethicscanvas.org/

during diagnostic imaging. We will use this because BODYTUNE was a technology idea in search of a validated clinical application need, while EASYJECTOR is a solution for an existing application that is well defined, but with a potential innovation need around reducing procedure complexity and cost.

Following a methodology like PLH is of course a PROCESS toward the advancement of a project. Important is that you have a guideline and you manage the project including the experiments and the learning.

27.7.1 Process Segment Tools

As mentioned above a very good process tool for innovation generation is the PLH methodology that we are presenting and describing in this chapter. Additional ones that we would recommend are OKRs—Objective and Key Results. An OBJECTIVE is what you want to accomplish, while the KEY RESULT should show and describe how you will accomplish the OBJECTIVE.

An example using BODYTUNE and EASYJECTOR could be:

BODYTUNE OBJECTIVE—Prove that Audio Auscultation can be used as a Biometric Marker

KR 1—Take >100 individual audio profiles and achieve a > 90% specificity.

KR 2—Extract >10 features from a signal of less than 10 seconds duration.

KR 3—Have 10 different users acquire the signals from 10 individuals with a 90% accuracy related to the classification of a particular individual.

EASYJECTOR OBJECTIVE—Prove that the EASYJECTOR produces equivalent contrast enhancement to a conventional Power Injector 100× the cost

KR 1—Create a laboratory setup and inject a liquid into a vascular phantom using a power injector and an EASYJECTOR prototype with less than 10% variation in flow rate.

KR 2—Calculate the cost of building an EASYJECTOR with economies of scale assumptions (e.g., 10.000 or 100.000 systems per year).

Another easy-to-use and implement process tool is the KANBAN board (see Fig. 27.17 for a simple example) that helps you manage your tasks.

There are many online programs and apps available for these essential process management jobs. The Kanban board is one of the agile project management tools used by remote teams and in-office agile teams. Putting your tasks on the Kanban board and envisioning/visualizing the workflow helps teams better understand the procedures and obtain an overview of the workload. It helps in removing useless and wasteful information. You can quickly identify the problematic work stages, and your team will work efficiently.

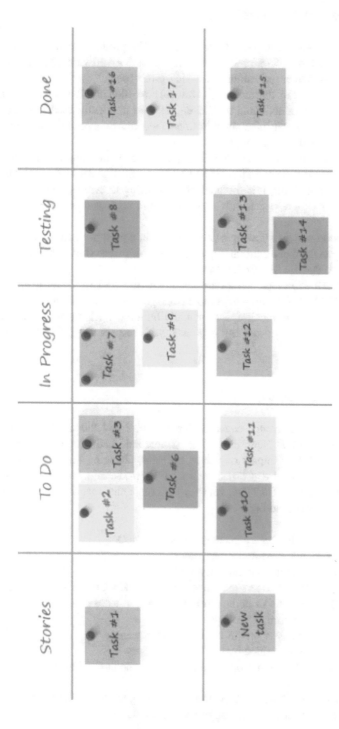

Fig. 27.17 A KANBAN board is an agile project management tool that helps you manage your tasks and To Do's

27.7.2 Product Segment Tools

With this EASYJECTOR project, we started in the product segment with a hand sketch, followed by 3D-printed demonstrators and a software app without real functionality, and then real minimal working prototypes.

These demonstrators (hardware or software) are called mock-ups. This is a model of what the final product could look like. They are frequently used to present and check a product in a real-life context or to get feedback on a product concept during the exploration and evaluation phases.

Prototyping is the process of building quick, cheap, and rough study models to learn about the DESIRABILITY, FEASIBILITY, and VIABILITY of ALTERNA-TIVE SOLUTIONS. So, it is the goal of prototyping to evaluate different designs and approaches with respect to solving the customer needs and providing the highest value.

A very good way of showing a product idea in a present or future context is to create a story in form of a cartoon. Figure 27.18 shows some stories that were created in a global FUTURE OF HEALTH event [6].

Figure 27.19 shows the prototype evolution of EASYJECTOR from using a HOW MIGHT WE question with a SO THAT goal and a simple drawing, creating hardware and software mock-ups, to actually building working prototype systems that were subsequently tested and evaluated with respect to solving the customer needs.

27.7.3 Metrics Tools

We suggest to use dedicated canvases for IMPACT and INNOVATION that are shown in Fig. 27.20. New innovation generation methodologies, exponential technologies and their impact on future business, changing business models and other factors also require a novel approach on measuring innovation outcome. INNOVATION ACCOUNTING Metrics for companies that are operating in the HEALTH segment could be:

- Annual R&D budget as a percentage of annual sales
- Number of patents filed in the past year
- Number of clinical users gained in a year
- Number of scientific papers that have used the product for clinical/research use
- Total R&D headcount or budget as a percentage of sales
- Number of active clinical projects
- Increase of active users on our clinical internet page
- Number of ideas submitted by employees or users
- Percentage of sales from products introduced in the past 5 year(s)

IMPACT ACCOUNTING refers to the purpose goals set in the MVMV canvas. Have we gotten closer on achieving our MOONSHOT?

Fig. 27.18 Cartoon stories—in this case future oriented—can be a great way to sow a product/service in actual use [6]. Originally published under CC-BY-SA license with permission

EASYJECTOR - PRODUCT SEGMENT

HOW MIGHT WE …
reduce complexity and
weight of MRI contrast
media application

SO THAT …
the cost of delivery
reduces dramatically
and the workflow
improves (time-
savings)

A

B

C

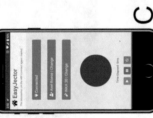

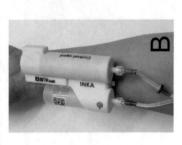

D

Fig. 27.19 Evolution of prototypes for the EASYJECTOR. (**a**) shows a first sketch that was drawn based on answering a HOW MIGHT WE question and a development goal—SO THAT -attached to. This was followed up by a hardware (**b**) and software (**c**) mock-up to show the setup to early potential customers and key opinion leaders. Real working prototypes (**d**) followed with the goal to evaluate alternative approaches

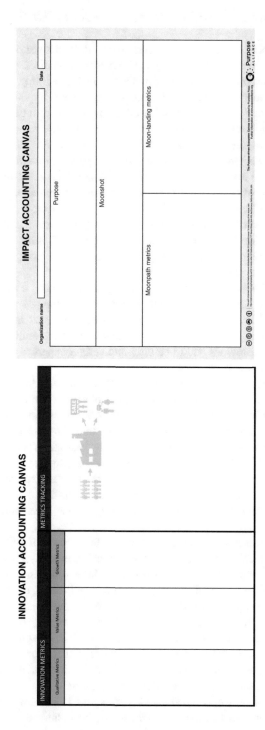

Fig. 27.20 the INNOVATION ACCOUNTING CANVAS (left) and IMPACT ACCOUNTING CANVAS (right) that can be used in the METRICS segment of the PLH. Originally published under CC-BY-SA license with permission www.purposealliance.org

27.8   Radar, Backlog, and PLH Results—BODYTUNE

The BODYTUNE project underwent 15 sprints so far over a period of 6 months. Initially, every week on a Monday starting with a presentation of the things that were attempted in the previous week and a discussion on the learning and validation (or not) of certain product, customer, or business model related hypotheses. This was typically a 20–30 minute Powerpoint followed by questions from the mentor and a check at the backlog that the team was supposed to work on.

Please note that is—just like in a SCRUM or other agile process—the responsibility of the team to decide which items are highest priority and therefore should be done first. They are even encouraged to add items that were not on the backlog TO DO list, but that just came up and required immediate attention.

Together with the Mentor, the individual Learnings are discussed and a point is awarded for every achieved new insight and two points for a validation of a hypothesis. A validation is here either positive or negative. It is also a very important, maybe even more important, learning if a hypothesis or an assumption turns out to be wrong. Embracing failure is important.

Figure 27.21 shows the Purpose Launchpad Progress (PLP) chart for BODYTUNE for the first 13 sprints with anywhere between 4 and 10 points of progress per sprint. The sprint duration as mentioned above was initially 1 week, which was gradually increased to 2 weeks and later to even 1 month due to the fact that minimal viable prototypes had to be built and certain experiments simply took more time to prepare and to conduct.

You also see in Fig. 27.21 that it took 5 sprints for the team to evolve from the Exploration phase and to enter the evaluation phase. Even after 15 sprints, BODYTUNE is still in this phase and most likely will remain there till first sizable revenues are being generated. This also requires that the project evolves to a commercial operation and that the university team evolves into a start-up company. In the case of BODYTUNE, it led to considering an alternative going to market strategy by working on a lifestyle product first that requires no regulatory approval and can be sold much faster than a certified medical product. That would also generate valuable datasets early on and possibly create loyal and supporting customers.

Also important is to monitor the team's mood. In the case of BODYTUNE, it started low (not clear on the process and expectations), was really good from sprint 3 to 9, and then dropped noticeably because of little progress. This was then addressed and the mood normalized again. A great validation of measuring the mood throughout a sprint process that goes over a longer period of time.

The next step is to then use the online assessment tool with the questionnaire for every one of the 8 segments (see Fig. 27.22 for the PURPOSE segment) to create a visual radar and to use that to determine the next things that should be targeted or resolved. It is also good practice to leave unresolved, but important items from the last week on the Backlog/To Do list for the coming sprint.

The radar and the newly developed backlog for sprint 15 is shown in Fig. 27.23. The visual radar shows the 4 action items that should be addressed in the upcoming

Sprint	Start	1	2	3	4	5	6	7	8	9	10	11	12	13
Progress	3,3	5	6	6	8	9	8	7	8	10	9	3	8	4
Total Progress	-	5	11	17	25	34	42	49	57	67	76	79	87	91
Mood	3,3	3,0	2,8	4,3	4,5	4,3	4,3	4,2	4,0	5,0	3,8	4,0	4,1	4
Evolution Phase	-	Explore	Explore	Explore	Explore	Evaluate	Evaluate	Evaluate	Evaluate	Evaluate	Evaluate	Evaluate	Evaluate	Evaluate

Purpose Launchpad Progres Graph

□ Explore □ Evaluate ■ Impact — Total Progress — Mood

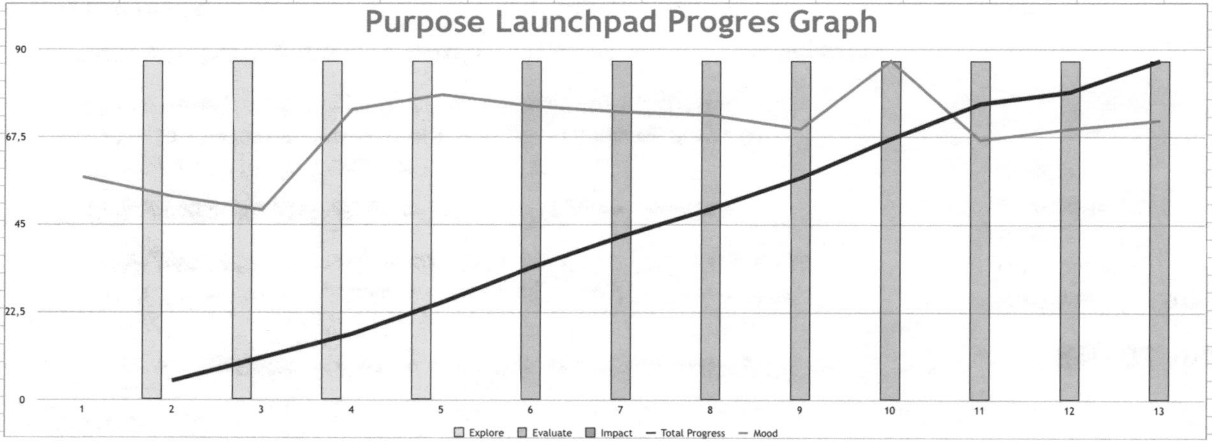

Fig. 27.21 The BODYTUNE BODYTUNE Purpose Launchpad Progress Graph. You see that it took BODYTUNE 5 sprints to evolve from the exploration to the evaluation phase. While there was continuous learning over the sprints the mood of the team varied slightly but noticeably. To advance to the Impact/Implementation phase will still take quite a bit longer, as this requires a validated business model and that on the other hand requires not a prototype, but a product. Especially in the healthcare field difficult to achieve quickly. But with alternative going to market strategies—e.g., creation of a lifestyle product first rather than a certified medical product—First revenues and business model validations achieved significantly quicker. Also something that BODYTUNE is considering now. Courtesy www.purposealliance.org and Francisco Palao

PURPOSE LAUNCHPAD

Purpose

Do you have a Purpose or Massive Transformative Purpose (MTP)?

○ No yet
○ Yes, but we have not defined our Moonshot, Vision, Mission and Values
○ Yes, and we have also defined our Moonshot, Vision, Mission and Values

Have you done any real progress in relation to your Moonshot?

○ Not yet, and we don't even know how to do it
○ Not yet, but we believe we know how to start getting real progress
○ Yes, we have started to get some real progress towards our Moonshot

Have you already generated a (massive) impact in relation to your Purpose?

○ Not at all
○ Yes, we are on track and moving with high speed towards achieving our Moonshot
○ Yes, we have already achieved our Moonshot and we may define the next one

Back   Next

Key Definitions

- Purpose and MTP
- Moonshot
- Vision
- Mission
- Values

Fig. 27.22 The Purpose Launchpad Assessment, sample questions for Purpose dimension. Every segment has such a questionnaire page that is subsequently used to create the visual progress radar (see Fig. 27.6). Source: https://members.purposealliance.org/assessment/new/- courtesy www.purposealliance.org and Francisco Palao

Evaluation phase – BODYTUNE – SPRINT #15	
Sprint Backlog	
Story	People assigneed
Three Customers - create 3 Elevator Pitches and validate the assumptions. Goal is to create the most important Value Propositions for each	Team assigns Individuals
Create a Marketing / Sales document (fake) for each of the customer segments and conduct experiments to find out more about the value that this product / service could have	Team assigns Individuals
Exponential Canvas - pick the 3-4 most important attributes (Engagement, Crowd&Community, Experimentation, ...) and use it for BODYTUNE. E.g. lifestyle will require Early Adopters that will help collect data, engage them and motivate them (tokens, gamification, rewards, shares, ...)	See above
Interfaces / Dashboard - how can you connect BODYTUNE to other devices / sensors / software for a DIGITAL TWIN. If our data is valuable then others must be interested. They may then do more then we can do (Staff on Demand, Experimentation). Think on how you can scale without adding expenses on the Bodytune side.	See above
Milestone Plan - when is what finished and when are products available. Make assumptions with respect to revenue streams, but think a little out of the box - not too far off from what we have, but a normal subscription model is probably not it for the lifestyle!	See above
Define the value for the Medtec product and the cost of the product. € 1.000 per device? Spot sales? Every ambulance should have one (22.000 in Germany alone)? Then we are talking several thousand per year in Germany alone.	See above
Finance Plan with Details on a quarter base. Gross Margin (product revenue - product cost) for each of the segments multiplied with the number of systems or subscription months or ... = revenue. Cost is (lets start with that) - staff (number of employees x € 75k - seperate for General Managment, Engineering, Product Support, Marketing & Sales, Adminsitration), office expenses (use 1k / month in year 1, and increase linear with number of employees), Patent and Legal expenses, Misc., Marketing expenses = total expenses (you also have depreciation and finance expenses, but that we can add later. Be ambitious, but realistic - it needs to be explainable. With that you can calculate the cash / investment needs.	See above
Do Innovation and Experimentation Accounting. Do something planned and measure the response compared to your expectations. Social Media posts .... measure increased traffic. And many more ...	See above

Fig. 27.23 The BODYTUNE Radar of sprint 15 showing clearly the segments that require further exploration and validation work. The color-coded backlog addresses the work for these individual segments

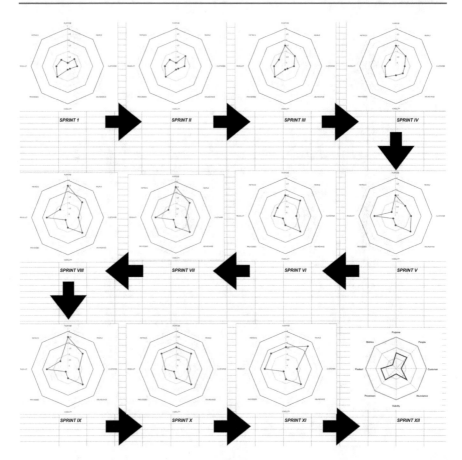

Fig. 27.24 The BODYTUNE Radar over the first 12 sprints. Visually shows the fast progress in the first 5 sprints and subsequently either no progress or even a setback. Sprint 12 looks different due to a change in the assessment tool

sprint. In this example, it is obvious that the customer (red), abundance (orange), viability (green), and metrics segments need to be worked on. The backlog list items and to-do's that should create learning and validate hypothesis.

Looking over the last 13 sprints (Fig. 27.24, but also see Fig. 27.21) the visual radar shows fast improvements in the first sprints, slower progress in the subsequent sprints and even some setbacks. The learning is however continuous and shows— also to the team—that invalidating assumptions is just as important as validating them.

27.9   PLH Summary

In this chapter, we explained the Purpose Launchpad Health methodology for defining Purpose and following methods for innovation and mindset generation from an initial EXPLORATION phase onward to an EVALUATION phase with the goal to validate the product, customer, business model, team, and scalability assumptions and hypotheses. A successful continuation in a subsequent IMPCT phase would typically result founding a start-up or funding a project venture inside a larger operation. Health-related innovations would need to now also work with regulatory issues and the national differences and put a lot of effort in creating an impactful product–market fit. Other helpful innovation tools for the IMPACT phase and additional agile methods will be presented in subsequent chapters.

Almost all presented tools are also available for download in [5], but will additionally be provided in the TOOL section of the book in Part 11.

The PLH should also enable the team to create a one-page summary page (already shown in Fig. 27.3) of the initiative as shown in Fig. 27.25 for the EASYJECTOR that highlights the customer and market need, the technology solution and the roadmap, plus describes the business opportunity and impact that is believed to be generated. It also lists the team members and an ELEVATOR PITCH sentence that is derived from the Value Proposition Canvas for the most prominent customer.

The PLH is continuously evolving and is undergoing subtle changes and adaptations as suggested by the community that drives the process. In the newest version, the VIABILITY segment is renamed SUSTAINABILITY and now ahead of the ABUNDANCE segment. This makes a lot of sense as the Business Model, showing that the project is sustainable, needs to be initially defined before you can use the attributes of the exponential canvas to find a connection to the ABUN-DANCE. Please also see Chap. 21 on that one. Figure 27.26 shows the newest layout of this innovation framework.

27.10   Remark with Respect to the Use of the Tools

The permission to use the presented templates has been obtained for this book. If no author or webpage is listed then the tools were developed or created by the Editor/Authors. Most can be freely used under Creative Commons license 3.0 or 4.0 BY NC SA. For license details, see https://creativecommons.org/licenses/by-nc-sa/4.0/.

Please apply the proper referencing when using them. The Editor and Authors of this book have obtained permission from the authors to use the tools in this book. For any other use outside the boundaries of the license, you may need to contact the authors directly and obtain permission.

The ones without the CC license can be freely used for anything provided that this chapter is properly referenced.

EASYJECTOR

Our ____
help(s) ____
who want to ____
by ____
and ____.
(unlike) ____

Team and Support
Lead: michael@easyjector.com thomas@easyjector.com
Member: stefan@easyjector.com axel@easyjector.com
Sponsor / Mentor : julia@easyjector.com isabella@easyjector.com

Elevator Pitch

"My team EASYJECTOR will develop A ONE-FLOW CM INJECTOR to help PRIVATE RADIOLOGISTS and RADIOGRAPHERS a HIGH STANDARD MRI IMAGING through SIMPLIFICATION unlike THE BIG AND COMPLICATED SYSTEM with their HIGH PROCEDURE COST and HANDLING ERRORS ."

Opportunity

Background
Contrast Media is applied (compliance and forensic issues) often by hand or with very expensive (CAPEX + OPEX), complicated power injectors that also take a lot of time to setup and service.

Customer
Hospital or Private Practice based Radiology centers with CT / MRI / … Inside these the main customer / decision makers are the radiographers and the radiologists together with the business and finance manager.

Solution
We provide a solution with a lightweight arm-mounted device that saves setup time (approx. 5-7 min per patient), reduces consumable waste significantly, and is dramatically cheaper (OPEX less than half, CAPEX 100x cheaper)

Technology

Key Technologies
Smartphone APP with an AI interface
3D printing (for the prototypes)
Nitinol based switch

Readiness Level
TRL 4 / TRL 5 - working prototypes tested in the MRI and CT, but not on patients.

Roadmap
File more patents for key technology aspects - now and continuous

Develop a device based on prefilled syringes for clinical studies (animals) - 6 Month

Acquire the needed funding (€ 1 Mio) to hire the team and get operation in place - 6-9 month

First product ready in 24 month (including certification)

Impact

Value for Customers
Much easier to use and to manage plus saves time and increases productivity

Revenue START-UP (Est. 1 yr / 5 yr / …)
1 yr: 0, 3 yr: 1 Mio, 5 yr: 12 Mio

Strategic Relevance
No significant invention in that space for several decades. Cost as a driver for innovation is getting more important.

Scaling Potential
Use for other modalities (Ultrasound) and other applications (Radiopharmaceuticals) and especially in Asian countries

Further Benefits
Ordering and inventory warehousing can be automated. Possibility to connect to existing billing system.

Fig. 27.25 A simple One-Pager for the EASYJECTOR project. It uses a simple ELEVATOR PITCH sentence derived from the VALUE PROPOSITION CANVAS (for the most important customer and the Value Proposition that is considered the most prominent one). With the three headlines OPPORTUNITY/ TECHNOLOGY and IMPACT it summarizes very well on one page the identified customer need, the way you want to solve that, and how you will achieve that plus the business opportunity behind that

Fig. 27.26 The newest Purpose Launchpad framework with the segment VIABILITY now called SUSTAINABILITY and being ahead of the ABUNDANCE

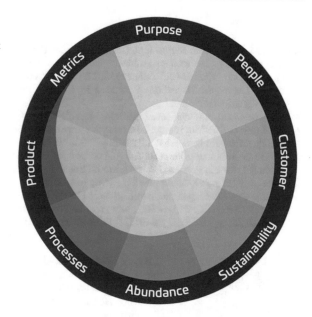

Take-Home Messages
- We need to invest a lot of time initially to understand the problem and define a purpose for our actions and ambitions.
- Only when we understand the depth of the problem for the community of users we can define a solution that solves the clinical issues.
- We often do not validate our hypotheses and too quickly jump to create solutions that will subsequently likely fail because the clinical and user need was not properly checked.
- PLH provides a set of tools for the 8 segments of the Purpose Launchpad for a first definition of the clinical needs, the health stakeholders, solution ideas based on the first Health-related innovations in an EXPLORATION phase.
- Following the method and using the tools will significantly decrease the risk of failure and help to gain innovator, health stakeholder, and investor confidence.
- The visual radar obtained after a questionnaire-based assessment helps to define the next steps for the innovation progress.
- After several sprints and iterations, you should be able to create a one-page initiative summary (with Elevator Pitch, Business Model, Customer Need, coarse Roadmap and technology solution), which is extremely valuable for the subsequent discussion to obtaining financial support.

References

1. Friebe M, Fritzsche H, Morbach O, Heryan K (2022) The PLH - purpose Launchpad health - meta-methodology to explore problems and evaluate solutions for biomedical engineering impact creation. IEEE EMBC conference 2022, Glasgow, conference paper, doi.org/XXX.YYY
2. Purpose Launchpad Repository download in English https://members.purposealliance.org/purpose-launchpad/resource/
3. https://members.purposealliance.org/purpose-launchpad/tools/
4. Ries E (2011). The lean startup: how today's entrepreneurs use continuous innovation to create radically successful businesses. ISBN-13: 978-0307887894
5. Massive Transformative Purpose, The guide to provide sense to your rojects and your life, Ángel Manria Herrera and Francisco Palao, Editorial Bubok Publishing, January 2021
6. Friebe M, Hitzbleck J, Merkel K (2021). Future of Health SCIFI Hive Event - Cartoon Book. https://www.researchgate.net/publication/351109123_FUTURE_of_HEALTH_%2D%2D_CARTOONBOOK_%2D%2D_story_results_from_the_global_SciFi_Hive_Event_March_20_2021_%2D%2D_PREDICT_PREVENT_PERSONALIZE_PARTICIPATE

Michael Friebe received a B.Sc. degree in electrical engineering, M.Sc. degree in technology management from Golden Gate University, San Francisco, and Ph.D. degree in medical physics in Germany. He spent 5 years in San Francisco, as a Research and Design Engineer with an MRI and ultrasound device manufacturer. He is a German citizen with expertise in diagnostic imaging and image-guided therapies, as a/an Founder/Innovator/CEO/Investor and a Scientist. He is also a Research Fellow with the Technical University of Munich, Munich; an Adjunct Professor with the Queensland University of Technology, Brisbane; and an honorary Professor of image-guided therapies with Otto von Guericke University, Magdeburg, Germany. Since 2022, he has been a Professor of biomedical engineering innovation with the AGH University of Science and Technology, Krakow, Poland. He is a listed inventor of more than 80 patents and has authored over 200 papers, and was part of over 35 Medical Technology Start-Ups. He is a Board Member of four medical technology start-up companies and an investment partner of a MedTec-fund. From 2016 to 2018, he was a Distinguished Lecturer of the IEEE EMBC teaching innovation generation and MedTec entrepreneurship. He is also a coach and trainer for OpenExO, and a master Launchpad mentor for the Purpose Alliance.

Oliver Morbach is Managing Director of pro.q.it Management Consulting GmbH. He is a coach, consultant, trainer, experience designer, and mentor working to enable co-creative processes and environments that unleash people's innovative potential in line with their purpose, by leveraging technology, lean-agile methodology, and exponential growth mindset. Over the last 25+ years, he held various senior and executive management roles in customer service and information technology organizations in the European ICT sector. Prior to that, he worked in defence software engineering. He holds a computer science degree (Diplom-Informatiker) from the University of the German Armed Forces Munich.

Part VIII

Health Leadership, Skills and other Methodologies

Soft Skills Needed for Problem Understanding and Innovation Generation

28

Sys Zoffmann Glud

Abstract

This paper seeks to illustrate what types of soft skills are needed in the different phases of the innovation process.

Attention here is directed mostly to the soft skills important for getting started with innovation and it is subsequently explained how the lack of certain soft skills, such as self-efficacy, can prevent people from engaging in innovative and entrepreneurial activities even if they clearly see an appealing problem to solve.

While it would become too comprehensive to provide a full list of soft skills needed for that, it will become clear that various skills are needed and that indeed innovation is a team act. *Not one person alone holds all the people skills needed.*

Keywords

Soft skills · Problem-solving · Empathy

> **What Will You Learn in This Chapter?**
> - Soft Skills needed for successful Health Innovation
> - Needs Driven Innovation as a structured and purposeful way of opportunity identification
> - How to practice empathizing.

S. Z. Glud (✉)
BioMedical Design, Institute of Clinical Medicine, Faculty of Health, Aarhus University Denmark, Aarhus, Denmark
e-mail: sys@clin.au.dk

28.1 Professional Expertise Is Highly Relevant, but Not Enough for Innovation to Happen

The ability to come up with great ideas new to the world (creativity) and the ability to develop and implement these ideas (innovation) are much studied phenomena that still leave researchers to debate the definitions of the words: creativity, innovation, and entrepreneurship.

In particular, the human component and the contribution of personality to successful outcomes still seem to be a bit of a black box [1].

Yet, everyone recognizes that the human ability to innovate and constantly solve ever more complex problems is the key to not only business success and national economies, but also for humanity to thrive and survive.

Gradually more understanding is gained about what personal skills successful innovators hold. There is a famous publication from Harvard Business Review 2009 titled "The Innovator's DNA" [2], which examines what makes some senior executives more innovative than others and the companies they lead more successful. The Innovator's DNA study concludes that five discovery skills distinguish the most innovative entrepreneurs from other executives. The authors used the metaphor of DNA to describe the connection between the five skills.

There are four patterns of action/doing behavior (questioning, observing, experimenting, and networking) that form the bases of the DNA. These are connected to each other via the DNA backbone which is a thinking behavior (Associate) that combines the information from the actions (questioning, observing, experimenting, and networking) into a synthesis that informs new insights and acts as fuel for innovation in the individual.

The authors claim that these skills can be cultivated and trained in people not born with them. While this still needs to be proven scientifically, the mentioned skills can be recognized as core elements of popularized innovation processes such as design thinking and the various needs-driven innovation programs. This will be elaborated below together with examples of other soft skills experienced to be critical for creativity and innovation.

Strong insights and yet unpublished empirical evidence of the importance of various soft skills in innovation are gained in the observation and follow-up study on participants in the BioMedical Design Novo Nordisk Foundation Fellowship Program from Denmark. The findings are underpinning existing published research, but in a more long-term perspective and in a much more interdisciplinary setting than usual for most research studies of innovation and entrepreneurial competences.

Figure 28.1 illustrates some of the many soft skills needed for at least the first half of the innovation pathway which is related to the more creative part. A range of additional soft skills are needed in the second half and are more related to the implementation of the innovation and integration into a running business/organization. The latter will not be elaborated on here.

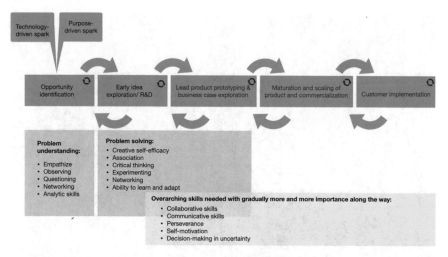

Fig. 28.1 Some of the soft skills needed for the Innovation process, especially for the initial creative part. For maturation different skills are required that are not part of this paper

28.2   Opportunity Identification

Opportunity identification is the ignition point for any innovation journey. It involves being receptive for new perspectives and ideas and being receptive for seeing the imperfections of current status, i.e., **noticing disharmonies and a willingness to question the status quo**.

Many people can identify things that could be better, but most just sweep away the thought by accepting that this is probably how it just happens to be.

The initial catalyst for eying an opportunity can be based on a technology-driven spark or a need-driven/Purpose-driven spark. In either case, uncovering an opportunity involves a process of building problem-understanding, that is, getting the problem definition right since this is key to getting a fruitful start of the innovation process and to optimize for success.

28.3   Needs-Driven Innovation Is a Structured and Purposeful Way of Opportunity Identification

At the core of needs-driven health innovation is the ability to **empathize** with patients, healthcare professionals, and secondary, yet critical, stakeholders of a given healthcare treatment and process.

Empathizing is also step one in the widespread methodology, Design thinking, used by design bureaus worldwide, and on which needs-driven innovation methodology is grounded.

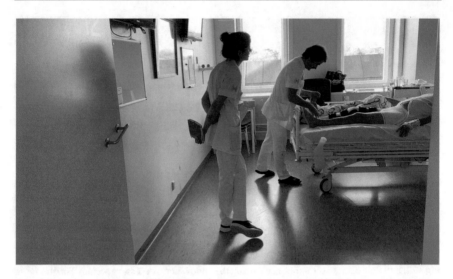

Fig. 28.2 A biomedical design fellow during clinical immersion at a hospital. (Photo: Katrine Lind Møgelgaard)

But Why?—why should we be able to understand and share the feelings and perspectives of patients and healthcare professionals?

Because no technology, simple or highly advanced, will work and create value any better than how it is being used by people. And humans are not rational (even though we like to think of ourselves as being so) and are very little compliant to whatever technology demands us to do.

Good technological solutions adapt to and fit with (work-)lives lived. And in situations where lifestyle or behavior needs to change, technology shall support in a motivating way.

To uncover lived (work-)life and motivations behind the multitude of different behaviors that patients and health professionals displays, one needs to study this by spending time alongside the involved people and try to understand (**empathize**) them (Fig. 28.2). Trying to see the perspective of others includes to also be sensitive to non-verbalized cues of likes and dislikes, taboos and not at least attention to own biased interpretations.

Note, when immersing yourself into a day in the life of a patient or following a healthcare professional it is important to hold a high ethical standard, e.g., make sure to have consent and use empathy to sense if you should withdraw from a vulnerable situation that suddenly arises or whether you can still stay to see the situation play out.

28.4   How to Practice Empathizing

For a start, empathizing involves **observing** the surroundings (listen —see —smell —feel, use all your senses), following what people do, say, do not say, how they look/react,—and gradually as it feels appropriate start asking naïve questions, "why is this and that important?" "Have you always done it like that?" "what else have you tried?", ask "how does it help you, in getting your jobs done when you do this task?"—observe differences in saying and doing as here lies often clues to what is actually needed and wanted.

As a natural elongation of empathizing follows **questioning** what is seen; asking "does this really create value for the people involved?" "is this satisfactory?", ask for people's personal experiences "What more could be achieved (better, faster, cheaper)?", then one **network** to get more answers and by applying **analytical skills** the unmet need to be solved is materializing.

28.5   Case Example

An observation in the pre-operative area: an elderly lady is being prepared to undergo acute operation for a broken hip caused by a fall in the home.

The anesthesiologist is troubled as she is lacking a good, precise, and easy measure of the patient's hydration level before administering the sedation.

The anesthesiologist is asking the nurses if the lady fell due to dizziness which could point to dehydration. But no one knows for sure and the lady herself is not providing reliable answers due to her acute situation.

Annoyed by the situation the anesthesiologist says out loud to the observer, "All kinds of interesting and advanced equipment have been developed, but none of them are intuitive in the daily work."

After the operation, the Observer pursues the observation about the lack of information around hydration status. Talks with geriatric doctors reveal that they have frustrations about restoring hydration levels after operations, because anesthesiologist tends to under-hydrate the patients to make the sedation work, which leaves the geriatrics with the task to make the kidneys work again after the surgery.

And down another pathway, the Observer investigates how often elderly people get dehydrated, the causes of dehydration, existing solutions gap, and how it is being handled by homecare people.

Then, working with patient journey mapping (see Fig. 28.3) and matching qualitative data as well as available numerical data and research literature, an opportunity of an important unmet need becomes clear and need-criteria that any solution shall meet can be defined.

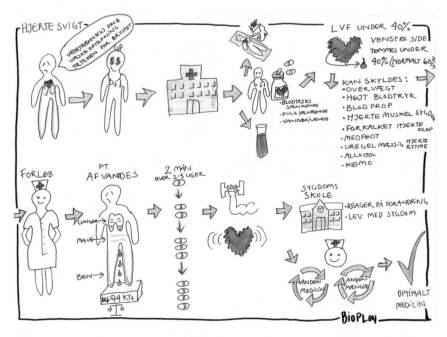

Fig. 28.3 An example of a patient journey mapping based on following heart failure patients. (credits: Kristine Kuni Buccoliero, BioMedical Design Fellow)

28.6   Knowing the Problem Is Not the Same as Daring to Solve It

However, being able to identify an opportunity is not enough in itself. People will not proceed with ideating and building a solution if they do not believe that they are capable of it. This is where **creative self-efficacy** (the belief one has in the ability to produce creative outcomes) is key. Many people find themselves full of ideas, but just as many find themselves not able to come up with new ideas. The latter can train this ability and improve their creative self-efficacy by involving themselves in creative process together with others and by learning different tools to stimulate creative thinking processes.

Idea novelty: The ability to **associate** seemingly distant information about how things work is a valuable ability to generate ideas of more radical character and true novelty (such as for instance the robotic pill from Rani Therapeutics—see Fig. 28.4).

The creative part of the innovation process is a frequent shifting between divergent and convergent thinking. As part of convergence **critical thinking** is key in order to select the right solution ideas to the problem. So is also, **networking** to collect the knowledge you do not have at hand and **experimenting** with many different solution ideas in rapid, iterative loops and combined with an **ability to learn and adapt** from both experimenting and the knowledge/information gained

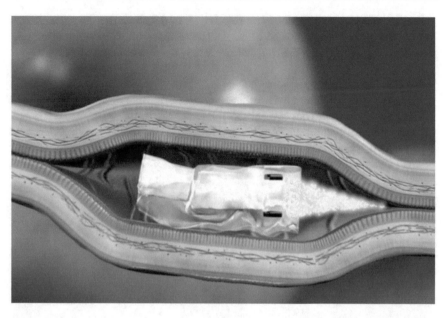

Fig. 28.4 A radical innovation from Rani Therapeutic which holds great potential to free thousands of patients from the pain and troubles of injectable drugs. (source: https://www.ranitherapeutics.com)

from the network,—i.e. also all the negative push back, rejections, failed experiments and so forth.

The process of innovating new solutions to complex problems in healthcare is a long and bumpy road and it is not completed with just inventing a cool device. It also has to be produced, sold, and implemented at the user-side.

Overarching soft skills that will increase your chances to make it all the way is a **collaborative mindset** in combination with skills and experiences of working in teams. Innovation is not a one-man job; hence, teamwork and collaboration are essential.

Perseverance is needed to overcome the many turndowns and obstacles that will come to you along the innovation journey. In that respect, **emotional intelligence** is likely to play an important role for success with managing an innovation process [3].

Good **communication** skills: in the first long period of an innovation project only you and your teammates believe in the idea and in the team, hence to convince other people to join, to help out, to invest, and to buy you need to be good at communicating what you aim to achieve, why and how, and why you and the team are the best bet. Good communication skills are also a critical competence for internal teamwork and for motivating employees.

Any innovation process is filled with ambiguity and uncertainty, the ability to not freeze up but continue to **make decisions** that moves you another step forward is critical, yet very difficult for many people.

Lastly, any innovator needs to be **self-driven and self-motivated** since there is no boss to tell you what to do and when. And there are no other than your teammates to pick you up when you are down, and likely they are down at the same time as you.

Reflection on Action Questions
- Where in your neighborhood are there innovation and entrepreneurship programs where you can train your soft skills?
- What is happening around you? When have you last been observing the world around you? Put the phone away and start seeing and listen to your surroundings and hereby stimulate your curiosity and your ability to question the status quo.

Take-Home Messages
Many diverse skills are needed during an innovation process. Some are related to:

- Abilities to interact productively with other people, teammates as well as externals
- While some skills are of more interpersonal abilities needed for obtaining, processing and applying information
- Pure subject professional knowledge and competencies are not enough.
- Being receptive for opportunities, and having the ability and willingness to learn from feedback is key to growing innovation skills.

Further, many soft skills can be stimulated and trained by participation in hands-on innovation and entrepreneurship courses and training camps/programs.

References

1. OECD Publication (2011) Skills for innovation and research. OECD Publishing, Paris. Accessible https://doi.org/10.1787/9789264097490-en
2. Dyer JH, Gregersen HB, Christensen CM (2009, Dec) The Innovator's DNA. Harvard Business Review. Free access via https://hbr.org/2009/12/the-innovators-dna
3. Hess JD (2014) Enhancing innovation processes through the application of emotional intelligence skills. Rev Public Adm Manag 02(01). https://doi.org/10.4172/2315-7844.1000143

Sys Zoffmann Glud, Managing Director of BioMedical Design Novo Nordisk Foundation Fellowship Programme, Sys Zoffmann Glud has graduated from Aarhus University with an MSc in Molecular Biology and PhD in Nanomedicine. Leaving the research career behind, Sys has since 2012 worked in the field of healthcare innovation and entrepreneurship applied through university courses and continuing education training programs. Through her work she has trained +300 students, healthcare professionals, and industry experts in needs-driven healthcare innovation and commercialization, resulting in more than 10 startups and attractive innovation talents who keep applying the methodology they have been trained in for years.

Leadership in Healthcare: A Novel Approach. Healthcare Executives' Traits, Styles, and Approaches

29

Beatrice Barbazzeni

Abstract

Leadership is discipline and a set of skills regarding the ability to lead individuals toward achieving goals and in which different styles of directing, guiding, deciding, and motivating a group are recognized. Considering healthcare, a dynamic system, to face the twenty-first-century challenges brought by Industry 4.0 and digital transformation, a novel and effective leadership approach to properly manage the system and guarantee high-quality care should be considered. Hence, a new set of skills is required which includes integrity, active listening, effective communication, technological expertise, teamwork and networking, collaborative and trusted relationships, management and business, creative thinking and shared vision, as well as, confidence and awareness. Moreover, revised educational and training programs are also needed to educate future leaders in responding and adapting to diverse healthcare situations with high efficiency, time management, and flexibility. Nevertheless, innovation has also to face challenges to become successful leaders in the context of digital healthcare.

Keywords

Healthcare leadership · Industry 4.0 · Digital transformation · Education · Effective communication · Team empowerment

B. Barbazzeni (✉)
ESF-GS ABINEP International Graduate School, Otto-von-Guericke-University, Magdeburg, Germany
e-mail: beatrice.barbazzeni@med.ovgu.de

What Will You Learn in This Chapter?
- Leadership is a research area but also a set of skills to influence, motivate, and inspire followers.
- Leadership styles indicate different approaches in directing, guiding, deciding, and motivating a group toward a goal-directed behavior.
- Considering healthcare, a complex system a novel and effective leadership approach should be developed according to Industry 4.0 and digital transformation.
- Integrity, self-confidence, awareness, active listening, team player, creative thinking, technical and financial are just a few skills that are needed to become a successful healthcare executive.
- Revised educational and training programs should be promoted to educate future healthcare leaders able to manage and overcome twenty-first-century challenges.

Leadership can be defined as a research area and a set of skills indicating the capability of an individual, a group, or an organization to lead and influence others. Hence, leadership represents a goal-directed behavior in which different styles, such as giving direction, planning actuation, inspiring, and motivating group activities, should be considered [1]. Indeed, different leadership styles were identified in several areas (e.g., military, political, business, government). Although only a few studies considered leadership in healthcare, most theories developed for business sectors were also adopted in healthcare. Therefore, innovative theories and styles suitable to face healthcare challenges are needed to improve patient care, system organization, and management.

If leadership styles indicate specific characteristics and behaviors of the leader in guiding, motivating, and managing the team to achieve goals, different theories were developed to identify these styles. Initially, the trait theory of leadership was proposed to detect recurrent personality traits associated with successful leaders when interacting with different situations and environments [1]. Moreover, besides the well-known leadership styles of Lewis [1] and its further extension [1], a recent approach was also developed [1] to identify those traits observed in high-quality leaders. These traits include adaptability and flexibility, assertiveness, capacity to motivate people, courage and resolution, creativity, decisiveness, eagerness to accept responsibility, emotional stability, intelligence, action-oriented judgment, need for achievement, people skills, self-confidence, task competence, trustworthiness, and understanding followers and their needs. Furthermore, different leadership theories were also developed over the last century to explain how certain individuals become successful leaders. From the early Great Man theory approach to different behavioral (e.g., authoritarian, democratic, and laissez-faire), situational, and contingency approaches, recent theories based on a supportive type of leadership were also proposed valuing interactions, trusted relationships, team empowerment, and job satisfaction [2].

In a context such as healthcare, composed of multiple interactions among multi-disciplinary professionals, departments, and specialists, leadership needs to manage and organize this complex system with high efficiency and effectiveness to guarantee the achievement of multidirectional goals and the avoidance of conflicts [2], while interfacing with exponential changes brought by Industry 4.0 and digital transformation. Thus, overcoming twenty-first-century challenges in the healthcare industry requires a novel leadership approach that effectively guarantees high-quality care services. In incorporating exponential technologies, revising ethical regulations, and delivering quality services more efficiently, the role of a healthcare leader is to plan, direct and coordinate health services, managing specific departments, and entire structures. In addition, healthcare leaders need to manage finances and interact with clinicians and key stakeholders; tasks require learning a novel set of skills. These include *integrity* to follow ethical standards when dealing with moral issues, *vision* to set and achieve goals, and *listening skills* to consider others' opinions while valuing their contribution [3].

Providing high-quality care that is patient-centered, cost-effective and efficient, reliable, and safe depends on specific leadership styles. In fact, increasing the quality of services would positively affect productivity, benefiting the entire healthcare system and its management. In this regard, effective leadership may be associated with patient satisfaction (e.g., better health outcomes, lower mortality rates); thus, healthcare professionals should employ adequate leadership styles accordingly with diverse challenges and responsibilities [3].

Transactional Leadership Valuing performance (e.g., at the group, organization, or supervision levels), it conceptualizes the relationship between healthcare executives and staff as a transaction, in which "reward and punishments" are implemented to sustain motivation toward goal achievements. Moreover, this type of leadership highlights respect for rules and procedures, clearly defined roles, and attentive supervision. This type of leadership results effectively in situations in which employees need a clear direction on how to proceed while following rules and decisions of their superiors, who can ensure the completeness of tasks. Nevertheless, this type of approach does not encourage, inspire, or support creativity toward problem-solving, although very effective in those situations that require focus and clear direction [3].

Challenges and Opportunities Digitalization of healthcare procedures, facilitating patient-provider communication and collaboration, might be affected by the lack of proper cybersecurity. Indeed, healthcare professionals and executives must adopt novel policies and procedures to guarantee data safety for the entire organization. Hence, a transactional leadership style may be suitable for defining roles, procedures, standards, and guidelines to execute a proper data-security plan while solving cybersecurity threats [3].

Innovative Leadership To bring innovation in the healthcare sector, leaders aim to approach situations with *innovative thinking*, particularly when dealing with uncertainty and unpredictability. This type of leadership overcomes obstacles with strategies, initiatives, ambitious goals, adaptability, and flexibility, leaving out conventional approaches to problem-solving. Innovative leaders aim to create a novel organizational culture where innovative thinking ("think out of the box") is the key to unlocking unlimited possibilities, resources, and potential while supporting and encouraging team members' creativity. However, innovative leaders need to manage, organize, direct, and grow this abundance for more innovation. Indeed, leaders have to supervise team members and their roles so that everything can properly function. Inviting and supporting employees to express their thoughts stimulates motivation, satisfaction, and productivity. Thus, innovative leadership may apply to those contexts in which a change and a novel team-building approach are needed, concrete decision-making, and effective management among different key stakeholders [3].

Challenges and Opportunities Digitizing healthcare services requires big data management and ease of the process. Hence, big data management would help healthcare executives improve the flow of operations more effectively and efficiently, eliminate administrative obstacles (paperwork), and promote a novel payment model (value-based payment). However, this outcome can only be reached by investing in advanced technologies and training on how to implement them for gaining insight into proper data management. In this perspective, innovative leadership results in the right approach in managing big data more efficiently. Indeed, an innovative thinking style is needed when investing in technologies and coordinating and training team members in their utilization to enhance an organization's outcome [3].

Charismatic Leadership Engaging in charismatic communication is a valuable way to empower visions while building trusted relationships, inspiring teams toward goal-directed actions. Hence, charismatic leaders support effective communication based on feelings expression valuing emotional connections, the importance of regulating emotions, social interactions, and sensitivity to interpret situations and social control. Leaders are mission-driven and aim to change healthcare toward valuable outcomes. Indeed, charismatic leaders consider loyalty and commitment in the workplace, generate innovation, transform challenges into opportunities, empower the team toward creative solutions and less risk-averse, increase high-quality productivity while valuing trust and respect. However, being strong, popular, personality-driven, and excellent speakers, charismatic leaders need to improve their listening skills and pay more attention to others' thoughts instead of compromising a professional business practice. Indeed, welcoming feedback would be highly recommended to promote further success in the healthcare system, as compassion is necessary to lead and support team members' skills and personal development [3].

Challenges and Opportunities Managing worker stress is another challenge to be faced when coping with long working hours, emergency and life-threatening situations, or any other healthcare routine. The effect on mental and physical stress of taking responsibility or coordinating activities can be highly demanding. In this case, a charismatic leader would be beneficial in dealing with stressful circumstances while encouraging and emotionally supporting the team to avoid burnout and risky behaviors [3].

Situational Leadership Situational leaders understand that a "one-size-fits-all" approach is not suitable when dealing with complex healthcare situations. Indeed, different circumstances need different approaches. Thus, they determine which leadership style is more appropriate when encountering a particular situation. In this regard, four approaches are adopted: telling (giving guidance and directives), selling (influencing others to adopt their vision), participating (encouraging and sharing decisions among members), and delegating (responsibilities and tasks). Assessing different situational factors, the nature, and capabilities of the team, as well as different phases of a project is prioritized when adapting different leadership styles (e.g., a selling leadership style may be adopted at the beginning of a project when team members are unfamiliar with the task or the environment. Whereas, during later stages of a project, a delegating leadership style can be applied when team members have already acquired confidence and needed skills to complete the task). Evaluating the team's "level of maturity" (i.e., members' competence and knowledge) is a central characteristic of a situational leader, supporting the team toward their growth and development. Indeed, this evaluation process goes from assessing the level of knowledge and skills (level 1), willingness and enthusiasm (level 2) regardless of limited ability, competence, and skills to achieve a task (level 3), advanced skills, and high motivation (level 4) [3].

Challenges and Opportunities Digital transformation and the implementation of advanced technologies in healthcare had the effect of revising regulatory procedures and ethical approvals. Hence, hospitals and any healthcare provider have to interface with different regulatory agencies (e.g., licensing, addressing frauds), investing time, money, and effort. The need for flexible, adaptable, and ready actions is the key to accomplishing this challenge. Thus, situational leaders' skills in encouraging participation, giving clear directions, and delegating responsibilities, are the key to accomplishing this challenge with flexible, adaptable, and ready actions [3].

Transformational Leadership Focused on empowering and encouraging the participation of team members, transformational leaders aim to transform the healthcare structure by sharing responsibilities and decisions among employees, independently of their roles in the organization. Healthcare executives value self-confidence and respect through participatory and motivating behavior while building a shared vision for a growing and enriched future. With goal-directed actions, performance and productivity are enhanced, as well as job satisfaction and a positive workplace environment. A few components have been identified as core concepts of transformational leadership: (i) intellectual stimulation (supporting creativity to generate innovation), (ii) individualized consideration (encouraging communication, shared ideas, trusted relationships), (iii) inspirational motivation (empowering the team with passion and motivation toward a vision), and (iv) idealized influence (act as role models) [3].

Challenges and Opportunities In shifting the value of payment toward health outcomes, the entire healthcare system needs to deal with an innovative payment method in which the possible threat of financial risk should be mitigated. Thus, a novel "value-based care" approach requires better synchronization and coordination with multiple health procedures and data acquisition. These measures need high accuracy and performance with a cost reduction. To overcome this challenge, transformational leadership may be suitable when dealing with disruption and exponential changes and managing costs and operations [3].

Furthermore, the need for great leaders to inspire and guide teams while increasing productivity and performance is high on demand nowadays to stimulate and support healthcare innovation. However, developing a proper set of skills and quality traits is necessary when leading organizations toward a successful outcome. In this regard, five leadership traits that healthcare executives should acquire were presented [4].

- *Leaders mentor others*: whether formal or informal, mentorship plays an important role when training and educating future generations to become public health leaders and professionals, guiding and inspiring them in building their career path.
- *Leaders face challenges*: overcoming difficulties while testing new approaches, with an innovative and growth mindset is a fundamental requirement to move out from a "comfort zone" while determining effective changes.
- *Leaders educate others*: similar to mentorship, education through awareness, communication, and empowerment are relevant in supporting public health and healthcare development.
- *Leaders practice humility*: accepting and welcoming different points of view or opinions is the key to development, change, and learning.
- *Leaders create opportunities for others*: mentoring, innovative thinking, educating, and an open-minded approach are the steps needed to greatly empower future generations, raising opportunities and space for growing toward innovation.

Effective leadership is needed in healthcare professionals to cope with new challenges brought by the modern healthcare system and increase efficiency and productivity while improving the quality and the management of healthcare services. Participation and engagement in novel training programs to develop further healthcare leaders are essential; indeed, learning to adopt diverse leadership styles facing different situations is a key point in fostering improvements and changes [5].

However, to become effective leaders, healthcare professionals need to develop new skills in responding to the increasing demand for quality health services. Therefore, based on the NHS, quality in healthcare can be measured in relation to patient safety and positive patient experience. Moreover, the Care Quality

Commission (CQC) identified quality as efficient and cost-effective services. For this reason, a good leader knows where to apply changes, adapt novel approaches when facing different environments and situations, and make the proper decisions to create a novel organizational culture. In this context, a transformational leadership approach would be the most successful in engaging and empowering the medical staff in decision-making, responsibilities, managing tasks, and contributing to the process of change [5].

Recently, besides the well-known Medical Leadership Competency Framework [6] supporting clinicians in developing leader's skills, to further grow healthcare workers into becoming effective leaders, a Healthcare Leadership Model has been proposed, designed over nine dimensions [7]:

- *Leading with care*: the recognition of team needs and behaviors while creating a caring and supportive environment
- *Sharing the vision*: communication based on trust to set and achieve long-term goals while inspiring confidence and credibility
- *Engaging the team*: the support of team members' participation
- *Influencing for results*: engaging in creating a collaborative environment, to properly commit
- *Evaluating information*: innovative and creative thinking approach to solve problems and to develop new concepts
- *Inspiring shared purpose*: taking risks and overcoming challenges based on NHS principles and values
- *Connecting our service*: the understanding of the system structure, organization, and politics, to connect with the inside and outside of the system
- *Developing capability*: seeking opportunities to support the growth of the team and to develop their skills further
- *Holding to account*: creating an innovative and growth mindset based on clear ideas and continuous improvement

Lastly, successful leadership programs aim to develop effective healthcare leaders, promote followership, and exploit changes and ameliorations. Hence, this approach would fit the concept of collective leadership for Healthcare [5]. Through an innovative culture, everyone would contribute to the organization's success in generating high-quality healthcare services.

29.1   Educating Future Healthcare Leaders

Leading healthcare innovation in the twenty-first-century requires effective and well-prepared executives managing the medical staff, key stakeholders, health providers, and distributors. Therefore, novel training and educational programs to prepare future healthcare leaders are necessary to empower professionals with a new role model balancing autonomy, accountability, teamwork in a value-based outcome, patient-centric approach, and interdisciplinary environment [8].

A few skills are indispensable in becoming an effective leader. Among these are organization and time management, building solid and trusted networks, and effective communication. Moreover, welcoming feedback and self-awareness are useful when adjusting behavioral habits toward dealing with different situations [8]. Furthermore, considering healthcare a complex system and organization, collaborations and teamwork are fundamental aspects. Indeed, developing proper communication skills while sending and receiving messages should be clear, congruent, stimulating, and reinforced by active talking and listening [8, 9].

Defying and planning novel curricula and training programs requires attention to encourage the development and assessment of future emerging leaders. Indeed, several experienced healthcare professionals and educators will retire in the following years [8], leaving the lead to new generations. Thus, promoting mentorship to support innovation should be considered to foster a novel organizational culture in healthcare, supervised by effective and responsible leaders [8]. A few practices suggested in supporting leadership development are reading more about leadership, attending training workshops and courses [8], joining mentorship programs, welcoming opportunities, and learning to take responsibilities.

A transformational approach based on teamwork is optimal to enable effective leadership and management, in which collaboration with multiple health professionals, educators, and providers leads to the achievement of goals. Moreover, leadership can be extended to the management and organization of different areas, from teaching to administration, research, and clinical setting. Training to use tools to stay organized while matching schedules or learning to delegate tasks can be useful when preparing future leaders. A few examples are defying milestones, a list of needed resources (e.g., staff requirements/qualities/skills), effective communication to set directions and complete tasks. Based on Oates (2012), a list of needed competencies to educate effective healthcare leaders is reported in Table 29.1 [8, 10].

Challenges A novel healthcare education for leaders that is extended among different professional areas, organizations (e.g., universities, hospitals, and other health services) might be influenced by the type of structure, policies, regulations, and culture within the organization. Thus, educators in healthcare leadership should define their approach to leadership (e.g., postgraduate, undergraduate education). To this, McKimm (2004) [9] (see Table 29.2) listed a few challenges that need to be considered when educating future healthcare leaders [8]. Learning novel leadership competencies become mandatory in any health sector, in which a multidisciplinary environment represents a new space of improvement based on mutual exchange across disciplines. In conclusion, to generate innovation in healthcare leadership, educators and revised training programs (or curricula) should promote teamwork, clinical skills, a patient-centered approach, responsibility, and autonomy [8] (see Fig. 29.1).

Table 29.1 Needed leadership competencies for educators, leaders, and professionals in healthcare (suggested and structured by van Diggele et al., 2020; revised and adapted from Oates, 2012) [8, 10]

Learn about leadership concepts, styles, and theories

Leadership theories and styles regarding the lead of an organizational structure, and the development of team members.

Become a motivator and mentor

- The focus is directed toward motivating and encouraging others.
- Role modeling and attentive delegation to develop future leaders

Effective communicator

- Effective communication can be developed based on different methods.
- Communication should be adapted following the organization's structure and team members (e.g., managers, administrators, medical staff, or patients).
- Communication should be respectful, supportive, and open to receiving feedback.
- Networking
- Communication means also the development of effective listening.

Learn to lead the team

Adapt the leading approach by understanding the atmosphere at the workplace. Hence, with a goal-oriented attitude, lead the team toward changes by managing and decision-making.

Leadership attitude and style

- Adapting leadership style in various environments and situations
- Express personal opinions with self-confidence
- Active and empathic communicator while engaging with the team

Learn to be a team leader, team player, and team-builder

- An effective leader should be a team leader and a team player to respect and consider the opinion of each team member and their role.
- Leaders have a role in teaching, coaching, and mentoring.
- Conflict resolution skills are necessary when leading a team or an organization.
- Problem-solving, management of conflicts, team growth, and development

Skills in healthcare education and research

A good leader engages him/herself in research and educational activities.

Skills in business and finances

- Human resources
- Management of business plans, workflows, optimal evaluation of budgets
- Minimization of inefficient processes and procedures
- Financial management to guarantee the best outcome for the healthcare providers

Self-management and self-development

Learn to effectively manage the time for a good work–life balance.

Become a teacher for others

Develop a role as a coach, motivator, and mentor to support team members in their self-development journey

Table 29.2 Challenges to be considered for educators, leaders, and professionals in healthcare (suggested and structured by van Diggele et al., 2020; revised and adapted from McKimm, 2004) [8, 9]

Personal challenges

Difficulty in maintaining a work–life balance

The complexity of managing a professional career (e.g., academic and clinical career)

Organizational, environmental, and cultural challenges

To succeed and achieve goals, leaders have to evaluate not only the workplace atmosphere but also the culture of an organization.

A few healthcare fields require the investment of time and resources in clinical and research activities.

Time management and high-demanding tasks

High demanding tasks in academic and in clinics would generate stress in health professional educators, besides efficient time management skills.

A leader and educator to foster innovation

Health educator leaders should be aware of the higher requirement in healthcare and education and stimulate interdisciplinary learning and collaborations. To foster innovative and effective approaches in leadership.

Fig. 29.1 Graphical representation of needed skills and competencies for innovative healthcare leadership

Take-Home Messages
- Healthcare is a complex and dynamic system and a novel leadership approach is needed to lead, actuate, and manage the process of health innovation.
- A novel set of skills is required when facing twenty-first-century challenges brought by Industry 4.0 and digital transformation
- Self-confident, aware, active listener, team player, creative thinker, technical and financial expert, visionary, and good communicator: the new profile of a successful healthcare leader
- Revised educational and training programs should be considered to educate future healthcare leaders to overcome challenges in any clinical situation.

References

1. Barbazzeni B (2021) Leadership: a born talent or a learned set of skills? Exo insight. Available at: https://insight.openexo.com/leadership-a-born-talent-or-a-learned-set-of-skills/ [Accessed on October 24, 2021]
2. Al-Sawai A (2013) Leadership of healthcare professionals: where do we stand? Oman Med J 28(4):285–287. https://doi.org/10.5001/omj.2013.79
3. AHU.edu (2020) 5 types of leadership styles in healthcare. AdventHealth University online. Blog. Available at: https://online.ahu.edu/blog/leadership-styles-in-healthcare/ [Accessed on October 22, 2021]
4. Burnham K (2020) Effective leadership in healthcare: 5 essential traits. Northeastern University Graduate Programs. Available at: https://www.northeastern.edu/graduate/blog/leadership-in-healthcare/ [Accessed on October 22, 2021]
5. Kumar RDC, Khiljee N (2016) Leadership in healthcare. Anaesthesia and Intensive Care Medicine, Management 17(1):64–65. https://doi.org/10.1016/j.mpaic.2015.10.012
6. NHS Institute for Innovation and Improvement and Academy of Medical Royal Colleges (2010) Medical leadership competency frame- work, enhancing engagement in medical leadership. 3rd edn. July 2010. Available at: http://www.institute.nhs.uk/images/documents/Medical%20Leadership%20Competency%20Framework%203rd %20ed.pdf (accessed 3 Sep 2015).
7. NHS Leadership Academy. Healthcare leadership Model (2015) The nine dimensions of leadership behaviour, Version 1.0, 2013. Available at: http://www.leadershipacademy.nhs.uk/wp-content/uploads/dlm_uploads/2014/10/NHSLeadership-LeadershipModel-colour.pdf (accessed 16 Sep 2015).
8. van Diggele C, Burgess A, Roberts C et al (2020) Leadership in healthcare education. BMC Med Educ 20:456. https://doi.org/10.1186/s12909-020-02288-x
9. McKimm J (2004) Developing tomorrow's leaders in health and social care education. Case studies in leadership in medical and health care education. Special report 5. Newcastle-upon-Tyne: Higher Education Academy, Medicine Dentistry and Veterinary Medicine, 2004
10. Oates K (2012) The new clinical leader. J Paediatr Child Health 48(6):472–475

Beatrice Barbazzeni is a proactive scientist and prolific author by *ExO Insight*. With a background in psychobiology, she is currently finishing a Ph.D. in cognitive neuroscience. She likes to challenge herself by joining different research institutes and interdisciplinary teams, with a particular interest in exponential medicine and health longevity. Beatrice is an effective communicator, fluently speaking several languages which allows her to share multicultural experiences and network globally. Last but not least, she has a fantastic passion for fitness, art, and music. The key to her success is discipline and grit; the values that inspire her to become a leader in science.

Future Skills Framework in Healthcare

30

David Matusiewicz and Jochen A. Werner

Abstract

The digital transformation will change the healthcare system in unexpected ways. Because the system as it was and is has no future. This great opportunity for improvement comes with three key challenges: (1) questioning of the self-images of the people who work in healthcare work, (2) the speed at which change is occurring, and (3) the uncertainty created by the intrusion of machines into fields formerly attributed to humans. The Future Skills presented here (Traditional Skills, New Work Skills, Digital Health Skills) can be used to address the change.

Keywords

Future skills · Digital transformation · Digital health skills

What Will You Learn in This Chapter?
- The needed skillset (abilities, competencies, and/or virtues) for a much more digital-based Healthcare.
- Increased speed of change leads to uncertainties—the human being has difficulties with that.
- New work and digital skills supplement the traditional skill set.

D. Matusiewicz (✉)
FOM Hochschule, Essen, Germany
e-mail: david.matusiewicz@fom.de

J. A. Werner
University Hospital Essen, Essen, Germany
e-mail: kontakt@medicalinfluencer.de

30.1 Introduction

"Nothing is as constant as change."—What Heraclitus of Ephesus observed some 2500 years ago still holds true today. The world of work is constantly changing and with it the demands on healthcare workers. What do employees need to bring to the table in the 2020s, both professionally and personally, to ensure that their company remains fit for the future?

The digital transformation is leading to completely new requirements on the part of employees in the healthcare sector. It is clear though that the future of healthcare will not work without people. And people need the ability to actively shape this future.

This development has an enormous impact on future requirements for the skills of employees and managers, but also for successful start-ups in the industry. This is a major challenge for the healthcare industry, which is faced with the conflicting demands of high-cost pressure and scarce economic resources, the need to bring innovations to healthcare, and yet continue the old core business ("innovation dilemma").

Acting professionally in a digitized world requires enhanced skills at the personal, social, and methodological levels. This includes a fundamental understanding of digitization, i.e., of drivers, contexts, and definitions, an entrepreneurial attitude with regard to opportunities and risks in one's own work context, and agile working in multi- and interprofessional teams. Be it the art of successful communication the ability of empathy toward the patient or today resilience in a dynamic world.

Education is essential in the course of digital transformation. Innovative technologies will accompany the healthcare sector in the future, and the digital transformation is being strongly driven by politics and business. In view of the far-reaching effects of digitization on thinking and acting, self-perceptions and job profiles, as well as work processes in the healthcare sector, education is essential. Employees need future skills to cope with change, and the healthcare player must empower employees through education so as not to miss the boat on digital transformation. The ability to digitally re-imagine the healthcare player will be determined in large part by a clear digital strategy supported by leaders who foster an open culture and are able to embrace innovation. What is different about digital transformation is that risk-taking is becoming a cultural norm as digitally advanced companies seek new competitive advantages. Equally important, employees of all ages want to work for companies that are committed to digital progress. Leaders must take this into account to attract and retain the best talent. Mature and successful digital organizations build competencies to execute strategy. Digitally mature organizations are four times more likely to provide employees with the skills they need than organizations at the lower end of the spectrum [1]. The ability to conceptualize how digital technologies can impact the health stakeholder is a skill that is lacking in many organizations in the early stages of digital maturity. Too often, there is a focus on single technologies.

On the one hand, fundamentals such as conceptualizations of digital health must be created, and on the other hand, topic areas such as digital business processes or design thinking must be anchored in the structures of the organization. Managers can act as multipliers, coaches and, not least, role models. In this context, continuing education is the key to success. In this article, the construction kit "future skills" with dimensions around education in the area of digital transformation is presented. This didactic concept gives the various stakeholders of healthcare institutions a well-founded view around the digital transformation in healthcare. The opportunities and limitations of digital transformation in healthcare organizations are then explained and a brief outlook is given.

With a new mindset in a new world of work, old values and virtues are reinterpreted and put into a new digital context. Pre-digital skills that will play a role today and still post-digital. Above all, it remains to be seen whether healthcare workers, on whether they are healthcare managers, doctors, nurses or others, will teach themselves the future skills autodidactically or on whether employers or associations will slowly but surely recognize the importance and take action.

This great opportunity for improvement comes with three key challenges:

1. Challenging the self-image of people who work in healthcare
2. The speed at which change is occurring
3. The uncertainty created by the intrusion of machines into areas once reserved for humans

30.2 Overview about Future Skills

Future Skills refers to abilities, competencies, and/or virtues that will be used in today's and the evolving new world of work, reinterpreting old values and skills, incorporating a new mindset, as well as being placed in a digital context in today's and tomorrow's organizations (see Fig. 30.1).

At the same time, companies should have the capacity to nurture and support their employees in terms of the necessary skills. This ensures that corporate success can be achieved in the long term through the application and improvement of skills. In the following figure, a systematization of the skills can be illustrated.

Traditional Skills
The more digitized medicine becomes in the future, the more likely it is that classic analog values, skills, and virtues will become obsolete. Classic skills or characteristics that everyone will need in their professional lives and for social participation in the future like creativity, entrepreneurial action, or stamina. People who master these skills can find their way in new situations and solve problems with their own ideas. These traditional skills will become even more important in the future, because task and job profiles are changing rapidly due to automation and digitization. The most important traditional skills are:

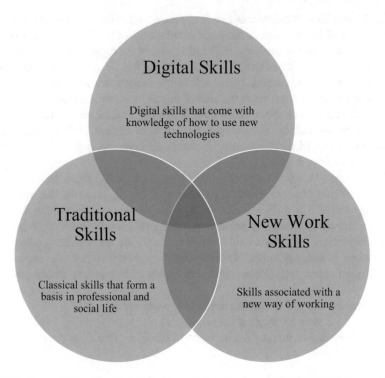

Fig. 30.1 Future Skillset for Digital Health at the intersection of Traditional, Digital, and New Work Skills

- *Charisma*
- *Creativity*
- *Empathy*
- *Honesty*
- *Humility*
- *Humor*
- *Inspiration*
- *Intuition*
- *Kindness*
- *Respect*
- *Self-reflection*
- *Serenity*
- *Trust*

New Work Skills

These are digital skills that everyone will need in their professional lives in the future. Those who master these skills will be able to work collaboratively and agilely as well as make critical decisions in an increasingly digital world. They include digital interaction and digital ethics, which will be required for everyone in the future. Whether in nursing, medicine, or management, a new mindset is important: error culture, speed, and risk-taking. Some new work skills are:

- *Adaptability*
- *Agility*
- *Attentiveness*
- *Awareness*
- *Collaboration*
- *Communication*
- *Critical Thinking*
- *Flexibility*
- *Health Literacy*
- *Innovative Abilities*
- *Interculturality*
- *Interdisciplinarity*
- *Internationality*
- *Lifelong learning*
- *New Leadership*
- *Qntrepreneurial Thinking*
- *Teamwork*

Digital Skills

Digital skills are shaping new professions and are being practiced by more and more employees. Those who have mastered these skills have the latest (information) technology expertise and can apply it. For example, this involves digital literacy (dealing with complex data sets), complex data analysis, or user-centric. These skills will create new job profiles across all organizations. They are often already shaping job profiles in startups and the smart healthcare organizations:

- *Augmented/Virtual Reality*
- *Cloud computing*
- *Cybersecurity*
- *Data Analysis*
- *Data Literacy*
- *Digital basic skills*
- *Digital Ethics*
- *Digital learning/microlearning*
- *Exponential Thinking*
- *Intelligent hardware/robotics*
- *Machine Learning/AI*
- *User centricity*

The biggest change in the healthcare sector is certainly digitization—it is now arriving globally. Paper and pencil, telephone, and fax are still widely present in 2022. But slowly, robots, digital health applications, interoperability, and other exponential technologies are being introduced and it is foreseeable that they will also exponentially grow. Are employees prepared for this—certainly not. And that is why the topic of individual education and corporate education plays a major role. However, the presented Future Skills Framework, which addresses the current skills needs of the healthcare industry, is similar to the skills needed in general business management.

We have the opportunity to creatively meet the challenges of tomorrow—today beside digital transformation also the long-term consequences of the corona crisis or climate change—with the right skills.

30.3   Discussion

Digitization offers opportunities for healthcare players to improve and automate processes so that resources can be saved. In a healthcare system that is still very analog and dominated by administration, in which, for example, around 40–50% of doctors' time and around 30–40% of nursing time is spent on bureaucracy, digitization alone can deliver a great deal of added value at this point. Digitization thus enables employees to spend more time on actual activities. Digital transformation can automate many processes and take over the requirements. The use of technologies also means that the collaboration and communication of all players will become more decentralized, networked, and direct. Collaborations of care networks are emerging so that health and research data can be actively used to ensure intelligent networking. Digital transformation means integrating digital assistance and database systems to ensure freedom of treatment for physicians and avoid one-sided control of care. The services and platforms created by digital transformation represent opportunities for strengthening and expanding care. Digital analysis tools and Big Data applications also enable IT-supported decision support systems that can support the transformation of organizations from administrative authorities to modern healthcare management companies and help them achieve a new level of quality in terms of patient outcomes. The digital transformation can also have a positive impact not only on the quality of healthcare but also on the development of healthcare costs.

Digital transformation is no simple task in healthcare. Legal, technical and financial framework conditions present challenges for implementation. Technical limits can arise from the clash of new technologies with existing systems or in the exchange of data between the parties involved. Linked to data exchange, data security also poses legal limits that must be considered in the digital transformation. At the same time, technological developments are difficult to predict and cannot be limited to specific applications. In addition, people set the limits of digital transformation. It is predominantly the employees who hinder the path to digital transformation. It is essential to create acceptance among all stakeholders for innovations for digital transformation of the organizations. Employees often associate the word digitalization with the replacement of humanity by robotics. This thought leads to employees not wanting to perceive digitization, not accepting it and refusing training courses. Those responsible must act as role models, multipliers, and coaches and learn a new leadership style. Curiosity and interest must be aroused from the very beginning in the company to understand digitization. Predominantly, leaders must accept the dynamics, be agile and flexible in dealing with new and unknown situations, and adapt to the new requirements of employees. It must be understood that employees are more motivated when they are involved in decision-making, co-development and are respected as individuals. Furthermore, success in digital transformation depends to a large extent on whether a culture of error is practiced in the company or promoted by managers. In addition, digital transformation fails due to a lack of corporate education strategy. If training courses are approved without a strategy, a patchwork quilt will develop in the company. What is needed is up-to-date continuing education and training in digitization, such as new technologies, digital processes, artificial intelligence, and other disciplines, in order to be able to successfully implement a digital transformation in healthcare facilities. Healthcare future skills can help to improve the knowledge base and help the organizations to meet a successful business model.

30.4 Conclusion

The healthcare sector needs more education regarding digitalization. Other countries show that digital transformation brings opportunities for patients, employees, and the company itself, in terms of structure, organization, and finances. Many potentials that are not yet exploited today need to be identified. It must be clearly delineated that digitization does not mean that a digital transformation is taking place. Digital transformation refers to the entire company; it means creating changes throughout the company. In order to further develop digital transformation in healthcare, various competencies/future skills and knowledge from different perspectives are required from those responsible and leaders. The management style or leadership and hierarchical positioning among each other must be adapted to the digital era, so that a leader must act as a coach and motivator. Personalization of patients and individualization of employees is becoming increasingly important for the healthcare players to stimulate creativity and identify innovations. Continuous machine learning must be established and used as an opportunity. The players need intelligent automation to save time and costs. Patient and employee experiences in the digital transformation are significant in order to be able to position themselves in the market in the long term. In addition, the healthcare players must implement design thinking methods to meet today's rapidly changing patient needs at all times. Pilot projects should identify errors and test phases should enable the implementation of innovative projects and processes without having to make high investments. In addition, a healthcare facility must continuously reinvent itself in the digital transformation because classic products quickly become obsolete and processes from today can be replaced by new technologies the next morning or may already be old.

The digital transformation of healthcare can only be driven forward through education or continuing education. This is the only way to develop a fundamental understanding of digitization and digital transformation in healthcare. Topics such as digital business processes, organizational change processes, disruptive business models in healthcare and global contexts are sometimes part of this. In this way, an up-to-date and well-founded overview of future skills in the national and international context is needed. In this context, a tool set is to be developed so that the healthcare player can manage digital transformation and strategy development in the healthcare sector.

But it is not technology that drives digitization, it is people.

Take-Home Messages
- *Future skills are essential in the course of the digital transformation in healthcare essential.*
- *Acting professionally in a digitized world requires enhanced skills at the personal, social, and professional levels.*
- *This includes a fundamental understanding of digitization and the personal leadership skills that are that are essential in the course of the digital depends on.*
- *It is not technology that drives digitization, it is people.*

Reference

1. Kane GC, Palmer D, Phillips AN, Kiron D, Buckley N (2015) Strategy, not technology, drives digital transformation. MIT Sloan Management Review and Deloitte University Press 14:1–25

David Matusiewicz is Professor of Medical Management at FOM University—the largest private university in Germany. Since 2015, he has been responsible for the university's Health & Social Sciences department as Dean and, as Director, heads the Research Institute for Health & Social Sciences (ifgs). In addition, he supports technology-driven start-ups in the healthcare sector as a founder or business angel. Matusiewicz is on various supervisory boards (advisory boards) as well as an investor in companies involved in the digital transformation of healthcare. www.david-matusiewicz.com

Jochen A. Werner, M.D., is currently Medical Director and CEO at the University Hospital of Essen, Germany. Prior to this role, Jochen Alfred Werner held the positions of Medical Director at the University of Marburg, Germany, Board Member of the University Hospital of Gießen and Marburg, as well as spokesman of the Medical Board (Rhön-Klinikum AG) until 30.09.2015. Since 1998, he has been Professor and Chairman for the Dept. of Otorhinolaryngology, Head and Neck Surgery at the Philipps University of Marburg. In May 2015 Jochen Werner became president of the German Society of Otorhinolaryngology, Head and Neck Surgery. www.jawerner.de

Porter's Five Forces Analysis: Quo Vadis Immunotherapy Industry

31

Joachim Maartens and Dietmar W. Hutmacher

Abstract

Investors and market analysts often seek different perspectives for market analyses of companies to gain a better picture of companies' positions and strengths within their industries. One tool for fundamental analysis that goes beyond just examining financial metrics such as the price-to-book ratio is Michael Porter's Five Forces Model. Therefore, the moment seems opportune to ask whether the five competitive forces remain adequate for characterizing the nascent cell therapy industry and whether the framework provides sufficient decision-making insight to stakeholders, be they insurers, providers, patients, employers, or policy makers. This article discusses the five competitive forces, (i) threat of new entrants, (ii) threat of substitute products or services, (iii) bargaining power of suppliers, (iv) bargaining power of buyers, and (v) rivalry among existing competitors for the nascent cell therapy industry. Factors influencing each of the forces provide insight into those areas most likely to impact industry profitability and therefore also stakeholder decision-making.

Keywords

Immunotherapy market · Porter five forces · Competition · CAR-T cells · Decision-making · Strategy

J. Maartens · D. W. Hutmacher (✉)
Queensland University of Technology (QUT), Brisbane, Australia
e-mail: dietmar.hutmacher@qut.edu.au

© The Author(s), under exclusive license to Springer Nature Switzerland AG 2022
M. Friebe (ed.), *Novel Innovation Design for the Future of Health*,
https://doi.org/10.1007/978-3-031-08191-0_31

> **What will you learn in this chapter?**
> - How the Porter Five Forces Framework can be used for a Health Industry analysis—here the Immunotherapy Industry as an example.
> - That the 5 Forces can be used to define a competitive advantage quantitatively and qualitatively.
> - Companies that are planning to enter the immunotherapy market must create a highly productive commercialization environment from research to proof-of-concept and early clinical trials.

31.1 Introduction

Four decades have passed since Michael Porter published his first-ever article in Harvard Business Review, entitled "How Competitive Forces Shape Strategy." Porter identified a framework of five competitive forces that define the nature and structure of any industry, including health care, with the purpose of informing the strategic decision-making process by industry stakeholders [1]. The intervening years have borne witness to a world, which has become a very different place not least due to the impact of the Digital Revolution and the concomitant Information Age. One could easily have added the subscript "mind the wave of disruptive technologies" as the Internet of Things continues to disrupt the world as we know it. During this period, as a particular case in point, biotechnology saw its definition being expanded to include new and diverse sciences such as genomics, recombinant gene techniques, applied immunology, and the development of pharmaceutical therapies and diagnostic tests. This has led to each of these new sciences contributing to the foundational "brickwork" of novel, breakthrough cell-based gene therapies.

A key milestone was reached on August 30, 2017, when the US Food and Drug Administration approved the first cell-based gene therapy (Kymriah, by Novartis) for the treatment of acute lymphoblastic leukemia (ALL) vastly improving the 1-year relapse-free survival to 60%, up from corresponding survival rates of 5–10% (adults) and 10–20% (children) under the best treatment options available. Despite many challenges, health care, as an industry, is being redefined.

Novartis, Kite (Gilead Sciences) and Juno (Celgene) are arguably regarded as the top three players in the Immuno-Oncology field, joined by Cellectis as the fourth member of this select vanguard, initially the only member of this select group to actively pursue an allogeneic or universal T-Cell (UCART) treatment approach. Several indicators are signaling that it may soon become significantly busier at the top. There are 200+ cell therapy companies pursuing dozens of technologies towards finding that breakthrough treatment. These activities are being complemented by mergers and acquisitions that already bear witness to a forty percent monetary increase over 2017 full-year figures and more than 300 active CAR-T clinical trials [3].

This being a significant anniversary of Porter's seminal treatise, the moment seems opportune to ask whether the five competitive forces remain adequate for characterizing, for example, the nascent cell therapy industry and whether the framework provides sufficient decision-making insight to stakeholders, be they insurers, providers, patients, employers, or policy makers.

31.2 Cell Therapy: Defining a New Industry

31.2.1 Casting the Die

In the late 1970s, a concatenation of political, social, and economic factors, complemented by strategic scientific, financial and business decisions shaped, stymied and encouraged Genentech's rise to the temporary pinnacle of its stock market debut. Genentech turned out well enough to serve as an adaptable template of a biotechnology industry about to emerge [4]. Although their stories are vastly different yet sufficiently similar, each of Novartis, Kite, Juno Therapeutics, and Cellectis have modeled their cell therapy journey from that template. However, the competitive environment within the industry seems to be in a continuous state of flux with a spate of Initial Public Offerings (IPOs) in the first half of 2018 already surpassing that of previous years and as a result, funding for start-up companies appears unencumbered [2].

A landmark paper published in 2011 in the New England Journal of Medicine, chronicling the journey towards remission for a patient diagnosed with Chronic Lymphoblastic Leukemia, is regarded by many cell therapy pundits as a signpost for readiness to deliver on earlier promises [5]. Fast-forward to the first quarter 2018 and many will agree that the nascent industry has reached an inflection point with cell therapy poised to become the fourth pillar of medicine after pharmaceuticals, protein drugs and devices. This notion is further supported by the global alliance of regenerative medicine companies, which includes gene- and cell therapies and tissue engineering therapeutic developers, committing significant resources, financial and otherwise, to more than 950 active clinical trials in 2018. In so doing, the industry has begun to embrace products that have already proven to be much more than mere "process." In fact, the final manufactured cell therapy product (CTP) is the cells themselves.

Viewed through the lens of cell therapeutics, the promise to transform the treatment of cancer and other debilitating diseases, by enabling mechanisms of action that small chemical compounds cannot provide, is being realized.

31.2.2 Industry Cost of Goods

If cells are to be used routinely for clinical or drug discovery applications, they need to be produced at an affordable cost of goods (COG). This implies scaling up the production of cells to a commercially viable level while optimizing the use of costly media and growth factors [6, 7]. Given these limitations, it is likely that manufacturing methodologies, spanning upstream (cell selection, purification and formulation) and downstream (preservation and distribution) processes should include some elements of automation [8–10].

This fact is strengthened by the growing market demand for larger cell quantities, and therefore, a concomitant step change in industrial manufacturing methods is required [11]. An additional obstacle for cell-based products that are largely manual in their manufacturing processes is meeting the regulatory requirements of process validation, quality control and reproducibility, which is extremely difficult to achieve [12]. As reported elsewhere, the large number of regulatory hurdles being faced in carrying out translational research is still regarded as one of the five Grand Challenges [13]. As a result, much of the current debate around pricing and reimbursement has been focussed on affordability and the associated value proposition cell therapies bring to industry stakeholders [14].

For example, a recent report on CAR-T therapies by the Institute for Clinical and Economic Review found that besides the treatment acquisition costs for Kymriah (USD 475,000) and Yescarta (USD 373,000), costs associated with

 (i) Hospitalization linked to CAR-T infusion and associated monitoring
 (ii) Hospital mark-up especially for the commercially insured population in the USA
 (iii) Managing B-cell aplasia and
 (iv) The potential cost of stem cell transplantation in CAR-T immunotherapy

if used as a bridge to transplantation, can be influential when determining the overall cost of the treatment.

Yet, the challenges encountered and previously described for the translation of cell therapy products to the market are not unique to this industry. In fact, the same happened in the early days of the semiconductor industry, which also started with unaffordable COG but today is considered mature (US$2 trillion in 2011) [15] with extremely low COG under the auspices of GMP and associated cleanroom technologies. In the early days, this industry also started with a lot of manual labor and the corresponding elevated costs associated with it. However, the cleanroom processes are nowadays fully automated, and constant innovation in the manufacturing technologies makes it possible for the prices of electronic devices to keep trending downwards. Although CAR-T immunotherapy adds another level of complexity by virtue of it comprising living cells, a plausible analogy with the semiconductor industry and the impact of Moore's Law has been proposed [16].

31.3 Applying Porter's Five Forces

Although different industries may differ at face value, the drivers that underpin profitability are the same. To gain an appreciation of the competition inherent to an industry, it must be analyzed through the lens of the five forces as depicted in Fig. 31.1.

Industry attributes become visible through its growth rate, technology and innovation, complementary products & services and regulations by virtue of the interplay between these five competitive forces.

31.3.1 Rivalry Among Existing Competitors

Kymriah, Yescarta, and more recently Tecartus and Breyanzi are described as being a CD19-directed, genetically modified autologous immunotherapy. These therapies were granted Orphan designation by the FDA and Novartis, Kite and Juno were able to take advantage of shorter FDA evaluation periods due to Breakthrough designations being granted. However, in the case of Kymriah, the intended treatment population includes pediatric and young adult patients 3 to 25 years of age with relapsed or refractory B-cell acute lymphoblastic leukemia (ALL). Both therapies have been approved for treating adult patients with relapsed or refractory diffuse large B-cell leukemia (DLBCL).

The scene appears to be set for classic rivalry in the form of price discounting, advertising campaigns and service improvements. Initially dearer by approximately USD 102,000, Novartis has since matched Kite's pricing for Yescarta to treat

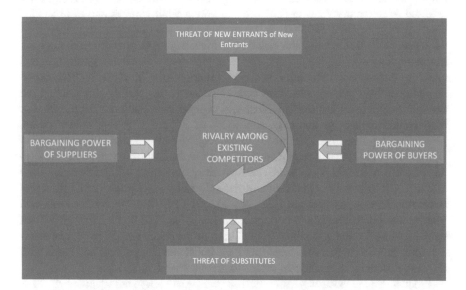

Fig. 31.1 Schematic representation of Porter's five competitive forces [2]

DLBCL. Novartis also seem to have the edge on the perception of being the front runner, i.e., first approval for a genetically modified cellular therapy by the FDA, first to obtain marketing authorization in the European Union, and the first product of its kind to be recommended as cancer treatment within the National Health Service (NHS) England.

Currently, Kite is pitching a shorter vein-to-vein time, largely driven by its cell expansion regimen. However, as discussed at the 2018 CAR-T Summit in Boston (USA), improving T-cell potency remains an area of intensive research, and there is an expectation that the industry may soon witness a step-change improvement in vein-to-vein time. Cold chain logistics, including custody control, are important adjuncts to cell therapy, especially for autologous products. As a result, Novartis, Kite (Gilead Sciences), and many others are investing strategically to ensure manufacturability of expected future commitments of providing worldwide treatment. Despite the small number of incumbents, the recognized complexity around patients and process will ensure rivalry continues to include training across sites, improvement of response rates, pricing, and market access. Due to tight regulatory control, the incumbents have a definite advantage, but a breakthrough could significantly alter the duration of this advantage. As a result, a qualitative assessment positions this as a WEAK force in its ability to influence industry structure and therefore the nature of competition within this industry.

31.3.2 Bargaining Power of Buyers

This is an important stakeholder group consisting of insurers, providers, patients, employers, and policy makers. Novartis has priced Kymriah at USD 475,000 for relapsed or refractory ALL in pediatric and young adult patients and has matched Kite's pricing of USD 373,000 for Yescarta for relapsed or refractory DLBCL in adult patients. This excludes associated costs of managing side effects resulting from the treatment over the short term, for example, cytokine release syndrome, and ongoing duty-of-care therapies such as immunoglobulin therapy. The initial response to these pricing structures echoed concerns of possibly jeopardizing the financial health of an institution willing to accept the associated risk for this kind of product and confirmed the price sensitivity of buyers as a stakeholder group. Innovative ways are being explored to reduce the acuteness of price sensitivity. This includes the calculation to determine whether the cost of treatment per quality-adjusted life year (QALY) falls between two threshold values of USD 50 k and USD 150 k. The comparative QALY figures determined by the Institute of Clinical and Economic Review (ICER) are USD 45,871 and USD 136,078 for Kymriah and Yescarta, respectively [17]. Other options being considered by Novartis include an outcomes-based pricing arrangement and in more desperate situations, patient assistance programs to help the under-insured, the uninsured and patients and families of lower socioeconomic status gain access to this treatment.

Product price sensitivity was further accentuated by the September 2018 announcement of NHS England's commercial deal with the manufacturer Novartis being the first in Europe and coming less than 10 days after the treatment was granted its European marketing authorization, especially since the agreement included an undisclosed discount off the wholesale acquisition cost. It represents one of the fastest funding approvals in the 70-year history of the National Health Service. The overseeing body, the National Institute of Health and Care Excellence (NICE) also green-lighted the treatment for entry into the reformed NHS Cancer Drugs Fund.

The notion that CAR-T immunotherapy is a game changer would have been foremost in the minds of the negotiators eager to establish the first win-win result for the buyer group and Novartis. Although leaders at NICE would have taken note of the earlier QALY calculation performed by their counterparts in the USA, they reached a similar conclusion that, initially, the cost per QALY for Kite's Yescarta exceeded the threshold values. Subsequent negotiations with Gilead Sciences resulted in NICE recommending CAR-T therapy as a treatment option for adults with certain types of non-Hodgkin lymphoma. This result has certainly raised the barrier to entry and has left the buyer stakeholder group with greater negotiating leverage for the future.

ICER and NICE are alerting all stakeholders that when a new treatment commands added health care costs, making it difficult for the health care system to absorb the same over a short time frame, possibly displacing other needed services or adding unexpected growth to health insurance costs, it can simultaneously limit access to high-value care for all patients. This remains a key challenge for the sector as it is linked to payer reimbursement and patient affordability. The short to medium-term outlook is that this is unlikely to impact the structure of competition and is considered a WEAK force.

31.3.3 Bargaining Power of Suppliers

Manifestation of the collaboration paradigm is strongly evidenced through the CAR-T cell therapy workflow, which encompasses upstream and downstream unit operations as well as analytics, quality, and logistics. The legacy of having to rely on significant manual intervention as part of the process is providing the impetus to seek opportunities to drastically shorten processing time, thereby enabling improved cost efficiencies and concomitant leverage at the bargaining table. Although not quite meeting the requirements of large-scale cell therapy manufacturing, the progress made by closed, automated cell processing systems such as CliniMACS Prodigy (Miltenyi Biotec), Cocoon (Lonza), and Beacon (Berkeley Lights) are sure to find their niche in the local manufacture of cells for early-stage clinical studies. The issue remains with the large disparity in required processing time for specific unit operations, such as transduction and cell expansion. This results in many of the upstream unit operations being idle, unable at being engaged for the next product and challenges the concept of scaling an integrated system.

Typical of a nascent industry, many aspects of the commercial landscape are unsettled, largely because the manufacturing platform is not yet standardized due to unanswered questions around robustness, COGS, scale and sustainability. It is thus quite understandable that both Novartis and Kite have been leaning towards a vertical integrative approach to secure the supplies needed to produce its product. In the early days of the first CAR-T clinical trials, Novartis signed an exclusive agreement with Thermo Fisher for the use of its proprietary magnetic bead platform for T cell isolation, activation, and expansion. This has since evolved into a non-exclusive agreement providing the opportunity for other CAR-T cell therapy developers to adopt the same platform. Both Novartis and Kite are using their own manufacturing facilities at Morris Plains (New Jersey, USA) and El Segundo (California, USA), respectively, with the parties taking active steps to extend their footprints into Europe.

The benefits of achieving Orphan Drug and Breakthrough designations from the FDA and their equivalent in the EU have chartered the course for other niche therapies to follow. The concomitant research effort to develop differentiated products and prove their safety and efficacy is significant. Hence the strong collaborative effort to align with the best tertiary academic institutions. Consequently, the rise in engagement with contract design and manufacturing organizations (CDMOs) and the importance of custody and control of the cold value chain have led to strategic alliances. For example, Novartis has signed an agreement with Cryoport to provide cryogenic logistics support, including chain-of-custody monitoring for its CAR-T therapy, Kymriah.

Trends in patents issued and clinical trials commenced are both strong indicators of research and development activity in the field, which would at the same time exert upwards pressure on R&D budgets. A concomitant trend has been outsourcing of R&D-related services. Key drivers for the increase in spending on contract research services include expanding R&D portfolios and a shift in company strategies towards a larger proportion of outsourced relationships in their supply chains. This is due to the need to address patent life issues, the need for novel delivery forms and other specialized capabilities, as well as a desire to increase decentralization for greater flexibility. Other reasons often cited for outsourcing are the desire to improve quality, increase efficiency, reduce costs, and time to market, improve processes and gain access to specialized expertise.

The global clinical research organization (CRO) market was estimated at USD 31 billion in 2016 and is expected to grow at a compounded annual growth rate (CAGR) of 6.9% to reach USD 38 billion by 2019. Key market drivers are (i) fifty percent of Phase II to IV activities are already being outsourced; (ii) an increased focus on large molecule therapy, rare diseases, specialty treatment, point-of-care assays and personalized therapy and devices; and (iii) investment in adaptive trial design which has an adoption rate of 50% as of 2016 following the FDA's emphasis on risk-based monitoring for clinical trial conduct.

In summary, the strong growth in outsourcing R&D-related services to CROs/CDMOs on the back of similar growth in patents issued and clinical studies started, the advances made in developing automated, closed-system cell manufacturing platforms, significant product differentiation through identified cell therapy targets, and the way collaboration has been embraced through strategic partnerships point to the current assessment of Supplier Bargaining Power as a STRONG force.

31.3.4 Threat of New Entrants

Barriers to entry are advantages that incumbents have relative to new entrants [18]. In an industry wrought with complexity, the issue of having access to a sufficiency of financial resources seems an obvious first hurdle to overcome. If the quantum of funds raised during Initial Public Offerings (IPOs) and subsequent Follow-Ons is accepted as measures of meeting capital requirements, this hurdle appears easily surmounted. The value of IPOs up to the third quarter of 2018 surpassed that achieved during the full preceding year five times over [19] (Fig. 31.2).

Next, scale economies can be found in almost all parts of the value chain. On the supply-side, Novartis and Kite are seemingly protected by scale economies in research, cell therapy manufacturing through vertical integration as well as the strategic location of logistics hubs and marketing to buyers. Matters appear to be less defined on the demand-side benefits of scale. Novartis has well-established credibility and respect in the pharmaceutical industry, but as a living cell therapy, Kymriah will need to "earn its stripes" as a preferred product. Although the approval of Yescarta by the FDA closely followed that of the Novartis product, Gilead Sciences (owner of Kite) is already facing some of the barriers. For instance, the issue of access to distribution channels. Besides preferential arrangements to ensure cold chain custody control, each living cell therapy has its own regimen that requires

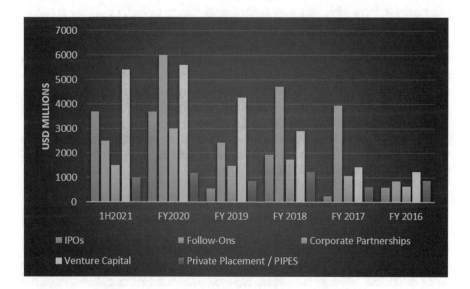

Fig. 31.2 Total global financings by type, by year [19, 20]

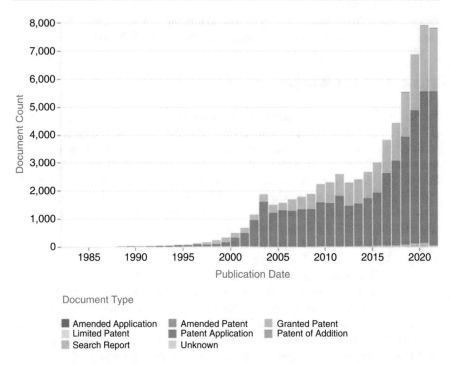

Fig. 31.3 Patent search using "chimeric antigen receptor" AND "immunotherapy" depicted by publication year. Source: lens.org

particular infrastructure and training to ensure consistency when administering. This is a highly regulated industry with limited short-cuts to market, i.e., those available through achieving Breakthrough and Orphan Drug designations from the FDA and similar terms from other recognized regulators such as the EMA and China FDA. Despite two strong incumbents, the threat is real and is being borne out by the value underpinning record IPOs and the concomitant investment in universal or "off-the-shelf" treatment options. The trend in patent registrations granted over the most recent decade bears witness to substantial preclinical activity and concomitant investigational clinical studies (Fig. 31.3).

Also progress made in the development of closed, single-footprint cell production technologies lowers the entry barrier for manufacturing cells for investigational clinical use [21]. This is a MODERATE force.

31.3.5 Threat of Substitute Products or Services

Considering that Kymriah and Yescarta were approved to treat patients with specific relapsed or refractory leukemic conditions, it would be incorrect to deduce that living cell immunotherapy is poised to replace treatments, for example, based on the use of monoclonal antibodies, checkpoint inhibitors, therapeutic vaccines, gene

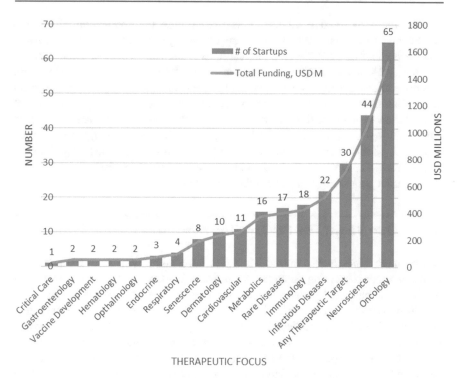

Fig. 31.4 AI-startups by therapeutic focus [22, 23]

therapies and the like. In many instances, the pharmaceutical drugs may remain the treatment of choice based on the cost per year of quality-adjusted life (QALY) gained and the associated benefits proffered by the healthcare system.

Furthermore, an intriguing situation is developing as biomedical tools and technologies rapidly improve, allowing researchers to produce and study an ever-expanding amount of complex biological data or "big data." AI-technology vendors are involved in almost every research aspect of modern drug discovery and development process—from data mining and biology research all the way to helping organize, manage, and improve clinical trials. Expect the benefits of data analytics to filter through to biologics via checkpoint inhibitors and monoclonal antibody specificity (Fig. 31.4).

Similarly, the Big Data to Knowledge (BD2K) program is a trans-NIH initiative that was launched in 2013 to support the research and development of innovative and transformative approaches and tools to maximize and accelerate the integration of big data and data science into biomedical research. While the impact of data analytics is expected in the realm of discovery, it is even more important with regard to post-market surveillance, where questions around long-term survival and efficacy are undetermined.

The number of active clinical trials that include the application of specific regenerative medicine or advanced therapy technology has been steadily increasing, and the number of CAR-T-based clinical trials is projected (at most recent CAR-T Summit, 2018) to increase from the current 311 to more than one thousand by 2020. The drive to contain costs of this treatment has ensured that significant resources are being committed to a universal or allogeneic therapeutic approach. Progression of clinical trials from Phase II to Phase III is perceived as slow, largely due to increased cold chain logistics requirements and the regulatory burden. Although a qualitative assessment of this force would describe it as MODERATE, it has the best probability of becoming STRONG soon. Its status as the second strongest force for this sector illustrates its potential for impacting the structure of competition in this industry.

31.4   Quo Vadis: The Patient, The Mission

Regenerative Medicine/Advanced Therapies have borne witness to extraordinary strides through pioneering outcomes for patients. Lessons continue to be learnt via previously unchartered pathways regarding regulation, reimbursement, legislation and manufacture of Kymriah and Yescarta as the first-ever genetically modified autologous T-cell immunotherapies approved by the FDA. This learning, unwittingly, has the benefit of impacting the configuration and therefore the strength of each of the five competitive forces, where the strongest of these forces has the best probability of impacting the industry's profitability. This knowledge should be integral to strategy formulation.

Rivalry among existing competitors. As the only incumbents and the clear industry leaders with approved CAR-T cell therapies, Novartis and Kite (Gilead Sciences) have the opportunity and resources to shape the industry structure by leading towards new ways of competing. This can be done by either redividing profitability in their favor or by expanding the overall profit pool, where rivals, buyers and suppliers can all share [18]. Given the nascent status of CAR-T cell therapy as a treatment modality, it is not surprising that Novartis and Kite initially appeared to favor a vertical integrative business model by aiming for more control over upstream suppliers and downstream buyers—recall Novartis' exclusive deal with Thermo Fisher regarding Dynabeads for activation and expansion of CAR-T cells. Pursuing improved economic efficiencies by lowering transaction costs between business units, synchronizing supply and demand along the chain of unit operations and seeking strategic independence considering, for example, highly sensitive reimbursement negotiations would, arguably, find plausible support in their respective boardrooms.

It is an open secret that manufacturing an autologous CAR-T cell therapy at scale is a significant industry-wide conundrum. While not a force, per se, it is a factor that impacts each of the five forces to a greater or lesser degree. The absence of a standardized manufacturing platform (i) limits incumbent's ability to reduce COGs and improve reimbursement outcomes, (ii) affects new entrants ability to meaningfully impact manufacturability of preclinical assay designs, (iii) cements manufacturing scalability advantages of alternative therapies such as monoclonal antibodies (iv) diffuses the focus and potential to optimize solution offerings of suppliers and (v) reduces industry's capacity to improve the therapy's value proposition to the buyer stakeholder group.

Novartis' response, through their global manufacturing strategy, has been to sign collaborative agreements for the development and manufacture of lentiviral vectors (Oxford Biomedica), for the manufacture of CAR-T therapies (CELLforCURE, France; Cellular Biomedicine Group, China) and for cryogenic logistics support (CryoPort). Kite has followed a similar strategy through collaborative manufacturing agreements (Fosun Pharma, China; Daichii Sankyo, Japan) and is well-represented in the US market through manufacturing sites in California (El Segunda and Santa Monica) and Maryland (Gaithersburg) with a new manufacturing facility planned for Europe (Hoofddorp, Netherlands). Neither of these approaches must be seen as favoring a particular business model, for example, centralized versus decentralized. Rather these regional manufacturing sites should be seen as a strategic move, on the one hand, to limit the use of cold and ultra-cold supply chains, which are significant contributors to COGs and, on the other hand, to enable faster response to therapy innovations which could favor the decentralized or point-of-care manufacturing model.

At the same time, this approach also seems to fit the alternative strategy of expanding the profit pool where opportunities abound for all competitors. The data capturing trends in CAR-T patent registrations and CAR-T clinical trials point to a growth industry that tends to mute rivalry through more opportunities for all competitors [18]. Casting the net to include the growth of the global immuno-oncology pipeline further highlights the position of cell therapy amidst the other classes of agents and their associated mechanisms of action [3].

While early results from clinical trials indicate that CAR-T therapy would extend life by an average of 4 years and 8 years for adult and pediatric patient populations, respectively, sales figures for Kymriah and Yescarta for the first 9 months of 2018 have been modest at USD48m and USD183m, respectively. Affordability, pricing models and reimbursement remain high on the agenda of discussion forums, also resulting in reports by ICER and NICE elucidating the value proposition that cell therapies are proffering [17]. It appears that the concept of value, and not price per se, is progressively gaining traction as the definition is extended to also include benefits to the medical community, health care system and society. However, competition between Novartis and Kite remains weighted in favor of pricing sensitivities, as emphasized by NICE's recent cost-effectiveness assessments for Kymriah and Yescarta, leading to both companies acknowledging that undisclosed discounts led to NHS acceptance, albeit with the aid of the Cancer Drugs Fund [25] (Fig. 31.5).

Threat of new entrants. The concept of tunable CAR-T cells where activation is dependent on the administration of a small molecule has already been demonstrated [26]. This would allow clinicians in the future to control the timing and dose of CAR-T cell activation. Immunogenicity of a universal CAR-T cell will soon be remembered as a minor obstacle in the path of the first approved allogeneic CAR-T

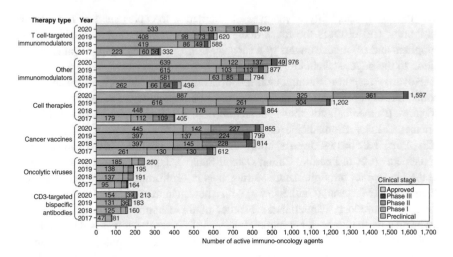

Fig. 31.5 Trends in the global immune-oncology pipeline [3, 24]

Fig. 31.6 Patient stakeholder group dimensions of value as gleaned from ICER report [17]

therapy. This will unlock enormous commercial potential when considering differentiators such as patient-centric factors, manufacturing scalability and business model adaptability. Its amenability to the mass-manufacturing model and associated lower cost of goods as well as to the off-the-shelf business model support this notion. This will be predicated on safety and efficacy data at least matching that of autologous therapy, possibly relegating the latter to thrive in niche or orphan type medical indications.

Threat of substitute products or services. Bispecific monoclonal antibodies and bispecific CAR-T therapies will enhance the potency of the CAR-T suite of therapies (Fig. 31.6).

Bargaining power of suppliers. Tangible progress on interface standardization between unit operations comprising cell therapy processes will precede maturation of point-of-care cell production technologies.

Bargaining power of buyers. It is incumbent upon industry participants to realize that the value proposition of this single-dose therapy will underpin its success as a treatment modality while simultaneously play a crucial role in determining the industry's structure and competitiveness. This should translate into improved value for buyers because of the greater focus on product features, support services and brand awareness.

It is evident that much would be gained by developing a value proposition for this therapy that is accepted by all stakeholders, i.e., value to the patients, value to the medical community, value to the health care system and value to society.

Industry structure grows out of a set of economic and technical characteristics that determine the strength of each competitive force. Competing to be unique is ultimately more sustainable than competing to be the best [18]. The industry has already passed the first inflection point. Add the latest industry trends, underpinned by peer-reviewed science and concomitant data, and the "extracellular matrix" that is collaboration will ensure a value proposition that is meaningful and supported by all stakeholders.

In the first nine months of 2018, Kymriah and Yescarta have generated sales income of USD48m and USD183m, respectively. These are modest figures by Big Pharma industry standards. The eventual industry structure will be determined by the interplay of the five competitive forces. The relative strength of these forces is certain to change over time. As a stakeholder in the cell therapy industry, what should your strategy be?

Information in Fig. 31.7 is but the first step towards answering this question. Here the authors provide a summary of key factors considered to impact the perceived strength of each of the forces. It is expected that individual weightings and industry scores could be vetted by stakeholders through appropriately structured surveys. On conclusion, this would serve as a valuable adjunct to the strategic decision-making processes of industry participants.

Take-Home Messages

- The five competitive forces of the P5F model are (i) threat of new entrants, (ii) threat of substitute products or services, (iii) bargaining power of suppliers, (iv) bargaining power of buyers and (v) rivalry among existing competitors.
- Growth rate, technology and innovation, complementary products and services and regulations interplay between these five competitive forces.
- The 5 forces can be categorized in WEAK, MODERATE, and STRONG.
- P5F requires an in-depth industry analysis that also provides in-depth insights.
- P5F can be used to compare different players and also different industry segments.
- P5F is an advanced tool, but to check the effect of disruptive technologies quite helpful.

		Criteria	Applied Weighting	Industry score	Subtotals	Total weighted Score
#1: Rivalry Among	Existing Competitors	Multi-site Treatment Uniformity	20%	59	11.8	**46.2**
		Vein-to-Vein Processing Time	20%	57	11.4	
		Market Access	20%	46	9.2	
		Cost-of-Goods	20%	33	6.6	
		Value Creation (proposition & chain)	20%	36	7.2	
			100%			weak

		Criteria	Applied Weighting	Industry score	Subtotals	Total weighted Score
#2: Bargaining Power	of Buyers	Reimbursement Options	20%	38	7.6	**46.2**
		Choice of Therapy	20%	43	8.6	
		Regulatory Pathways & Surveillance	20%	47	9.4	
		Urgency / Lead Time to Consumption	20%	45	9	✓
		Treatment Value	20%	58	11.6	
			100%			weak

		Criteria	Applied Weighting	Industry score	Subtotals	Total weighted Score
#3: Bargaining	Power of Suppliers	CROs / CDMOs	25%	77	19.25	**65.75**
		Strategic Partnerships / Acquisitions	25%	79	19.75	
		Supplier Substitutes	25%	37	9.25	✓✓✓
		Product differentiation	25%	70	17.5	
			100%			strong

		Criteria	Applied Weighting	Industry score	Subtotals	Total weighted Score
#4: Threat of New	Entrants	Supply-side Economies of Scale	20%	47	9.4	**51.6**
		Demand-side Economies of Scale	20%	48	9.6	
		Government Policy	20%	51	10.2	
		Incumbency	20%	61	12.2	✓✓
		Distribution Channels	20%	51	10.2	
			100%			moderate

		Criteria	Applied Weighting	Industry score	Subtotals	Total weighted Score
#5: Threat of	Substitute Products	Late Phase Clinical Trials	20%	56	11.2	**53.8**
		Alternative Therapies	20%	48	9.6	
		Buyer Risk Profile	20%	51	10.2	
		Buyer Switching Costs	20%	57	11.4	✓✓
		Buyer Price-point Sensitivity	20%	57	11.4	
			100%			moderate

Fig. 31.7 Identification and weighting of factors contributing to the five forces in the immunotherapy market

References

1. Porter M (2008) The five competitive forces that shape strategy. Harv Bus Rev 86:78–93
2. (2018, 5 November 2018) ClinicalTrials.gov. Available: https://clinicaltrials.gov/
3. Tang J, Pearce L, O'Donnell-Tormey J, Hubbard-Lucey VM (2018) Trends in the global immuno-oncology landscape. Nat Rev Drug Disc 17:783, 10/19/online 2018
4. Hughes SS (2011) Genentech: the beginnings of biotech. University of Chicago Press
5. Porter DL, Levine BL, Kalos M, Bagg A, June CH (2011) Chimeric antigen receptor–modified t cells in chronic lymphoid leukemia. N Engl J Med 365:725–733
6. Prestwich GD, Bhatia S, Breuer CK, Dahl SLM, Mason C, McFarland R et al (2012) What is the greatest regulatory challenge in the translation of biomaterials to the clinic? Sci Transl Med 4: 160cm14
7. Paull D, Sevilla A, Zhou H, Hahn AK, Kim H, Napolitano C, et al. (2015) Automated, high-throughput derivation, characterization and differentiation of induced pluripotent stem cells. Nat Methods 12:885+, 2015/09//
8. Rowley JA (2010) Developing cell therapy biomanufacturing processes. Chem Eng Prog 106: 50–55
9. Hümmer C, Poppe C, Bunos M, Stock B, Wingenfeld E, Huppert V et al (2016) Automation of cellular therapy product manufacturing: results of a split validation comparing CD34 selection of peripheral blood stem cell apheresis product with a semi-manual vs. an automatic procedure. J Transl Med 14:76
10. Haddock R, Lin-Gibson S, Lumelsky N, McFarland R, Roy K, Saha K, et al. (2017) Manufacturing cell therapies: the paradigm shift in health care of this century. 13. Available: https://nam.edu/wp-content/uploads/2017/06/Manufacturing-Cell-Therapies.pdf
11. Lysaght MJ, Hazlehurst AL (2004) Tissue engineering: the end of the beginning. Tissue Eng 10:309–320
12. Vaes B, Craeye D, Pinxteren J (2012) Quality control during manufacture of a stem cell therapeutic. BioProcess Int Suppl Cell Ther Anal 10:50–55
13. He B, Baird R, Butera R, Datta A, George S, Hecht B et al (2013) Grand challenges in interfacing engineering with life sciences and medicine. IEEE Trans Biomed Eng 60:589–598
14. Regenerative Medicine Opportunities for Australia (2018). Available: https://www.mtpconnect.org.au/images/MTPConnect%20Regenerative%20Medicine%20Report.pdf
15. Mack CA (2011) Fifty years of Moore's Law. IEEE Trans Semicond Manuf 24:202–207
16. Maartens JH, De-Juan-Pardo E, Wunner FM, Simula A, Voelcker NH, Barry SC et al (2017) Challenges and opportunities in the manufacture and expansion of cells for therapy. Expert Opin Biol Ther 17:1221–1233
17. I. f. C. a. E. Review (2018, December 7, 2018) Chimeric antigen receptor T-cell therapy for B-cell cancers: effectiveness and value [Onliine]. Available: https://icer-review.org/wp-content/uploads/2017/07/ICER_CAR_T_Final_Evidence_Report_032318.pdf
18. Porter ME (2008) On competition. Harvard Business Press
19. (2018, 3 August 2018) Alliance for regenerative medicine data report Q1. Available: https://alliancerm.org/publication/q1-2018/
20. Regenerative medicine in 2021: a year of firsts & records. Alliance for regenerative medicine, September 2021. Available: https://alliancerm.org/sector-report/h1-2021-report
21. Tang J, Hubbard-Lucey VM, Pearce L, O'Donnell-Tormey J, Shalabi A (2018) The global landscape of cancer cell therapy. Nat Rev Drug Discov 17:465, 05/25/online 2018
22. BiopharmaTrend (2018, 13 November 2018) A landscape of artificial intelligence (AI) in pharmaceutical R&D. Available: https://www.biopharmatrend.com/m/charts/

23. (2021, December 14, 2021). *BiopharmaTrend.com* [online]. Available: https://www.biopharmatrend.com/topic/industry-trends/
24. Upadhaya S, Hubbard-Lucey VM, Yu JX (2020) Immuno-oncology drug development forges on despite COVID-19. Nat Rev Drug Discov 19:751–752
25. N. I. f. H. a. C. Excellence (2018, December 10, 2018) Tisagenlecleucel for treating relapsed or refractory B-cell acute lymphoblastic leukemia in people aged up to 25 years. Available: https://www.nice.org.uk/guidance/gid-ta10270/documents/final-appraisal-determination-document
26. Wu C-Y, Roybal KT, Puchner EM, Onuffer J, Lim WA (2015) Remote control of therapeutic T cells through a small molecule-gated chimeric receptor. Science 350:aab4077

Joachim Maartens, Post-Doctoral Fellow at the Centre of Orofacial Regeneration, Reconstruction and Rehabilitation (COR3), School of Dentistry, The University of Queensland, Brisbane, Australia. There he is part of a team promoting the translation from preclinical to clinical application of novel scaffold-guided bone and soft tissue regeneration. He is part of the CCRM Australia sponsored international Industry Mentoring Network in STEM (IMNIS) program in regenerative medicine.

Dietmar Hutmacher, PhD My scholarly track record illustrate successful mastery of a major challenge in an interdisciplinary field: the ability to transcend traditional disciplinary boundaries, to initiate and nurture excellent research and educational programs across different disciplines. This was achieved through an interdisciplinary research program via convergence of science & engineering (bioengineering, biomaterials science, computational modeling, chemistry and nanotechnology), the life science (molecuar & cell biology, stem cell research, genomics, proteomics, bioinformatics), and clinical research (orthopedics, plastic and reconstructive surgery, radiology). I am also an academic entrepreneur and one of the few academics who successfully translated tissue engineering research programs from fundamental research to routine clinical application.

Part IX

Health Entrepreneurship

Global Health Markets and Their Different Needs

32

Jörg Traub

Abstract

Healthcare, digital health and medical technology is global, but in each country, there are different entry barriers, challenges, and opportunities due to the individual healthcare system and society. It is a strategic decision in which region the market entry should take place and in which time frame and with which strategy the internationalization takes place. The market for medical device sales varies greatly from region to region. The states have different sizes, different social structures, and different healthcare systems. This brings individual opportunities for medical devices and health tech, but also challenges for market access and success. Specifically, this relates to market size, routes to market, regulatory burden and costs, and other compliant issues, as well as how reimbursement is achieved. These analyses and considerations, along with the regional marketing, sales, and service approach, are key success factors for sustainable market access and healthy growth. In general a product and service shall be designed to be suitable for all markets, the challenges are regional in the fit to the local markets and thus require dedicated know how even for exponential, scalable solutions.

Keywords

Markets · International · Access · Opportunities · Sales · Marketing · Certification · Reimbursement

J. Traub (✉)
Forum Medtech Pharma & Healthcare Bayern Innovativ, Munich, Germany

© The Author(s), under exclusive license to Springer Nature Switzerland AG 2022   411
M. Friebe (ed.), *Novel Innovation Design for the Future of Health*,
https://doi.org/10.1007/978-3-031-08191-0_32

What will you learn in this chapter?
- Different countries come with different healthcare business models.
- Analyze the individual healthcare markets with respect to your offering ideas.
- Sales and service issues are important as well.
- A regional fit of your product/service may be a solution.

32.1   Introduction

The global market for medical devices, excluding in vitro diagnostics, was valued at around USD 422 billion in 2018 (SPECTARIS Jahrbuch 2019/2020 [1]). In 2017, the USA ranked first with 38.8%, ahead of Germany with 9.9%, followed by Japan and China. In 2020, there are a total of 194 states.

It is important to have a sound product or service, that is addressing a burning problem with highest technical excellence and a robust and unbreakable solution. This also applies to digital health solutions and great use cases, and a study shows how to succeed in regional markets (PWC [2]).

The factors for an international market access assessment are listed below to provide a basis for decision-making in prioritizing market access and the associated probabilities of success. Developing any market ties up resources and requires investment. These must be compared to the likelihood of success in a country-specific commercialization plan and individual entrepreneurship activities. Also digital products that seem to have better opportunities for scaling need the regional fit for success. New products and solutions need to master the local boundaries of integration in the healthcare systems and in particular getting the appropriate reimbursement.

32.2   Market and Competition

The market and the estimation of its potential depend on the category of the product or service. For example, the number of clinics or special departments may be the relevant value for capital goods or the number of patients or procedures for a particular disease, diagnosis or therapy. This will be a good initial estimate of the market size. The data are in general available in regional reports or reports of the world health organization (WHO) in the world health statistics [3].

Market sizes vary enormously. Population size can be used as one possible indicator. Here, too, the range is wide: from about 800 inhabitants in the Vatican City State to about 1.4 billion inhabitants in China in 2019. Other indicators are the wealth of society in the states and per capita spending on healthcare. Already in the OECD countries, an enormous range is again evident: from the USA with USD

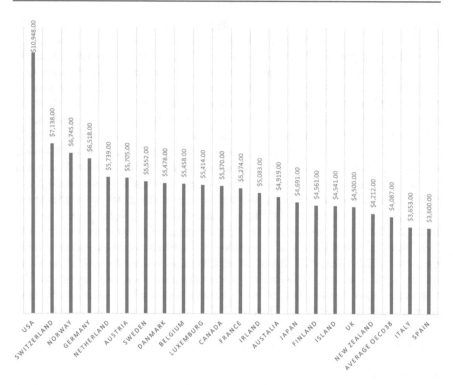

Fig. 32.1 Per capita expanses per country are an indicator on the spending in healthcare per person. Data from 2019 from Statista (Statista) [4]. Source: Own graphic

10,586 per inhabitant in 2018 to USD 209 in India in 2018 (Statista) (Fig. 32.1). The use of this parameters and indicators allows the selection of the priority for market access and gives a focus to your venture into internationalization.

In addition to size and wealth, however, the local competitive situation also plays a role in the decision to enter the market. For example, how high is the penetration of the market by competitors? Sometimes it can make sense to establish the product and service in a niche market without competition, as part of a so-called "blue ocean" strategy [5]. Cultural factors in the respective target markets can also play an important role here. For example, goods from the target country may be preferred or "Made in Germany" may be advantageous for a medical product, especially selling to international markets.

32.3   Reimbursement Options

It is important to understand that a product or service is only applicable if there is a clear strategy for reimbursement, either by the state, by the medical doctor, by the healthcare system or private payer. Like your own business model, you need to create a business model for your partners, e.g., hospitals or medical doctors in order to have sustainable success beyond research interest. Precondition is that it is a painkiller for a burning problem, and there is a clear patient or business need.

A wide variety of healthcare systems exist, which generally form a closed market. When billing for preventive services, diagnoses, and therapies in the inpatient area, so-called CPT codes (Current Procedural Terminologies) are used in some regions, e.g., in the USA, among others, by the insurance companies Medicare and Medicaid. In this case, the procedure used and thus also the technique used are reimbursed. It is important to obtain a reimbursement code for your procedure, either using an existing one or establish a reimbursement code yourself. Later is in general in need for clinical studies and large upfront investments in the market. In contrast, in other regions, e.g., in Germany, billing is based on diagnosis-related case groups (DRG). This means that reimbursement depends on the diagnosis but is independent of the procedure chosen for treatment. These different systems have a decisive influence on the likelihood of success of health technology business models. Similarly, reimbursement for capital expenditures varies across states. In some cases, these are made from capital budgets, i.e., as a one-time investment amount; in others, models such as leasing or financing purchases are required, as reimbursement is offset against per-case fees per procedure. In many states, private payers are the main component of the health care market. Either you gain deep knowledge of the reimbursement options of the market you want to enter, or you need to collaborate with a trusted partner supporting that you create a lasting success in the market.

32.4 Regulation and Certification

Prior to market entry, you need to match local regulations of the state. In almost all states, certification is required to market and sell medical devices in the local market. In some states, if CE certification or FDA approval has already been obtained, only a registration with valid certificates is required. Going through the necessary approvals before market launch is essential. Resources can be bundled here through strategic planning. Testing protocols for certification, e.g., according to the IEC CB Scheme, an international system for mutual recognition of test results and certificates according to harmonized standards, to register electrical and electronic products for approval in various countries without renewed test protocols. This can save costs and time. This is because, among other things, the waiting times for test laboratories that meet the necessary legal framework of international requirements are sometimes three to nine months in care all test scenarios are passed without major rework.

Clinical data usually play a major role in the approval process. These can occasionally be obtained from experience in other countries, but as a rule new clinical studies are required in the target country, especially for products and services higher risk classes. This then means an enormous expenditure of resources. In addition, for successful approval, important product documentation must be written in the language of the country or in general at least in the language that is understandable by the user and operator of the product or service.

32.5   Legal, Political, and Social Framework Conditions

In addition to regulatory issues, export and import and the associated taxes play an important role and need to be understood before a country is entered. They can have an influence on the pricing of products. In some cases, there may be restrictions on exports (e.g., embargoes), but there may also be restrictions on imports due to proof of the raw materials used (e.g., REACH) or import bans to protect the domestic market. In digital health also data privacy regulations need to be considered. This also means that patient data needs to be stored on servers in the country or countries that are legally accepted by country. The EuGH/Shrems II, i.e., the privacy shield regulations for data between EU and US prohibits since July 2020 the storage of EU data on servers outside the EU and also access of any services on the servers storing these data. Legal consultation of an expert is required to ensure that you comply with the legal framework of your target country.

When entering prospect target markets for the sales of your products and services, the respective social and cultural framework conditions should also be taken into account. These can play a significant role in the acceptance of a product: for example, the technology affinity of the population when marketing high-tech products, or when establishing stable business relationships. Knowledge of local traditions and customs can be an invaluable advantage in this respect. Gender awareness and religious issues as well as naming might also play an important role to avoid crucial and predictable failures of your products and services.

32.6   Market Entry Plan and Market Entry Opportunities

There are several ways to enter the international market (see Table 32.1). In each country, it has to be evaluated individually which distribution strategy is the best choice for the product [6]. While in the core markets, in some cases, the establishment of a subsidiary is the best solution, in other countries, it may be a good choice to cooperate with an exclusive dealer or a dealer network or several non-exclusive ones.

In addition to the market entry strategy, the training and service concept for the respective country must also be established. Depending on the complexity and size of the market, dedicated personnel must be trained and further educated here to be able to ensure the required quality in the long term. It is strongly advisable to have a direct and local presence in your key market, considering also moving the location of a start-up venture to the market with highest business opportunity to ensure also the linkage to regional politics and providing fast and reliable service.

Table 32.1 To scale to regional business, there are different models applicable, from direct sales with high risks and costs but full control to agent models with low upfront investments but dependency and even lock-in effects depending on the license holder for the sales permission

Model	Advantage	Disadvantage
Direct sales	Direct feedback, highest control, highest margin	Expensive and partly not possible from a regulatory point of view
Local branch/ subsidiary	Direct control, partial disclaimer, positive image of the company	Very expensive and complex, very high upfront costs and high risk
Agents	Inexpensive, no obligations and usually no exclusivity, no fixed costs, good market knowledge and regional network	No control, agent commission due after project completion, liability risk remains, lots of support from application specialists, high costs for market introduction
Distributors/ resellers	No fixed costs, long-term exclusive commitment, partial disclaimer, existing network, and regional market knowledge	No control, high distributors margin on project realization, a lot of support from application specialists

32.7   Conclusion

Entering new markets requires a differentiated consideration of risks and opportunities, costs and realizable reimbursement and sales opportunities. A structured consideration of the characteristics listed here, market, market size, competition, approval, framework conditions, reimbursement, and sales strategy enable prioritization of target markets and strategic business development in internationalization. Scaling from a regional market in direct sales to a global market is a major challenge, especially for young ventures and for small companies with new products or services. Expansion must be well planned, and the various options must be carefully weighed and evaluated. It is essential for success in expanding into other markets to understand the target market, determine the best strategy based on one's own resources and strategic objectives, and then implement it. This also applies to digital products and disruptive business models. It needs a regional fit to be successful, and even so a digital solution is ready to scale, the regional and country-specific success depends on the precise modeling of the culture and healthcare system fit and its reimbursement.

> **Take-Home Messages**
> - Healthcare markets are working based on different models and you need to analyze the market dynamics and stakeholder decision-making—what works in place may not work in another and vice versa.

(continued)

- Growth for your product/service can be achieved by entering new regional markets.
- Smart choice of the regions for market entries are key for success.
- Entering new markets requires understanding of the opportunities, overcoming market entry hurdles and country-specific business models and therefore, precise knowledge and trusted partnerships.
- A systematic analysis of the markets supports strategic decisions for market development.
- Your venture should be located in your key market, at least have a strong presence there.
- Digital health models in general address global business, but need regional fit to scale.
- Local and regional business models based on reimbursement or the willingness to pay are success factors for market access.

References

1. The German Medtech Industrie (2021) Die deutsche Medizintechnikindustrie: SPECTARIS Jahrbuch 2019/2020 auf Basis von Daten des Beratungsunternehmen Frost & Sullivan; see https://www.spectaris.de/fileadmin/Content/Medizintechnik/Zahlen-Fakten-Publikationen/SPECTARIS_Jahrbuch_2019-2020.pdf
2. Isler C, Jeandupeux M, Nye W (2021) From pilot to scale – How to make digital health stick; PWC; see https://www.pwc.ch/en/insights/digital-health.html
3. World Health Statistics (2021) Monitoring Health for the Sustainable Development Goals; see https://www.who.int/data/gho/publications/world-health-statistics
4. Statista (2019) Annual per capita health expenditure in selected OECD countries in 2018. (Jährliche Gesundheitsausgaben pro Kopf in ausgewählten OECD-Ländern im Jahr 2018); see https://de.statista.com/statistik/daten/studie/37176/umfrage/gesundheitsausgaben-pro-kopf/
5. Chan Kim W, Mauborgne Renée (2005) Blue Ocean Strategy. How to Create Uncontested Market Space and Make the Competition Irrelevant, Boston, Massachusetts: Harvard Business School Press, ISBN 978-1591396192
6. Traub J, Friebe M (2018) The difficult way from the regional to the international market – Case Report of a company in the field of interventional imaging; Der schwierige Weg vom regionalen zum internationalen Markt – Case Report einer Firma aus dem Bereich der interventionellen Bildgebung, p. 279–289 from M. A. Pfannstiel, P. Da-Cruz, V. Schulte (Editor). Internationalization in Healthcare – Strategies, Solutions, Practical Examples; Internationalisierung im Gesundheitswesen – Strategien, Lösungen, Praxisbeispiele. Springer Gabler

Jörg Traub received his PhD in Medical Computer Science. He has more than 15 years experience in the medical device start-up industry and was as founder and CEO responsible from the idea generation, through the implementation, certification, internationalization and multiple financing round all the way to the exit of the medical technology company and university spin off SurgicEye. Following this he was business development director and coordinating the strategic marketing and sales activities at a smart medical robotics solution company, resopnsible also for strategic corporate partnerships.

Since 2020 he is director healthcare section at Bayern Innovativ and is Managing Director of Forum Medtech Pharma e.V., responsible for innovation incubation, infrastructure projects in healthcare innovation and networking.

Patient: Health Relation and Digital Health Entrepreneurship

33

Thorsten Hagemann

Abstract

The easiest benefit to achieve using digital solutions is a patient's adherence to the disease. By engaging with the disease through an app, patients learn about themselves in dealing with that very disease. You need data to create benefit, but you only get the data if you create enough value. Think about the societal challenges and meet them not only by digitally transforming existing processes, but also by redesigning them. Healthcare markets are over-regulated and full of vested interests. Unfortunately, motivation is maximized only where these lobbies can make or save money. However, at the heart of the matter is the relationship between the patient and the (health) care provider, which is many generations old. Think of the whole person. Think of the patient's ecosystem and the market structure in which that ecosystem swims. Where the data is, that is where the market will develop the strongest momentum in the future.

Keywords

Health data · Electronic patient record · Patient-doctor relation · Health ecosystem · Regulatory approval

What will you learn in this chapter?
- Health needs and market size correlation.
- The big 5 innovation reasons and needs for Health Entrepreneurs.
- The patient ecosystem and how to follow the data.

T. Hagemann (✉)
adesso SE, Dortmund, Germany
e-mail: info@thorstenhagemann.de

33.1 Needs: It Is All About the Pains and Gains—Or Is It About Data?

33.1.1 Value Creation Path of Data

Solutions will only survive in the market if they generate sustainable benefits. In the digital world, benefits can only be generated through data. We have known about benefit-centric innovation for quite some time.[1] But we want to think beyond banalities like diary functions, calendars, and specific reminders. By purely increasing adherence, you do generate some benefit,[2] but today we understand that the magic of a digital solution is not technicality.

This technicality is precisely why it will be about data processing—not about mere delivery or presentation.

Such dashboards will be limited and can only generate very limited benefits. Only the digital value chain of data creation, data integration, data processing and result at the point of care forms the base of successful solutions. This does not mean that a single solution has to accompany the entire path of the data as mentioned. On the contrary: By choosing the right partners, the value-added steps that are missing in one's own organization should be opened up and made accessible in shared business models.[3]

33.1.2 User Benefit First

It seems like a vicious circle (chicken or egg): the digital solution needs data to be able to generate a (medical or care) benefit. Similarly, this digital solution only gets data from users the moment it generates sufficient benefit to motivate users to share their data. Without a perceived or demonstrated use: no success!

Think carefully, if the benefit will be sufficient in the way that it is at least enough to donate data (compare e.g., German jurisdiction[4]). It is permissible, especially in shared business models, to initially aim solely for data donation, if parts of your partner network want to participate in the data itself or in the opportunities that arise. This is also a proven method in the context of studies, for e.g., approval or certification.

---

[1] BILGRAM, Volker; BREM, Alexander; VOIGT, Kai-Ingo. User-centric innovations in new product development—Systematic identification of lead users harnessing interactive and collaborative online-tools. International journal of innovation management, 2008, 12. Jg., Nr. 03, S. 419–458.

[2] SABATÉ, Eduardo; SABATÉ, Eduardo (Hg.). Adherence to long-term therapies: evidence for action. World Health Organization, 2003.

[3] Baron, C. (1999). Synergien durch innovationsorientierte Partnerschaften. In: Public-Private-Partnership-Konzepte für den IT-Markt. Deutscher Universitätsverlag, Wiesbaden. https://doi.org/10.1007/978-3-663-08228-6_7

[4] § 327 BGB bzw. in den §§ 327a-u BGB sowie in §§ 455c BGB und 475a BGB

33.1.3 A Must: Think in New Dimensions—Think Disruptively!

> *"If you digitize a shitty process then you have a shitty digital process."*
> — Thorsten Dirks, former CEO Telefónica, 2015.

Healthcare processes and solutions should also be thought of disruptively. The healthcare sector is facing a terrific shortage of personnel in the face of various demographic challenges. According to latest appraisals, there could be a shortage of up to 500,000 nursing staff in Germany alone by 2035.[5] This likely is also the case worldwide or in other European countries.

Also relevant are the infrastructure changes in healthcare, some of which are politically desired like the formation of treatment centers for specific indications . These certainly increase patient safety and quality of care in the context of the treatment taking place there. What is challenging, however, is how providers in general will deal with the geographically extended distance between them and their patients. Digital solutions will be able to help in the vast majority of cases.[6]

Today, we tend to focus on the chronic diseases of Western civilizations. The reasons for this are as easy as they are sad. Of course, many people are affected by these diseases over long periods of time, but at the same time, these diseases devour millions of health insurance premiums, which should certainly be sufficient motivation for payers to invest in this area.

Processes must be rethought and redesigned: every process can be hybrid controlled and thus globally usable and effective.

For entrepreneurship activities, this also means: a "disruptor" is waiting around every corner. So always think disruptively for yourself!

33.1.4 Market Rules Rule

In many healthcare systems around the world, especially in European countries, there are no classic market rules. Sometimes the possibilities of a clever business model are limited by governments and health related regulatory process and approval peculiarities.[7]

In addition, there are always challenges and issues arising regarding data security and data protection, which prevents openminded thinking in many places. The smart

---

[5] Seyda, Susanne / Köppen, Robert / Hickmann, Helen, 2022, Großer Fachkräftemangel in den Altenpflegeberufen in Nordrhein-Westfalen, Gutachten gefördert durch das Ministerium für Arbeit, Gesundheit und Soziales des Landes Nordrhein-Westfalen, Köln

[6] BERTELSMANN STIFTUNG (HRSG.): Faktencheck Krankenhausstruktur, www.faktencheck-gesundheit.de/de/publikationen/publikation/did/faktencheck-krankenhausstruktur/

[7] Kloiber, Otmar; Flenker, Ingo, Grenzen des Wettbewerbs im Gesundheitswesen, Dtsch Arztebl 1996; 93(38): A-2381 / B-2054 / C-1906.

buzzword of the business model generation: "privacy is dead," which startups in other target markets are allowed to use, will be much more difficult to establish in the healthcare sector, depending on the ethical values of a company and the already mentioned laws with respect to health data privacy.

The ecosystem of a healthcare system is also riddled with many complex conflicts of interest.[8] Alone, the idea of how a resident in Central Europe gets a drug describes it quite precisely:

> The patient asks (explicitly or implicitly through their condition) the doctor for a prescription. This prescription is delivered through a long value chain of pharmaceutical research, development, marketing and pharmacy distribution. Pharmacies engage in commerce. Health insurance systems and insurance companies—historically strongly represented in political lobbies—however, have to pay for the prescriptions. It will certainly not be disputed that each stakeholder has its own interest and definition of benefit here. And of course they all want to help the patient—but only in the case that enough money can be earned.

So choose the beneficiary of your solution wisely. Create cross-stakeholder benefits. Find the approaches to your business model that bring users and payers "together" and harmonize their respective Pains and Gains.

33.1.5 Chase the Big Five!

1. Above the dynamics of the individual disease of a person will always be the force of the societal needs. In Central Europe, it is mainly the **demographic change of our society with increasing aging** and—not surprising—strongly growing incidences of certain diseases associated with age.[9] Dementia (and other neurodegenerative diseases), tumors and chronic "diseases of affluence" (e.g., obesity related) in western countries are among those.[10] Above all, it will demand solutions that can counteract or at least counteract this movement. This has implications for the design of the interface as well as for appropriate market and business model strategies.

2. **Distances and infrastructures**: Close and bridge the distance between a patient and his practitioner in a hybrid world. Think global. No process remains local.[11] Not only so-called telemedicine will be the answer, but a general new thinking between local needs and global answers.

3. **Save costs.** Whatever your new business model or digital solution will involve, it should save someone's money. The healthcare market is so saturated that our

---

[8]Prof. Michael Simon, Lobbyismus in der Gesundheitspolitik, Bundeszentrale für politische Bildung, www.bpb.de/themen/gesundheit/gesundheitspolitik/200658/lobbyismus-in-der-gesundheitspolitik/

[9]Gesellschaft, Demografie, Demografischer Wandel in Deutschland, Statista, First published 4. Dezember 2020

[10]Prognose zum Anstieg von Zivilisationskrankheiten bis 2050, Statista, First published 09.04.2010

[11]Michael E.Porter, James E. Heppelmann, How Smart, Connected Products Are Transforming Competition, Harvard Business Review, First published November 2014

healthcare spending will grow at a very limited rate relative to other challenges. Think, for example, of environmental catastrophes, general social services, and, unfortunately, more recently, defense spending, which will significantly alter the cost structures of economies. If a system does not get fresh money to flow, then of course those solutions will sell, which bring a saving potential with itself.

4. **Relieve medical, supplying, and nursing personnel.** Relieving the burden on those who care for our patients and the elderly remains one of the most important issues of the solutions we want to successfully bring to market. Either you enable them to do things easier, faster to learn or safer, or you replace parts of their activities. From purchasing, to supply, to care, to medical intervention, to just monitoring, there are countless possibilities open to you.

5. **Pandemics and disasters:** The saddest point of the "big five". We will not be able to arm ourselves, but in these cases, solutions will be demanded of the digital enablers and developers in the long term.

33.2 The Patient (Experience) Ecosystem

It is our western health care, diagnostics and therapy ecosystem that has been saturated over millennia. "Healthcare" is simply one of the oldest industries known to man.

Historically, our health care and health care delivery consisted virtually exclusively of the relationship between the individual, his or her caring immediate family, and a medical practitioner.[12]

Even today, some surgeries look like their historical predecessors back in the Middle Ages (see Fig. 33.1: cataract surgery then–now–abundant sources). In some countries of the world, even today, unfortunately, the core relationship between doctor and patient remains the expression of all available health care. The family has now been replaced by professional nursing care. New treatment methods required new infrastructures. Infrastructures wanted to be paid for. Medical care was to become available to everyone, which entailed strict political regulation with regard to diagnostics, therapies and interventions, nursing care and their cost structures.

So the challenge is to re-understand patient-doctor relationships and patient's ecosystem relationships and to create digital benefits within them. Ideally, this benefit leads to a positive care effect and thus to relief along the patient care processes. These benefits can serve the system as a whole or individual stakeholders. Knowledge of the market to be addressed and its various dynamics and partial motivations is therefore indispensable!

---

[12] Schüler, Dietz: Medizinische Psychologie und Soziologie. 1. Auflage Thieme 2004, ISBN: 3-131-36,421-1

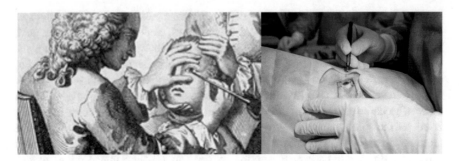

Fig. 33.1 Therapies developed over hundreds of years. Of course with better devices, higher sterility, ... Foto Wikipedia

> "Health is a state of complete physical, mental and social well-being and not merely the absence of disease or infirmity."
> — WHO Constitution, 1948.

We also learn from classic industrial design thinking approaches[13] that it is about much more than the digital answer to a (medical or care) customer need, but about an inclusion of the human psyche and its effects on the individual. If we now trigger parts of the doctor-patient relationship through digital services, psychological elements of being sick should urgently be considered as well. Talcott Parsons (1951), for example, describes the complex events of being sick–both on the doctor's side and on the patient's side. From primary and secondary disease gain to patient coping mechanisms, elements will also be reflected in digital interaction with patients.

Today, the combinatorics of connectivity, edge computing, miniaturization, and sensor technology make every single digital step of care globally available. Ordering a nursing device becomes processable anywhere—the measurements for a medical support stocking are suddenly available worldwide. Conversely, global medical knowledge becomes accessible for the smallest process step in medical care (see Fig. 33.2.).

33.3 The "Ecosystem of the Ecosystem" of the Patient

Once you have understood the ecosystem of an individual patient, you need to understand the "ecosystem's niche market," *thus market structure in which that ecosystem can be found.*

---

[13] Hasso Plattner, Christoph Meinel, Ulrich Weinberg: Design-Thinking. Innovation lernen – Ideenwelten öffnen. mi-Wirtschaftsbuch, München 2009, ISBN 978-3-86,880-013-5.

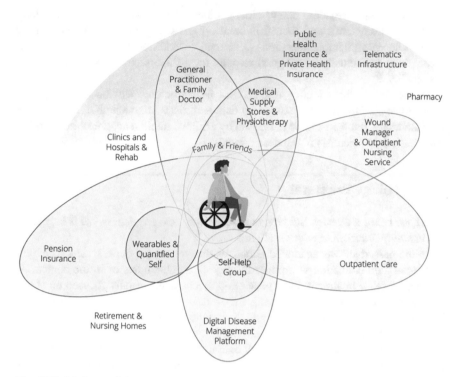

Fig. 33.2 Modern patient ecosystems

Thus, the personal network of the patient's relationships is found within a fabric of further systems: for example, self-help groups, symptom-related communication platforms, disease-specific chats in social networks, so-called disease management platforms, such as those recently also offered by health insurance companies, and many more.

Structured and unstructured data usually converges here. The knowledge gathered here is hardly available to the general society, let alone to the data provider "patient."

The politically regulated health care system also produces a wide variety of other structures. In these various structures, qualified medical information is sometimes already available in the form of structured data. Other structures, such as a physician information system of an outpatient physician, has a data quality that can hardly be reused by anyone or anything else. Other, partly emerging, data exchange platforms (for example, the electronic patient file of the telematics infrastructure, gematik[14]) over-regulate access to data in order to keep only structured data. The delays in the rollout of the electronic patient file in Germany, for example, (for which there are

---

[14] www.gematik.de/anwendungen/e-patientenakte

certainly multiple reasons) show, among other things, that an excessively long dispute over standards and their fulfilment criteria can paralyze and significantly delay a successful implementation—although, in addition to a legally anchored lever, there are actually many good reasons for an intersectoral exchange of data via an ePA.[15]

In the future, the winning systems and infrastructures will be those that can consolidate structured and unstructured data and make it comparable. They collect the data today that then can be harnessed by artificial intelligence, machine learning and clever data scientists in the near future.

33.4 Follow the Data!

You have heard it already, but here one more time, *"the physical world is becoming increasingly intertwined with the digital world."*

In the past, data was generated almost exclusively about our diseases. The only structured data collection was either through research studies or in the context of inpatient disease treatment. Data generation mechanisms usually focused on short, disease-related time periods, leaving the same—to be treated—individual unobserved in the convalescence phase and during the healthy periods.

New technologies, through miniaturization and edge computing, allow us to move the sources of structured medical data from the large data centers pharmaceutical companies or health care providers to each individual willing to have data collected on their person.[16]

The direct interaction between direct human body measurements and a physical device has recently become feasible. Self-regulating insulin pumps that automatically adjust their personalized medication delivery on the basis of measured values (and the feedback-loop of analyzing delivery and body response) are probably just the beginning (see Fig. 33.3).

Today, we still need high-performance computing power for the use of artificial intelligence. Tomorrow, however, these algorithms will also be available as local edge intelligence. The magic will no longer happen in the data center, but on a smartphone or a sticky electronic device, for example.[17]

---

[15] Reisepass und Gesundheitskarte mit Bits und Bytes, Handelsblatt, First published 22.03.2004 www.handelsblatt.com/technik/it-internet/schmidt-will-mit-elektronischem-rezept-milliarden-einsparen-reisepass-und-gesundheitskarte-mit-bits-und-bytes

[16] Stefanie Duttweiler, Robert Gugutzer, Jan-Hendrik Passoth, Jörg Strübing (Hrsg.): Leben nach Zahlen. Self-Tracking als Optimierungsprojekt? transcript, Bielefeld 2016, ISBN 978-3-8376-3136-4.

[17] Janine van Ackeren, Edge AI: Künstliche Intelligenz der nächsten Generation, www.iis.fraunhofer.de/de/magazin/kuenstliche-intelligenz-ki-serie/edge-ai-uebersicht.html, First published August 2020

Fig. 33.3 Closed-loop systems consist of an insulin pump, a sensor for continuous glucose measurement in the subcutaneous fatty tissue, a blood glucose meter for calibrating the sensor, and a computer program that automatically controls the insulin pump. Picture: Medtronic

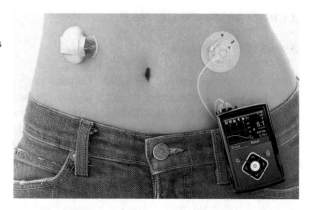

If we mentally couple this development with the miniaturization of medical technologies (mini bots, nanotechnology, mRNA vaccines, genetic cell therapies, etc.), suddenly an infinite number of new use cases can be addressed.

For example, think of vascular implants that react to blood pressure in a self-regulating (= automated) way to, for example, relieve a defected heart, or electrophoresis patches containing active substances that react specifically to certain sensor readings in such a way that they automatically regulate the release of active substances.

The limits of our imagination and that of medical technology developers and researchers will become more and more remote. The associated ethical discussion will be conducted by researchers and experts. The decision about social or personal acceptance, however, will be made by patients and the "market."

33.5   Use Real-World Data!

Always assume in your considerations that the solution you are thinking about today will already be able to draw on the information of a digital twin of a human being to the greatest possible extent.

People will be individually measurable—unlike in the past—not only at the moment when they fall ill, but always and in every life situation. Creating an incentive to have the data provided by a "customer" will have to be the subject of your business model. The data is a critical supplier for your emerging solution. It is the currency, the gasoline that powers your engine.[18]

---

[18]Gilbers, O., "Real-World-Data und -Evidenz in der Regulatorik" in "Monitor Versorgungsforschung" (03/19), S. 14–17; doi: 10.24945/MVF.03.19.1866-0533.2140

33.6 The Way Is the Goal

33.6.1 Think Simple

Disease education, health counselling, wellness programs, and patient group support: surely the most accessible target segment is healthy lifestyle support. Here, digital solutions range from offerings we call infotainment to evidence-based therapy approaches for behavioral management of specific diseases. Benefits can be felt directly through human interaction with the digital application. Be it guidance on a special diet or instructions on gymnastic exercises in the orthopedic field: such solutions have a limited complexity, require above all subject-related content and precise expertise, and can quickly lead to success. The price for this simplicity is a completely saturated market with low chances to differentiate and establish yourself as a follower.

So look for the market that is big enough to make money, but small enough that the niche has not yet been discovered. Mostly a niche will be discovered—but there is also unfortunately no large enough market opportunity in that niche.[19]

If digital solutions are limited to the conventional so-called user interfaces we are familiar with today, we are talking about smartphones, monitors, or voice interfaces. If a digital solution is limited to such feedback loops back to the human, you will always be limited to diagnostics and behaviorism.

Behavioral changes can generate medical benefits that can sometimes be superior to, for example, medications.[20] Digital coaching on diet, motivation, and information can improve a diabetic condition in such a way that blood glucose-lowering drugs can be omitted. We should therefore not underestimate this aspect. In the case of coaching in particular, human–human interaction has been shown to deliver better results than human–computer interaction.

33.6.2 Hybrid Things

Other effects that could happen to or within the human body are typically require some sort of hardware, which in some cases will be extremely small. Here it will be a matter of developing marketable solutions from research topics with high investment requirements, coupled with the highest level of expertise. Simply put, the more specific the medical challenge, the more complex the requirements for a hybrid cyber-physical solution.

---

[19] Sam Atkinson, "The Business Book: Big Ideas Simply Explained," ISBN 978-1,465,415,851

[20] Albrecht, Urs-Vito: Rationale. In: Chancen und Risiken von Gesundheits-Apps (CHARISMHA). Braunschweig, Peter L. Reichertz Institut für Medizinische Informatik der TU Braunschweig und der Medizinischen Hochschule Hannover (2016).

33.6.3 Big Data

Data consolidation is also the most important asset in medical applications. Only if I can achieve a critical mass of data, will I be able to draw sufficiently valid conclusions from this data. In recent years, especially in Europe, much discussion has been invested in the challenges of interoperability. Only those who collected structured data could make it available to other sectors with the necessary medical precision. If today, as before, it is primarily up to large research platforms to pull together many different data sources on a common infrastructure, tomorrow it will be about the availability of the interpretation results in the periphery. In the near future, we will hear from virtually all places about data lakes that will be able, thanks to artificial intelligence, to use even unstructured data for precise questions and algorithmic considerations. Structured and unstructured data will be able to coexist and be available in knowledge bases and information repositories for medical purposes.

Systems here will be able to generate binary approaches to medical decision-making equally as approaches that will be supported by artificial intelligence, along a medical care process.

The regulatory intricacies, particularly in medical device approval, will want to distinguish between the two approaches. The approval requirements of a medical device with artificial intelligence and machine learning will be much higher than the approval of a product with fixed and explainable algorithms.

Take-Home Messages
- You will always be able to increase adherence to a disease. Be better than that.
- Choose the right partners and think in terms of shared business models.
- Always center the benefits to be created.
- Consider the grand societal challenges.
- Find your way between the lobby interests of the market.
- Cut costs—follow the relevant areas.
- Think disruptively and think holistically.
- Consider the complex coexist of different ecosystems at different levels.
- Follow the data streams and the big data.
- Make the data results available in the periphery.
- Use modern technologies like miniaturization and edge intelligence.
- Think simple, hybrid and collect structured but also unstructured d.

Thorsten Hageman MD is a trained medical specialist and has more than thirty years of experience in healthcare. He advises MedTech, HCP and payers on the digital transformation of their IT architecture, products and processes. At adesso SE, one of the leading holistic IT service providers in Germany, he is currently Head of Business Development for the Line of Business Health and LifeScience. Together with his partners and customers, he digitizes patient care processes and is shaping new business approaches from mobile health apps to interoperable platforms.

Health Start-Up: Create Impact and be Investment Ready Intra- and Entre-Preneurs

34

Michael Friebe

Abstract

This chapter will provide some advise for the Intrapreneurs turned Entrepreneurs, because their concept is too far away from the current business model of the mother company and for the typical start-up Entrepreneur. The needs and requirements are generally the same with a slight advantage for the Edge solution that may come with some initial financial and/or other support from the mother company. This book is not a start-up book. But all the methodology—if properly applied—will eventually lead to a dedicated entity for the transformation from the idea and prototyping concept towards creating a positive impact on healthcare. Funding the development, business creation activities and needed staff is one of the core problems at this stage. While this chapter cannot provide detailed and dedicated entrepreneurial advice and information, it provides some valuable general comments and introduce tools to increase the odds of creating a successful setup for your health start-up.

Keywords

Pitch deck · One-pager · One page strategic plan · Purpose launchpad health · Start-up funding · Venture financing · OPSP · TRL

M. Friebe (✉)
AGH University of Science and Technology, Krakow, Poland

Otto-von-Guericke University, Magdeburg, Germany

IDTM GmbH, Recklinghausen, Germany

FOM University of Applied Science, Center for Innovation and Business Development, Essen, Germany
e-mail: info@friebelab.org

What will you learn in this chapter?
- Some—hopefully—valuable advise on how to prepare and present your start-up for the purpose of obtaining external funding.
- What you should have done in terms of validation before you make a decision—the Health Venture Requirements.
- The importance of a great vision combined with realistic milestones.

34.1   Introduction

You hopefully applied some of the methodologies presented in this book to advance from an initial Exploration phase towards an Evaluation phase. You should now have a clear understanding of the problem space that you are attempting to address with your innovation.

You also have validated the needs and the technological developments in that space, you have defined and started to validate the value propositions of your solution concept and have already created some crude prototypes for a Product-Market fit and checked Desirability, Feasibility, and Viability of the alternatives. And you also have invested some time to define the purpose of your activities, a longer-term vision, shorter term mission, and the values that your entity is based upon.

Figure 34.1 shows some of the different starting points. You may have gotten exposed to a health-related problem or issue because you believe you have found a health problem that needs to be urgently solved. It could also be that you have a personal motivation to address health challenges or issues. You find yourself in the HEALTH INNOVATOR/ENTREPRENEUR part on the left of Fig. 34.1. You would dive deep into the problem in a PATIENT/USER EMPATHIZE phase and come up with hypotheses and many ideas, that need to be (in)validated. You would run many customer experiments with crude prototypes. That process goes relatively fast, is focused on iterative learning using agile methodologies, and does not cost much. Many project ideas and initiatives do not survive that phase because the problem is not really a problem, the motivation has dropped, or there is not enough support to continue. The next phase requires a team, a good enough project plan and value propositions packaged in a future vision to convince early investors. The goal would be to further validate the ideas and create solutions that allow you to find early "paying" customers/users. "Crossing the chasm" is the title of a book that describes the need and presents concept on how to move from a very small number of "crazy" early users to create an attractive product and offering for the majority customers. This phase requires much more capital, a larger team and many start-ups and projects will not reach that phase. And then you need to scale your operations and maybe even create new markets and additional products.

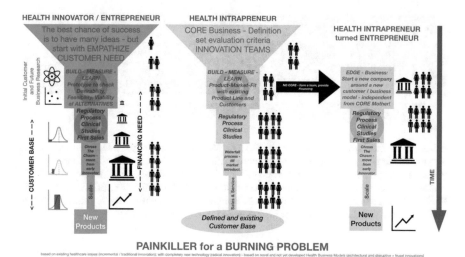

PAINKILLER for a BURNING PROBLEM

Fig. 34.1 Three possible scenarios of health innovation starting points. An innovator (person/team) that will become an entrepreneur on the left (RED). Many ideas are reduced in agile and iterative learning and validation processes, further reduced in a subsequent start-up phase. At the same time, the staff and financing needs go up to eventually achieve a sustainable operation. You could also come from within an existing organization (INTRAPRENEUR (ORANGE) with a goal to develop a new product or product improvement that fits to the existing offering (CORE business compatibility). The development and translation process is typically following a waterfall principle. And in case the new concepts require a new business model or address a customer that is not being served by the organization (EDGE business), this may then lead to a spin-out with much the same problems and issues as the ENTREPRENEUR

The difference for an intrapreneur/innovator within an existing, most likely successful medical technology or health delivery organization is that it comes with an established business and customer base. The new idea needs to fit into this existing operation (CORE). The innovation is probably with respect to improving the current offering or by extending the product to new markets or customers. These companies have existing expertise in developing, producing, marketing and will likely follow a typical waterfall approach to translate the ideas to their existing customer base. For many product improvement ideas, this is the most reasonable method.

And in case that the intrapreneurial activity turns out innovation approaches that are not in line with the existing business model and operational setup of the organization (EDGE) then there is still the option of spinning a company out. Important difference here, this new company needs to act and develop like a real start-up and needs to be detached from the mother organization. If that is not the case, it will act, behave, and argue just like the mother. It could be a little easier for these companies to get started, as they may get initial funding from the mother. We typically distinguish here between LINKED EDGE (disruptive and scalable business model that uses the mother company's assets—e.g., manufacturing capabilities, some engineering, research cooperations) and PURE EDGE (completely

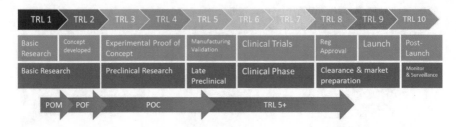

Fig. 34.2 Technology Readiness Levels (TRL) for Medical Technology Products. A Start-Up and with that funding/financing activities make sense starting at TRL 4–5 in the Proof of Concept (POC) Phase. For that you need to have completed the Proof of Market (POM), Proof of Feasibility (POF) Phases. Or in other words, you should have validated the DESIRABILITY, FEASIBILITY, and VIABILITY. The first 3 phases typically can be completed in just a few month by yourself or a very small team and with little (personal or institutional) funds [graphic from 1]

interdependent with respect to the mother company). Also the CORE opportunities can be further classified into EDGE-CORE, BLUE CORE, and PURE CORE depending on whether it is unique to the mother company, can be used in other organizations or is a product that highlights one or few specific features while reducing others. Something that is called VALUE INNOVATION in a BLUE OCEAN strategy (Value Innovation = simultaneous pursuit of differentiation and low cost leading to increased value for both the buyers and the company).

But all concepts/methods that we present should be applicable for all these scenarios as they are all based on some important core principles.

If you used the Purpose Launchpad Health (PLH) methodology for that process, you will have extremely valuable and convincing data available that supports your start-up plans.

Using Fig. 34.2, and the defined Technology Readiness Level as a reference, you are now at TRL 4 or 5 in the Proof of Concept phase, but have already checked the market and analyzed the on whether the planned technological features are actually feasible.

The results up to now can be summarized and presented in a ONE-PAGER document.

An earlier and simpler version (one that has already been presented and discussed in the PLH section) is shown in Fig. 34.3, and a more detailed one in Fig. 34.4, both for the EASYJECTOR case study project.

And Yes, you should have invested some thoughts in the Viability check:

- Is there a potential business model?
- What would that look like?
- How would I be able to make money (Business Model Canvas plus financial planning Details)?
- What would I need in terms of finance resources to pay for the development and translation to the patient/market?

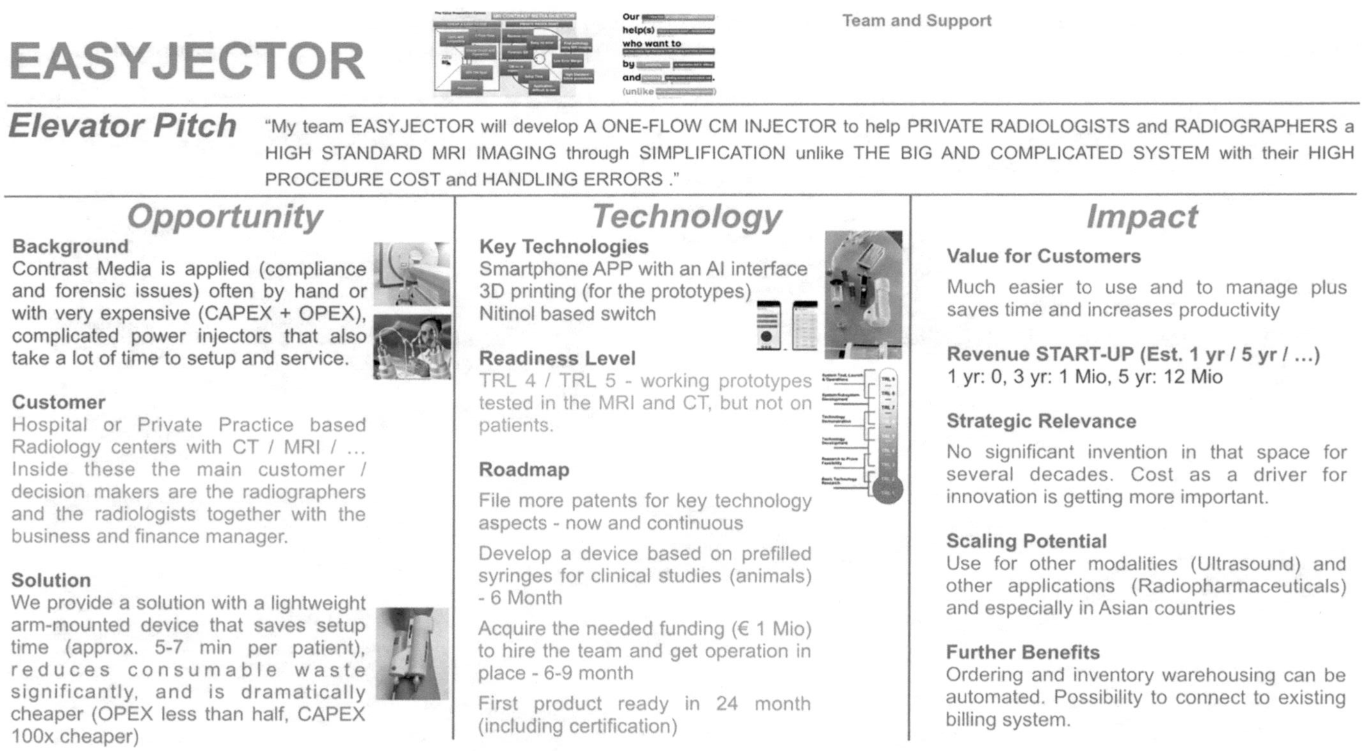

EASYJECTOR

Team and Support

Elevator Pitch

"My team EASYJECTOR will develop A ONE-FLOW CM INJECTOR to help PRIVATE RADIOLOGISTS and RADIOGRAPHERS a HIGH STANDARD MRI IMAGING through SIMPLIFICATION unlike THE BIG AND COMPLICATED SYSTEM with their HIGH PROCEDURE COST and HANDLING ERRORS ."

Opportunity

Background

Contrast Media is applied (compliance and forensic issues) often by hand or with very expensive (CAPEX + OPEX), complicated power injectors that also take a lot of time to setup and service.

Customer

Hospital or Private Practice based Radiology centers with CT / MRI / ... Inside these the main customer / decision makers are the radiographers and the radiologists together with the business and finance manager.

Solution

We provide a solution with a lightweight arm-mounted device that saves setup time (approx. 5-7 min per patient), reduces consumable waste significantly, and is dramatically cheaper (OPEX less than half, CAPEX 100x cheaper)

Technology

Key Technologies

Smartphone APP with an AI interface
3D printing (for the prototypes)
Nitinol based switch

Readiness Level

TRL 4 / TRL 5 - working prototypes tested in the MRI and CT, but not on patients.

Roadmap

File more patents for key technology aspects - now and continuous

Develop a device based on prefilled syringes for clinical studies (animals) - 6 Month

Acquire the needed funding (€ 1 Mio) to hire the team and get operation in place - 6-9 month

First product ready in 24 month (including certification)

Impact

Value for Customers

Much easier to use and to manage plus saves time and increases productivity

Revenue START-UP (Est. 1 yr / 5 yr / ...)
1 yr: 0, 3 yr: 1 Mio, 5 yr: 12 Mio

Strategic Relevance

No significant invention in that space for several decades. Cost as a driver for innovation is getting more important.

Scaling Potential

Use for other modalities (Ultrasound) and other applications (Radiopharmaceuticals) and especially in Asian countries

Further Benefits

Ordering and inventory warehousing can be automated. Possibility to connect to existing billing system.

Fig. 34.3 One-Pager Idea Summary. The OPPORTUNITY is described with background on the problem, a customer definition and a short solution idea. In the TECHNOLOGY section, you can describe the technologies that you plan to integrate, how far you are with the development and market/customer validation (TRL). And finally, the IMPACT part shows (also for intrapreneurial projects) what the customer and economic value of the proposal could be and on whether there are adjacent opportunities and further benefits

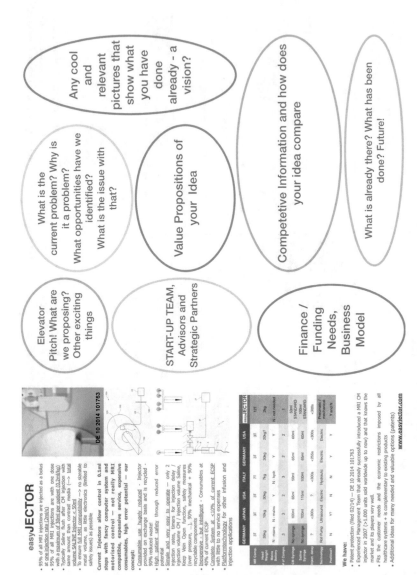

Fig. 34.4 A more comprehensive One-Pager. This is just a proposal layout that some will find too compacted and too busy, others appreciate the information content. So also here difficult to suggest a general template. On the right side, you see the rough content of the one-pager segments

This one-pager can then be used for building a more detailed presentation deck, also called a pitch deck. This can follow a similar sequence than the one-pager and should not take more than 10–15 minutes to present (so max. 15–20 slides).

Very often the time slots for presentations at investor conferences are only 5 minutes (or 5–7 slides—see below), which means you need to identify the key elements and convincingly package them together (Slide 0. Project Name and a Teaser Picture, 1. Vision, 2. Core Problem Identified, 3. How is it done right now and why is it a Pain and why should it be solved and for whom, 4. What is your proposal and your main Value Propositions, 5. How and with Whom will you solve that, 6. And what is the Timeline and Intermediate Steps to get that done, 7. What is the business Model and what do you need … plus a short repeat on why you are the coolest project).

34.2 Health Entrepreneur Checklist Towards a Start-Up

The following bullet points come with little comments and are meant to be used as a checklist. If you have done all that and have the information presentably summarized (do not forget to list and mention your learning process and the times you were wrong), this will give you a base for a convincing pitch and increase the chances of being successfully presenting your ideas and convincing investors to fund the future development:

- You completed the Definition of the HEALTH problem and have some validated NEEDS results.
- You have followed an iterative and agile methodology and have (in)validated your hypothesis and applied the learning to define the value propositions of your solution idea(s)—you have that documented can show where you were wrong. Investors like to see that Start-Up leadership is eager to learn!
- You have build some Prototypes to check the value propositions and find the ones that are most important to the customer and that create the highest impact.
- Your start-up concept has defined an initial Purpose (Massive Transformative), Moonshot, Vision, Mission and Values.
- You have formulated development steps and meaningful and measurable milestones and have an idea what it will require to complete them (Staff and Qualifications, Investments, Office/Lab space other Infrastructure, Funding to reach certain Milestones).
- You have checked on whether IP protection is possible (if you have not filed for patent protection, for example, do not disclose too many details/if you have filed then please do only show the first page).
- You have identified (for disruptive innovation formulated a business model vision) how the business model could look like and a good definition of the current and future market that your idea/product could address.
- You know the pathway and what it takes to obtain regulatory approval (if needed) for use on or with a patient.

- You have established a network of contributors, opinion leaders (not always good, because they are often part of the established business … but essential for incremental/traditional innovation), and other supporters and collaborators.

34.3 One Page Strategic Plan (OPSP)

This is a work canvas developed by Verne Harnish from the Growth Institute/ Gazelles, Inc. that can be downloaded from the institutes webpage including some explanations on how to use it [2]. This is an incredible tool that list all the core values, the companies purpose, and makes you think about meaningful metrics for the targets and goals. For me this (or such a) tool, used and updated as a living document, is all you need to combine a long-term vision with meaningful shorter term goals and to also continuously think about the PEOPLE (Employees, Customer, Shareholders) as REPUTATION DRIVERS and attached PROCESSES as PRODUCTIVITY DRIVERS.

This agile worksheet can easily be adjusted based on new learnings and new developments and is able to present all relevant planning and future vision information on one page. This is a blueprint of all activities and goals and with that is also well suited to update collaborators and investors (see Fig. 34.5 for an OPSP of EASYJECTOR).

34.4 Additional Advice

- Continuously learn, especially stay up to date with technology and market developments.
- In your team … be as diverse as possible with respect to age, gender, past experience.
- Stay close to a university till you reach TRL 4/5, then physically separate.
- Present your Vision and idea initially often and learn to become more and more convincing.
- Do not try take all feedback serious. You will often get 5 opinions from 4 people that are not congruent to each other.
- The vision is of utmost importance to get people/investors excited, but only the realistic informations make you a credible partner.
- Health innovations typically take longer and cost more than expected. You need to be patient and also need to look for investors that understand that.
- Never lie, never make things up … admit when you do not know the answers.
- Be realistic with your projections—if you are too low and conservative you will be considered.

Strategy: One-Page Strategic Plan (OPSP)   Organization Name: EasyJECTOR GmbH

Your Name: Michael Friebe, PhD   Date: 25. January 2022   Gazelles

People (Reputation Drivers)

Employees
1. Innovation_KPI > 1 prototype
2. Customer_Active_KPI > 4 Le
3. Daily_Huddle_Score > 80%

Customers
1. NPS_Radiographer KPI > 8
2. Number of REPEAT custome
3. UTI-KPI=CM pat. EJ/total C

Shareholders
1. NCF-KPI ≥33% of Revenu
2. #Hospital users > +20% p.a.
3. Gross Margins - KPI > 60%

Process (Productivity Drivers)

Make/Buy
1. B: Molds for Base Componer
2. B: Engineering Services for F
3. M: 3D printed prototypes

Sell
1. Licenses for certain Geograp
2. Licenses for certain applicati
3. Base product and consumab

Recordkeeping
1. Number of Consumables per
2. Online order turn-around incl
3. Regular weekly and monthly

COREVALUES/BELIEFS (Should/Shouldn't)	PURPOSE (Why)	TARGETS (3–6 YRS.) (Where)	GOALS (1 YR.) (What)	ACTIONS (QTR) (How)	THEME (QTR/ANNUAL)	YOUR ACCOUNTABILITY (Who/When)
- Make Contrast Media Injections really EASY! - Provide solutions that make SENSE and improve the workflow - Focus is on EASE, AFFORDABILITY, PATIENT SAFETY - We want to be known as partners that are open to customer ideas - We are listening! - We can build a customer specified prototype for validation in 1 month - We attempt to predict future market developments through the use of agile and iterative innovation methods(revalidated every 3 month). - We will intensively communicate with our stakeholders	Create HealthCare Technologies with the main purpose to provide AFFORDABLE solutions.	*Future Date* 2025 Revenues 5 Mio Profit 2 Mio Mkt Cap/Cash 25 Mio **Sandbox** MRI/CT/US injection tech in E = € 500Mio + € 300Mio drug delivery	*YR Ending* 2022 Revenues 100k Profit -300k MKT Cap 3 Mio Cash 0.75 Mio Gross Margin A/R Days Inv. Days Rev./Emp.	*Qtr #* ending 31.12.2022 Revenues 150000.- Governme Profit less than (- 200,000 Mkt Cap 3.5 Mio Pre-Money Cash 500,000 still availab Gross Margin n.a. A/R Days n.a. Inv. Days n.a. Rev./Emp. n.a.	*Deadline:* 31.12.2022 *Measurable Target/Critical #* Generate more than 12 different additively **Theme Name** DOZEN_IP_GEN	*Your KPIs* *Goal* 1 Members LinkeIn group > 500 2 Returned Customer Questionnaires > 50 3 Inventions identified > 5

Actions
To Live Values, Purpose, BHAG
1. Build MVP's to prove the point!
2. Have continuous interaction with the
3. Be in an experimental and
4. Think about GLOBAL solutions.
5.

Key Thrusts/Capabilities
1. Turn idea into MVP... file IP
2. Have a global innovation network
3. Open database for Unmet Clinical
4. Provide scientific proof and publish
5. Win Innovation Prices for most

Key Initiatives
1. Receive growth funding -- create
2. Hire dedicated CEO -- growth
3. Create a one sentence Elevator
4. Create and motivate clinical
5. Create an ideation to Innovation

Rocks *Who*
1. Investment Pitch for Series A — MF
2. Second Patent filed — SH
3. State Government Funding proposal for — MF
4. Employees (1 x Software IG Engineer, 1 x Mechanical — SH
5. Webpage and LinkedIn group started and — SH

Scoreboard Design
Describe and/or sketch your design in this space
- Drive Mechanism - Spring, Air, Osmotic, ...
- Syringe over syringe
- Bag on Valve
- Communication Options (IR, BT, WIFI)
- Start/Stop Options

Your Quarterly Priorities *Due*
1. Develop a Customer Questionnaire for the — 03.22
2. Develop a validated test protocol /lab — 03.22
3. Identify potential international — 06.22
4. Create a digital collection of — 06.22
5. Identify and secure members of an — 06.22

Strengths/Core Competencies
1. Close contact to the clinical users.
2. Know the market.
3. Innovation Generation

Profit per X
X = Installations
2024 = € 2.500 per site per year
2025 = € 2.750 per site by 50%

BHAG
Change the akward and complicated way of CM Injection for Tomographie

Weaknesses:
1. Dedicated 150%, CEO missing
2. Need young, motivated team
3. Too convinced that the concept is i

Brand Promise KPI's
□ CEO plus BD Directc
□ Motivated CEO in ple
□ no dedicated CEO

Brand Promises
Fall in love with the Ease and Cost savings in the first day of full use

Critical #: People or B/S
□ 500 LinkenIn Group Members
□ 400 LinkenIn Group Members
□ <250 LinkenIn Group Members

Critical #: Process or P/L
□ Calculated Volume Cost
□ Calculated ... < 40€
□ Calculated ... > 100€

Critical #: People or B/S
□ > 100 Customer Talks and
□ > 75 Customer ...
□ < 50 Customer

Critical #: Process or P/L
5 -- IP Filed or Journal
3 -- IP ...
2 or less -- IP

Critical #: Process or P/L
□ 75% GM
□ 70% GM
□ <50% GM

Celebration
River ELBE Boatrip Celebration on a Friday Morning ... with Dinner in the Evening.

Reward
An Innovation Trip -- Sightseeing, Drinking, Eating, Learning -- to Scotland (4 days) for all

Trends
1. Dramatic cost reductions in Health
2. IoT Devices that allow for easy ord
3. Portable and easy to use meeting

4. Contrast Media and injection devic
5. (Semi)-Personalized packaging
6. Low waste

To download more copies and to get help implementing these tool

To download more copies and to get help implementing these tools, plea

Copyright 2015 Gazelles, Inc.

Fig. 34.5 The ONE PAGE STRATEGIC PLAN from Verne Harnish, Gowth Institute/Gazelles, Inc. [2] combining short-mid- and long-term goals, with vision and purpose and many other relevant planning details. The OPSP is an agile blueprint of the company and can easily be updated when new insights have been generated

34.5 Summary

Every start-up is different, and every product/service/workflow/process improvement or novel approach needs to be differently presented. There are some general things that you should have completed before a start-up financing has realistic chances. And remember that the Patient and especially a future Empowered Patient will change everything with respect to a Future Health business model [3, 4].

Take-Home Messages
- You can be over-prepared! Preparation takes time and effort! You need to find a balance between what you need to know before and what you will still need to find out once you have started your venture.
- Enthusiastically present a future-oriented vision about the impact of your start-up combined with realistic and measurable goals towards that vision.
- Especially for disruptive innovations, you cannot create a credible business model assumption. So be honest and say that you do not know.
- Never make things up or clearly present them as unvalidated assumptions and hypotheses.
- Finance numbers and business assumptions are not as important as you think. An investor wants to see your ambitions, that you are motivated, realistic (neither too low nor too high), know your deficits, that you are willing to check and re-check your propositions based on your learnings, and that you are an open communicator.
- Show that you understand the UNMET CLINICAL NEED, the problem definition, the current alternatives, the market around that problem, and future development/ideas trying to address that problem–as an investor, you cannot ask for much more than that.
- You should have made it—using the Purpose Launchpad Health or other methodologies—to at least TRL 4/5 before you are somewhat ready to get external funding. The funds needed to get to this point have to be paid by yourself/your team/your community or by applying for government grants.
- Create a one-pager summarizing the problem, your solution, the market and vision—this called a Teaser Document, because it should be interesting enough to ask.
- Healthcare needs, and business models, are different in just about every country—what does work in Germany may not work in India or vice versa. Invest time to check and identify the markets that are best suited for your product and company. Also, check Reverse Innovation opportunities.
- Think about how you can use exponential technologies and an exponential mindset to connect your product to the "abundance."

(continued)

- Using an agile planning tool like the One Page Strategic Plan is incredibly helpful for you and your interaction with your investors, customers, and employees.
- The single biggest opportunity for health innovations is by EMPOWERING PATIENTS!

References

1. From https://growmed.tech/wp-content/uploads/2019/04/Grow-MedTech-Call-Info-Sheet-FINAL.pdf viewed and downloaded March 20, 2022
2. Harnish V—One Page Strartegic Plan. Available via the Growth Institute https://blog.growthinstitute.com/scale-up-blueprint/opsp-one-page-strategic-plan viewed and downloaded March 22, 2022
3. Deloitte Consulting—on the Future of Health: https://www2.deloitte.com/us/en/insights/industry/health-care/forces-of-change-health-care.html, viewed and downloaded March 22, 2022
4. Future Today Institute—Future Health Trends—available https://futuretodayinstitute.com/mu_uploads/2022/03/FTI_Tech_Trends_2022_Book06.pdf, viewed and downloaded March 22, 2022

Michael Friebe   received the B.Sc. degree in electrical engineering, the M.Sc. degree in technology management from Golden Gate University, San Francisco, and the Ph.D. degree in medical physics in Germany. He spent 5 years in San Francisco, as a Research and Design Engineer with an MRI and ultrasound device manufacturer. He is a German citizen with expertise in diagnostic imaging and image-guided therapies, as a/an Founder/Innovator/CEO/Investor and a Scientist. He is also a Research Fellow with the Technical University of Munich, Munich; an Adjunct Professor with the Queensland University of Technology, Brisbane; and a honorary Professor of image-guided therapies with Otto von Guericke University, Magdeburg, Germany. Since 2022, he has been a Professor of biomedical engineering innovation with the AGH University of Science and Technology, Krakow, Poland. He is a listed inventor of more than 80 patents and has authored over 200 papers, and was part of over 35 Medical Technology Start-Ups. He is a Board Member of four medical technology start-up companies and an investment partner of a MedTec-fund. From 2016 to 2018, he was a Distinguished Lecturer of the IEEE EMBC teaching innovation generation and MedTec entrepreneurship. He is also a coach and trainer for OpenExO, and a master Launchpad mentor for the Purpose Alliance.

Regulatory Issues for Health Innovations

35

Axel Boese

Abstract

There is a hurdle between the creative innovator and the successful business creator in the medical device industry: the standards and laws regulating market access!

As the names already reveal, standards and regulations do not correlate with free-minded and creative acting. Nevertheless, innovation is only an innovation when it arrives at the market. The good thing about regulatory is that you have to follow the rules. The bad thing is that you have to follow many rules, and they differ from country to country. For the medical device industry safety and performance of the products are crucial. Every one of us is also a patient and needs to rely on and trust the diagnostics and treatments for which these products are used. To ensure that, there are regulatory frameworks that guarantee a high level of quality, safety, performance, traceability, reproducibility, and define responsibilities. As a real "Innovator," you should be willing to claim this for your cool idea and start the transfer into a real product. One more motivation statement before you start to read our short summary on regulatory issues: Others have done that before, it is doable! Yes, you have to learn a lot, yes, it is paperwork, yes, it will maybe make you crazy from time to time, and yes, you will make decisions that need "corrective actions" from time to time. But, if you have managed it in the end, you can be proud, and even if you are not successful, you will earn experience and knowledge.

Keywords

Regulatory · Certification · Intended purpose · Quality management · Product development · Documentation · Go to market

A. Boese (✉)
MEDICS GmbH, Magdeburg, Germany
e-mail: boese@medics-md.de

What will you learn in this chapter?
- Medical Products are governed by lots of regulations—but not all is bad . . . and not all is bad for your product and company.
- Trust in the product performance and high safety standards are expected . . . think as a patient.
- It is a lot of paperwork, but doable . . . especially when you start early and document everything you do . . . regulatory approval will likely boost the attractiveness of your company.

35.1 Regulatory Framework

Bringing ideas into the healthcare market requires the consideration of a strict regulatory framework. Healthcare products have a significant impact on patient's life and health. The safety and efficiency of these products have to be self-evident. The regulatory framework defines the rules to place medical devices and services on the market to ensure that. In the European Union (EU), the Regulation (EU) 2017/745 of the European Parliament on medical devices (MDR) [1] builds the basis of this framework that is transferred into national law.

In the USA, the rules are defined and controlled by the Food and Drug Administration (FDA) [2].

Other countries have different but comparable regulations. Even though these systems have their differences, the basic concepts are similar. The regulations define the minimum standards on good practice, traceability, documentation, safety testing, performance evaluation, convenience and a defined procedure and documentation to track responsibilities.

For innovators and especially startups, the task is to arrange their innovative ideas and work with this regulatory framework. This sounds not like fun and creates a lot of extra work, and some may say that this is the killer of creativity. However, good ideas can only turn into innovations when they arrive on the market, and no one wants to be subsequently treated by a creative but not tested medical device. The strict regulations in the healthcare business can be seen as a chance for making innovative product ideas more realistic and marketable. The strong focus on quality, documentation, user integration, and evidence increases safety and performance and can turn a visionary idea into a safe and functional product.

For startups, mastering the regulatory requirements can also demonstrate the willingness and professional abilities to bring an idea to the market and increase attractiveness with investors.

To transfer an idea into the market and overcome the regulatory hurdles, some key activities are needed on three different levels: people, company, and product.

35.2   Preparing the People

The people are the main drivers behind the translation process from ideas into products. Nevertheless, this sometimes comes with a change of roles and tasks. For example, scientists and creative minds have to take over straightforward and regulated jobs in a company team.

As we have learned, medical product innovation is challenging, and if you are not an expert in this field, defined training on this is mandatory. In addition, a basic understanding of the regulatory framework should be made available for all team members if you work in a team. Reading the MDR or FDA requirements and applicable standards could be one way, but these documents are hard to understand if you are a newbie.

Nowadays, there are many consultancy companies offering dedicated training courses. They offer several levels of teaching, starting from an overview up to a deep dive into the regulations. Usually, these courses come with many examples and explanations and can create a network of fellows dealing with the same problems. So invest the money and take these teachings to an early stage. Of course, the quality of these courses varies, but it is much easier to learn and discuss in a group of like-minded people. For the EU, the training should include at least the basics of the MDR, the quality management standard ISO 13485 [3], and the risk management standard ISO 14971 [4].

With a basic knowledge and understanding of what is needed, the team responsibilities have to be defined. Pushing the idea in the direction of a certified medical product is an essential activity and has to be taken seriously from the beginning. The whole team has to accept and support that. Some of the essential roles requested by the MDR and the connected standards are:

- General Management GM: Strategic decisions and responsibility.
- Quality Management QM: Supporting the GM to build up a QM system documented in a QM Handbook including a QM policy, the responsibilities, company structure, all QM objectives, company processes, standard procedures, work instructions. . . .
- Risk Management RM: Product-based risk identification, prevention, and documentation.
- Regulatory affairs RA: Checking for news and changes on regulatory and compliance with the regulations of the country where the product is placed (person responsible for regulatory compliance).
- Development team: for Hardware/Software following GM, QM, RM, RA.
- Resources (finances, human, materials, partners, suppliers, ordering, invest).
- Studies and Testing: study design, study partner management, statistics, documentation, design input.
- Sales and Marketing: public relations, contracting, delivery, teaching and information, user contact (Medical representative).

The assignment of these roles can be connected with personal liabilities, and some of these roles are essential for the survival and success of a company. This has to be considered when assigning the roles to the team members. Defining the roles is also connected with a gap analysis. If an expertise gap in your team is identified, it has to be filled with additional expertise or education. For some of these roles, an external subcontracting could—and in an early stage should—be an alternative.

35.3 Preparing the Company

Besides all necessary structures and procedures to run a successful business, a quality management system (QMS) is essential for medical companies. This is required by the MDR but also by the FDA, and its structure is defined in the quality management standard for medical devices ISO 13485 [3].

The idea of a QMS is to have defined and traceable procedures, responsibilities, and goals and to follow a continuous optimization process. Although defining the roles and responsibilities is the first step towards a QMS, creating a company structure covering these roles is the key to defining the company processes and procedures.

Therefore, a good starting point is a sketch of "things to be done" inside the company and all connections to the outside world. The processes usually are structured as core processes of the company (e.g., production chain), management processes (e.g., accounting), and support processes (e.g., machinery maintenance).

The structure of the company and all processes are documented in the Quality Management Handbook (QMH). This acts more or less as a guide for everyone and everything that happens inside the company and how the company interacts with the outside world. It is crucial that the QM handbook is known and accepted by the team and all members are living the defined rules. For startups, it could be beneficial to create the QM handbook with the whole team and to start the process very early on.

The QMS is a living system and can be optimized or adapted if needed. If your company wants to act on the international market, it would be a good idea to create all QMS documents in English, but be aware that all team members have to understand and follow the QMS. Before the official certification of your QMS, it is helpful to have an evaluation and optimization phase where all defined procedures and associated documents are tested and adapted if needed. Role games (Empathize) to take over the position of customers, partners, suppliers, or stakeholders in this test phase can help to identify definition gaps in the processes or inconsistencies.

35.4 Preparing the Product

Medical products need to be certified as such. The rules depend on the country's regulations where the product is planned to be placed on the market. We will give a short summary of the documents needed for the certification. The essential document for a medical product is the so-called "Technical Documentation" (TD). This is development documentation that describes the idea all the way to an evaluated and certified product on the market. The preparation of this documentation is mandatory and has to start as early as possible, the latest with the shift from creative research to a structured development process. The starting point for the TD is the definition of the intended purpose of the product and, according to that, the risk classification based on the classification rules of your target market.

- What exactly is the medical application that the product addresses?
- Who will use this product?
- How will this product be used?
- For which procedure or patient cohort is the product suitable?
- Besides the medical application, are there other intended uses like storage, interconnection, transport, cleaning . . . .

The intended use serves as input for the risk classification of your medical product. In the EU, the 22 classification rules are listed in Annex VIII of the MDR [1]. The risk classes are subdivided into class I for lower risk, IIa, IIb, and III. The risk class depends on several aspects such as invasiveness, duration of use, or if the device is "active."

The FDA defines the risk classes I, II, and III. A guide on how to classify your product can be found on [5].

[1] and [5] provide an assessment tool for determining the risk classes of your individual product.

The definition of the intended purpose has to be carefully balanced with the evaluation of the size of the addressed market and the risk classification. A wide range of an intended purpose can open up a larger market but can create more testing efforts to prove the proposed value. The same counts for a higher risk classification that again can lead to higher development, testing, and certification costs.

The second step in setting up the TD is the description of the use case and, based on that, the definition of the basic functionalities of the product. This leads to the definition of product requirements and the first system architecture and is the starting point for the ongoing risk analysis. All possible risks have to be identified and documented. During the development process, these risks have to be prevented or reduced by appropriate solutions.

The structured development process transfers the requirements and functionalities into realizable solutions, including the determination of all components, production processes, suppliers, sterilization strategy, packaging, delivery, user education, and so on. In parallel, a testing and evaluation strategy has to be designed. Depending on the complexity and the intended purpose of the product, the features, components, and overall product has to be tested.

It has to be demonstrated that the performance of the device fulfills the intended purpose and the product use is safe. All steps in the development process, including decisions, changes, and test results, have to be documented in the TD and comply with the actual regulatory standards and knowledge. The remaining risks of the product are opposed to the expected medical benefit in a Clinical Evaluation Report

(CER) that is also part of the TD. This analysis should show a clear advantage of the application over the remaining risk. Another part of the TD is the description of the post-market activities to observe the use of the product and identify possible failures or miss-use. This should offer the possibility for the provider to react fast if any unwanted or unexpected event occurs in the real-world application.

The goal of the described structured and documented development process for medical products is to achieve a high level of quality, traceability, and control and thus an increased product safety. For products that should arrive on international markets, again an English documentation would be beneficial. Regarding product testing, it could be beneficial to collaborate with certified bodies (CB) following the CB scheme. This offers manufacturers a simplified way of obtaining multiple national safety certifications for their products based on the same testing.

In the EU, the TD of the product and the QMS of the company are, depending on the risk class, assessed by "Notified Bodies" (NB) for conformity with the regulatory. An NB is an organization designated by an EU country carrying out tasks related to a conformity assessment as a third party [6]. If the conformity assessment is successful, the product gets certified and can be placed on the market. Not every NB is assessing every kind of product. The NB in the EU and their focus are listed in the "New Approach Notified and Designated Organizations" (Nando) Database [7]. In the USA, the FDA is responsible for the assessment of the company and the product documentation.

The success of a medical product is highly related to the "go-to-market" strategy. Since regulations and reimbursement differ from country to country, the related costs and time-to-market can vary. Sometimes it might be of advantage to start with a lower risk class product or a non-medical lifestyle product to evaluate user acceptance or collect first data. This counts especially for digital applications or wearable products in the home care market. But as soon as your product fulfills the definition of a medical device of your target market, it is a medical device!

For the whole development, risk management and testing procedures, several standards and guidance documents are available. It might be challenging to keep an overview of all these documents, but the positive message is: There is guidance and documents that help to manage the task!

35.5 Checklist for Starting the Regulatory Journey

- Do you know the regulatory and the go-to-market processes of your target market?
- Do you have your team ready to fill the roles needed for the regulatory journey?
- Does your company fulfill the requirements to act as an economic operator on the medical device market?
- Does your company have a Quality Management System and is that alive?
- Did you define an Intended Purpose of your product idea?

- Did you establish a structured development process, including strict documentation?
- Do you have access to the right consultancies and assessment organizations?

Take-Home Messages

People:
1. Stop being a team of scientists/researchers, start to be a company team.
2. Connect to professionals for guidance, exchange, and communication— build up dedicated regulatory knowledge.
3. Start working, make decisions, fill out the documents. Regulatory affairs and all related activities can only be successful if you start to work on it.

Company:
1. A Quality Management System is essential for medical device companies.
2. Continuous optimization of a QMS is a chance for improvement and efficiency.
3. A QMS has to be alive and accepted by the team.

Product:
1. A structured product development process has to be documented in a Technical Documentation.
2. The definition of the right range of the Intended Purpose of the product is crucial for costs and market.
3. For successful certification, conformity with the regulations has to be assessed.

References and Further Reading

1. Regulation (EU) 2017/745 of the European Parliament and of the Council of 5 April 2017 on medical devices. 2020. Zugegriffen: 9. Februar 2022. [Online]. Verfügbar unter: http://data.europa.eu/eli/reg/2017/745/2020-04-24/eng
2. Mallis E. An introduction to FDA's regulation of medical devices, S. 40
3. "ISO 13485:2016", ISO, 2016. https://www.iso.org/cms/render/live/en/sites/isoorg/contents/data/standard/05/97/59752.html (zugegriffen 13. Februar 2022)

4. "ISO 14971:2019", ISO, 2019. https://www.iso.org/cms/render/live/en/sites/isoorg/contents/data/standard/07/27/72704.html (zugegriffen 9. Februar 2022)
5. Health, Center for Devices and Radiological. "FDA: Classify Your Medical Device". FDA. FDA, 22. Oktober 2020. https://www.fda.gov/medical-devices/overview-device-regulation/classify-your-medical-device
6. Decision No 768/2008/EC of the European Parliament and of the Council of 9 July 2008 on a common framework for the marketing of products, and repealing Council Decision 93/465/EEC (Text with EEA relevance), 218 OJ L § (2008). http://data.europa.eu/eli/dec/2008/768(1)/oj/eng
7. "EUROPA – European Commission – Growth – Regulatory policy – NANDO". https://ec.europa.eu/growth/tools-databases/nando/ (zugegriffen 9. Februar 2022)

Dr. Axel Boese is a mechanical engineer and an expert in medical device development and clinical application with years of experience. Besides several developments, publications and patents in the field of image-guided therapies and instruments, he is a certified Quality Manager for ISO 13485 medical devices, an investigator for clinical studies and medical device testing, evaluation and GCP. Axel Boese is collaborating with clinicians and medical professionals to create solutions for real clinical needs. Beside his scientific career he is co-founder of three companies and founder and CEO of MEDICS GmbH, Germany.

Innovation Think Tank Frameworks for Resolving "The Innovator's Dilemma" in Healthcare

Sultan Haider

Abstract

Innovators are constantly seeking the right challenges, stakeholders, and environment for implementation of novel ideas. Technological advancements in healthcare and their implementation require identification, development, testing, and validation of high-impact use cases. Timing is crucial for successful market positioning and commercialization, which is highly dependent on (1) customer insights during requirements generation and (2) co-creation and validation of minimum viable prototype (MVP). One constraining factor is trust building with various healthcare system stakeholders as well as the current limited understanding of the entire product life cycle. The purpose of this paper is to firstly present Innovation Think Tank frameworks developed in the last 17 years at its various locations for knowledge reuse and opportunity and impact analysis for accelerating decision-making, and secondly to elaborate the success factors and their role in the innovation life cycle. Three hundred project modules were analyzed. Creativity, tools, and methods as well as trusted partnerships for real-time customer feedback were identified as key success factors. The framework can be used for identification and evaluation of the opportunities and accelerating MVP creation, testing, and validation in customer environment.

Keywords

Innovator · Innovation Think Tank · Healthcare system · Co-creation · Innovation indicators · Siemens healthineers

S. Haider (✉)
Siemens Healthineers AG, Innovation Think Tank, Erlangen, Germany
e-mail: sultan.haider@siemens-healthineers.com

© The Author(s), under exclusive license to Springer Nature Switzerland AG 2022
M. Friebe (ed.), *Novel Innovation Design for the Future of Health*,
https://doi.org/10.1007/978-3-031-08191-0_36

What Will You Learn in This Chapter?
- The Innovation Think Tank framework
- The results of decade-long global projects with healthcare partners
- And how it can be used for identification and evaluation of health innovation opportunities

36.1   Introduction

Innovators are individuals or groups who create products and solutions for addressing societal needs, by pushing boundaries, challenging the status quo, and achieving growth. They can be entrepreneurs, but they can also be active and employed in companies and act as intrapreneurs. Some of the challenges faced by these innovators are:

(a) Difficulties in acquiring a project mandate,
(b) They often overlook the big picture that includes scope, stakeholders, pain points, etc., and thus experiencing analysis paralysis.
(c) They fail to identify and prioritize ideas holding the biggest impacts.
(d) They lack guidance on how to use appropriate mechanisms, and
(e) They often do not have access to environments to engage different entities for implementation and commercialization.
(f) Or they lack resources, strategy, transdisciplinary, and time.

A number of these challenges contribute to the "The innovations dilemma" which spans from (1) selecting the right problem for research, (2) making a right decision for acquiring skills and education, (3) how much creativity is needed at a certain step considering the customer needs and available resources, (4) defining the starting point, (5) Opportunity/impact evaluation, to (6) choosing the right stakeholders [1]. Innovators need to know when and how to change traditional business practices [2].

Healthcare systems worldwide increasingly require innovators to come up with holistic and cost-effective solutions to address the needs of patients, providers, and payers due to limited budgets and rising patient populations. Healthcare in many countries is not providing care in a sustainable manner, they are largely based on leveraging values for reactionary and acute care, which is not well suited to constantly increasing chronic disease burden on aging population. COVID 19 pandemic has revealed the need for disrupting healthcare delivery and adaptation of the societies with new technologies (e.g., home care, remote diagnosis, education, etc.).

Innovation Think Tank (ITT) is a global infrastructure of co-creation programs and labs established at Siemens Healthineers and several of its partner universities and hospitals. Its methodology includes elements ranging from acquiring a mandate for initiating projects to commercialization, covering the entire innovation lifecycle [3].

The Healthcare System Frameworks proposed by ITT engage stakeholders and decision makers to get insights into the trends and solutions [4] and with that may help to overcome the innovators' dilemma of large companies.

ITT offers Expertise Development Programs (EDP) worldwide with a focus on innovation methodologies, entrepreneurship, and commercialization. ITT EDPs are experiential learning programs based on the experience of successful implementation and management of Innovation Think Tank programs and labs at Siemens Healthineers and several prestigious institutions worldwide. The interactive program is designed to develop creative pioneers capable of delivering innovative and customer-centric solutions to the world's greatest challenges in healthcare, in their own field of profession.

The program strongly focuses on making use of intrinsic creativity and various tools and frameworks for addressing customer needs. ITT EDPs have been offered at several prestigious institutions in the form of Innovation Think Tank Certification Programs, Innovation, Management, and Leadership Certification, fellowship programs, etc., in reputable global institutions such as Peking University, China; Imperial College London, U.K.; Technical University of Munich, Germany; Otto von Guericke University Magdeburg, Germany; Oxford University Hospitals, U.K.; FAU Erlangen Nürnberg, New York University Abu Dhabi; HS Coburg, Germany, University of Freiburg, Germany; OTH Amberg-Weiden, Germany; Vellore Institute of Technology, Texas A&M, Bogazici University, Acibadem Hospital Group and Hospitals, Georgia Institute of Technology, ERA Lucknow Medical College and Hospitals, Baskent Hospital Group and University, Fakeeh University Hospitals, Western University, Canada, University of Evora, Portugal and various Siemens Healthineers global locations [3, 5].

36.2   Material and Methods

Three hundred project modules were analyzed from a period of 10 years in 15 ITT locations with a team of 100 engaged employees [Table 36.1]. Project modules were categorized into three main levels: healthcare system level, departmental level, and system and component level. The weekly outcome reports, including project data, stakeholder identification, and analyzed content, were clustered and key performance indicators (KPI's) were monitored and validated with the annual targets.

36.3   Results

The ITT framework has been visualized with the core modules as shown in [Fig. 36.1] and could be used during various stages of the innovation life cycle. Similar to the ITT Healthcare System Framework elaborated in [4], clinical workflow, technology assessment and business model modules support capturing and validating trends, visualize its stakeholders, facts and figures, challenges, identifies opportunities, best practices, and interdependencies. Several example pain points and solutions for 22 disease pathways have already been created and presented in [5].

Table 36.1 Project data

Project module Category	Project data categories	Content analyzed	Project outcomes analyzed
Healthcare system level	Hospitals and healthcare systems of the future. Healthcare system analysis for Turkey, USA, India, China, South Africa, Egypt, Canada, Germany, South Korea, Austria, Italy, UK, UAE, Saudi Arabia, France	Patient journey, total healthcare expenditure, reimbursement model, healthcare policies, current trends, pandemic management, disease pathways	# Savings # Additional revenue potential created # Time to market # Number of customer interaction sessions # Project modules reused # Project participants, expertise
Departmental level	Radiology infrastructure of the future, radiology user interfaces of the future, laboratory workflow analysis for digitalization, other workflows (operating room (OR), emergency room (ER), intensive care unit (ICU), cardiology, catheterization laboratory)	Business models, process steps, current trends, KPIs, success factors	
System and component level	Computed tomography (CT), medical resonance imaging (MRI), X-ray, molecular imaging (MI), fluoroscopy, ultrasound, patient table	Modality workflows, business models, lifecycle visualization, technological and clinical trends, KPIs, success factors	

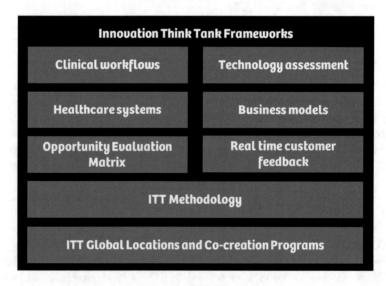

Fig. 36.1 Innovation Think Tank Frameworks overview

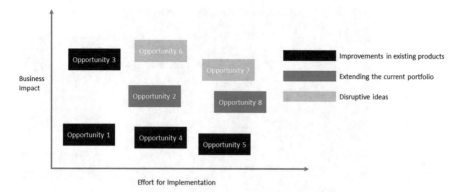

Fig. 36.2 Innovation Think Tank Opportunity Evaluation Matrix (ITT OEM) supports assessment of business impact vs. effort for implementation. Business impact consists of additional revenue potential created due to new features, savings, and disruption potential. All opportunities are relative to each other

Fig. 36.3 Innovation Think Tank success factors

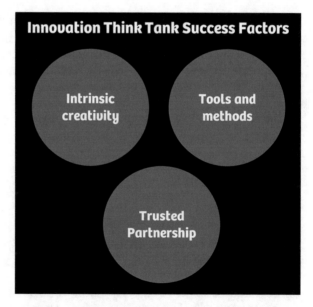

An Innovation Think Tank Opportunity Evaluation Matrix (ITT OEM) supports assessment of business impact vs. effort for implementation [Fig. 36.2]. Business impact consists of additional revenue potential created due to new features, savings, and disruption potential. All opportunities are relative to each other.

Intrinsic creativity, tools and methods and trusted partnership for real-time customer feedback have been identified as key success factors resulting in annual savings and additional revenue potential with new offerings and shortened time to market [Fig. 36.3].

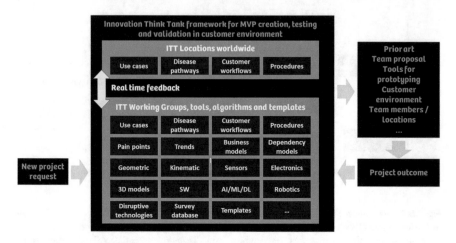

Fig. 36.4 Innovation Think Tank knowledge reuse leading to accelerated decision-making and shortening time to market and savings success factors

The Innovation Think Tank framework connects all ITT locations worldwide, where open innovation and co-creation programs result in the generation of knowledge (Technology trends, pathways, procedures, product requirements, etc.).

For a new project request, a semi-automated ITT Framework proposes reusable assets as well as competencies, tools, customer environment, templates, and locations.

The new project outcomes are curated and indexed within the ITT framework [Fig. 36.4].

36.4 Discussion

The ITT Frameworks aid in resolving the innovator's dilemma by

(a) Providing a systematic approach for acquiring mandate and accelerates decision-making by utilizing prior knowledge and its transdisciplinary global infrastructure.
(b) Real-time customer insights support the creation of a big picture and support engagement with the users and decision makers.
(c) The opportunity evaluation matrix supports the business impact calculations with respect to a required effort for implementation.
(d) The clinical workflow, healthcare systems, and technology assessment modules in combination with ITT methodology provide guidance throughout the innovation lifecycle.

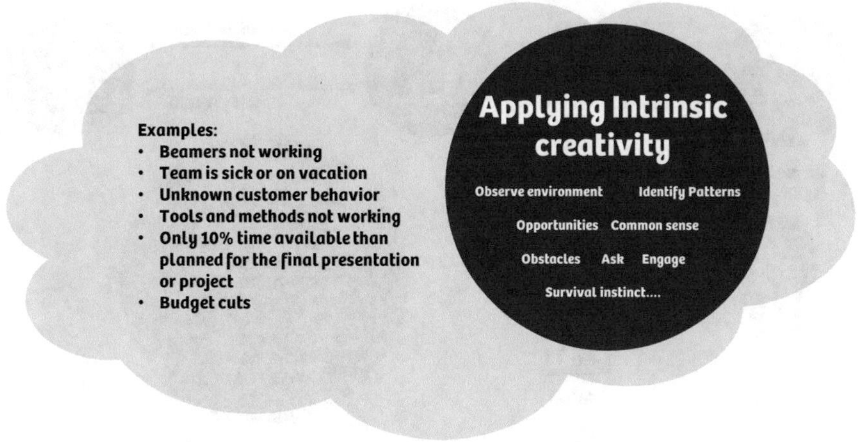

Fig. 36.5 Dealing with imperfect situations—Practicing for making use of intrinsic creativity

(e) The globally connected ITT locations and co-creation programs offer access to environments for implementation and commercialization (ITT Labs, Expertise development Programs, ITT Investor ecosystems, etc.)

Some limitations of the innovation approaches have been observed, when individuals or groups:

(a) Focus too much on tools and methods rather than making use of intrinsic creativity (human ability to identify patterns, trust, relationships, decision-making, risk taking, etc.)
(b) Do not invest enough on partnerships and remain in silos

This is the reason, the three-success factors were highlighted [Fig. 36.3].

Here is a list of key recommendations identified by the ITT Framework during the analysis of the projects, the challenges that arose and the success factors identified to optimize the product lifecycle.

- **Deal with imperfect situations by making use of intrinsic creativity:** While dealing with imperfect situations, which could happen during the various project phases—changing environment, customer expectations, team changes, one should not only rely on standards processes, which might not be for imperfect settings. Here making use of intrinsic creativity is extremely beneficial, which includes abilities and instincts of human beings. Some examples are illustrated in Fig. 36.5.
- **Be careful with the creativity trap and not losing the customer focus:** Innovators (also the aspirants) should pay special attention to the creativity trap and not losing the customer focus. This means not only choosing the most innovative solutions but the appropriate one with sensitivity to available

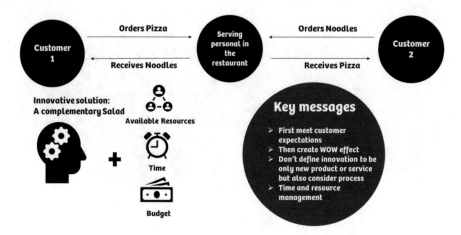

Fig. 36.6 Be careful the creativity trap and customer focus—Not only choosing the most innovative solutions, sensitivity to resources and time

resources and time. Figure 36.6, depicts example for a customer ordering food in the restaurant who gets something else by mistake or because of unavailability of the option. If this comes as a surprise and the customer does not want it, would result in disappointment. Even worse would be a highly unusual situation where a creative chef, delivers completely unexpected food to the customer and which is not desirable/acceptable for the customer. This could result in waste of creativity and resources. One solution to pursue creativity further would be first delivering what customer asks for and prime with new feature/product to get feedback, as shown in Fig. 36.6.

- **Giving options to decision makers creates flexibility and could widen the scope:** Addressing customer requirements is one of the integral parts for innovation process. It becomes important to look through different perspectives and alternatives for identifying key value propositions. Offering options to decision makers can strengthen the deliverable according to their focused preferences. Figure 36.7 outlines the ideology of ordering food in restaurant, no single customer can deny ordering at least one item from the list of food mentioned in menu. The menu provides range of choices through which customer's preference can be satisfied. Unavailability of choices can be one of the causes for customer dissatisfaction. Providing options assures that decisions are not only restricted to certain aspects like just yes or no, but also gives flexibility in decision-making process, which increases the scope of creative outcomes and optimize the time and resources spent by focusing on different possible solutions instead of wasting all the energy and organizational assets into one solution.

- While being attentive to customer requirements, the innovators should also emphasize on consolidating different phases of process to capture and visualize given efforts and resources during each phase. Figure 36.8 portrays an example

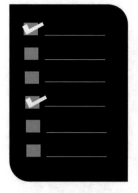

Fig. 36.7 Giving decision makers options—Creating flexibility and introducing wide area of scope

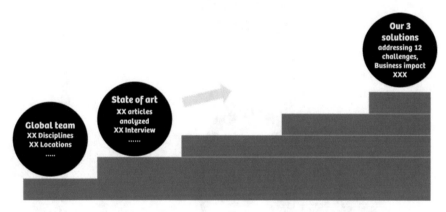

Fig. 36.8 Quantifying efforts, showcasing roadmap and abilities—Increases credibility of the solutions and appreciation of efforts

template for quantifying efforts at every step of process starting from how many resources from different backgrounds and locations are allocated followed by numbers of articles and contents analyzed to define state of art, similarly other phases can be quantified which results in consolidating key outcomes at the end. Here showcasing roadmap and utilized abilities during process not only increases credibility of the solutions but also acknowledge and bring a sense of appreciation to the given efforts.

• When identifying different solutions and value propositions within the ecosystem we usually end up analyzing a portion of problem space for specific areas. Innovations from the lens of big picture consider focusing on key root cause analysis, which includes not only identification of various challenges, trends, vision/mission, or organization goals but also highlighting detailed factors such as resource planning, communication, and interactions. Here Fig. 36.9 illustrates that for creating a holistic approach in innovation one must connect the dots and

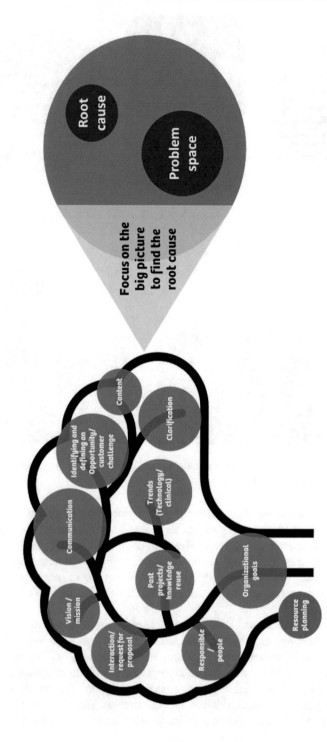

Fig. 36.9 Connecting the dots and keeping focus on the big picture—Across the innovation lifecycle

keep focus on the big picture to identify possibilities across the innovation lifecycle.

- **The starting point—Creating capacity within the existing environment:** Often new ideas stimulated by inspirations, customer insights, interests, skills, etc. for their implementation require changes and additional resources in the Existing Environments (EEs). Disruption or transforming EEs is not always possible in short term due to unclear impact and could discourage innovators and aspirants. A starting point could be creating capacity within EEs by (a) understanding the EEs overall organizational/societal goals, (b) Identifying dependencies and interfaces, (c) Appreciation of success factors, (d) Reuse knowledge, and (e) Partnering with the stakeholders f) Defining contributions and acquiring resources for implementation [Fig. 36.10].
- **Overcoming bias—Creation of interest in the field:** Biasing act as a stumbling block for innovation process and could take place during interactions with customers, suppliers, investors, within own teams, and against oneself. Being aware of it and of others could help in identifying ways to address it. Some of the recommended traits which could help the innovators overcome the bias are mentioned in Fig. 36.11.
- **Keeping and nurturing the innovation spirit:** Being an attractive field, innovative environments attract attention of individuals who sometime underestimate the continuous learning requirements, operative tasks efforts for sustaining or do not fully utilize the possibilities. This could lead to deterioration of the creativity, quality of the outcomes, attractiveness of the environments, and productivity loss. To keep the motivation on the same or even on higher level, an innovator should keep focus on increasing trust with stakeholders leading to customer pull, portfolio expansion, adjusting innovation processes, upgrading skills, and more importantly interest in the field [Fig. 36.12].

36.5 Conclusion

An innovator in the healthcare domain creates and commercializes products, solutions, and services for addressing needs of its stakeholders, such as patients, care providers, payers, etc. The role is crucial in disrupting healthcare systems direly in need of cost-effective and sustainable solutions due to the increasing patient population worldwide. The Innovation Think Tank framework could be used for identification and evaluation of the opportunities and accelerating MVP creation, testing, and validation in customer environment.

Aspiring individuals and organizations could go through the ITT white papers, frameworks, lab tours, disease pathways, co-creation program outcomes and choose the elements according to their requirements. The open innovation approach offers participation possibilities in various ITT programs [6].

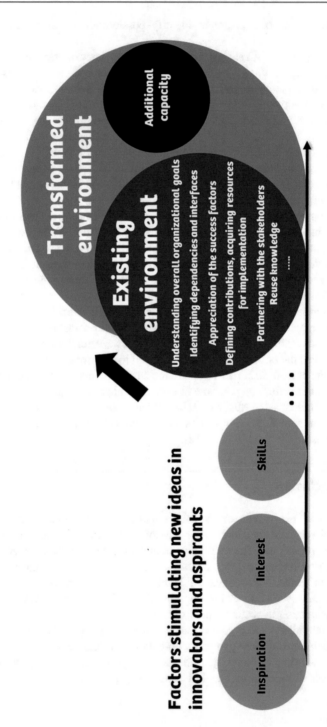

Fig. 36.10 The starting point—Creating capacity within the existing environment

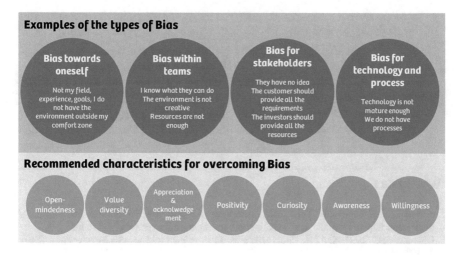

Fig. 36.11 Examples of bias and recommended characteristics for overcoming them

Fig. 36.12 Keeping and nurturing the innovator sprit

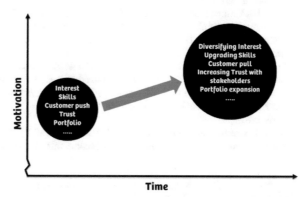

Take-Home Messages
- Healthcare systems worldwide increasingly require innovators to come up with holistic and cost-effective solutions to address the needs of patients, providers, and payers due to limited budgets and rising patient populations.
- "The innovations dilemma" spans from (1) selecting the right problem for research, (2) making a right decision for acquiring skills and education, (3) how much creativity is needed at a certain step considering the customer needs and available resources, (4) defining the starting point, (5) Opportunity/impact evaluation, to (6) choosing the right stakeholders.

(continued)

- The ITT frameworks provide tools to evaluate identified problems, possible solutions and recommendations for various stages for innovation lifecycle.
- Healthcare comes with regional differences and with that requires regionally adapted innovations. A global network of innovation labs helps to identify these and additionally creates learning with respect to delivery and policy changes that could affect other geographies as well.

References

1. Haider S (2022) Innovation Think Tank Frameworks for resolving "The innovator's dilemma" in healthcare. https://www.siemens-healthineers.com/at/careers/innovation-think-tank
2. Christensen C (1997) The Innovators Dilemma, 1st edn. Harvard Business Review Press. ISBN-13 978-1633691780
3. Haider S (2021). Addressing healthcare needs with Innovation Think Tank global infrastructure and methodology. https://www.siemens-healthineers.com/careers/innovation-think-tank/siemens-healthineers-itt-white-paper.pdf
4. Haider S, Vasavada JNN, Goenka A, Hassan D, Azem G (2022). Healthcare System Framework by Innovation Think Tank for understanding needs and defining solution requirements. https://www.siemens-healthineers.com/careers/innovation-think-tank/white-paper-system-framework.pdf
5. Fritzsche H, Barbazzeni B, Mahmeen M, Haider S, Friebe M (2021) A structured pathway toward disruption: a novel HealthTec innovation design curriculum with entrepreneurship in mind. Front Public Health 9:1300. https://doi.org/10.3389/fpubh.2021.715768
6. Innovation Think Tank (2022). https://www.siemens-healthineers.com/at/careers/innovation-think-tank
7. Annual External Innovation Think Tank Exhibition (eITT), Disease pathways, 2021 https://www.siemens-healthineers.com/news-and-events/conferences-events-new/eitt

Sultan Haider is the Founder and Global head of Innovation Think Tank (ITT) at Siemens Healthineers, which he established in 2005. His inspiring vision of innovation culture formed ITT to become a global infrastructure of 85 activity locations (Innovation Labs, ITT certification programs, ITT Incubation Centers etc.) in Germany, China, UK, India, USA, UAE, Turkey, Canada, Australia, Egypt, Saudi Arabia, Portugal, Switzerland, Brazil, Jordan, and South Africa. Prof. Haider has generated more than 440 inventions which resulted in more than 150 patent filings. Under his leadership, ITT teams have worked on over 2500 technology, strategy, and product definition projects worldwide. He is a Principal Key Expert at Siemens Healthineers (SHS), a title awarded to him by the SHS Managing Board in 2008 for his outstanding innovation track record. Furthermore, Prof. Haider has been awarded honorary directorships, professorships and has developed innovation infrastructures and implemented innovation management certification programs for top institutions.

A Primer on Patents and IP for Health Innovations

Michael Friebe

Abstract

In this short paper, some basic rules and to-do's are listed with respect to securing the intellectual property (IP) of your innovation. This is not a comprehensive summary, but it should help to make some initial decisions. IP is not limited to patents, but includes—especially in the world of digitization and data—software code, algorithms, blockchains, process documentation, clinical trial data, hardware sourcing information, a unique business model, logo/brand/copyrights, and other design creations. Exponential growth often requires that relevant information is created or distributed openly and for everyone to use and share and especially in the health sector and for democratization efforts, the value maybe in the data collected and used. IP is still—especially when external investors need to get involved—a very important topic. The paper will focus mostly on patent rights, where we will also talk about limitations and opportunities, licensing (in- as well as out licensing), and some basics with respect to costs /duration and content.

Keywords

Intellectual property · Patents · Licensing · Digital health · Health innovations

M. Friebe (✉)
AGH University of Science and Technology, Krakow, Poland

Otto-von-Guericke University, Magdeburg, Germany

IDTM GmbH, Recklinghausen, Germany

FOM University of Applied Science, Center for Innovation and Business Development, Essen, Germany
e-mail: info@friebelab.org

What Will You Learn in This Chapter?
- What Intellectual Property (IP) and associated Rights (IPR) are
- What you can and cannot patent and what the typical layout of a patent is
- Some commercial use of IP (sale, bargaining, license, investor relations, marketing)

37.1 Introduction

Health Innovations can be assigned to at least seven categories, whereby some innovations may be at the intersection of several of these [1].

These could be:

- THERAPEUTICS for prevention or complete treatment of a disease of health condition (e.g., molecule-based drug, vaccines)
- MEDICAL DEVICES or instruments for treatment, diagnosis, prevention … of a disease or health condition (e.g., surgical tools, robots, implants)
- DIAGNOSTICS, devices, or process techniques used for a clinical assessment or diagnosis (e.g., diagnostic imaging systems, biosensors, stethoscope, pregnancy test)
- DATA SCIENCES, data-based solutions for interpreting, collecting, and storing of medical data for a specific purpose (e.g., cancer registries, genetic information database, artificial intelligence/machine learning applications)
- DIGITAL APPLICATIONS are products in the form of an specific and dedicated software (e.g., smartphone apps, medical records)
- RESEARCH TOOLS describes and categorizes the set of tools that are needed to conduct health-related research (e.g., PCR, antibodies, reagents, animal models, drug models, simulation tools)
- HEALTH SYSTEM INNOVATIONS describes measures that improve, alter healthcare delivery (e.g., policies, programs, external funding, support systems)

Any of these innovations is attempting to be successful on the health market with a more or less pronounced commercialization element (e.g., being bought by a patient or other health user) and often is based on a unique and new research finding. Intellectual human activities lead to that discovery and the ideas, inventions, and creations in the industrial, scientific, literary, and artistic fields (including symbols, names, images, designs, and models) are part of intellectual property rights (IPR) granted to creators and owners of this work.

With protected IP you would technically be able to create a limited duration monopoly in a certain field or for a certain offering. How valuable that really is needs to be answered individually. That, however, also means that if you are not protecting your technology or you are limiting it to only few geographies then you are basically

Table 37.1 The different types of Intellectual Property, short explanation, and examples from the Health Segment

COPYRIGHT	Exclusive legal right to produce, reproduce, publish or perform an original work of art—health data analytics and digital apps, clinical questionnaires
TRADE SECRET	Formulas, process steps for manufacturing, delivery chains, ingredients, supplier sourcing—drug design, special manufacturing, experimental setups
TRADEMARK	Design, symbol, logo, phrase, word/picture combination that is recognized for the association with a brand—e.g., BAYER or GE or SIEMENS Logo and colors
PATENT	Exclusive legal rights granted in exchange for public disclosure of an invention, a product or process that solves a technological problem—therapeutics, medical devices, diagnostics

allowing that the technology can be copied or reverse engineered either in general or for any of the countries that you have not filed for protection.

A very quick rule of thumb therefore is that NO PROTECTION ALLOWS EVERYONE TO COPY YOUR WORK.

This is not completely true with respect to design and art work or with respect to some other written and documented work.

Another important—actually the most important—rule is that YOUR CANNOT FILE FOR PROTECTION AFTER YOU HAVE ALREADY MADE IT PUBLICLY AVAILABLE as part of a publication or presentation. Especially for the scientific community that is a very important statement. A researcher is interested to publish their work and get credit for being the first one that showed functionality or a new approach. Having published that without having filed a patent protection beforehand means that no more patenting is possible for that!

In the case of a possible patent application at a university the entire process could take quite a long time however. Invention needs to be disclosed, the university (or employer) needs to accept that as valuable to them, then the patent filing is actually prepared and possibly iterated several times, and finally filed with the local patent office. At least 6 month in some instances that the author was involved in 24 month. For the researcher, that means no public presentation or publication at that time.

There are several other IP rights, like Trademarks (e.g., unique Names, Word Combinations), Designs (e.g., Logo, Architectural work), and Copyrights (Books, Films, Music). And there are internal intellectual knowledge (e.g., sourcing of supplier, special manufacturing techniques, recipes, detailed software code, …) that is important to an operation, but never externally published or shared (see Table 37.1). That "secret sauce" does not have a limitation with respect to duration. Sometimes it is better to just keep those internals a secret!

In the digital Health world, we will increasingly deal with dedicated and personal health data and their use in combination with deep-tech. This data undoubtedly has a value, but who is the owner of it? Is this considered IP? Is the software and the algorithms IP? Are the calculation and use of data in a novel and unique way patentable (IP they are for sure)? These questions need to be individually answered and are therefore not an in-depth part of this write-up.

What needs to be considered in this case is the importance of the technology for the companies operation, the uniqueness of the approach, on whether the application possibilities are broad or rather limited, value towards the external world, and several more. Important is to think and discuss that early on.

If you believe that your company/product deserves a unique name, you should certainly protect that and immediately secure the internet name domain rights.

IP also describes the data and information from clinical trials and drug research from pharmaceutical and biotechnology developments. Healthcare providers and entities also may have copyright protections on their publications, protocols, and policies and procedures.

37.2 Why Do We Need to Protect Our Intellectual Property?

It is the core of our activities and describes the uniqueness of our operation and technological approach. Protection could be achieved by keeping it a secret (software only ideas and concepts), by moving fast and capturing a large enough share from which to grow and scale further (not likely happening or an option for Health Innovations), and/or by filing for patents. An excellent summary on Health-related IP, the Why and How, can be found in [1].

37.3 Patents

A patent for an invention is granted by a government to the inventor. When patent protection is granted, the invention becomes the property of the inventor, which like any other form of property or business asset, can be bought, sold, rented, or hired. Patents are territorial rights: German patents will only give the holder rights in Germany and rights to stop others from importing the patented products into Germany.

To be patentable your invention must:

Be NEW—never been made public in any way, anywhere in the world, before the date on which the application for a patent is filed.

Involve an INVENTIVE STEP compared to with what is already known that also not be obvious to someone with good knowledge and experience of the subject.

Be capable of INDUSTRIAL APPLICATION, which means that you need to convincingly demonstrate that the invention is capable of being made or used in some kind of industry. By the way, you do not need to provide physical evidence that you have actually build something that can actually do what you are claiming.

This means that the invention must be of physical nature in the practical form of an apparatus or device, or consist of or be a new material or an industrial process or method of operation.

An invention is not patentable if it is:

- *An accidental discovery.*
- *Describes a scientific theory or mathematical method.*
- *Is characterized as an aesthetic creation, literary, dramatic or artistic work.*
- *Is a method for performing a mental act, playing a game or doing business.*
- *Just presents information or is a computer program.*
- *However, if it involves more than these abstract aspects and comes with novel physical features, then it may be patentable.*
- *Important for HEALTH INNOVATIONS:* **a method of treatment of the human or animal body by surgery or therapy or a method of diagnosis is not patentable.**

A typical patent filing layout (you can actually write it yourself—you do NOT need a patent attorney, even though it may be advised) could look as shown below, but the best is that you check existing patents in your field of expertise or even in the particular segment that you believe your invention would fit it. You can easily search and find existing patent documentation on dedicated search engines. A good way is to either look for patents filed by companies that are already in that field or by looking for a particular inventor that you know is active in that field. The results give great cross-references and may also help to finetune your own application. Please be aware that only documents that are older than 18 month are publicly available, as the first 18 month from filing should allow the inventor/patent owner to keep it a secret and be able to work on introducing the idea to the market.

That also means that you—as inventor/patent owner—should not share the patent documents in that time without a confidentiality agreement.

Typical layout would be—researchers will see that it is following the typical journal paper layout for presenting research results:

- **Front page**—used in the same way a book will have a title page, a patent will have a front page that gives useful bibliographical details.
- **Abstract**—(compiled by the applicant) and may contain an illustration. This layout is becoming increasingly standardized, but the information given can vary from country to country.
- **Opening statement**—usually states the problem that the invention addresses.
- **Background information**—"state of the art" with key references showing the nature of the problem.
- **Description of the invention**—Explains the proposal/improvement/alternative, the inventive step and how it works.
- **Claims**—numbered claims that cover the legal aspects of the "monopoly." The first being the main claim and the later dependent claims referring back to earlier claims in describing what is new about the invention. Be careful not to add several "inventive" claims to one patent application. Every invention requires an individual patent application.
- **Illustrations**—as many as required showing the invention.

37.4 Are There Any Disadvantages Filing Patents?

Yes there are! Following is just a short and incomplete list of some of the drawbacks:

- Patent filings come with high cost and administrative effort to maintain them (ideal case without much legal interaction: € 3.000 for the initial filing in your country, € 3.000–5.000 for any additional country, € 200–1.000 annual fees per country and patent, with 15 years of actual patent issuance and three countries that would total at least € 40.000).
- The require that you invest time and effort in checking possible infringements and possibly take legal action against infringers that cost more money, take time with an unclear outcome.
- With a patent filing, you actually show and describe your technology in the patent documentation, which also means that you show what could be done to circumvent the patent protection.
- Patents have a maximum duration of 20 years from the time of original filing. While this looks like a long time, it really is not knowing that a medical product often takes 10 years and more to actually create revenue. This also means that you need to continuously file new patents.

37.5 What Is the Value of IP?

That is a very good question and cannot easily be answered. It may be key to you particular offering and it may be a base for having convinced investors that your idea should be funded. It also may be a stopping point for others to get into that particular field. In case of licensing, there is a clear monetary value associated with it. If your technology is sold to another company, then patents likely increase the value significantly. They could also be used as bargaining tools, when dealing with other competitors based on the principle "you allow me to use your patents and you can use mine for this COUNTRY and/or this TECHNOLOGY SEGMENT." Of course, the accountants see that differently as they normally determine the asset value based on the monies that already have been invested for filing and the upkeep.

37.6 Licensing (In/Out) and Patent Ownership

The patent list the inventors and the owner of the technology. The inventors will never change and are always natural persons, while the owner often is a legal entity. The owner has to have an executed legal agreement with the inventors about the transfer of the patent and associated rights. The details do not need to be publicly disclosed, and in certain countries the inventors maintain a non-exclusive right to use the described technology for their own non-commercial or limited commercial use.

We have talked a lot about incremental versus disruptive innovation for Health. If you—as an example—develop a new algorithm that would improve the signal-to-noise ratio of a diagnostic image (MRI or CT) by 20%, then this would certainly be noticeable and could in return lead to a reduction in scan time, which may be more important than the improved signal to noise. This is difficult to sell and promote as a standalone product for all the different MRI and CT systems out there. An alternative option would be to license the technology to the manufacturer of MRI and CT systems. This can be done EXCLUSIVE to a segment, geography, or to one particular vendor. Exclusive use in that respect would mean that you have a contractual agreement that only one user (e.g., specific Diagnostic Imaging manufacturer) can integrate and sell your technology on their MRI system, or only in one country, or use it for all their offerings. The contractual agreement could be based on a fee per installed system, a revenue share of the increase or total sales price and be combined with a minimum annual royalty fee. It is recommended to make the agreement as simple as possible, especially from an auditing point of view. It is much easier to estimate and verify the number of systems than to calculate or verify the actual revenue that a company has achieved with the patent protected product.

NON-EXCLUSIVE means that you can license the technology rights to more than one user in any of the combinations mentioned above. Transparency is important here. Depending on the product and offering, it may make sense to get into a non-exclusive agreement from the licensor and/or the licensee's position.

37.7   Future of Patenting and Digital Health Inventions

With further developments and exponential improvements using artificial intelligence/machine learning-based technologies, it could soon happen that the AI system (unsupervised-, deep-, federated-learning) with no or minimal human involvement. Who would then be the inventor?

As we have learned mathematical models (algorithms) and associated computational methods for data analysis (so also machine learning) are not patentable in most countries with the purpose to provide a method of treatment of the human or therapy or a method of diagnosis.

It is possible however to patent and secure intellectual property on hardware configurations, use of sensors and sensor data, and other aspects. A consultation with an expert on that matter (Healthtech/Digital Health patent attorney) would be advised.

HEALTH INNOVATIONS in the current healthcare system and most likely also for a FUTURE OF HEALTH will benefit from IP protection, especially when there are business models to be protected or new ones developed. Any investor will likely require some form of protection. Blockchain-based IP may possibly replace or change the way we currently apply and maintain patent and other IP rights.

37.8 Patent Recommendations

Please see Fig. 37.1 for a short patent recommendation flow chart. If you are dealing with an exclusive software product, it is difficult to file a patent, however, you could describe and patent a new process and in combination with a hardware. It is recommended to patent in your local country (e.g., Germany—Deutsches Patentamt) and also patent as many as possible incremental steps to increase your coverage. The cost for these local patent applications is actually quite low (without Patent Attorney, but with examination < € 500 in Germany, with a Patent Attorney—depending on your negotiation skills—around € 2.500 per patent), but obviously you will only ever have a protection for that local market. If that market is large enough, it still may have a high value. You then have 12 month before you have to initiate the next step. Either file for a PCT (roughly € 5.000) that buys you another 18 month before you have to decide on the individual countries or file directly. Recommendation here is to file initially only in the three–five countries with the largest potential market.

37.9 Summary

IP rights and in particular patenting activities will still—and probably remain for the next decades—be very important for health innovations and with that for Healthtech companies or other stakeholders. They can be used to create revenues for the company, secure a technology from copying and use by others, or prevent others to secure a technology. Once you have a technology secured you do not have to enforce, if you do not want to, but at least you avoid allowing others to create "monopolies."

DO NOT PUBLISH ANYTHING PUBLICLY BEFORE FILING A PATENT!

The author would certainly recommend to patent as early, file as many as possible applications, and subsequently decide on which of the secured technologies are more important than others. Continue these and file them in other countries, but maintain at least in your home country even the ones that you do not want to spend the extra resources.

You can learn a lot about patenting by analyzing filed and issued patents, and you can also write a patent application on your own without involving a patent attorney. The filing and examination costs are then rather low (< 1.000 USD/€). A good reference can be found in [3].

With respect to making healthcare available for everyone, there are discussions about the enforcement of patent rights and making patented technologies available for economies that cannot afford but need the concepts to battle inequalities and difficult access to health services. Please see [4, 5] for interesting articles on the issues between patenting and the right to health. Even the World Intellectual Property Organization (WIPO) *"seeks to raise awareness and understanding of the complex linkages between global health and access to medical technologies, innovation, technology transfer and trade. The goal is to leverage intellectual*

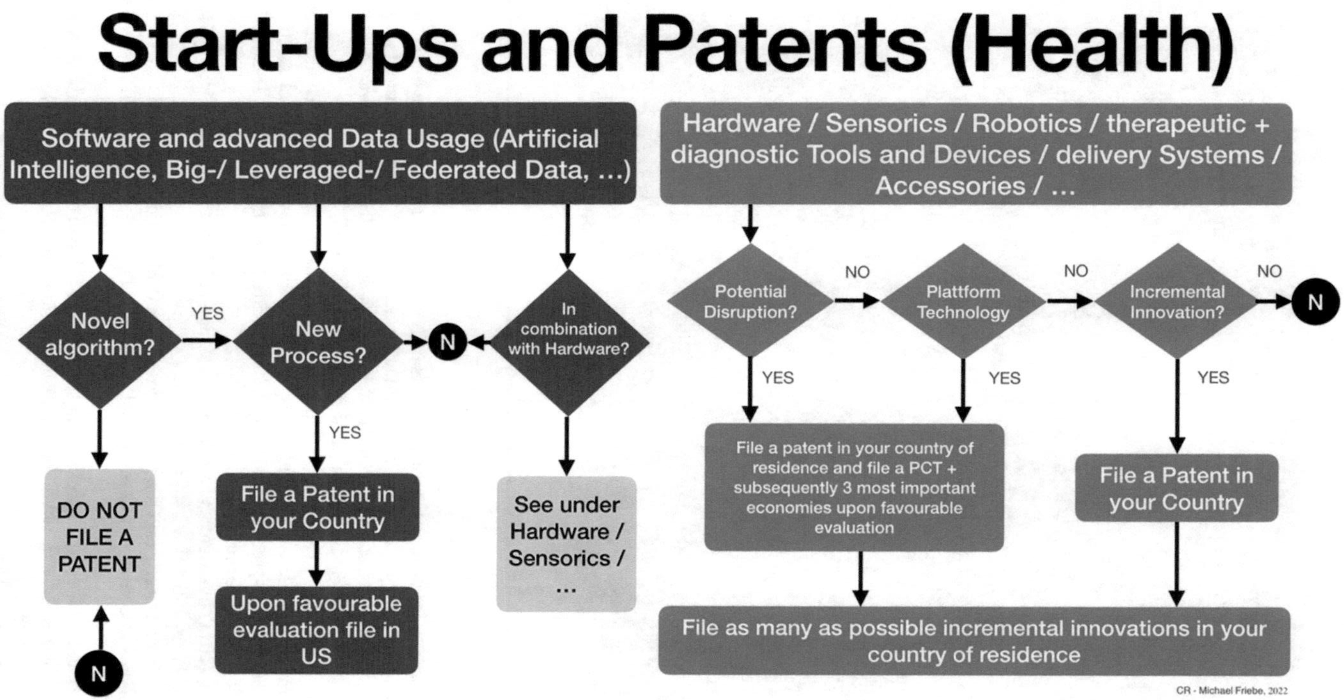

Fig. 37.1 Recommendation on patenting for Start-Ups in the Medical Field. Distinguishes between SOFTWARE/DATA heavy or HARDWARE heavy inventions. Quality is always important, but quantity as well, especially when you want to cover many aspects. Here it might make sense to file incremental steps in your country with limited expenses. An International PCT makes senses when you are still not sure on whether the patent application will have a value for your product, market, business model. You can rely the national patent filing by 18 month with that step. From [2]

property (IP) as a tool that contributes to meet the world's most pressing health needs."[6]

Take-Home Messages

- Do not publish anything about your idea before you have decided to file or not file a patent.
- Not patenting in essence means that you allow everyone to copy your technology.
- There may be good reasons not to patent, but then publish quickly so that others do not patent your idea and prevent you from following up on it yourself.
- Patents can be written by yourself—difficult, but not impossible.
- Patents can also be used for bargaining purposes.
- File as many patents for the same topic as you can—continue only the ones that are important.
- IP—intellectual property—is not only patents, but also is non published and disclosed Know-How, software code, design and art work.
- Methods for treatment of the human or animal body by surgery or therapy or a method of diagnosis cannot be patented. But devices for treatment, therapy, surgery, diagnosis, prevention can be patented.
- Software and algorithms, data analysis can typically not be patented, but in combination with novel sensors they may also be patentable.
- Patent protection is for 20 years from the time of initial filing . . . means you need to continuously file additional patents.

References

1. Intellectual Property & Regulatory Considerations Handbook. Michael G. DeGroote Innovation, Commercialization & Entrepreneurship Programming in partnership with the McMaster Industry Liasion Office (MILO). https://research.mcmaster.ca/app/uploads/2019/09/IP-and-Regulatory-Handbook-October-18-2018.pdf - downloaded and viewed March 12, 2022
2. Immer Ärger mit den Patenten!? Wie den IP-Transfer aus Hochschulen und Forschungseinrichtungen für Ausgründungen optimieren? Available for download at https://www.business-angels.de/wp-content/uploads/2022/03/BAND-Thema-im-Fokus-01-2022-IP-FINAL.pdf - viewed and downloaded March 23, 2022
3. Reingand N (ed) (2011) Intellectual Property in Academia - A Practical Guide for Scientists and Engineers. CRC Press. ISBN 978 1439837009
4. Salazar S. Intellectual property and the right to health. Available for download at https://www.wipo.int/edocs/mdocs/tk/en/wipo_unhchr_ip_pnl_98/wipo_unhchr_ip_pnl_98_3.pdf - viewed March 12, 2022
5. Cullet P. Patents and medicines: The relationship between TRIPS and the human right to health. Available for download at https://library.fes.de/libalt/journals/swetsfulltext/17639153.pdf - viewed March 12, 2022
6. WIPO - Global Health and IP. https://www.wipo.int/policy/en/global_health/ - viewed March 12, 2022

Michael Friebe received the B.Sc. degree in electrical engineering, the M.Sc. degree in technology management from Golden Gate University, San Francisco, and the Ph.D. degree in medical physics in Germany. He spent five years in San Francisco, as a Research and Design Engineer with an MRI and ultrasound device manufacturer. He is a German citizen with expertise in diagnostic imaging and image-guided therapies, as a/an Founder/Innovator/CEO/Investor and a Scientist. He is also a Research Fellow with the Technical University of Munich, Munich; an Adjunct Professor with the Queensland University of Technology, Brisbane; and a honorary Professor of image-guided therapies with Otto von Guericke University, Magdeburg, Germany. Since 2022, he has been a Professor of biomedical engineering innovation with the AGH University of Science and Technology, Krakow, Poland. He is a listed inventor of more than 80 patents and has authored over 200 papers, and was part of over 35 Medical Technology Start-Ups. He is a Board Member of four medical technology start-up companies and an investment partner of a MedTec-fund. From 2016 to 2018, he was a Distinguished Lecturer of the IEEE EMBC teaching innovation generation and MedTec entrepreneurship. He is also a coach and trainer for OpenExO, and a master Launchpad mentor for the Purpose Alliance.

Successfully Implementing Ambidexterity in the Medical Industry

38

Gunther Wobser

Abstract

Disruptions are fundamental, destructive changes in markets can ruin entire businesses. The Innovator's Dilemma sums things up unequivocally: Even though managers do everything right, one day, they get swept away. The concept of ambidexterity as a solution for the dilemma is presented. It requires an integrated solution with an individual organizational unit and three essential success criteria: Identification of the strengths, capabilities of the organization for exploration, and support and control from top management. Successful implementation requires an exceptional degree of complexity management from the leaders though. The newly developed Nine-field Ambidexterity Matrix makes it possible to classify innovations in the core business, in business development, and beyond. This way, an ambidextrous strategy is not just visualized; instead, opportunities and risks are revealed.

Keywords

Innovator's Dilemma · Ambidexterity · Disruption

What Will You Learn in This Chapter?
- The concept of Ambidexterity
- The Nine-Field Ambidexterity Matrix
- Core business innovation while monitoring and working on disruptive opportunities

G. Wobser (✉)
LAUDA DR. R. WOBSER GmbH & Co. KG, Lauda-Königshofen, Germany
e-mail: gunther@wobser.family

© The Author(s), under exclusive license to Springer Nature Switzerland AG 2022
M. Friebe (ed.), *Novel Innovation Design for the Future of Health*,
https://doi.org/10.1007/978-3-031-08191-0_38

38.1 Introduction

Risks lurk everywhere, especially where you do not suspect them. Disruption has become a buzzword that is often misused to describe even smallest changes. That is a shame, because it cheapens the dramatic effect of "real disruption," which happens unexpectedly and can be life-threatening.

Initially, disruption in the narrower sense arises from more straightforward products, which are acquired by overwhelmed customers or those who have not been approached at all yet.

However, in the broader sense, disruptions are fundamental, destructive changes in markets (not just developments or innovations), for instance, through new technologies (keywords: Artificial Intelligence, Blockchain), new business models (keywords: Platforms, Pay-per-use), or unpredictable events (keywords: COVID pandemic, war in Europe).

38.2 The Innovator's Dilemma

I'd like to introduce the Innovator's Dilemma by former Harvard professor Clayton M. Christensen [1, 2]. His insight in a nutshell:

Even though managers do everything right, one day they get swept away.

Why is that? Concentrating on the core business, increasing modularization, and the performance competition of ever newer products, frees up gaps that are cleverly filled by intruders. This initial "tempest in a teapot" can quickly erase an entire industry, in other words, cause disruption. Although established companies were sometimes the first to have an innovative idea and recognize the threats, they are often powerless because, for example, established factories and their workforces prevent them from changing course, unlike a fresh new start-up.

The Innovator's Dilemma of Christensen, who died on January 23, 2020, is especially significant in the context of innovation. This US scientist describes a situation in his theory, which was published in 1997, where business leaders act completely correctly from their perspective, and yet are nonetheless threatened by newly established competitors, and in the worst case, swept away. This behavior, which is rational in and by itself, but devastating, relies first and foremost on the successes achieved in the past.

"Successful firms that are disrupted are not complacent or poorly managed. Instead, they continue on the path that brought them to success."

Due to the increasing cost pressure, companies seek to develop ever higher quality products in small innovation steps and to market these in order to record attractive profits according to shareholders' expectations (resource dependence). Adding ever more functions leads to an oversaturation of the customer (performance oversupply). As a result, there is a shift of the purchase motivation from functionality to reliability to convenience, and finally to price. Due to this market development, manufacturers neglect entry-level segments, which are then occupied by new

Fig. 38.1 S-curve courses of two technologies (cf. Christensen [1], p. 40, own diagram)

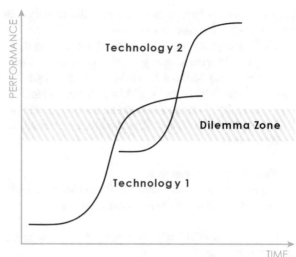

companies with newly developed products based on innovative technologies. These risks cannot be detected with classic market research.

Another point of attack are simplified products for previous non-consumers. These products continue to develop and later draw customers away from established companies.

Interestingly, the established companies often developed the attackers' new technologies themselves, but then decide against a market launch. The existing customers do not value the new technology enough (high-end mentality, especially in medicine), and the hoped-for sales, revenue, and most importantly the gross margin are too small.

If the attackers occupy these uncultivated segments, they will attack the established companies in their core segments one day and perhaps even force them out of the market (see Fig. 38.1).

Christensen characterizes the life cycle of technologies as a typical S-curve, which initially demonstrates a weak, then a strong, and in the mature phase of development, once again a weak performance. In the dilemma zone, the new technology 2 on the rise suddenly pops up in the established company's market. The leaders then face a dilemma: They do not know whether and how they should react. In the worst case, from the established company's point of view, the replacement of technology 1 comes with the new technology 2.

38.3 Ambidexterity

The answer to the described problems is a two-handed action called ambidexterity. Successful companies are balanced. They master their core business and at the same time explore opportunities beyond it. Threatened companies need to defend their core business as best as they can in order to build new businesses as quickly as possible with diminishing revenues.

The concept of ambidexterity was formulated as a direct response to the Innovator's Dilemma and goes far beyond it. The two US scientists Charles A. O'Reilly and Michael L. Tushman flesh out Christensen's original insight and suggest establishing an individual organizational unit to develop disruptive innovations [3]. They recommend an integrated solution.

It includes, in addition to the establishment of an individual unit, three essential success criteria:

- Identification of strengths
- Capabilities of the organization for exploration, and
- Support and control from top management

For the successful practical implementation, attention must be paid to four fields of action in descending order:

1. **Clear strategic goal**, which justifies the necessity of exploitation and exploration, including the explicit identification of the strengths of the CORE organization and the possibilities to take advantage of these through an exploration unit (EDGE).
2. **Obligation of the management** to encourage the EDGE company, to finance it and protect it against its enemies.
3. **Sufficient separation from the CORE organization**, so that the new EDGE company can develop its own architectonic orientation and the careful design of the organizational interfaces can develop. These are necessary to take advantage of the critical strengths and the company's potential. In addition, the definition of clear criteria for the decision to shut down the EDGE unit or to integrate it into the CORE organization.
4. **Vision, values, and a culture**, which ensure a common identity for the EDGE unit and the CORE organization and thus bind all participants to one another.

Ambidexterity requires an exceptional degree of complexity management from the leaders [4]. Inconsequential action can have serious consequences. With an integrated, structurally not separated organizational model, leaders and employees must increasingly develop capabilities in both hands.

Ambidexterity as a method puts high demands on the modern leader and can quickly cause "inner conflict" after a phase of "welcome change" and enriching experiences [5]. That is why ambidexterity should be anchored transparently in the organization through a clear concept—regardless of individual people.

38.4   The Nine-Field Ambidexterity Matrix

In my company group LAUDA, we are applying the Nine-field Ambidexterity Matrix I developed to navigate the tension between Exploitation, harvesting of current business areas, and Exploration, the examination of potential future business areas (see Fig. 38.2). This is where product management and our separate innovation unit *new degree* play an important role, requiring close, dynamic coordination and adjustment.

The classic product market matrix, also known as the Ansoff matrix, is included in this. In addition to customers, however, skills for developing products and services are considered rather than products and services themselves. What is completely new is the consideration of people who have not previously purchased anything at all and skills that no company has had before, which means they are truly novel and innovative.

On the customer side, markets can be expanded with existing skills, or markets can be developed with new skills. On the skills axis, skill expansion and skill development are mirror images. The pure form of exploration is when completely new customers are acquired with completely new skills.

LAUDA is the world leader in precise temperatures. The temperature equipment and systems are at the heart of important applications, contributing to a better future. As a full-service provider, the family-owned company guarantees the optimum temperature in research, production, and quality control. LAUDA is the trusted partner for electromobility, hydrogen, chemicals, pharmaceuticals/biotech, semiconductors, and medical industries. With competent advice and innovative solutions, around 530 employees have been inspiring customers worldwide every day for over 65 years.

The first innovation project at LAUDA, a mobile fridge to carry temperature-sensitive insulin, viewed from today's perspective with this systematic approach, was precisely in this high-risk, but also opportunity-rich area.

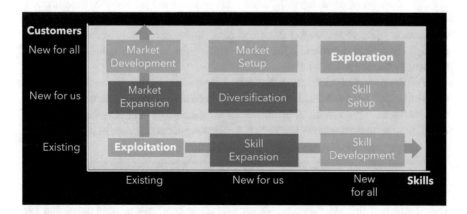

Fig. 38.2 Nine-field Ambidexterity Matrix (authors diagram)

Unfortunately, it failed spectacularly. If I had this systematic approach at the time, I might have been more cautious. We have been much more successful with our long-standing activities on the medical technology market and our newly established subsidiary LAUDA Medical (2021). We have served as a contract manufacturer of hypothermia devices since the early 1990s. These are used to cool the human body down during a heart operation in order to put the body's metabolism in a resting state. After the operation, the patient is brought back up to body temperature.

Currently, we sell these devices via a specialized medical technology manufacturer. Here, we are working with existing skills in an existing market (exploitation). With the revenues from this business, we are financing the establishment of the company LAUDA Medical and a complete new development of this device. For this, we have to build up new skills, however also especially serve existing customers in the future. As a result, we are moving in the skill expansion field.

Interesting are new digital products that are supposed to ensure the reliable, optimal operation of the device. According to our market knowledge, there is no such digital support thus far. Therefore, we are establishing new skills for the development of the digital products, which we will present to customers for the first time worldwide. On the medical technology market, there are already suppliers of digital products, however not in our product category. As a result, there will be a market set-up, the establishment of a new market, which is formed from previous non-users.

Ultimately, we have determined that an organized exchange of experiences among the users of hypothermia devices, the perfusionists, could result in significantly better handling of the devices during health-critical heart operations. Previously, operation, including the setting of the parameters, was handled primarily in isolation by locally knowledgeable specialists, who in turn worked in multi-shift operation without a systematic exchange of experiences. The establishment of an online community for the exchange of knowledge would significantly improve the work of every perfusionist around the world. We are absolutely certain that no manufacturer relevant for our application has successfully established such a community. In addition, there are no or hardly any experiences or skills for the establishment of this community. As a result, we classify this ambitious project as exploration. The classification does not have to be completely selected, because perhaps individual approaches for communities exist in the manufacturer sector of medical technology as a whole.

LAUDA's medical technology business is a good example of ambidexterity since a business that is new in various dimensions is being established in addition to the core business. This produces an excellent balance of current and future fields of business, one that represents the potential of a long-term, successful strategy.

38.5   Conclusion

Disruption and the Innovator's Dilemma threaten established companies and can fundamentally change entire industries.

The concept of ambidexterity enables action with both hands in order to bring various options into balance today and tomorrow. In addition to an individual innovation unit, a clear management strategy is required for this.

The newly developed Nine-field Ambidexterity Matrix makes it possible to classify innovations in the core business, in business development and beyond, enabling not only clear systematic ambidextrous action, but also an assessment of the resulting opportunities and risks. In particular, taking into account customers who are not activated and who are new to everyone allows for a graduated use of blue oceans based on the company's skills.

Likewise, unique skills that are new to everyone can be used to reach not only the company's own customers or its competitors' customers, but also, at the highest level of exploration, previously unnoticed passive customers. Accordingly, Exploration is the riskiest option, but it also offers opportunities for long-lasting profitable growth.

Take-Home Messages
- Disruptions are major change events (new business models, new technologies completely replacing existing ones, or unpredictable events causing significant shifts)
- Company leadership often fails, because it does not recognize disruption or acts too late, and because it ignorantly continue their past success route
- Established companies typically prefer to implement improvements (incremental innovations) to products they have been successful with and with that push the customers' purchase motivation from functionality to reliability to convenience, and finally to price
- New market contenders can target the entry-level (low-cost) segment and eventually pull customers from higher segments over (e.g., a Point of Care Ultrasound system for 20% of the cost of a mid product range ultrasound—https://clarius.com)
- Or they provide simplified products for users that were not customers before (e.g., see above example and also portable low-cost MRI system for use in intensive care of for pediatric applications—see https://hyperfine.io)
- Successful companies need to improve their Core technologies and at the same time invest in exploring technologies that could replace their CORE
- Start and finance an Exploration Unit or EDGE Unit that is sufficiently separated from Core, but with similar vision, values, and culture
- While Exploitation (Core) uses existing skills for an existing market, an Exploration (Edge) requires new skills for development and market understanding

References

1. Christensen CM (2016) The Innovator's Dilemma. When new technologies cause great firms to fail. Harvard Business School Publishing, Boston
2. Christensen CM, Raynor ME (2003) The innovator's solution. Creating and sustaining successful growth. Harvard Business Review Press, Boston
3. O'Reilly CA, Tushman ML (2016) Lead and disrupt. How to solve the innovator's dilemma. Stanford University Press, Stanford
4. Wobser G (2022) Agiles Innovationsmanagement. Dilemmata überwinden, Ambidextrie beherrschen und mit Innovationen langfristig erfolgreich sein. Springer Gabler, Berlin
5. Wobser G (2020) Neu erfinden. Was der Mittelstand vom Silicon Valley lernen kann. Beshu Books, Berlin

Gunther Wobser has been running LAUDA (www.lauda.de) for more than 25 years and is the CEO and main shareholder of the group. LAUDA employees 530 people and generates approximately USD 100 million in annual sales. He is also an active angel investor (www.mango.ventures) and member of the Band of Angels, Silicon Valley's oldest seed funding organization. Recently, he wrote about his findings of Innovation in the books "Neu Erfinden" and "Agiles Innovationsmanagement", both published in German language.

Gunther holds a Master's degree (Diplom-Kaufmann) in Business Administration from the University of Bamberg, Germany, and a PhD (Dr. rer. pol.) in Economics from the Technical University of Freiberg, Germany. He regularly contributes to conferences, panel discussions and to professional social networks.

Reverse Innovation: Circumvent Digital Health Transformation Issues

Michael Friebe

Abstract

Exponential technologies are generally described as something that will in a given time period double data generation/evaluation and/or half the associated cost with it. A Medical technology example is the cost of genome sequencing that has dropped from millions to thousands and now to under USD 1.000 within a little more than a decade. In many other areas, there is a potential and hope that certain technologies could lead to significant clinical knowledge gains and procedure improvements combined with dramatic cost reductions and increased clinical and patient satisfaction. Specifically the increasing life expectancy and the aging societies in combination with less and less available healthcare staff, ever-increasing cost associated with healthcare delivery/products and services, or the inequalities between rural and urban areas, particularly in developing nations, need to be addressed urgently. This paper will discuss some developments with a particular focus on reverse innovation, where new technologies and delivery approaches will be implemented in areas with existing health access inequalities or limited access, low quality of health service, and other urging needs (e.g., general availability and cost issues). A desire of a local government to increase population-, public-, and individual- health by allowing access of new tools despite missing regulatory approval would additionally be helpful.

M. Friebe (✉)
AGH University of Science and Technology, Krakow, Poland

Otto-von-Guericke University, Magdeburg, Germany

IDTM GmbH, Recklinghausen, Germany

FOM University of Applied Science, Center for Innovation and Business Development, Essen, Germany
e-mail: info@friebelab.org

Keywords

Reverse innovation · Digital transformation · Health innovation · Health entrepreneur

> **What Will You Learn in This Chapter?**
> - The concept of Reverse Innovation
> - How to circumvent digital health transformation issues by introducing technologies to areas that have a much higher need and little funding
> - Different geographies have different health innovation needs requiring different value propositions for the development

39.1   Introduction

In low-income nations (LIN), healthcare is—with very few exceptions—dealing with sick people within a complex, expensive, and overly bureaucratic environment. There is very little focus and money spend on preventing people from getting sick. Avoiding medical problems has a potentially huge negative financial impact on healthcare companies, clinical service providers, and everyone involved in healthcare delivery [1]. There is also evidence that a significant percentage of clinical procedures currently performed are not necessary and done just to avoid malpractice and liability issues [2].

At the same time, there is a growing concern that the healthcare industry's focus is at the moment mainly on developing and advancing existing technologies that only help the relatively small population in the high-income nations with developed healthcare infrastructures of the developed world rather than to invest money into simplifying and reducing complex systems and procedures to make them available for everyone.

There is an understandable logic behind that if you empathize with the business models of current healthcare stakeholders, including the medical technology providers.

A short calculation will illustrate the problem: let us assume an MRI system brings in USD 1 Mio revenue with a gross margin of 50%. That would mean that the company can spend USD 500k on development activities, marketing, support, shareholder profit . . . a 10x reduced cost for an affordable MRI (it is possible already with reduced applications and reduced image quality) with the same Gross Margin would require 10x as many MRI systems to be sold to come up with the same absolute number. Is that possible? And is that something that a company wants to support or actively invest in? I guess the answer to the first question is Yes, but only if the MRI is made available to other professions but the radiology department. This may very likely not help to reduce the total healthcare expenses in the short term. Answer to the second question is that the manufacturer should stay up to date and watch the developments that happen on the edge to not get disrupted and if they are smart be prepared to offer systems on their own when the market requires it.

Simplification and associated cost reductions could potentially lead to truly innovative and disruptive technologies that would benefit most people on this planet. They could also be the base for a shift towards preventive and personalized solutions.

39.2   Digital Transformation

In the next decades systems that combine advanced hardware with Artificial Intelligence (AI) will be able to provide diagnostic information that should be equally available to the poorest and wealthiest on Earth.

The patients genomic sequence and machine learning will allow us to understand the root cause of many non-communicable diseases like cancer, cardiovascular and neurodegenerative problems, and also provide information and advise on what to do about it. Many more technologies will be introduced and eventually be operated semi-autonomously or even fully autonomously in the near future, like surgical robotic systems. With 5G connections, these will then even allow these systems to be operated from a distance without having an expert on site and by being available on a 7/24 schedule.

We will collect and be able to use all individual health data to create a personalized and holistic view on the health status, analyze and predict health changes.

Other transforming technologies are 3D printing (3D cell models of organ tissue or bone implants can already be printed).

One of the pre-requirements for all these technologies is digitization. Processes and services that cannot be digitized will for the time being be impossible to grow or improve exponentially.

The associated digital data content is key to analysis, evaluation, learning, drawing conclusions, and subsequently providing informations for personalized and individualized therapy decisions and outcome improvement. It will also be essential for a transformation from a reactive to a pro-active healthcare system.

We have already started that process with the large number of healthcare apps that are available and by recording all kinds of personal data with the support of wearable sensors. While most of these sensors and apps are for personal use only at the moment and are not used in the official clinical diagnosis yet, that will change for sure.

Healthcare participants (so everyone) will use sensor systems for blood pressure -, glucose -, behavior - monitoring and many other applications.

They will use their Internet of Things (IoT) devices to connect directly or via smartphone to cloud servers that make the data available for researchers, regulatory offices, and also healthcare providers (doctors, hospitals, ...).

These home-based and personally used technologies and the data generated by them will eventually lead to providing fast, and personalized healthcare status information directly to the patient.

The patient does not have to go in person to see a doctor for a majority of causes, but instead gets detailed feedback and even a prescription digitally. In case of urgent or immediate clinical intervention a clinical expert would be contacted and an appointment made or an ambulance dispatched.

This is the promise, but the reality is different. Digital Health is not the provision of tools and devices, but it comes along with a mindset transformation. This does not happen over time. We have an established, grown, and quite well-working healthcare delivery setup in the western world. But the healthcare provision is not determined by the patient! The patient is not an equal partner in the process, does not even have access to his/her own patient data. And the typical human is not investing as much as they should in health (or actually "sick") prevention.

Prevention is currently mainly limited to exercise and nutrition and only complemented by a voluntary clinical feedback (check-up) every other year or so. Prevention is also very closely linked to the individuals education and socioeconomic standing. The better you are off, the more you invest in good food, good living environment, sports, exercise and relaxation (body and mind). Not everyone can afford that!

Home-based or personally worn sensors and other technologies would be able to give much more detailed, fast, and regular feedback on the current healthcare status to everyone and for pretty much no cost. Improvements or situation worsening could be monitored and combined with follow-up clinical treatments were needed and indicated. So instead of an early and reactive transition from home to the healthcare system, as is typical for today, a future healthcare system will stimulate pro-active prevention, provide personalized healthcare status and recommendation, and transfer the patient relatively late to become a real patient.

Many challenges and issues remain like regulatory permissions, innovation funding, large-scale acceptance, reimbursement and service payments, data privacy, and of course general adoption of such a setup. It can already be safely forecasted though that some home-based diagnostic platforms will soon arrive and with that start to establish medical care at home [3]. But the "soon" cannot really be defined at the moment, especially with a missing business model and trust issues of the established parties with respect to efficacy and data usage. When you have a working business model within a capitalistic economy, your goal is to grow your billable services and increase revenues and profits. A new model that challenges and disrupts that established model is initially always facing fierce opposition, as the successful players do not know how to deal with all the unknowns and are afraid that they will lose their position and status. So for the healthcare business model that pays for diagnosing and treating sick people, a shift towards a home-based and (partly) doctorless medicine with highly affordable systems and a data-based prediction is a complete paradigm shift.

This is combined with a regulatory process that takes many years, requires significant funds and is based on evidence. That means that you need to prove over years and hundreds of patients that a certain clinical process is better than the current setup. So in essence you will always compare a novel technology to the performance of a human. The clinician and clinical process is the Gold Standard. Let us assume that you use a very large number of multi-dimensional digital health data (e.g., lab tests, genetic information, current disease information, diagnostic imaging information, general health data, ...). You then let an unsupervised artificial intelligence algorithm/network use that data to extract features and attempt to cluster them. The result may (no, will) show information that will identify a certain combination of data points to be associated with a certain disease. The result is not really explainable and cannot be tested, as there is no Gold Standard available. So, should we discard these results as untrustworthy?

39.3   Reverse Healthcare Innovation

One of the major problems for the implementation and acceptance of these new technologies in the healthcare systems of high-income nations (HIN) is the current setup and associated adoption issues. Also, the regulatory environment in Europe, Northern America, and Japan is already very complicated and slow and, with the implementation of the new Medical Device Regulation (MDR), this will get even more complicated and slower in the near future.

There is only little interest from the established healthcare providers to rethink the current healthcare delivery process and to start thinking about more cost-efficient solutions that would replace their own existing business model.

The current healthcare delivery solutions have been developed by the developed world for the developed world. But little has been done to develop cost-effective solutions for the developing nations that address vital local problems there.

So, developed nations have an issue with acceptance and developing nations are missing essentials. Reverse Innovation is combining these two needs.

Up to now high-priced ("scarce") and complicated medical equipment was produced by developed nations for developed nations, installed there and used for 5–10 or even 15 years. After that it was exchanged by a Next-Generation Product that more or less did the same task, but better, faster, with more features—something called sustaining or incremental innovation. The old equipment was then often sold as used or refurbished to emerging or low-income nations. But it was often still too expensive and too complicated to install /operate or may only be useful/accessible to a very small part of the population there.

An alternative is to build a much lower specified system that more closely fits the requirements of the local healthcare system and clinical needs. The result is a dedicated system for the specific needs that is quite simple, very efficient, and can be manufactured for a fraction of the cost of the advanced technologies.

Yes, these products most likely do not provide the same features, come with a lower quality and may be slower ... but they may be applicable for a majority of the cases even in the developed world and/or at rural areas.

These point of care and local technologies therefore will have a valuable use in our healthcare systems as well. This process of adopting low-cost efficient solutions from developing nations is called Reverse Innovation and highlighted in Fig. 39.1.

Another slightly changed version of Reverse Innovation could be the introduction of novel, cost-effective solutions that are currently outside the existing healthcare business model in these areas. Using disruptive and radical health innovation developments could solve health-related challenges there and the data points and results could then be used for proving efficacy of that approach and helping to build trust in the developed world [4]. This does not mean that the low-income nations should become the test environment for unproven technologies. It should rather solve local healthcare delivery and address local problems (an AI-powered doctor that is available at little cost is much better than NO doctor).

A desire of a local government to increase population-, public-, and individual-health by allowing access of new tools despite missing regulatory approval would additionally be helpful.

This could help to circumvent some of the digital health acceptance and mindset transformation issues that we experience from the healthcare providers/stakeholders in the higher income countries. If it works well and solves problems plus provides valuable health information there, then how can you object to the use and here?

39.4 Health Innovations from Outside the Western World

It has been proven that low-income nations can come up with and subsequently develop and produce extremely efficient products addressing local problems that could substitute expensive products in the HIN world and therefore could easily be transferred [5].

Companies located in the HIN world have a hard time accepting that, however, as it would stipulate a complete rethinking and reorganization process. Only very few companies have therefore invested in reverse innovation. But the ones that did were rather successful [6].

A solution could be a combined learning process between organizations/companies in developing and developed nations specifically looking at creating cost-effective, easy, small and with that valuable products and processes that are helpful for a global health systems [7].

And there is plenty of data supporting such joint activities and the value of a development focus on creating less expensive devices. While only 13% of the medical technology manufacturers are located in lower income countries almost two-third of the incremental health care spending is used for affordable technologies. The products that come from these low-income nations very often are quite disruptive and creative with a technology basis that is significantly simpler than competing technologies from developed nations.

A recipe for joint developments between LIN and HIN could be:

1. What medical needs are common to LIN and HIN?
2. Start the innovation process in the LIN

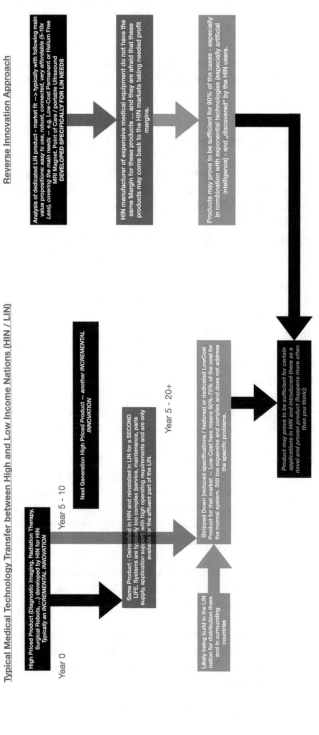

Fig. 39.1 The current use of high-tech, scarce, and expensive medical equipment. Developed for high-income healthcare systems and eventually replaced and reused in lower income areas. But these devices were never made for the needs of these areas/countries. If devices are made dedicated for LIN needs, they may prove to be sufficiently good for most needs in HIN as well. Also, the need to come up with solutions that make basic (homecare based) and affordable healthcare available to everyone in LIN is much higher, as there often is nothing available or only available large distances away that comes with very high costs [adapted from 3]

3. Initiate the Reverse Innovation process and introduce to HIN with proven concept in LIN
4. Implement and establish the technology in HIN

But this process and the focus on disruptive value-based clinical innovation can also be initiated and managed by entrepreneurs through dedicated start-up companies in LIN or HIN [8, 9].

39.5 Importance of Entrepreneurship for the HC Transition

Without entrepreneurs that challenge the current setup and that come-up with alternative and new systems, methods and processes, a healthcare change as outlined will not happen soon. The large medical device companies and healthcare providers do not have a large enough incentive to start a rethinking their process of creating sustaining/incremental innovations immediately and to a full extend. Another reason for this slow and little radical innovation process is regulatory issues and the high cost of entry.

But we are now on the way to a much later transition from home care to professional healthcare and from a reactive healthcare delivery process to a more prevention-oriented and personalized pro-active one.

There will for sure be entrepreneurs that will tackle and address the opportunities that come with such a transition, and that will nevertheless put the patient in the midst of their thinking while maintaining the regulatory rules that come with new healthcare products and services [10].

These entrepreneurs (and the more progressive established medical technology suppliers and healthcare providers) will also learn that valuable products and services can be developed and introduced adapted to the local needs, with the main key value propositions being inexpensive and easy to use.

These local needs include special diagnostic and/or therapy systems to address different disease patterns, dedicated and simple technical support and training, availability of inexpensive spare parts, affordability of system and related medical service for all not just a minority, and fulfilment of all applicable regulatory issues. These systems could also open opportunities for local manufacturing, servicing, and support companies with that creating employment and tax income in these countries [11].

Entrepreneurs in that segment will understand that success is predominantly based on the knowledge of the local markets and customer needs. These needs are locally different and may require completely different product features when HIN and LIN are compared, for example.

Finding these differences and designing the product/service that fits to the local market requires dedicated need finding in a combination between technical and clinical innovators. And it also requires a different skill set that is commonly not provided as part of a university-based training.

To address the mentioned healthcare innovation, reverse innovation and entrepreneurial opportunities, a special skill set and dedicated knowledge are required that can only be taught within a new interdisciplinary purpose-oriented health innovation program.

Current education offerings will make you a good doctor or an excellent engineer or possibly a great economist—depending on your choice of subject. But it does not prepare you for being a future leader in healthcare innovation. The skill set needed for that—in an interdisciplinary triangle environment between technical possibilities, medical needs and economic realities—is quite different from the either technical /clinical/or economic foundation that is provided as part of a university training.

You need to be creative and visionary and also be trained in personal skills like leadership, social responsibility, and in an empathic understanding of problems and concerns of any participant in the healthcare system, including (but not limited to) patients, doctors, and other clinical staff.

Entrepreneurial basics with a focus on healthcare also need to be addressed and taught. Future entrepreneurs that can collaborate with medical device or healthcare provider industry, researchers, clinical staff and public organization's will have a distinct advantage and will be able to more easily identify real innovations solving medical needs.

A new study program, incorporating knowledge of healthcare system and provision, innovation challenges and strategies, basic knowledge of clinical problems and future-oriented technical possibilities, interdisciplinary and international teamwork, combined with management and entrepreneurial skills, is needed [12].

39.6   Discussion and Conclusion

Exponential technologies could lead to a dramatic change in the way that healthcare is delivered. Currently almost all of the national healthcare systems treat sick patients rather than to prevent people from becoming patients. The future will be a pro-active predominantly data-based healthcare that will collect, process, and analyze information providing a personalized report that would also allow preventive measures.

Digitization is a key requirement for these developments. With that real disruptive medical technologies could be developed that would change diagnosis, treatment, billing, financing, and many other aspects of healthcare delivery [2].

These disruptions may very likely come with easier to use, less complex, and much more affordable solutions—some of them originating from low-income nations that are subsequently adapted by the developed world as they have proven to provide adequate or even superior solutions for a fraction of cost. Disruptions developed could also be used to improve the healthcare setup in low-income countries dramatically and lead to subsequent adoption in high-income countries that are currently still stuck in the digital transformation issues.

So it may be a good advise for health innovators to check the needs, the regulatory environment, and the willingness to introduce frugal, disruptive, or radical health innovations to solve local/regional issues in other countries and other innovation environments.

Entrepreneurs will play a vital role in the future change of healthcare, as they will develop and introduce solutions that are based on addressing local clinical needs.

But these changes in technology leading to completely new business models will also require innovation leaders that have a skill set that fits the twenty-first-century healthcare challenges that are not taught in conventional university-based study programs [12].

> **Take-Home Messages**
> - Value propositions that are focusing on low cost and ease of use are very attractive to solve the health needs of low-income countries.
> - They typically do not fit into a business model of established healthcare delivery systems . . . yet!
> - Validating a product in a low-income market may solve many issues related to acceptance, trust, and proof of efficacy.
> - Starting in a country or area that combines actual health needs with willingness and openness to accept a new technology may be a good strategy for a product ideas that are too disruptive or radical for the established health care systems.
> - Health Innovators should check Reverse Innovation opportunities.
> - Health Innovation design and entrepreneurship need to be taught in dedicated programs.

References

1. Christensen C, Bohmer R, Kenagy J (2000) Will Disruptive Innovations Cure Health Care? Sept.-Oct. 2000 issue Harvard Business Review. Available at https://hbr.org/2000/09/will-disruptive-innovations-cure-health-care
2. Diamandis P (2016). Disrupting todays healthcare system. http://www.diamandis.com/blog/disrupting-todays-healthcare-system
3. Friebe M (2017) Exponential technologies + reverse innovation = solution for future healthcare issues? What does it mean for University Education and Entrepreneurial Opportunities? Open J Bus Manage 5:458–469. https://doi.org/10.4236/ojbm.2017.53039
4. Bottles K (2012) Reverse Innovation and American HealthCare in a Time of Cost Crisis. PEJ July/August 2012:S18:20. http://ldihealtheconomist.com/media/reverse_innovation_and_american_health_care_in_a_time_of_cost_crisis.pdf
5. Immelt J, Govindarajan V, Trimble C (2009) How GE is disrupting itself. Harv Bus Rev 87:56–65
6. Govindarajan V (2012) A reverse-innovation playbook. Harv Bus Rev 04/2012. Available at https://hbr.org/2012/04/a-reverse-innovation-playbook
7. Shamsuzzoha BS et al (2012) Developed-developing country partnerships: Benefits to developed countries? Global Health 8:17 https://doi.org/10.1186/1744-8603-8-17
8. DePasse J, Lee P (2013) A model for 'reverse innovation' in health care. Global Health 2013(9): 40. https://doi.org/10.1186/1744-8603-9-40

9. DePasse JW, Caldwell A, Santorino D et al (2016) Affordable medical technologies: bringing value-based design into global health. BMJ Innov 2016(2):4–7. https://doi.org/10.1136/bmjinnov-2015-000069
10. Hendricks D (2016) Why Entrepreneurs Are the Future of Healthcare. http://www.inc.com/drew-hendricks/why-entrepreneurs-are-the-future-of-healthcare.html
11. Glifford G (2016) The use of sustainable and scalable health care technologies in developing countries. Innov Entrep Health 3:35–46
12. Fritzsche H, Barbazzeni B, Mahmeen M, Haider S, Friebe M (2021) A structured pathway toward disruption: a novel HealthTec innovation design curriculum with entrepreneurship in mind. Front Public Health 9:715768. https://doi.org/10.3389/fpubh.2021.715768

Michael Friebe received the B.Sc. degree in electrical engineering, the M.Sc. degree in technology management from Golden Gate University, San Francisco, and the Ph.D. degree in medical physics in Germany. He spent five years in San Francisco, as a Research and Design Engineer with an MRI and ultrasound device manufacturer. He is a German citizen with expertise in diagnostic imaging and image-guided therapies, as a/an Founder/Innovator/CEO/Investor and a Scientist. He is also a Research Fellow with the Technical University of Munich, Munich; an Adjunct Professor with the Queensland University of Technology, Brisbane; and a honorary Professor of image-guided therapies with Otto von Guericke University, Magdeburg, Germany. Since 2022, he has been a Professor of biomedical engineering innovation with the AGH University of Science and Technology, Krakow, Poland. He is a listed inventor of more than 80 patents and has authored over 200 papers, and was part of over 35 Medical Technology Start-Ups. He is a Board Member of four medical technology start-up companies and an investment partner of a MedTec-fund. From 2016 to 2018, he was a Distinguished Lecturer of the IEEE EMBC teaching innovation generation and MedTec entrepreneurship. He is also a coach and trainer for OpenExO, and a master Launchpad mentor for the Purpose Alliance.

Part X

Health Innovation Education and Incubation

Health Innovation Design: "CAMPing" for the Unmet Clinical Need

Jörg Traub

Abstract

Entrepreneurial education and design thinking start at the university. The subject domain courses educate students on the hard skills required for their professional career. At TU Munich (TUM) there is a great tradition to foster and incubate innovation and business creation, not only by ideation and creation of patents and other intellectual property (IP), but also by stimulation of an entrepreneurial spirit and a link to executors, business angels and supporters in the particular domains. At the Chair of Computer Aided Medical Procedures (CAMP) this is a great tradition with an incubation of ideas and subsequent creation of start-ups. There are several elements that make this environment unique, i.e., focus on applied science and industry collaboration, encouragement of boldness and risk taking, and the education in innovation generation in multidisciplinary setups between engineers, computer science, medical, and business students.

Keywords

CAMP · Medinnovate · Spin-Off · University · TUM · UnternehmerTUM · Entrepreneurial Education · Real-World Labs · Networks · Soft skills · Mentors

J. Traub (✉)
Forum Medtech Pharma and Healthcare Bayern Innovativ, Munich, Germany

M. Friebe (ed.), *Novel Innovation Design for the Future of Health*,
https://doi.org/10.1007/978-3-031-08191-0_40

> **What Will You Learn in This Chapter?**
> - An example of a university-based Innovation Generation setup for Healthtech.
> - Innovation Generation is a mindset process and starts during university education.
> - Multidisciplinary setups, hard work, and a lot of fun are ingredients of a successful recipe.

40.1 The Setup

Created in 2003, the Chair for Computer Aided Medical Procedures (CAMP) by Prof. Nassir Navab enabled a series of applied research and translational projects within an open innovation setup. Coming from industrial research in medical image processing, computer vision, and augmented reality, it was important that besides creating impact research it also make impact to healthcare delivery. This included a focus on IP and translational (from bench to bedside) projects. All researchers, postdocs, and PhD students were deeply involved in clinical projects with real data generation and applications in real lab environments or even in clinical studies. This also lead to early spin-offs from the chair, for example SurgicEye, in intraoperative molecular imaging; MicroDimensions, in digital pathology; ImFusion, a software platform for medical image processing and computer-assisted procedures; Piur Imaging, 3D ultrasound imaging and AI-based image processing; and OneProjecs, a platform for improvement in treating cardiac arrhythmias.

This open space combined with the drive to implementation of ideas, created a unique ecosystem in computer science in medical applications. Along these activities also educational programs involving the practical experience of the start-up founders were designed to share and multiplicate the success scaling ideas to the market, i.e., innovation generation in the healthcare domain.

With this spirit Michael Friebe, as a guest lecturer, created a lecture series with the title "Medical Imaging Entrepreneurship" in the winter semester 2010/2011. Following the success of the first lecture, three additional lecture blocks "Image-Guided Surgery," "Innovation Generation in the Healthcare Domain," and "Medical Entrepreneurship" were subsequently created and are until today in the curriculum of the computer science faculty as interdisciplinary education.

40.2 The Role Models and Early Leaders

CAMP was set up by Prof. Navab as a collaborative, interdisciplinary and applied science laboratory and teaching environment. One of the activities was the encouragement of the researchers to work in close collaboration with clinical partners and industrial collaborators. This was also manifested by the creation of applied research labs at the two university hospital in Munich. At Klinikum rechts der Isar the IFL[1] (interdisciplinary research lab) was established with many clinical collaborators and other science institution with an focus on navigation solutions, intraoperative imaging, workflow assessment, artificial intelligence, and medical robotics. At Klinikum Innenstadt, the city hospital of the Ludwig Maximilian University Hospital the NARVIS lab[2] was established with the focus in augmented reality, simulation and training, artificial intelligence as well as advanced intraoperative imaging solutions in close collaboration with the surgery department of the hospital.

Being present in the hospitals supported a deep understanding of the problem solution domain and formed compelling tandems and teams between engineering researchers and medical doctors. It supported a deep understanding of the engineering disciplines of the application and problem domain the so-called unmet clinical need or the burning problem that needed to be solved. On the other side, the medical disciplines got a profound understanding of the technical capabilities and potential and thus were able to help to design and execute the creation of the painkiller for these burning problems.

This ideation and innovation generation melting pot created joint visions and one result was that five of the first ten PhD graduates from the CAMP chair between 2007 and 2010 were co-founders of a health tech-related start-up, actively executing and implementing the research conducted during their research assignments.

40.3   Educational Courses Supporting the Innovation

The lecture series started in the winter semester 2010/2011 with the course "Medical Imaging Entrepreneurship" by Michael Friebe. This provided the students a deep dive into entrepreneurship in the healthcare domain with many real-world entrepreneurs as guest lecturers. The most important message was that it is not only the excellence of the idea and the solution and product, but its need and thus acceptance outside of research hospitals.

In summer term 2013, this lecture was extended with the lecture Image-Guided Surgery—from bedside to bench that focused on the initial challenge, finding the burning problem. Within this class groups of students were sent to the different hospital departments, emerging into the clinical routine, observing the daily routine, and using a structured process to address the challenges and the shortcomings and potential improvements.

This was only possible with the great support of progressive medical professions from the discipline's trauma surgery, visceral surgery, radiology, pneumology, gynecology, and interventional radiology. The setup at the affiliated university hospitals allowed this setup and the medical professional dedicated valuable time

---

[1] https://www.in.tum.de/campar/labs-locations/ifl-lab/.

[2] https://www.in.tum.de/campar/labs-locations/narvis-lab/.

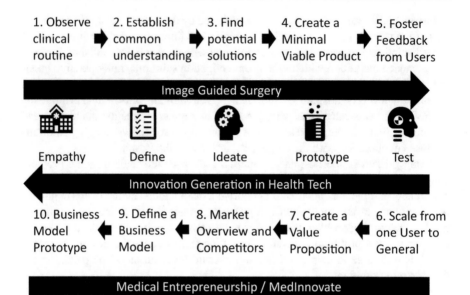

Fig. 40.1 The different lecture blocks that support finding the unmet clinical need and the incubation of business creation along the design thinking process adapted to the medical and healthcare application space

in the solution domain understanding of the talents in the course. The lecture series was completed by the creation of the course "Innovation Generation in the Healthcare Domain" that emphasized on the structured assessment of the market needs and potential, scaling an idea from a problem solution fit in one setup to a business model and a general market. Therefore, methods of the value proposition design and business model design [2–4] were applied and groups of students validation the initial hypothesis in this canvas in the real-world setup, accessing further key opinion leader and user in the field.

Figure 40.1 provides an overview of the lectured along the iterative design thinking process.

Inspired by this series and Yock et al. [1] CAMP alumni Christoph Hennersperger participated in the BioInnovate program in Irland, the course "BioInnovation: from unmet clinical needs to solution concepts" with the concept that the entire design thinking process along the Stanford Biodesign concept was used in one semester in selected interdisciplinary teams with great support by the incubator UnternehmerTUM. This was also led to the creation of the MedInnovate community in Munich.

Besides the methodology in health tech design thinking, the soft skill training was an important part of all courses of CAMP as well as the link to the real world. The link to the real world was achieved by invited presentation throughout the lectures of active entrepreneurs and founders of health tech start-ups. The courses also focus on teamwork and solving challenges as the following picture series demonstrated:

Group assignments and teamwork with value modal canvas, business model canvas and other design thinking methods were elementary parts of the course.

Final demonstrations were done with minimal viable products, presentations, and illustrative models. Most final team presentations were done during a 2–3 day retreat with the entire innovation generation class, forming a great alumni team with over 600 students that completed these classes between winter semester 2010/2011 and winter semester 2021/2022.

During the final retreat, the creation of a health tech business was linked to many real-world examples. One example was mountaineering, hiking, and rock climbing.

While the route ahead looks initially challenging and unsolvable, it can be successfully tackled with the right preparation and training, the right equipment, mentoring and expert advice. The lecturers (Friebe and Traub) were always ensuring the correct material was available to execute and perform, and they belayed the group to master the challenges and push their limits outside the comfort zone.

Mastering the challenges was very rewarding. During the lecture, this was the generation of a minimal viable product and a business case with market validation, creation of impact solutions addressing burning problems of patients or medical staff. During the retreat, the challenges included reaching the summit of Drachenkopf (middle image) or the sledge riding challenge from Rotwandhaus and Hintere Firstalm during the night (images on the right and left).

All lectures were enriched with invited presentation (left image), e.g., with Dr. Martin Groher, former CEO from MicroDimensions and field trips (right image) here visit to the innovation labs of Siemens Healthineers in Forchheim.

The final reward was the graduation of the courses and the fun all participants had during the hard and time-consuming course assignments creating a viable product and a competitive business model. Within the alumni of the course, there were more than ten founders of health tech start-ups getting inspiration and practical hints how to execute a business creation with higher odds for success in this domain.

40.4 The Current Lecture Setup for "Innovation Generation in the Healthtech Domain"

Over the past 10 years, the course graduated more than 600 students amongst them the co-founders of various companies with novel healthcare solution, e.g., Resmonics, Orbem, or Custom Surgical. The lecture format and its content are constantly refined to create the highest educational impact for the participating students while also finding sound modes and business cases to solve burning problems. Within one semester, about 25–30 students are working in teams of 4–5 students on one challenge, mainly on digital technology in healthcare. This includes the definition of the problem, field work in refining the problem and validating the value proposition, creation of a minimal viable product and pivoting to find the perfect problem solution fit. Within six lecture blocks, each 3–4 h, the groups get knowledge and perform exercises. In the 2–3 weeks between the lecture blocks, there are field assignments to validate the generated hypothesis and to present the result to the group. In every semester, 3–4 invited lecturers are enriching the format with real-world experience and provide direct entrepreneurial spirit and advise from their daily work. The assessment by the students of the course is normally one of the top-ranked lectures in computer science and won already two times the golden punch card as the best lecture award in computer science at TUM.

> **Take-Home Messages**
> - Innovation is not created in the lab but in the application space.
> - Incubation Ecosystem supports the translation from ideas to business.
> - With early innovators the CAMP Chair at TUM enabled various spin-offs.
> - Mentoring is an intrinsic DNA of the CAMP alumni and affiliate.
> - Creating something that has a real value for a patient or a clinical user is a huge motivation for students and lecturers alike.
> - Having fun is essential.

References

1. Yock PG et al (2015) Biodesign: the process of innovating medical technologies. Cambridge University Press. ISBN: 9781107087354, Online resources: http://ebiodesign.org/
2. Osterwalder A, Pigneur Y, Smith A, Etiemble F (2020) The invincible company: how to constantly reinvent your organization with inspiration from the world's best business models. John Wiley & Sons
3. Osterwalder A, Pigneur Y, Bernarda G, Smith A (2014) Value proposition design: how to create products and services customers want. John Wiley & Sons, Online resources: https://strategyzer.com/books/value-proposition-design
4. Osterwalder A, Pigneur Y (2010) Business model generation: a handbook for visionaries, game changers, and challengers. John Wiley and Sons, Hoboken, Online resourced: http://businessmodelgeneration.com/

Jörg Traub received his PhD in Medical Computer Science. He has more than 15 years experience in the medical device start-up industry and was as founder and CEO responsible from the idea generation, through the implementation, certification, internationalization and multiple financing round all the way to the exit of the medical technology company and university spin off SurgicEye. Following this he was business development director and coordinating the strategic marketing and sales activities at a smart medical robotics solution company, resopnsible also for strategic corporate partnerships.

Since 2020 he is director healthcare at Bayern Innovativ and is Managing Director of Forum Medtech Pharma e.V., responsible for innovation incubation, infrastructure projects in healthcare innovation and networking.

Health Technology Innovation Generation (HTIG) Lecture and Project Classes at AGH University

41

Katarzyna Heryan and Michael Friebe

Abstract

HTIG was initiated at AGH in the autumn/winter semester of 2021/2202. The project classes HTIG is a micro-scale experience of healthcare innovation generation where students learn dedicated tools and techniques the hard way (Learning by Doing, Problem-Based Learning, Design Thinking). To reinforce this educational experience useful tools to maintain student excitement, engagement, and focus are presented. Creating the innovation apart from deep technological knowledge, awareness of healthcare socio-economic aspects, understanding of clinical needs, and the inclusion of patient perspective, require specific soft skills (learning and life skills). These soft skills are especially hard to master for technical students, therefore, the course curriculum is adjusted to maximize the learning outcomes in this area.

Keywords

Healthcare 4.0 · Engineering education · Biomedical entrepreneurship · Twenty-first-century skills · Healthcare innovation · Healthcare start-up

K. Heryan (✉)
AGH University of Science and Technology, Krakow, Poland
e-mail: heryan@agh.edu.pl

M. Friebe
AGH University of Science and Technology, Krakow, Poland

Otto-von-Guericke University, Magdeburg, Germany

IDTM GmbH, Recklinghausen, Germany

FOM University of Applied Science, Center for Innovation and Business Development, Essen, Germany
e-mail: info@friebelab.org

What Will You Learn in this Chapter?
- How to teach students to become a MedTech innovator.
- Why soft skills are so hard for technical students, and why are they so important.
- What is the AGH <u>H</u>ealth <u>T</u>ech <u>I</u>nnovation <u>G</u>eneration curriculum content?
- That to create innovations ideating around medical needs (not technology) and verifying the proposed solution with early adopters is essential.
- How to maintain student engagement, and excitement throughout the course. Valuable methods, mechanisms, and tricks.

41.1   Introduction

Are entrepreneurship and a knack for innovation generation something you can learn or something that you have in your blood or genes? Do we even need to answer this question?

During the HTIG classes, the professor supports and supervises the group work indicating that an interdisciplinary team that consists of diverse personalities can handle the complex development of business ideas in healthcare likely more successful. The course starts with a short online assessment test taken by the students that reveal their business DNA [1]. Builder, opportunist, innovator, and specialist, which one are you? This exercise is not done to select teams, but rather to show what strengths and deficits the team currently has. Incredibly important for a start-up.

Someone with highly developed empathy and emotional skills that cares about team mood, someone focused on the final goal that drives people further, a person who is a specialist dedicated to ensuring the quality of each tiny detail, and a person who sees the whole picture and is capable of generalizing the idea and expressing it to the stakeholders and investors; a pessimist who by discovering the product's deficiencies creates opportunities for improvement, and an optimist who motivates the rest in need. All are desired and once well managed synergize work.

One should be aware that students who attend the lecture come from different backgrounds and have different educational experiences, and therefore their expectations differ. For some of them, it is difficult to find themselves in this novel educational setting that is focused on encouragement and development of personal skills rather than technical knowledge that often can be gained with or without a teacher.

Hence, the HTIG is designed to be a process of changing the paradigms and mindsets to become empowered, to leave the individual comfort zone and start to think out of the box, and to be able to communicate that. For some people, this is a huge challenge, since especially technical universities are rather focused on the students' individual work and learning technologically challenging topics. The necessity to publicly argue their ideas and convince an audience of their solution is a brand new experience that often comes with significant personal stress. While these are very important soft skills necessary for professional adult life, they are often neglected in the current university education or not treated with the same importance. Mastery of these skills would allow students to play a more prominent role in healthcare innovation generation. Whoever teaches that class should understand these limitations and introduce students step by step to the exciting tools and techniques that may be used for creating MedTech innovations.

The classes themselves are a micro-scale experience of medical innovation development that enables students to immerse in this process. Students learn tools and methods to create innovation in healthcare (translation, entrepreneurship, exponential development) by actually doing these things, experiencing them, and reflecting iteratively over the validation results (lean). Focus is given to the ability to present and discuss these outcomes on a regular basis.

Learning Box

The HTIG teacher, based on hers/his expertise and using the tools of positive psychology, should supervise this work, focusing not only on the final result but also on reflecting upon a road taken and team collaboration issues. She/He should indicate strong aspects of individual and group assignments and room for further improvement. The provided feedback needs to be honest and constructive. Students' strong skills should be identified and appreciated. Time dedicated to analyzing these issues is as important as dedicated to technical content evaluation.

41.2   Organizational Information on HTIG Course Run at AGH Autumn/Winter Semester 2021/2022

Course Title: Health technology innovation generation (120-INTCOURSE-xS-186)
Course groups: Information and Communication Technologies - Interdisciplinary Courses
ECTS credit allocation: 5.00
Number of students: 27
Type of class: lecture, 30 hours; project classes, 30 hours
Examination: Course - Final grade, lecture - Exam grade, project classes - Graded credit
Final grade is calculated as follows: 0.6 project classes and 0.4 examination
Participation in all project classes in compulsory

The course HTIG is a part of the university elective course database available to AGH-UST students regardless of faculty, a field of studies, and experience (Bachelor, Master, and Ph.D. students). This interdisciplinarity gives students the chance to

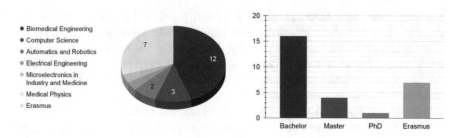

Fig. 41.1 Diversity of students participating in HTIG 2021/2022 (faculties, degree, respectively)

synergize their knowledge and experiences, which is a requirement in healthcare innovation and which is also exciting to the teaching staff.

In the 2021/2022 winter semester, 27 students were participating HTIG (Fig. 41.1), including 7 ERASMUS students, mainly from Portugal.

41.3 Curriculum Content

This chapter presents the individual topics raised during the classes in such an order so that to maximize the learning outcomes (Fig. 41.2). The first part (3×W) is dedicated to provide students with the required knowledge and shape their mindset so that they became aware and engaged in healthcare problems and solving unmet clinical needs (lectures mainly). The second part (2×W) provides information on the healthcare innovator profile (knowledge, skills required) as well as tools and methods useful for innovation generation (learned the hard way, mainly project classes—team assignments). The given content is suitable for direct use. Practical advice and tools to strengthen the learning process are provided. In addition, the HTIG syllabus can be found in [2].

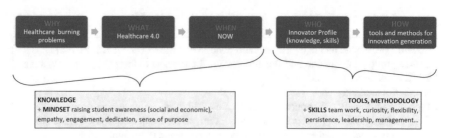

Fig. 41.2 HTIG curriculum overview

41.3.1 Raising Students' Awareness on Current Healthcare Problems and Challenges (Awaking = WHY)

This section explains problems that form the basis for further development of healing solutions and innovations in healthcare.

> **Learning Box**
> An important starting point for raising students' awareness of healthcare problems.
> What are the current healthcare issues? What are the solutions?
> Tools: sticky notes and flip-chart brainstorming (problem/solutions), grouping items related to similar issues, discussion, see Fig. 41.12.

Two aspects of healthcare problems can be distinguished: cost and availability. Both are closely related and strengthen each other.

Costs

Cost is one of the most burning healthcare problems. Indeed healthcare is really **expensive**—it takes up approximately 10% of GDP worldwide. Most are spent to follow highly regulated procedure guidelines rather than actual patient needs (fear of overlooking something, redundant confirmation, overtreatment, a sense of liability). And yet more and more is invested in MDs insurance. Underinvestment is one thing, but misallocation is another. Of course, most students have no idea about that and most—being young and healthy—have not experienced the sick care side of healthcare delivery.

Availability

Availability that can be understood on many levels.

First of all, the **long waiting time** to access diagnostics and therapy: from the first symptoms to the first visit to the GP, specialist, to the actual treatment. It means more expensive treatment of a more severe or spread disease with a worse prognosis instead of the provision of precise and minimally invasive treatment.

But are you sure that the clock is ticking from the first symptoms? Or maybe from the first predictions? Imagine what if, we could **predict** a disease before the first symptoms appear. To notice the unnoticed, to measure and quantify the silent early development phase indicators. Instead of treating a patient with advanced symptoms, we could have a chance to **prevent** the disease from developing. Not to mention—treatment costs that would be significantly lower—if we could only predict this. Saving money here allows us to allocate it wisely elsewhere.

Additionally, the predictive model could take into account the key influencing factors such as genetics, other comorbidities, and lifestyle to choose what is the best for this particular patient and include his/hers expectations—**a patient-centric personalized approach** [3].

Second, **equality of access to health care is not ensured for everyone.** Does every person have the same chances of survival and optimal therapy choice regardless of their wealth? Will she/he receive the same quality of care? Unfortunately, no. Each of them, after all, pays health insurance and taxes. But still, there are huge disparities in the quality and availability of healthcare between people, rural/urban areas, and economies.

> **Learning Box**
> An important starting point for raising students' empathy towards patients' needs.
> Working around patients, doctors, and stakeholder perspectives. Put yourself in different economical or geographical settings.
> Tools: team up over particular perspective, identify problems.

Moreover, this different economical perspective and related different needs and expectations have to be taken into account when creating innovations [4]. There might be a significantly bigger market for some innovations in developing countries than in developed countries, and vice-versa. In addition, developing countries may be in greater demand, and for this reason, it is worth considering dedicating efforts towards them. Understanding these differences makes it easier to create/adapt solutions to the particular community (Fig. 41.3).

Third, developing a new drug or medical device takes a very long time (between 3 and 12 years) and can cost many millions to bring to the patient. The reason of this is that healthcare is a very **rigid system** and there is a lot of stakeholder **resistance** to accept new solutions.

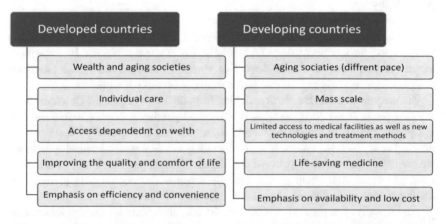

Fig. 41.3 Understanding the differences between developed and developing countries, and related needs and expectations, makes it much easier to create/adapt solutions to the particular community

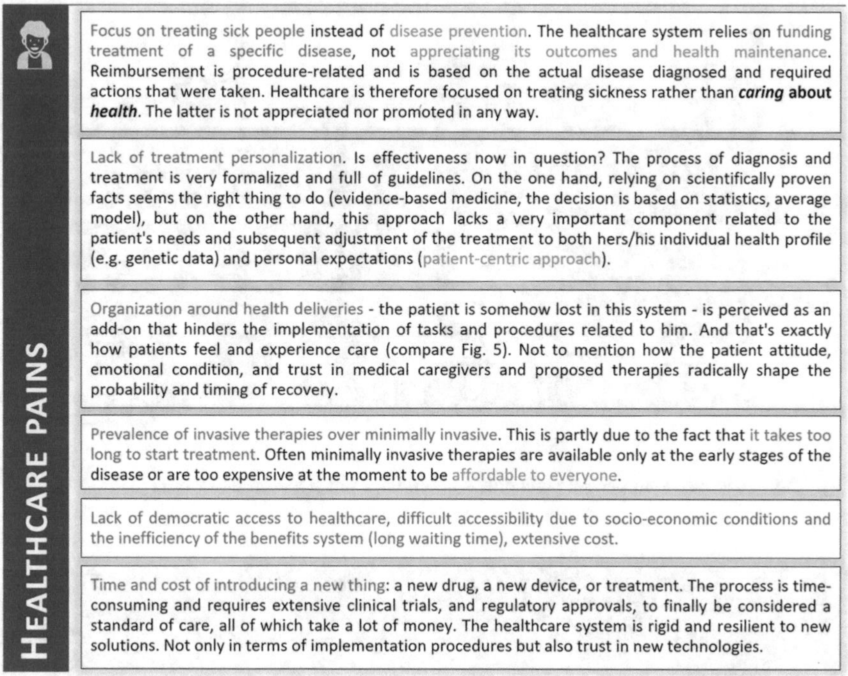

HEALTHCARE PAINS

Focus on treating sick people instead of disease prevention. The healthcare system relies on funding treatment of a specific disease, not appreciating its outcomes and health maintenance. Reimbursement is procedure-related and is based on the actual disease diagnosed and required actions that were taken. Healthcare is therefore focused on treating sickness rather than *caring* about *health*. The latter is not appreciated nor promoted in any way.

Lack of treatment personalization. Is effectiveness now in question? The process of diagnosis and treatment is very formalized and full of guidelines. On the one hand, relying on scientifically proven facts seems the right thing to do (evidence-based medicine, the decision is based on statistics, average model), but on the other hand, this approach lacks a very important component related to the patient's needs and subsequent adjustment of the treatment to both hers/his individual health profile (e.g. genetic data) and personal expectations (patient-centric approach).

Organization around health deliveries - the patient is somehow lost in this system - is perceived as an add-on that hinders the implementation of tasks and procedures related to him. And that's exactly how patients feel and experience care (compare Fig. 5). Not to mention how the patient attitude, emotional condition, and trust in medical caregivers and proposed therapies radically shape the probability and timing of recovery.

Prevalence of invasive therapies over minimally invasive. This is partly due to the fact that it takes too long to start treatment. Often minimally invasive therapies are available only at the early stages of the disease or are too expensive at the moment to be affordable to everyone.

Lack of democratic access to healthcare, difficult accessibility due to socio-economic conditions and the inefficiency of the benefits system (long waiting time), extensive cost.

Time and cost of introducing a new thing: a new drug, a new device, or treatment. The process is time-consuming and requires extensive clinical trials, and regulatory approvals, to finally be considered a standard of care, all of which take a lot of money. The healthcare system is rigid and resilient to new solutions. Not only in terms of implementation procedures but also trust in new technologies.

Fig. 41.4 Important healthcare pains

Students should also be aware that these important healthcare pains could be a base for Health Innovation Generation, including the patient—clinician relation (see Figs. 41.4 and 41.5). And that globally, there are a lot of unmet clinical needs.

41.3.2 Healthcare 4.0, Technical Readiness, and Anticipated Technological Development (WHAT are the Expectations?)

In the last few decades, we have experienced many dynamic developments as part of the industrial revolution 4.0 (communication and connectivity, information spread, process automatization). A subset of the presented technologies (Fig. 41.6) is exponentially growing (for example, 3D printing, nanotechnology, AI, IoT, cloud computing). Employing these can lead to exponentially improving advancements in quality, size, speed, and manufacturing/delivery cost reduction.

We use these technologies every day and yet they have not been properly employed in healthcare. At least not to such an extent as it could be expected. Why? Healthcare system is highly formalized and conservative, and therefore resilient. For this reason, it has only improved in incremental steps—not at the same speed as other segments. However, these incremental innovations do not provide a solution to current burning problems.

Fig. 41.5 A word cloud patient perceptions about their physician in two words [5]

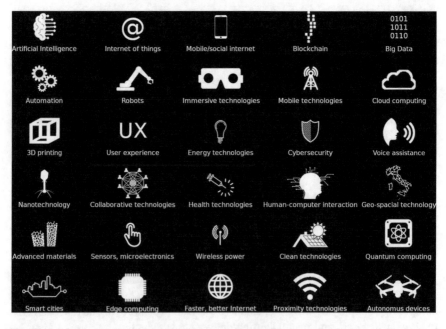

Fig. 41.6 MedTech innovations come from the synergy of many technologies and the work of interdisciplinary teams: doctors, IT specialists, engineers, lawyers, economists, and sociologists. To a greater or lesser extent, each of these exponential technologies indicated as most prominent in the next decade, is/will be present in healthcare [6]

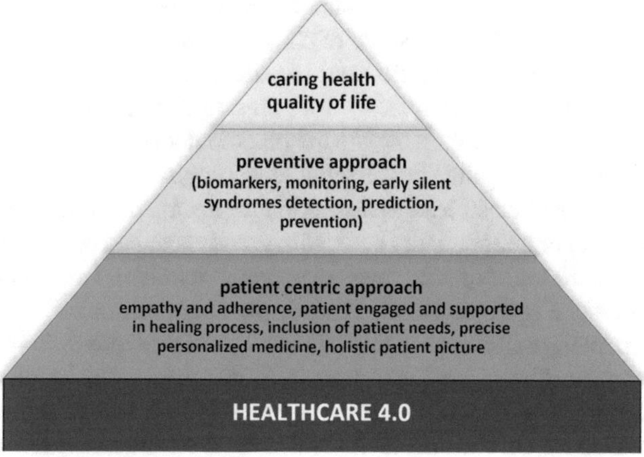

Fig. 41.7 Key aspects of Healthcare 4.0

Next Generation Healthcare or Healthcare 4.0 is putting patients first. Instead concentrating over the healthcare providers, focus on specific patient needs and requirements. So a goal of the lecture here is to introduce the students to the anticipated shifts in healthcare delivery and how novel technologies (machine and deep learning, big data analytics, IoT, mobile apps, advanced sensors, . . .) approach prevention, prediction, personalization, and a physical shift from professional healthcare to a homecare setting.

> **Learning Box**
> An important starting point for raising students' empathy towards patients' needs.
> Working around patients, doctors, and stakeholder perspectives.
> Put yourself in different economical or geographical settings.
> Tools: team up over particular perspective, identify problems.

The Healthcare 4.0 pyramid in Fig. 41.7 is a good base for describing the upcoming paradigm shifts. It will be the patient, not the healthcare provider, that is most important, and the focus will be on a preventive approach towards *caring* about the *health* and maintaining a longer and healthier life.

Such a healthcare system could approach the patient with greater empathy and devote enough time for conversation, support, and education. A patient conscious of his healthcare status reinforces the treatment outcome. Moreover, shifting healthcare provision from hospital to patient home supports this process even further by ensuring patient well-being in comfortable conditions, making him more eager to disease control and participative in therapeutic processes supervised by professionals. It is especially important today in the face of aging societies, understaffing, and the increasing share of chronic diseases.

The healthcare paradigm change—putting the patient first (shift from provider-centric towards patient-centric system), focusing on disease prevention (shift from sick care to health care), reimbursing the treatment outcome (not the procedure), and personalized approach requires exponential technologies adaptation. Their application in healthcare can indeed lead to disruption (replacing a current technology with something of equal or better quality), but with greatly reduced cost (demonetization) and size (dematerialization) that may eventually increase the healthcare global availability—despite socio-economic barriers (democratization).

However, paradigm shifts—actually another word for disruptions—are not in line with the current healthcare business models. While it is not quite clear how these new business models will look like, it is important to be aware of the developments and trends and to be able to use tools dedicated to healthcare innovation generation—the learning goals of HTIG.

41.3.3 The Future of Healthcare Starts Today in Front of Our Own Eyes (It Is Already Happening = WHEN)

Patients are getting more and more aware of new health developments and they are bringing that information to the treatment discussions with the clinical staff. They ask questions, compare data, and choose the doctors they prefer (scores and opinions). Patients are engaged and eager to understand the nature of their well-being and factors that disturb it and want to have a choice of what is best for themselves among the offered treatment options. The traditional healthcare model is no longer aligned with patient mindset and expectations. This enforces the shift from a system to a patient-centric approach with a digitally empowered patient as prominent driving force of healthcare transformation (Fig. 41.8).

For that, innovative and in many cases, disruptive developments are needed, and innovation education needs to be adjusted to meet upcoming needs.

Both the technological readiness brought by the fourth industrial revolution (Big data, AI, disruptive technologies) and patients' increased expectations reinforces the healthcare system disruption (Fig. 41.8). This future for biomedical engineering is happening before our very eyes. And it is time to provide required training for students to be able to face these challenges. To ensure that their contribution is significant and to increase the likelihood of commercial success. For that, a knowledge of innovative tools learned during HTIG is essential. So the question is not about when or if it will happen, but whether we will participate in this.

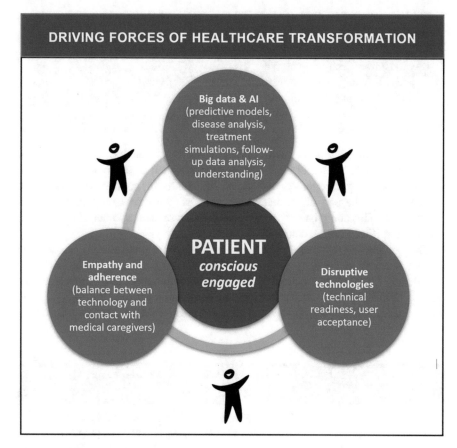

Fig. 41.8 Driving forces of healthcare 4.0 transformation among which—an empowered patient plays the most prominent role

41.3.4 Conscious Role of the Biomedical Engineer in Innovation Design (What is Expected and How It Can Be Learned = WHO)

Despite the fact that innovation and commercialization are currently not a part of the science education curriculum, a growing need exists for professionals who specialize in interdisciplinary innovation generation and technology transfer that can bridge the gap between medicine and technology and who can manage tasks effectively and efficiently within a socio-economic context, and with respect to legal determinants. Therefore, it is necessary to revise and adapt the traditional educational curricula in many aspects of the public health system and the way that we approach future issues.

> **Learning Box**
> An important starting point for students to understand what is needed to become a MedTech innovator.
> Imagine your future professional career. What knowledge and skills the MedTech innovator should have?
> Tools: group work, brainstorming, discussion.

Heath Tech innovators are people that liaise between technologies, clinical needs, and the patient. They should be able to identify problem areas, look for disruptive solutions, put a team together, and iterate the results for maximum impact. The engineer should be part of the innovation process as a developer, innovator, and entrepreneur. This could be only achieved through proper comprehensive education and training (Fig. 41.9)—such as HTIG.

The path leading to meaningful healthcare innovations generation is presented in Fig. 41.10. It is initiated by awareness of healthcare problems and is driven by knowledge of exponential technologies reinforced with trained tools and techniques for innovation generation and empowered by demanding patients. It is only possible with a special persona (John—the innovator) mindset (curiosity, empathy) and skills

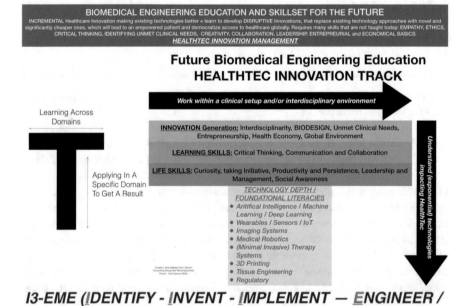

Fig. 41.9 Knowledge and skills set of healthcare innovators [7]

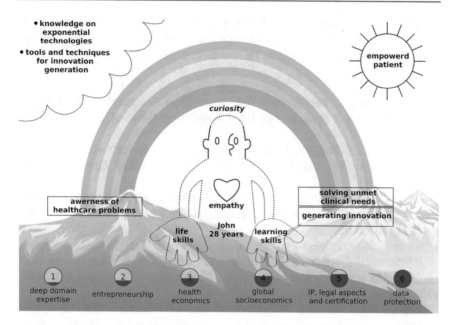

Fig. 41.10   The innovator persona and the path towards innovation generation

(life skills and learning skills) build on solid grounds of deep domain expertise and knowledge of associated aspects (1–6).

The comprehensive education and training on HTIG should include and follow the order:

- First—focus on solving unmet clinical needs, which requires a solid understanding of current health problems (regional, global), and domain knowledge (available solutions and their limitations), further being able to notice, identify, and define specific unmet problems.
- Second—understanding of the possibilities of exponential technologies in the health domain, and training on tools and techniques for innovation generation.
- Third—being empathetic with respect to the clinical stakeholders and the patients—being aware of their concerns and needs. To create meaningful innovations leading to healthcare democratization, a starting point is to empathize with patient-centric benefits.
- Fourth—despite deep domain expertise—entrepreneurship, awareness of socioeconomic aspects, IP, regulatory approvals (e.g., CE, FDA), certification process, and data protection are essential and therefore symbolically located as the innovation's grounds in Fig. 41.10 1–6.
- Fifth—being armed with essential learning (critical thinking, communication, working in an interdisciplinary team) and life skills (curiosity, flexibility, productivity, persistence, taking initiative, leadership and management, social

awareness, and responsibilities). Those are the most important twenty-first-century skills [8]—crucial catalysts of disruption—that allow the biomedical engineer to play a prominent role in healthcare innovation generation.

41.3.5 Tools and Techniques for Innovation Generation (Building Your Own Path to Health Technology Innovation = HOW)

Health innovation should be based on meaningful solutions that solve meaningful problems (including availability, excessive cost, procedure time, or difficult handling).

To design an innovation (= invention × commercialization), one starts with identifying real clinical needs or unsolved problems (always ideating around a problem, not a technology).

For that, this unique training uses an adapted **Stanford BIODESIGN** approach of IDENTIFYING UNMET CLINICAL NEEDS, IDEATE solutions for these needs, iterate with the stakeholders, and subsequently work on IMPLEMENTATION as start-up companies.

This combined with other VALIDATION models and experiments enables us to properly understand the problem, the future opportunities and to create and validate ideas through Minimum Viable Prototype (MVP). MVP is a method for quickly creating, launching, testing, and improving a product before perfecting all the fine details to validate the first assumptions as soon as possible and at the lowest possible costs. Also, the product (prototype) that fulfills this description is called the MVP.

Figure 41.11 presents the outline of innovative technologies taught during HTIG classes.

To effectively and clearly present the innovation to the stakeholders, investors, and clients Scientific Writing, and Pitching strategies are presented to students. Emphasis is put on differentiation of the presented content depending on who is a recipient, deciding what is really important and what is unnecessary so that the

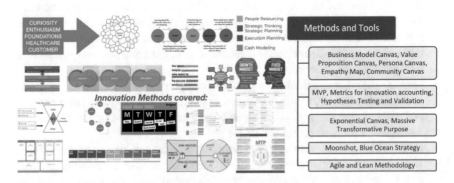

Fig. 41.11 Overview of the innovative methods taught at HTIG

delivered message is clear and meets the recipient's expectations on content. Also, the use of appropriate language, graphics, and quality in general, are discussed.

> **Learning Box**
> An important starting point for students to learn innovation generation. This exercise can be continued through subsequent classes while new methods and tools are introduced step by step (e.g., BIODESIGN, BMC, VPC, ExO Canvas,. . .).
>
> Think about a very simple innovation, e.g., completely new design of a water bottle or rehab in virtual reality.
>
> Tools: group work, brainstorming, discussion, pitching.

41.4   Tools for Keeping Students Engaged During and Between Classes

The following teaching methods and techniques are applied during HTIG: Case study, Problem-based learning, Design Thinking, Project-based learning, Gamification, Mind mapping, Concept mapping, Peer assessment, Socratic questioning, Debriefing Group, Flipped classroom, Project assignments, and Discussion.

41.4.1 Communication Channels and Engagement Tools Between Classes

WhatsApp group	Keeping students in touch, assigning work systematically between onsite classes, announcements, sending deadline reminders, and auxiliary materials related to HTIG
Mail	handling individual issues, and grading explanations
MS Teams	literature and classes recordings storage, announcements on dates and location of the upcoming classes
Zoom	hosting additional consultations and online lectures, encouraging discussions
Homework	Finishing a lecture with a question / an issue for further consideration before the next classes. For example what will happen in biomedical engineering in next 10 years (10x better, 10x cheaper, 10x more users)

Fig. 41.12 Brainstorming during HTIG classes. First identification of healthcare problems, and then designing innovations

41.4.2 Keeping Interest During the Classes

- Engaging in discussions, asking puzzling/rhetorical questions, pointing out one individual if no one was willing to answer.
- Embedding the presented content in the current reality by giving examples with which students can identify.
- Brainstorming (Fig. 41.12)—everybody gets sticky notes and needs to answer a question. Similar answers are grouped and discussed, e.g.:
  - What innovation really is.
  - Healthcare problems/solutions.
  - Ideas for live longevity improvement.
  - (System) problems observed when you or someone close to you was ill.
- Innovation game (15 min of discussion in groups, 5 min for presentation outcomes of discussion to other groups):
  - Design a surgery room 2035.
  - Cancer (lung, prostate, breast) imaging + therapy, how it will look like in 2035.
- Regular assessment of teamwork innovation with Purpose Launchpad [6] and backlog, joint discussions, and feedback for other teams.

> **Learning Box**
> Keeping the mind fresh:
>
> - Quick team up for ideating tiny problems (e.g., invent a completely new water bottle), towards collaboration, competition, gamification.
> - Providing short, out-of-the-scope breaks. For example, guess what is in the picture. For a while, students discuss what they see in the picture, and after a while, the answer is revealed.

41.5   Team and Individual Assignments, Grading

In 2021/2022 edition, most of the lectures and project classes were carried out in person. However, the HTIG program is flexible enough to be fully held online with more extensive use of the aforementioned communication channels. For example, group work can take place in the separate *zoom* or *gmeet* rooms, project assessments with Purpose Launchpad and backlog are anyway made using the online tools, and the final exam was also held online (*Kahoot*). See Fig. 41.13 for details.

In 2021/2022, students worked on the following innovative ideas: SURGERY AR, HEART BIT, CLEVERCARE, ACCELERATOR, ASSISTANT, CELLMED, MEDNOW. Students delivering final presentations were captured in photos (Fig. 41.14). The highest grades were given to team assignments (team core project and its presentation, and team paper, Fig. 41.13). This reflects students higher enthusiasm and engagement towards teamwork, and shared responsibilities, in comparison to individual assignments.

41.6   Conclusion and Lecture Summary

Healthcare innovations come from synergized work of interdisciplinary teams. HTIG classes aim at preparing engineers to find their place in these teams. For that a significant change in the current education curriculum was required to offer future healthcare innovators a comprehensive education and training to bridge the gap between medicine and technology, and provide consciousness of socio-economic context, and legal determinants.

First familiarized with current burning problems in healthcare and a holistic view of its future, students learn to empathize with clinical perspective, analyze workflow, and subsequently identify unmet clinical needs. Then, equipped with the innovation generation tools and strengthened with the required life and learning skills, they are ready to participate in a meaningful process of generating clinically relevant and affordable innovations, leading to healthcare democratization. Factors like keeping the students engaged and dedicated to the case and actual learning by doing—eagerness to find the right solution, immersing in innovation generation, were of the most importance in developing an impactful learning outcome and students satisfaction.

Individual assignment 1 (15pt)	**Future Technology Aspect in Healthcare** • Proposal on Technological or Process Solution • Should include problem description, what is the technological solution to the problem that is not currently available, but will be by 2035, what it does, and what benefits it brings. • Graded on content, value, enthusiasm, methodology, and scientific writing. • Form: use a scientific template - 4pg IEEE EMBC conference, provide references and figures.
Individual assignment 2 (15pt)	**How your job will be performed in the future** • How it would differ from right now, and why, what tools are likely to be used in 2041, what innovations would you suggest, • Form: use a scientific template - 4pg IEEE EMBC conference, provide references and figures.

Team Assignment = paper + presentation (50pt)	• First, **a short presentation** of an initiative taken: team composition (Bosi DNA, team canvas), What problem will be solved? How is it done so far? What are the limitations? Who do you believe your customer is? How do you want to solve the problem? What is the value proposition? First assumptions and their verification. The project is then discussed in project classes, the first PL evaluation is performed and the backlog is defined.
	 • Getting deeper into the initiative - throughout the semester students work on the following (project development):
	    → Unmet clinical need definition (problem statement, How is it done so far? What are the limitations? What are the alternatives? - come up with 3-5 ideas).
	    → Stakeholder perspective - problem definition, ideas on how to disrupt that - prove why the problem is worth solving.
	    → Customer perspective - persona canvas, empathy map - ask questions, do the interviews.
	    → Potential value propositions - evaluation scheme, develop VPC for the customer, validate, correct, prioritize, ITERATE.
	    → Team building, tasks assignment.
	    → Developing a BMC, and ExO Canvas, observing the relation between VPC, BMC, ExO.
	    → Creating an MVP.
	    → Summarizing the aforementioned data for final evaluation in the form of a 4-page paper (IEEE EMBC template).
	 • During classes, the results are discussed together, students exchange experiences with other teams, receive feedback from the teacher, and clues on further steps. The project is regularly evaluated with PL, and the backlog is refined.
	 • Second, **a final presentation** (summarizing the innovative project results and perspective) is delivered by the team members. Again, it is publicly discussed and PL evaluated.

Extra points	• Extra points for taking the online ExO Foundation course (3pt) • Preparing a personal MTP canvas (3pt)
Exam	• **Final exam** - online Kahoot (20pt) - multiple questions

Fig. 41.13 HTIG Students deliverables and grading—mix of individual and group assignments

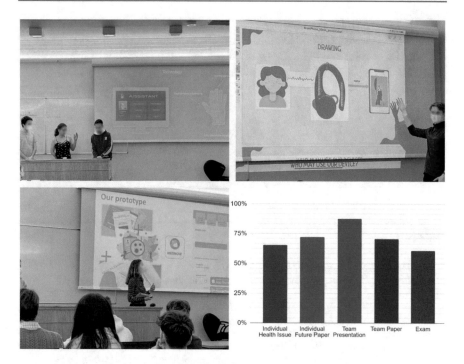

Fig. 41.14 Project final presentations and student grades

Take-Home TIPS and Messages TEACHERS
- Take into account that students have different educational backgrounds, expectations, soft skills and emotions.
- Adjust the content to bridge the educational gap at your university.
- Be a servant leader, facilitate discussions, support teamwork, guide, and supervise.
- Chose innovative tools or techniques you would like to teach and prepare exercises (learning by doing).
- Introduce as many as possible different games, experiments, and activities to empower creativity and reinforce the learning outcome.

Take-Home TIPS and Messages STUDENTS
- Collaborate with clinicians and patients, be an empathetic observer in a clinical setup, and invest time in a proper understanding of the clinical needs.

(continued)

- Always innovate around the problem, not around the technology. So make sure that there will be a market for your solution.
- Taking into account regional needs and healthcare economics, try to reduce errors' costs keeping the quality and availability of the solution at the same time.
- Remember to test your hypotheses and iteratively validate your MVP.
- When preparing a pitch or project documentation, tailor the content to the actual recipient (stakeholder, patient, doctor), and respect hers/his time with only relevant and clearly stated information.

References

1. https://bosidna.com/assessment. Accessed 3 Apr 2022
2. https://sylabusy.agh.edu.pl/en/document/510f3563-4c2b-4bca-b1dc-741476d4b0e5.pdf. Accessed 3 Apr 2022
3. Era Per Med (2022) Prevention in personalised medicine. Call text, updated on 16 Feb 2022. https://erapermed.isciii.es/joint-calls/joint-transnational-call-2022/. Accessed 3 Apr 2022
4. European Commission (July 2014) Advice for 2016/2017 of the Horizon 2020 Advisory Group for social challenge 1, "health, demographic change and wellbeing". https://ec.europa.eu/transparency/regexpert/index.cfm?do=groupDetail.groupDetailDoc&id=15073&no=1. Accessed 3 Apr 2022
5. Singletary B, Patel N, Heslin M (2017) Patient perceptions about their physician in 2 words. The good, the bad, and the ugly. JAMA Surg 152(12):1169–1170. https://doi.org/10.1001/jamasurg.2017.3851
6. https://medium.com/@seanmoffitt/the-top-30-emerging-technologies-2018-2028-eca0dfb0f43c. Accessed 3 Apr 2022
7. Friebe M (2020) Healthcare in need of innovation: exponential technology and biomedical entrepreneurship as solution providers (Keynote Paper). In: Proc. SPIE 11315, Medical Imaging 2020: image-guided procedures, robotic interventions, and modeling, 113150T (16 March 2020). https://doi.org/10.1117/12.2556776
8. Bailey A, Kaufman E, Subotic S (2015) Education, technology, and the 21st century skills gap. World Econ Forum Geneva Switz 16(03)

Katarzyna Heryan is a passionate biomedical engineer (graduated with distinction from Computer Science and Medical Electronics in 2014), employed at Department of Measurement and Electronics, AGH-UST Krakow, Poland. Following her desire to create meaningful innovations in healthcare, she took the training at the University of Lund, Sweden on soft skills and entrepreneurship, design thinking, business model canvas, coaching, and teaching in higher education—Train the Trainers Transformation.doc. She also studied biomedical signal processing and medical image analysis, ML/DL at ETH Zurich (School on Biomedical Imaging) and Masaryk University (Advanced Methods in Biomedical Image Analysis). She is the author of more than 25 original research papers and four patents.

Michael Friebe received a B.Sc. degree in electrical engineering, an M.Sc. degree in technology management from Golden Gate University, San Francisco, and a Ph.D. degree in medical physics in Germany. He spent 5 years in San Francisco, as a Research and Design Engineer with an MRI and ultrasound device manufacturer. He is a German citizen with expertise in diagnostic imaging and image-guided therapies, as a/an Founder/Innovator/CEO/Investor and a Scientist. He is also a Research Fellow with the Technical University of Munich, Munich; an Adjunct Professor with the Queensland University of Technology, Brisbane; and an honorary Professor of image-guided therapies with Otto von Guericke University, Magdeburg, Germany. Since 2022, he has been a Professor of biomedical engineering innovation with the AGH University of Science and Technology, Krakow, Poland. He is a listed inventor of more than 80 patents and has authored over 200 papers, and was part of over 35 Medical Technology Start-Ups. He is a Board Member of four medical technology start-up companies and an investment partner of a MedTec-fund. From 2016 to 2018, he was a Distinguished Lecturer of the IEEE EMBC teaching innovation generation and MedTec entrepreneurship. He is also a coach and trainer for OpenExO, and a master Launchpad mentor for the Purpose Alliance.

Health Innovation Design at a University: INKA INNOLAB at Otto-von-Guericke-University

42

Holger Fritzsche

Abstract

Incremental innovation is the constant and gradual improvement of existing products, services, and processes: making them better, cheaper, or more effective; this is also the standard innovation process for medical product development. Here, disruptive technologies are seen as a simple alternative to existing products with reduced functions and limited performance. Innovation is not just something new but an invention multiplied by a commercial translation. There must be added value for the clinical user and the patient. New ideas and concepts require new technologies in interdisciplinary cooperation between engineers, scientists, economists, and the medical user. Biomedical engineering students, in particular, must understand that only through correct observation, process know-how, and subsequent analysis and evaluation can clinically relevant and affordable innovations be generated and possibly used for entrepreneurial ventures. The laboratory for innovation, research, and entrepreneurship—Innolab IGT (Innovation laboratory for Image-Guided Therapy) is run by a research group of engineers (INKA Application Driven Research) at the University Hospital of the Otto-von-Guericke-University in Magdeburg. It forms a network node between medicine, research, and business/industry. Here, students work in a focused manner from the innovation process to technology transfer with the user. At the same time, they should stimulate startup activities in this area. The concept of the Innolab IGT is simple but very effective and follows the innovation methodology I^3 EME (Identify, Invent, Implement together with Engineers, Medical Users, and Economists), a fusion of the Design Thinking approach and Stanford Biodesign. First, needs are identified together with the clinical users. For

H. Fritzsche (✉)
INKA Application Driven Research, Medical Faculty, Otto-von-Guericke-University Magdeburg, Magdeburg, Germany
e-mail: holger.fritzsche@ovgu.de

these, initial solutions are being developed within the framework of the Innolab IGT and checked for commercial exploitation. The implementation of the ideas is then carried out by a startup-focused transfer downstream of the Innolab IGT in cooperation with the transfer structures of the University and broad network activity. Especially for the continuation of startup ideas beyond the Innolab IGT, early involvement and support of the Key Opinion Leaders and Stakeholders are needed.

Keywords

Stanford biodesign · Innovation generation · Startup · Medical technology transfer · Entrepreneurship · Medical Research Laboratory

What Will You Learn in this Chapter?
- Multidisciplinarity is key to Health Innovation.
- The INNOLAB uses a methodology of Design Thinking combined with the Stanford Biodesign approach.
- The ultimate goal of such a lab is the translation of validated concepts to a Startup—but the mentorship of entrepreneurial activities continues.

42.1 Introduction

The potential for innovation generation and subsequent translation in startup activities are rarely part of academic training. Biomedical engineering courses, in particular, impart knowledge from the natural sciences in a technical context but innovation, creativity, and implementation are rarely part of the curriculum [1]. However, there is a high potential for collaboration between clinicians and engineers, especially in medical technology [2]. In developing product ideas, this collaboration is new. It has developed into an essential part of medical technology training in Magdeburg, with an established clinical network focusing on minimally invasive surgeries and image-guided procedures. From a series of lectures, a network of clinicians and engineers has emerged in a short time. This network produces numerous inventor reports, patents, publications, startups, and works on joint research projects with industrial partners. The interdisciplinary approach of the research group is based on the identification of clinical needs, the implementation of product ideas in cooperation with medical users, and the transfer to industrial partners. The focus is on image-guided, minimally invasive diagnostic and therapeutic procedures and the necessary medical technology systems. The approach for interdisciplinary cooperation with the medical user is pursued in research and development, as well as in student training.

42.2   Focus

The focus of the IGT incubator is to directly generate innovations in the field of image-guided surgeries where they can be used together with the actual users. These innovative processes and projects are for therapeutic tools and systems (e.g., tumor removal under imaging, lymph node biopsies, catheter, and delivery systems, endoscopic components, and much more) for the clinical area of interventional radiology, neuroradiology, urology, and ENT.

42.3   Design and Structure

At its core, the Innolab IGT is intended to be a "think tank" for entrepreneurial thinking and action within diagnostic and therapeutic medical technology. The target group is those interested in founding the company, consisting of students, experienced clinicians, and engineers. The research group uses the I³EME (*I*dentify—*I*deate—*I*mplement with a team of *E*ngineers, *M*edical Staff, and *E*conomic know-how) innovation concept [3] as a solid basis to raise awareness. Interdisciplinary collaboration between physicians and engineers is explicitly required. According to this maxim, qualification and support offerings are tailored to the users of the Innolab.

42.3.1 Creative Space

The laboratory includes a creative area for generating ideas, concept studies, and meetings with individual and group workplaces. Bright and colorful setup invites students to work creatively. The flexible furniture offers a variety of possibilities to meet the group's needs. Cork walls, whiteboards, a smartboard, and mobile projectors offer plenty of opportunities to find, discuss and develop ideas (see Fig. 42.1).

Fig. 42.1 The Creative Space of the INNOLAB, a university hospital-based Healthtech innovation environment to stimulate inter-/multidisciplinary exchange

42.3.2 Prototyping and Electronics Laboratory

The adjoining prototype laboratory is used for invention and technical implementation. Smaller and larger product ideas are checked for technical feasibility and built directly as prototypes. Various 3D printers and many precision tools are available for machining the finished parts and creating and implementing electronic components.

42.3.3 Simulation Surgical Room

The simulation SR is used to implement and verify the developed prototypes. A minimally invasive surgical setup with a patient table, ultrasound tomography system, endoscopy tower with HF generator, ultrasound systems, navigation/tracking devices, and various phantoms offered the opportunity to test with the user in a simulated clinical environment (Fig. 42.2).

42.4 Student Training

The lecture series Innovation Generation and Entrepreneurship in the Healthcare Domain, Image-Guided Surgeries, Translational Technology Entrepreneurship and Health Tech Innovation Design (Fig. 42.3) were offered as a training basis for the students, mainly in the master's seminars. Above all, students with interdisciplinary approaches from the respective master's degree courses participated (especially integrated design engineering, medical systems engineering, mechanical engineering, electrical engineering, software engineering and medicine).

Interdisciplinary student teams (3–5 members) are formed every semester to identify clinical needs when visiting live surgeries and develop many ideas for each problem. The ideas are then regularly passed back to the clinicians, who see and discuss the developed prototypes.

Fig. 42.2 A simulated Surgical Room is the centerpiece of innovation attention. Robots, therapy systems and diagnostic + navigation equipment is available as well as a kind of gold standard environment

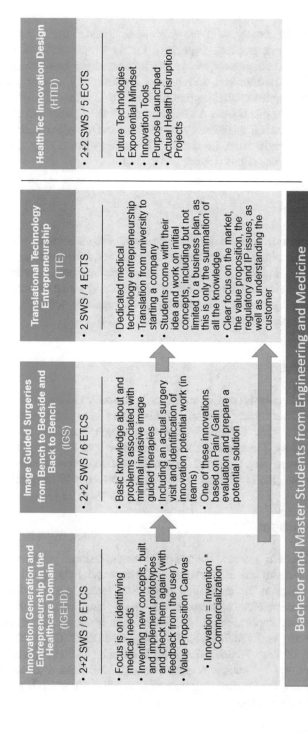

Fig. 42.3 The series of lectures on innovation design and entrepreneurship conducted by Prof. Michael Friebe [4, 5]

42.5 Network

The Innolab IGT works closely with the established transfer structures of the University and its network partners. Especially for the continuation of startup ideas beyond the Innovation process within the Innolab, early involvement and support of Key Opinion Leaders and Stakeholders are necessary. Strong partners provided practical and content-related support for the "identify, invent, implement" development approach [6] mentioned. A clinical and industrial panel was created, and the medical and electrical engineering faculties are currently involved in the innovation processes (identification, invention). The industry board helps find and identify technology transfer options (for implementation).

42.5.1 Clinical Panel

Current clinical cooperation partners are ENT, urology, neuroradiology, radiology, nuclear medicine, dermatology, vascular surgery, orthopedics, and cardiac surgery at the University Hospital Magdeburg. In addition, interdisciplinary student teams are formed each semester to identify clinical needs while visiting actual surgeries and finding many ideas per problem. The ideas are then regularly fed back to the clinicians, who come, see, and discuss the developed prototypes.

42.5.2 Industry Board

The industry advisory board with several small, medium-sized, and large companies from Saxony-Anhalt and other German locations was also established. The companies come from the most diverse medical technologies branches and follow different business models. Therefore, the board should be available as an advisor, especially for questions relating to market design, production, and medical technology certification. In total, the Innolab IGT received support from 19 companies.

42.5.3 Graduate School T^2I^2

The INNOLAB IGT acts as a central contact point for the doctoral program of the graduate school "Technological Innovations in Therapy and Imaging—T^2I^2." A structured doctoral program for innovation generation, technology transfer, and medical technologies' commercial implementation. The training includes a technical understanding of medical application and consideration of economic factors industrial board. Furthermore, the INNOLAB IGT is represented in networks such as the VDI (Association of German Engineers), the IEEE Branch (Institute of Electrical and Electronics Engineers), or maintains close contact with BMEidea (Biomedical Engineering–Innovation, Design, and Entrepreneurship Alliance), which has a wide range of industrial partners, clusters and research networks.

42.6   Outlook

The Innolab IGT has created an innovation and idea generator in which clinicians and engineers work closely together in a simulated clinical setup on the university hospital campus. As a result, the engineer better understands everyday clinical processes and can identify problems and deficiencies in clinical workflow or technical products. Another unique selling point of the Innolab IGT is the close cooperation between students, doctors, scientists, and industrial business partners. Solutions and innovative ideas are developed, implemented, changed, or discarded in constant consultation with doctors. Test and evaluation by the clinical user and requests and suggestions from business partners are continuously integrated into the individual development phases of new medical products and guarantee market-oriented product development. Of course, these products/services cannot be used on patients due to a lack of regulatory recognition. However, the Innolab IGT itself represents an ideal development environment for a future translation in the field of image-guided therapies, thanks to the existing therapeutic and diagnostic infrastructure and the training concepts.

The aim is to involve students in real-life projects early in their academic training and teach them innovation management and entrepreneurial thinking to enable technology transfer. The biomedical engineer gets a fundamental understanding of the problem through live medical intervention observation, good communication and an empathic understanding of the user (doctor, patient, nurse), which forms the basis for the innovation process within the Innolab IGT. Creative work produces many ideas and possibilities to solve the unmet clinical needs quickly checked and validated by the related innovation concepts for their usability in terms of technical implementation, market launch, reimbursement, and regulatory obstacles. The validated ideas can be quickly converted into prototypes with good laboratory equipment and broad support from network partners. Furthermore, students receive hands-on training on real-life problems and applications while working on the project. As a result, all stakeholders involved benefit from scientific recognition of economic translation via patents and startup generation, education and knowledge transfer, or economic stimulus.

Take-Home Messages
For the successful development of innovative ideas as a basis for technical translation and startups, the following points must be intensified or newly created:

- Closer networking between university medical technology and clinics
- The necessity of short distances on site
- Development at the clinic with direct and rapid input from clinicians
- A well-equipped prototyping infrastructure

(continued)

- Cooperation between engineers with an understanding of the medical/therapeutic problems and physicians with innovation and technical know-how
- Mutual recognition, trust, and empathy as the basis for cooperation
- Empathy toward clinical users and patients
- Understanding of the value chain in healthcare (national and international)
- Recognize potential for exploitation/develop business strategy (lean startup approach)

References

1. Traub J, Ostler D, Feussner H, Friebe M (2019) Globale Innovationen in der Medizintechnik – Interdisziplinäre Ausbildung an der Universität. In: Internationalisierung im Gesundheitswesen. Springer Fachmedien Wiesbaden, Wiesbaden, S 265–278
2. Boese A (2016) Lösungsfindung mit dem Endnutzer, ein neuer Ansatz in der methodischen Produktentwicklung am Beispiel der Medizintechnik, 1. Aufl. Shaker, Herzogenrath
3. Friebe M, Traub J (2015) Image guided surgery innovation with graduate students - a new lecture format: basics - experience - observe - analyse = innovation generation recipe and motivation boost for biomedical and medtec master students. Curr Dir Biomed Eng 1:475–479
4. Friebe M, Boese A (2016) I3 EME — innovation awareness für Medizintechniker. Magdeburger BEITRÄGE ZUR HOCHSCHULENTWICKLUNG. pp 24–27
5. Friebe M (2020) Healthcare in need of innovation: exponential technology and biomedical entrepreneurship as solution providers (Keynote Paper). In: Medical imaging 2020: image-guided procedures, robotic interventions, and modeling
6. Yock P, Zenios S, Makower J (2015) Biodesign: the process of innovating medical technologies, 1st edn. Cambridge University Press, Cambridge

Holger Fritzsche is currently working at the Otto-von-Guericke-University Magdeburg (OVGU) where he is part of the INKA research group. He takes care of the design and organization of the Innovation Laboratory for Image-Guided Therapies, which is funded by the EFRE (Europäischer Fonds für regionale Entwicklung) and coordinates the Graduate School T^2I^2—Technology Innovation in Therapy and Imaging.

Clinical Innovation at Acibadem Biodesign Center

43

Ata Akın, Erdi Dirilen, Kutalp Kurt, and Esra Bal

Abstract

Acibadem Biodesign Center (ABC) was founded in January 2017 with a joint funding from the state and university. Its mission is to provide innovative solutions to clinical problems, ranging from medical devices to healthcare delivery processes. Over the years we completed academic and industrial projects as well as provided innovative designs for entrepreneurs admitted to the Acibadem Incubation Center. ABCs design approach rests on three main key concepts: (1) design thinking, (2) need-based innovation, and (3) purpose-focused design. One other major goal of the center is to cultivate the employability skills in the students through promoting a transdisciplinary approach.

Keywords

Biodesign · Design thinking · Clinical innovation · Medical device design

A. Akın (✉)
Department of Biomedical Engineering, Acibadem University, Istanbul, Turkey
e-mail: ata.akin@acibadem.edu.tr

E. Dirilen · K. Kurt
Acibadem Biodesign Center, Acibadem University, Istanbul, Turkey

E. Bal
Department of Psychology, Acibadem University, Istanbul, Turkey

43.1 The Need

Problem Statement: Can we minimize the uncertainty in diagnosis and improve clinical outcome through engineering tools?

Medicine is both a science and an art and there is a good reason for it. It is a science because it relies heavily on scientific methods in diagnosis and therapy. Physicians and healthcare practitioners are committed to provide healthcare based on established results. It is considered an art because physicians also rely on their intuitions when delivering a decision. This insight and intuitive reasoning may sound a bit unreliable, but it is developed only after years of continuous reading and practice.

Performance measurement in healthcare services has been ongoing for over several decades now. Factors contributing to performance have been included in mathematical models to develop strategies. Due to its complex nature, it has been quite difficult to accurately model the performance of healthcare services for optimization procedures.

In the early 1980s, a more profit maximization approach was undertaken while recently more patient-focused approaches have become favorable [1].

The performance of a healthcare service provided by a hospital (P(t)) can be formulated as below:

$$P(t) = w_d \times D(t) + w_s \times S(t) + w_h \times H(t) + w_o \times O(t) \qquad (43.1)$$

This equation is an additive objective function that is a linear sum of several of the factors. A physician's performance ($0 < D(t) < 100$) is multiplied by a weight (w_d), the contribution of a physician to the hospital's overall performance. The nursing services performance ($0 < S(t) < 100$) is multiplied by a weight (w_s), their contribution to the hospital's performance. The level of infrastructure of the hospital ($0 < H(t) < 100$) is multiplied by a weight (w_h), the contribution of this infrastructure to the overall performance of the hospital. Lastly, the suitability of the sociopolitical and physical environment ($0 < O(t) < 100$) is multiplied by a weight (w_o), the contribution of the outside environment to the performance of the healthcare service.

This is a time-dependent equation, hence, any change in the number and quality of the physicians, or the support teams will greatly affect the healthcare performance. Similarly, an improvement of infrastructure or the lack of it will lead to drastic changes in performance.

Clearly being good in one of these is not enough to maximize this equation.

We need methods and technologies to back the healthcare delivery up where the equation is failing. Several universities as well as healthcare providers grasped the necessity of continuous innovation required to push the performance bar higher. This paper will outline how we at Acibadem University decided to handle this endeavor, namely via the establishment of the *Acibadem Biodesign Center*.

43.2   This Is the Way!

November 8, 1895, was a very important day for all (bio)medical engineers. That date late afternoon and in the following days, Wilhelm Conrad Röntgen discovered that a strange light (X-ray) can pass through most material and even the human body. His work was published on December 28, 1895, and in a couple of weeks he became famous.[1] He was awarded the first Nobel Prize in 1901. A great start for all the biomedical engineers worldwide!

Röntgen was a mechanical engineer and a physicist, yet with his discovery and invention, he was awarded an honorary Doctor of Medicine degree. This brilliant mind was the embodiment of modern (bio)medical engineers. He was educated in the ways of engineering and basic sciences and had a deep interest in applying his theories to medicine.

Nearly around the same years another inventor, the Dutch Willem Eindhoven, a doctor, a physiologist, and inventor, showed an early interest in the heart's electrical currents. He imagined a device that could measure the electrical activity of the heart without the need to stick electrodes to the patient. Coming from medicine, he was aware of the great need to monitor this electrical activity in a noninvasive manner. He worked for several years to convert a galvonometer into an electro-cardiogram (ECG) instrument. His efforts were again praised with the 1924 Nobel Prize.

It seems anyone working to develop some sort of a medical device is a potential Nobel prize candidate. These two stories are examples of what is now called the biodesign approach [2–7]. The term was coined by two Stanford professors to imply the innovation and creativity process that leads to the development of a novel technique or an instrument in clinical settings. This process starts in the clinical setting, continues in a design thinking environment, and ends in a prototyping machine shop.

A biodesigner should be trained in the ways of medicine and engineering to be able to fulfill her destiny. The training includes both theoretical and applied components. A good engineer should also have the necessary communication skills

---

[1] Wilhelm Röntgen: https://en.wikipedia.org/wiki/Wilhelm_Röntgen

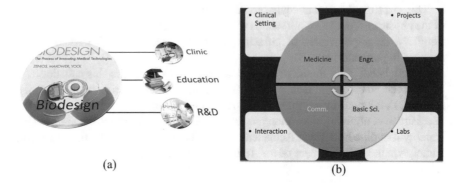

Fig. 43.1 (**a**) The Biodesign Book cover, (**b**) Transdisciplinary approach

to guide customers in their definition of their problem and need, to integrate in a teamwork and exchange ideas, to present her ideas and convince the managers to support her design.

So, the Biomedical Engineering Program at Acibadem University was developed 8 years ago with these expectations (Fig. 43.1).

43.2.1 Biomedical Engineering at ACU

The curriculum of Biomedical Engineering undergraduate program at Acibadem University rests upon three main pillars:

1. Biodesign Perspective
2. Employability Skills
3. Transdisciplinarity

43.2.1.1 Biodesign Perspective

This field not only deals with patient care or trying to make the hospital more technological but also examines workflow processes and offers optimization suggestions with engineering approaches [8–11]. We continue our projects at the Acibadem Biodesign Center, which we have established to examine the problems encountered during health services and to produce solutions for need, together with healthcare professionals and other health stakeholders [4–7, 12].

The ABC houses a 60 m^2 innovation room, a 40 m^2 prototyping workshop with 3-D printers, a mechanical workshop, an electronic work bench with the necessary prototyping tools, design computers and software and bench space for individual projects. We currently employ two full-time industrial designers, one patent advisor, and one entrepreneur mentor. The center is located within the Acibadem Incubation Center where the entrepreneurs can also access the facility. This way the ABC acts as a conduit between the industrial innovation and academic resources, both human resources and the rest of the R&D Labs at the university.

ABC WAY

There are three main elements that drive medical design: (1) Psychological/administrative expectations of the need's stakeholders, (2) Physiological/physical expectations of the need, (3) Treatment/application algorithm requirements. Hence, a careful analysis will yield a solution for the intersection of user clusters with different needs, consisting of the patient, patient relatives, health professionals, hospital management, regulations, and laws.

A good example is the "Nurse Call System" which is a communication system that enables patients to call nurses from patient rooms. Currently, the fact that the nurses cannot determine the level of the emergency with the existing alarm systems and the noise pollution on the floor constitute a problem.

NCS is one of the most common elements in the treatment process for in-patients. As a result of our research and approach, we discovered that by replacing the alarm sounds in existing systems with therapeutic sounds, the anxiety and fatigue of the problem stakeholders were reduced and the treatment process was supported, and we designed the therapeutic voice-assisted nurse call system as in Fig. 43.2.

Fig. 43.2 A nurse call button with a soothing sound designed by ABC

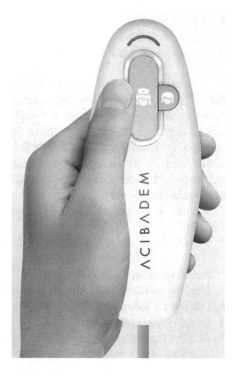

The center receives most of its project ideas from the clinicians, nurses, and managers of Acibadem Health Group and recently from other national and international customers. We have helped design and develop special syringes for neonatal use, mechanical instruments to aid aortic valve replacement, kiosk for COVID mask disposal, a make-up for a rapid COVID testing instrument and very recently worked with surgeons in creating a 3-D structure of the bones, nerves, muscles of a Siamese twin for surgical planning.

The ABC also plays a crucial role in the senior design projects of biomedical engineering students. The team at ABC provides design recommendations and prototyping facilities during the projects. ABC also is the place where students meet the entrepreneurs for their internships who would perhaps become their future employers. For more on ABC please visit the ABC LinkedIn page.

43.2.1.2 Employability Skills: Mastering Core Competencies

Studies show that soft skills play a critical role in hiring decisions [13] and a lack of soft skills is a major barrier for young people's employability [14].

Recently in Turkey, 295 decision-makers and HR executives from a variety of industries participated in the survey of the "Human Resources Trends and Salary Report" prepared by the multinational human resources consulting firm Randstad. The results showed that 35% of the barriers in the selection process are due to soft skills deficits [15].

Engineering education is currently based on three main areas:

1. Natural Sciences
2. Engineering Sciences
3. Humanities and general sciences

The goal of this three-legged education is to guarantee that the graduate possesses the skills and competencies sought after in professional life. Besides providing a strong engineering background, the engineering curriculum is designed to include several courses in humanities, and social sciences. Nevertheless, the demand for outstanding competencies outweighs the search for the brightest engineer. Hence, we decided to design a parallel education platform (very much like a parallel universe) within our curriculum to cultivate these competencies that are highly sought after in business life.

We call this approach "mc^2: Mastering Core Competencies" and the universe the ABC:

The offerings of this universe to the habitants are to enable students to:

1. Function on multi-disciplinary teams
2. Identify, formulate, and solve engineering problems
3. Display an understanding of professional and ethical responsibility
4. Communicate effectively
5. Acquire the broad education necessary to understand the impact of engineering solutions in a global, economic environment, and societal context

6. Recognize the need for and engage in lifelong learning
7. Exhibit a knowledge of contemporary issues
8. Use technique, skills, and modern engineering tools necessary for engineering practices
9. Design a system, component, or process to meet desired needs within realistic constraints such as economic, environmental, social, political, ethical, health and safety, manufacturability, and sustainability

These expectations are addressed in courses at each level by hands-on experience where the students work in groups to (1) solve a real-life clinical problem, (2) analyze and report their findings and advancements, and (3) present their results and discuss the outcome of their projects.

The outcome of this study will either be (1) a prototype, (2) a system, (3) or a project proposal, while the success can be defined through (1) the effectiveness of the solution, (2) the funds raised, or (3) the strength of ties with the industrial or academic partners.

The main purpose is to integrate the students vertically from different years within each research axis on a group basis and coach them on their quest to solve a problem that will be carried out for several years. Hence, the atmosphere of this universe is basically a PBL (project/problem base learning) environment, and the students are grouped with respect to projects but not to years.

We have been testing this method for the last 5 years and observed that the students who have captured the value of this approach were employed even before graduation at very prestigious international companies. Even more motivating for us was the demand from the companies who have hired one of our graduates to send them even more graduates.

43.2.1.3 Transdisciplinarity

Clinical problems come in a variety of sorts. The solutions therefore cannot be limited to one field of study, like electrical engineering or mechanical engineering. A biomedical engineer will need the background of electrical, computer, chemical, and mechanical engineering to be able to address the problems delivered by the healthcare practitioners. They will also need to be equipped with a good understanding of biology, biochemistry, and human physiology so that they can converse with medical experts. However, all these uploads will not be enough to make a biomedical engineer a good engineer. The future biomedical engineer (in fact any engineer) will need to possess several skills and competencies that are highly sought after in industry. In a meta-review by Passow, he shows that contrary to common belief by academicians, the industry seeks very different competencies [16]. The meta-analysis reveals that, contrary to the beliefs and obsessions of engineering professors with math and science courses, the most widely used competency turned out to be problem solving, communication, ethics, and lifelong learning skills. Specific to biomedical engineering programs, the students must also be taught how to interact with healthcare providers. This can only be possible when the students are immersed in those environments and interact with clinical experts in their own habitat. This type of approach, where the student is immersed in different environments like clinical settings, engineering facilities, design thinking workshops calls for a transdisciplinary education.

In a paper by Ertas et al., transdisciplinarity is explained as:

> Transdisciplinarity can be defined as the practice of acquiring new knowledge through education, research, design, and production with a broad emphasis on complex problem solving. The goal of transdisciplinary practice is to improve students' understanding of complex issues by extracting the valuable aspects of typical academic disciplines and thereby generate both a more integrative and universal solution to support an issue of importance to society. [17]

This elegant integration of academic disciplines for the holly pursuit of acquiring new knowledge to attack complex problems shall be the essential and major goal of all engineering programs. In a paper, Nicolescu mentions that:

> The report to UNESCO of the International Commission on Education for the Twenty-first Century, chaired by Jacques Delors, strongly emphasizes four pillars of a new kind of education: learning to know, learning to do, learning to live together with, and learning to be. In this context, the transdisciplinary approach can make an important contribution to the advent of this new type of education. [18]

This very straightforward recipe provides the program chairs with how and what should be done to convert our dull education approach to the transdisciplinary one. Fischer and Redmiles guide us by posing that:

> If the world of working and living relies on collaboration, creativity, definition and framing of problems and if it requires dealing with uncertainty, change, and intelligence that is distributed across cultures, disciplines, and tools, then education should foster transdisciplinary competencies that prepare students for having meaningful and productive lives in such a world. [19]

Fischer and Redmiles correctly warn us of the bumps on the road in designing the transdisciplinary curriculum and the challenges lying ahead for our transdisciplinary engineers.

43.3 Conclusion

Innovation in clinical settings is far different from any industry. The clinical environment does not enjoy the so-called disruptive innovations, since disruption means the loss of sequential healing practices that have been perfected over the years. So, one cannot propose a novel technique that will replace all of a sudden all the sequence of events that lead to the therapy.

The Biodesign approach is a unique innovation method tailored specifically for clinical problems. To be able to fully appreciate and use it in excellence, one needs a persevering education environment where the apprentice will be exposed to this way of thinking continuously.

We try to provide such an environment at Acibadem University with the Acibadem Biodesign Center at its core.

Take-Home Messages
- Disruption is not appreciated by the clinical profession.
- Transdisciplinary work is at the core of biomedical engineering innovation.
- The BIODESIGN approach is specifically geared for tackling clinical problems.
- The design approach should account for all the needs and expectations of the stakeholders.
- Mastering Core Competencies of the Future need to become part of university education and are particularly important for creating health innovations.

Acknowledgments   ABC was established funds in part by a grant from Istanbul Developmental Agency.

References

1. Zelman WN, Pink GH, Matthias CB (2003) Use of the balanced scorecard in health care. J Health Care Finance 29(4):1–16
2. Brinton TJ, Kurihara CQ, Camarillo DB, Pietzsch JB, Gorodsky J, Zenios SA, Doshi R, Shen C, Kumar UN, Mairal A, Watkins J, Popp RL, Wang PJ, Makower J, Krummel TM, Yock PG (2013) Outcomes from a postgraduate biomedical technology innovation training program: the first 12 years of Stanford biodesign. Ann Biomed Eng 41:1803–1810
3. Herr GL (2010) Biodesign: the process of innovating medical technologies. Biomed Instrum Technol 44:388
4. Wall J, Hellman E, Denend L, Rait D, Venook R, Lucian L, Azagury D, Yock PG, Brinton TJ (2017) The impact of postgraduate health technology innovation training: outcomes of the Stanford biodesign fellowship. Ann Biomed Eng 45:1163–1171
5. Wall J, Wynne E, Krummel T (2015) Biodesign process and culture to enable pediatric medical technology innovation. Semin Pediatr Surg 24:102–106
6. Wynne EK, Krummel TM (2016) Innovation within a university setting. Surgery 160:1427–1431
7. Yock PG, Zenios S, Makower J, Brinton TJ, Kumar UN, Watkins FTJ, Denend L, Kurihara CQ, Krummel TM (2015) Biodesign. Cambridge University Press
8. Braithwaite J, Glasziou P, Westbrook J (2020) The three numbers you need to know about healthcare: the 60-30-10 challenge. BMC Med 18(1):102
9. Grigoroudis E, Orfanoudaki E, Zopounidis C (2012) Strategic performance measurement in a healthcare organisation: a multiple criteria approach based on balanced scorecard. Omega 40(1):104–119
10. Omachonu VK, Einspruch NG (2010) Innovation in healthcare delivery systems: a conceptual framework. Innov J 15(1):1–20

11. Zambuto RP (2004) Clinical engineers in the 21st century. IEEE Eng Med Biol Mag 23(3): 37–41
12. Yock PG, Brinton TJ, Zenios SA (2011) Teaching biomedical technology innovation as a discipline. Sci Transl Med 3(92):92cm18
13. Hurrell SA (2016) Rethinking the soft skills deficit blame game: employers, skills withdrawal and the reporting of soft skills gaps. Hum Relat 69(3):605–628. https://doi.org/10.1177/0018726715591636
14. Clarke M (2016) Addressing the soft skills crisis. Strateg HR Rev 15(3):137–139
15. Randstad (2019) HR trends and salary report 2019. https://www.randstad.com.tr/en/workforce360/archives/hr-trends-and-salary-report-2019_185/
16. Passow HJ (2007) What competencies should engineering programs emphasize? A meta-analysis of practitioners' opinions informs curricular design. In: Proceedings of the 3rd International CDIO Conference, MIT, Cambridge, Massachusetts, USA, June 2007
17. Ertas A, Frias KM, Greenhalgh-Spencer H, Back SM (2015) A transdisciplinary research approach to engineering education. In: Proceedings of the 2015 ASEE Gulf-Southwest Annual Conference. The University of Texas at San Antonio
18. Nicolescu B (1993) On transdisciplinarity
19. Fischer G, Redmiles D (2008) Transdisciplinary education and collaboration. In HCIC-2008: "education in HCI; HCI in education". MIT, 2008

Dr. Ata Akin received his Ph.D. on Biomedical Engineering from Drexel University in 1998, his MS on Biomedical Engineering and BS on Electronics and Telecommunication Engineering both from Istanbul Technical University in 1995 and 1993, respectively. He is currently a faculty member at the Department of Biomedical Engineering at Acıbadem University and serves as the Dean of Faculty of Engineering and Natural Sciences. Dr. Akin's research interests are in the field of functional neuroimaging. He is also the founder and director of the Acibadem Biodesign Center.

Erdi Dirilen is an industrial designer by training. He is the Projects Manager at Acibadem Biodesign Center.

Kutalp Kurt Graduated with a degree on Industrial Design from Karabuk University in 2017. He has worked as a designer in HK Muhendislik and at Aşut Fiberglass Inc before joining the Acibadem Biodesign Center in 2020. He has helped design many medical technologies during his employment and participated in several patent and design awards during his time at the center.

Dr. Esra Bal After receiving her bachelor's degree from the Sociology Department of Boğaziçi University, Esra Bal completed her MBA Degree at Northeastern University (USA) in 2002. Afterwards she worked at Yeditepe University until 2007 first as an instructor of advanced English and then as a lecturer at the Communication Faculty. During this time, she completed her Ph.D. in Organizational Behavior at the Business Administration Department of Marmara University. After completing her Ph.D, she started working as a Human Resources Consultant at the Istanbul partner office of the global human resources consulting firm Development Dimensions International (DDI). As a consultant, she took part in the assessment and development center activities of various local and multinational clients, conducted competency profiling sessions and delivered targeted selection trainings. During this time, she continued her academic studies and delivered courses in the area of Industrial and Organizational Psychology as a lecturer at Acıbadem University's Psychology Department where she has become a full-time faculty member since 2021. Dr. Bal's main research interests focus on the development of students' employability related competencies as well as occupational health pyschology concepts. She is a member of Human Resources Management Foundation (PERYÖN) and European Association of Work and Organizational Psychology (EAWOP). She also serves on the board of YEKÜV (21st Century Education and Cultural Foundation), an NGO supporting socio-economically disadvantaged students. Dr. Bal is married and has two children.

Example of a Needs-Driven Innovation Training Program: The BioMedical Design Novo Nordisk Foundation Fellowship Program

44

Sys Zoffmann Glud

Abstract

This paper will present the Danish BioMedical Design fellowship program, which is an interdisciplinary fellowship that goes through a full cycle of a health-related innovation process over 10 months. This very successful needs-driven approach with the goal to actually detect unmet clinical needs, validate them, and build a transformational business model approach for a subsequent implementation is presented.

Keywords

Biomedical design fellowship · Needs-driven innovation

> **What Will You Learn in this Chapter?**
> - What the needs-driven innovation model of BioMedical Design is.
> - What the primary learning outcomes are
> - Why a program such as BioMedical Design is relevant

S. Z. Glud (✉)
BioMedical Design, Institute of Clinical Medicine, Faculty of Health, Aarhus University, Aarhus, Denmark
e-mail: sys@clin.au.dk

© The Author(s), under exclusive license to Springer Nature Switzerland AG 2022
M. Friebe (ed.), *Novel Innovation Design for the Future of Health*,
https://doi.org/10.1007/978-3-031-08191-0_44

44.1   Introduction

The Danish BioMedical Design Novo Nordisk Foundation fellowship is an elite postgraduate 10-month program, where interdisciplinary teams of talented and experienced professionals with diverse academic backgrounds are trained in a full cycle of a health-related innovation process. It combines lecturing sessions with clinical immersion, teamwork, and hands-on work such as prototyping in maker-spaces, workshops, or tests with end users and relevant stakeholders.

The cornerstone is a needs-driven innovation approach (as opposed to a technology-driven innovation approach) mediated by an 8 weeks long clinical immersion where the fellows are provided access to a clinical department in order to identify unmet needs for technological solutions.

The needs-driven approach in combination with other methodological principles has the potential to lower the risk of a poor product–market fit because the problem to be solved is witnessed in real life at the hospitals and further researched and validated through multiple channels before solutions are ideated.

Moreover, it is a fundamental part of the BioMedical Design that the participants (fellows) are organized in multidisciplinary teams. This is to train the participants' collaborative abilities and to leverage on the creativity that can arise in a multidisci-plinary group.

Figure 44.1 illustrates the overall layout of the program with the different phases that the fellows are taken through. The foundational knowledge for beginning a needs-driven innovation journey in healthcare is established in the bootcamp (in the first 4 weeks) followed by 8 weeks of clinical immersion.

After intense observations in the clinic and validation work, the creative skills phase is started, based on 2–3 identified and prioritized needs. Large amounts of ideas are generated, and some are prototyped, and at the end, one need with one

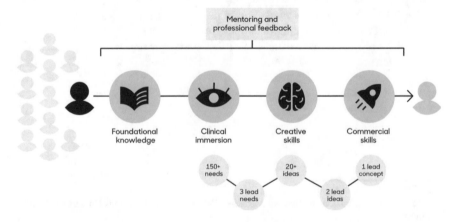

Fig. 44.1 Biomedical Design Fellowship layout from lecturing foundational knowledge toward a validated concept in a 10-month multidisciplinary team setup

product concept solution makes it to the last phase: the commercial skills phase where a business model and a strategy for business implementation are drafted.

44.2 The Methodological Framework of BioMedical Design

In short, BioMedical Design is best described as **a blend of disciplines** to fuel creative problem-solving and innovation. Several theoretical concepts addressing innovation methods and personal inventory are applied. The most central ones are described in the following.

The BioMedical Design program has a dual learning and training focus:

- Cognitive skills: concrete, applicable knowledge, tools, and methodologies in the field of innovation and commercialization to accomplish the entire innovation process.
- Non-cognitive traits: development of personal behavioral skills to enable ongoing successful application of the concrete competences.

To obtain lasting competences, ongoing and conscious attention is paid to the interconnection between cognitive skills and non-cognitive traits. The cognitive skills are trained through relating theory and methods with hands-on tasks and assignments in a domain-specific and real-world context.

The non-cognitive traits development is trained by having consistent supervisors mentoring the fellows with feedback related to their task accomplishment and behavior (adaptability, social sensitivity, critical thinking, networking skills, and communication).

44.3 Harvesting Tacit Knowledge and Sticky Information for Problem-Solving

The needs-based innovation approach builds partly on the widely acknowledged **Design Thinking** model for new product development (empathize, define, ideate, prototype, test), which is an attractive innovation model for many businesses. Particularly, the **empathize and define** elements are key to the needs-driven innovation approach as it allows for a systematic decomposition of the complexity of healthcare problems for the identification of information that can be tacit and sticky to obtain.

When empathizing and defining is applied with a 360-degree stakeholder perspective, as done in BioMedical Design, this enables solution ideation to take place on a more informed level.

Furthermore, in the healthcare sector, there is also a high level of hidden information at the systemic level, which can hinder the adoption of a new technology. The ability of the individual to identify the location of this information as well as the capability to decompose complex innovation problems are needed to succeed in the health innovation process.

44.4   Understanding the Process of Commercialization

What Design Thinking lacks is a focus on business implementation and outside market and industry dynamics that impacts the possibilities for what can be invented and what can be marketed.

Hence, a range of strategic business elements such as intellectual property rights, regulatory approval, health economics, business models, reimbursement and so forth are necessary constituents to shape and foster entrepreneurs to launch new ventures with their eyes open to what comes ahead.

However, without prior market knowledge and professional experience, such strategies become theoretical and far from reality if only performed as desktop work. Therefore, the principles of get-out-of-the-building (adapted from Lean Startup [1]) are applied throughout the fellowship. The methodology is basically an iterative process of formulating very concrete assumptions, identifying people and stakeholders who may be able to qualify the truth of the assumptions and conducting semi-structured interviews to gather proof or disproof of the assumptions.

The approach provides a fast-paced and efficient way to acquire learning and insight that many experienced entrepreneurs often state that they would like to have had beforehand.

44.5   Creative Self-Efficacy

The focus on developing personal behavioral skills at BioMedical Design specifically addresses and increases individuals' creative self-efficacy since this is a necessary precursor of creative efforts [2].

Creative self-efficacy means bringing fellows to a level where they believe in their own abilities and resources to act creatively and produce creative outcomes in the medical device/health tech arena. Consequently, high self-efficacy is critical for motivation and actual capacity to engage in the pursuit of certain tasks and to have the perseverance to endure obstacles and long processes [2]. These are elements that are equally important for entrepreneurial activities—not to mention innovation in itself [3].

Therefore, being able to produce creative outcomes requires:

(a) Self-efficacy, i.e., a belief in task capacity as a requirement for task accomplishment
(b) Domain expertise and knowledge, i.e., personal input for the production/making of ideas, actions, approaches

(c) Skills specific to creativity, i.e., tools, methodologies and processual understanding and managerial capacity

The fellows bring with them domain expertise, i.e., the subjective input for creativity, while the domain context (the healthcare setting) and the training format of the program add the tools and knowledge to develop creative self-efficacy.

44.6 Interdisciplinary Collaborative Abilities and Team Skills

The interdisciplinary focus of BioMedical Design is relevant because it drives the ideas generated to a higher inventive level and because it trains the fellows' ability to interact with and communicate with all the necessary intellectual disciplines required for succeeding in health innovation.

It is well-established knowledge that inter- and multidisciplinarity in teams enriches and heightens the innovative work [4]. However, fruitful teamwork and interdisciplinary collaboration are not a given result merely on the basis of bringing competent people together in a project.

Therefore, it is a key objective to strengthen the individual's aptitude to take part in an innovation team and to collaborate with specialists of diverse disciplines.

Research demonstrates that communication strategies, clarity of vision, respect and understanding of team role differences lead to better teamwork in interdisciplinary settings.

In order to promote good team development, the Tuckman model [5] for team development (forming, storming, norming, performing) and the results of Project Aristotle [6, 7] (Google team study) are used as a framework for guiding the BioMedical Design teams in combination with fairly intensive mentoring.

Innovation teams not only need to be good at executing, they also need to be proficient in group learning to integrate the constant flow of new knowledge and realizations into the teamwork. Good group learning promotes adaptation to change, greater understanding, and improved team performance results.

It is the intention that a focus on learnings about team dynamics will leave the fellows with an ability to promote group learning in future innovation teams.

44.7 Fostering a Pipeline of Talented Innovators to Shape the Future of Healthcare

The Novo Nordisk Foundation-sponsored BioMedical Design Fellowship Program plays a relevant role in the Danish innovation and entrepreneurship system of life science because it empowers and stimulates subject-matter experts and experienced professionals to create and commercialize novel healthcare solutions that can benefit patients and prepare our healthcare system for the future.

Denmark is a highly developed welfare society with a big knowledge-intensive life science industry that is among the best performing in the world. However, too

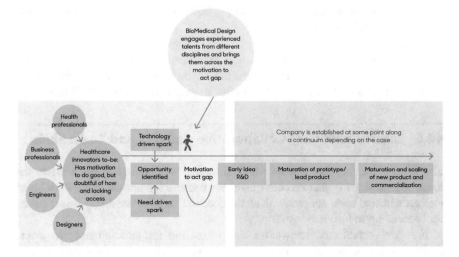

Fig. 44.2 BioMedical Design seeks to bridge the motivation gap. When fellows have completed the 10-months fellowship they have gained a solid knowledge-base, key insights and network to either take a whole new innovation cycle or launch a startup journey based on the projects developed in the program

few people are taking the opportunity to invent new healthcare solutions and pursue a startup career. Despite a generous and well-branched innovation and business growth support system, the pipeline of talents and qualified cases is particularly dry in medtech and healthtech.

To provide the tools, the motivation and a solid innovation opportunity BioMedical Design is a hands-on experience of acting as entrepreneur, but in a safe environment. The needs-driven approach is a focal point around which the whole process is revolving and which at the end of the fellowship year has provided the participants with a **compelling unmet healthcare need and a strong purpose to continue forward** (Fig. 44.2).

> **Take-Home Messages**
> – BioMedical Design is focusing heavily on both individual learning as well as team-level learning. The aim is to empower and motivate more people to pursue ideas that will benefit patients and healthcare.
> – All participants begin the process with a blank sheet of paper, no ideas and no clear mind of what problems to solve. They leave with a well-described unmet need, a novel invention and a clear project plan to continue on.
> – BioMedical Design trains the participants in a needs-driven approach to health innovation that builds on design thinking and combines it with biomedical engineering and a life science commercialization framework.

References

1. Steve Blank and Bob Dorf (2012) The Startup owner's manual. K & S Ranch Publishing Division
2. Bandura A (1997) Self-efficacy: the exercise of control. W.H. Freeman and Company, New York
3. Amabile T (1988) A model of creativity and innovation in organizations. Research in organizational. Behaviour 10:123–167. http://emotrab.ufba.br/wp-content/uploads/2019/06/AMABILE-Teresa-A-model-of-creativity-and-innovation-in-organizations.pdf
4. Hill L, Brandeau G, Truelove E, Lineback K (2014) Collective genius. Harvard Business Press
5. Tuckman BW, Jensen MC (1977) Stages of small-group development revisited. Group Organ Stud 2(4):419–427. https://doi.org/10.1177/105960117700200404
6. Duhigg C (25 Feb 2016) What Google learned from its quest to build the perfect team. NY Times. Hentet 10. October 2017. https://www.nytimes.com/2016/02/28/magazine/what-google-learned-from-its-quest-to-build-the-perfect-team.html?_r=0. Accessed 27 Feb 2022
7. Google Project Aristoteles. https://rework.withgoogle.com/print/guides/5721312655835136/. Accessed 27 Feb 2022

Sys Zoffmann Glud, Managing Director of BioMedical Design Novo Nordisk Foundation Fellowship Programme, Sys Zoffmann Glud has graduated from Aarhus University with an MSc in Molecular Biology and PhD in Nanomedicine. Leaving the research career behind, Sys has since 2012 worked in the field of healthcare innovation and entrepreneurship applied through university courses and continuing education training programs. Through her work she has trained +300 students, healthcare professionals, and industry experts in needs-driven healthcare innovation and commercialization, resulting in more than 10 startups and attractive innovation talents who keep applying the methodology they have been trained in for years.

Addressing the Healthcare Needs with Innovation Think Tank Global Infrastructure and its Methodology

45

Sultan Haider

Abstract

There is a growing need for self-sustaining innovation infrastructures on healthcare providers and medical device manufacturers to meet their customer-centric requirements. The Innovation Think Tank (ITT) addresses this with its global footprint and innovative framework methodology. The purpose of this paper is to firstly outline the value propositions derived from best practice ITT implementations at Siemens Healthineers, universities, and hospitals and secondly to describe the ITT methodology developed and optimized based on experiential learning. A total of 150 projects from R & D, product definition, lab designs, and co-creation programs at 20 ITT global activity locations were selected for this paper. Projects and locations were chosen to represent ITT's global footprint, portfolio, and programs. Further, project outcome data was analyzed and clustered to optimize and validate the ITT methodology. The database of ITT co-creation programs and lab framework consisting of best practices from various locations was updated, which can be used to plan, build, and operate ITT locations and projects. The global ITT infrastructure with its portfolio provides opportunities for healthcare system stakeholders to engage in a sustainable way. The ITT methodology is independent of the type of projects and is applicable across the innovation lifecycles and industries.

Keywords

Innovation Think Tank · ITT methodology · Innovation lifecycle · Co-creation · Product and portfolio definition · Siemens Healthineers

S. Haider (✉)
Siemens Halthineers, Innovation Think Tank, Erlangen, Germany
e-mail: sultan.haider@siemens-healthineers.com

What Will You Learn in this Chapter?
- The Innovation Think Tank methodology of joint health innovation creation.
- How a Global Medtech company connects universities, hospitals for co-creation.
- Global health improvement potential analysis on the base through a world-wide network of dedicated—mainly university-based—innovation labs.

45.1   Introduction

Healthcare systems globally continue to increasingly face highly complex challenges. In addition to generic solutions, e.g., in the field of digitalization and automation, there is a trend that healthcare institutions worldwide are looking toward establishing their own innovation and incubation structures locally to create tailored solutions to the needs of their patients and workforce [1].

ITT [2], founded in 2005, is a global infrastructure incorporating 85 activity locations (co-creation programs conducted/established and labs) at Siemens Healthineers (SHS) and several of its partner universities and hospitals, e.g., [3–8]. ITT methodology has been developed based on the experience of implementing and managing ITT locations and corresponding projects. Over the past 16 years, it has evolved and applied to accelerate healthcare innovation lifecycles.

ITT empowers partner healthcare institutions to re-invent themselves and find customized solutions to their challenging problems. The ITT Labs offer a sustainable co-creation environment to work on (1) the deep dive projects e.g. disease pathways [2], radiology workflow optimization [9], digital twin [10], etc., and (2) ITT Certification programs focus on engaging stakeholders and open innovation projects [2, 4, 11] (3) for developing patent and invention disclosures, e.g., [12–15] etc. and (4) generating savings due to synergies, and (5) creating additional revenue potential stemming from new product offerings. ITT programs have been progressively transforming into incubation centers funded by joint public funding programs.

45.2   Materials and Methods

Inputs from 150 project outcomes from 20 ITT activity locations in Erlangen, Kemnath, Forchheim, Magdeburg, Munich, Wuxi, Shanghai, Princeton, Gurgaon, Bengaluru, Dubai, Abu Dhabi, Cairo, Riyadh, Cape Town, London, Houston, Columbia, Istanbul, and Ankara have been analyzed for this methodology overview. The projects ranged from designs of the ITT labs, product definition and R & D to co-creation and ITT Certification Programs (Table 45.1). Table 45.1 also illustrates the consolidation of different backgrounds ranging from students and fellows to

Table 45.1 Example projects content analyzed during weekly project review meetings. The project categories are created by clustering the projects and are selected to represent the global portfolio from the last year. Highly diverse—both interdisciplinary and intercultural teams show a pattern of influencing each other when placed with common goals and examples from the previous projects, shortening on-boarding times and implement cycles. KPIs have been listed cumulatively in the last column for all the categories

Category	Locations and project examples	Participants, project teams	Projects content analyzed	Impact KPIs # = Number of
ITT Co-creation and Certification program	Western University, Ontario, Canada, NYU—Abu Dhabi, SHS South Africa, ITT Texas A & M, Acibadem University, SHS Saudi Arabia, Baskent University and Hospitals, ITT University of South Carolina, SHS ITT India in collaboration with SHS ITT China, Georgia Institute of Technology, ITT Magdeburg in collaboration with the Technical University of Munich, ERA Lucknow Medical College and Hospitals, etc.	465 interdisciplinary participants, 60 institutions, 95 projects, 30 countries	Business models, technology, disease pathways, presentations and feedback, challenges, and breakthroughs in digital hospital of the future, theragnostic and molecular imaging, robotics and interventional radiology, and technological advancement in medical imaging, healthcare system of the future	# Institutions engaged # Key Opinion Leaders engaged # Projects # Revenue potential due to new product offerings # Public funding applications # Savings and cost avoidance # Number of inventions and patents filed # Hospital workflows and customer needs analyzed #Time to market
ITT activity locations	ITT Headquarter, SHS Erlangen; Mechatronics Products, SHS Kemnath; Computed Tomography Performance Product Line, SHS Forchheim; Medical Electronics, SHS Erlangen, ITT SHS Shanghai and ITT SHS Wuxi, ITT SHS UAE, ITT SHS Saudi Arabia, ITT SHS Egypt,	60 participants, 40 projects	Customer insights and feedback, product definition, product roadmap, strategy workshops, technology development and validation	

(continued)

Table 45.1 (continued)

Category	Locations and project examples	Participants, project teams	Projects content analyzed	Impact KPIs # = Number of
	ITT SHS USA, ITT Bangalore and ITT Gurgaon, etc.			
Product and portfolio definition	Technologies and business models which can disrupt healthcare; disease pathway elaboration; requirement collection and MVP creation for the medical systems etc.	20 projects, 55 participants, 15 ITT SHS teams and partner universities and hospitals	Business models, technology, disease pathways, competitor analysis, presentations and feedback, customer workflow analysis, pain-points, stakeholders, decision propositions, outcome presentations	

experts and physicians. Employees, projects, and locations from these different backgrounds were chosen to represent ITT's global footprint, portfolio, and programs. The projects and their life cycle stages were clustered to identify synergies and improvement potentials to maximize the business impact. The best practices were added to the ITT database and the ITT methodology was refined for guidance to the teams. The value propositions were revised.

45.3   Results

It was observed while analyzing the sample project data for this paper, that ITT infrastructure, methodology, and portfolio offerings have further evolved respectively to the changing healthcare needs. The new definitions and facts and figures are now included into its global strategy and communication [2].

The key value propositions for ITT labs for addressing healthcare institution challenges were clustered into following: (1) Creation of intrinsic innovation infrastructures to address their particular organizational challenges and needs, (2) Access to international research facilities and databases to position them in the research communities, (3) Innovative organizational change to enable conducting specific research and expertise development programs, (4) Continuous customer insights across the product life cycle to shorten TTM and to meet customer requirements and (5) Key differentiator in strategic projects (Fig. 45.1).

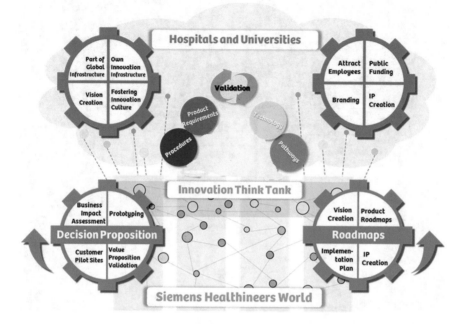

Fig. 45.1 Innovation Think Tank value propositions are presented here for the host organizations. ITT supports different departments within Siemens Healthineers by creating decision propositions and thus shaping the roadmaps. For hospitals and universities, it creates an own innovation infrastructure connected to global ITT locations and addressing organizational needs

The database of ITT co-creation programs and labs framework consisting of best practices from various locations was updated. The database is used for planning, building, and operating new ITT activity locations and co-creation programs, e.g., ITT Lab design (Fig. 45.2) and 3-D model designs for the customer world (Fig. 45.3).

The ITT methodology is presented into four steps: Acquire Mandate, Big picture creation, Co-creation on decision propositions and Deploy commercialization (Fig. 45.4). The definition of the ITT methodology steps was created based on (1) the importance of the project stages with respect to the KPIs including revenue potential and saving etc. and (2) the simplification of process to ensure tangibility for the users. KPIs (Table 45.1) could be adjusted based on the project goals. It has also been observed in various projects that the steps could be overlapping or unessential due to the organizational maturity and degree of trust with stakeholders.

The project clustering showed the key portfolio elements of ITT that enable host institutions to (1) Create their own innovation infrastructures which address their project requirements, (2) Fashion decision propositions that include business impact assessment, identification of pilot sites and supports shaping of the roadmaps and (3) Fosters innovation culture and expertise development via its ITT Certification program (Fig. 45.1).

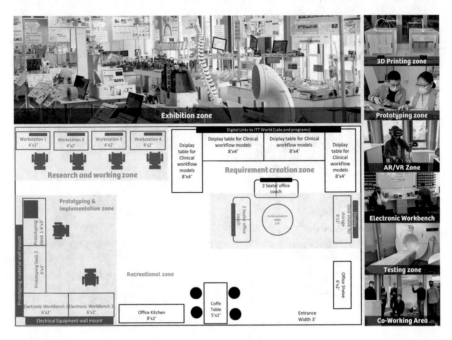

Fig. 45.2 ITT lab design tool and database has been built with inputs from various ITT labs and programs worldwide which focus on different scopes and goals and thus need various resources (HW/SW/competence) to fulfill them. The labs have a customer requirement zone showcasing the workflow, pain-points and best practices, research and working zone and prototyping and implementation zone. The proximity of the zones is carefully designed to simulate creativity and optimize resource usage

45.4   Discussion

Oversimplification of the innovation process and structures, without considering the customer expectations and importance of persistent efforts to gain the trust of the stakeholders has been observed as a limitation for their successful implementations. Another challenge is the lack of business models to sustain and deficient measurable KPIs. ITT addresses these gaps with its methodology and global footprint.

Being applied to healthcare projects, ITT intrinsically addresses complexity of the changing environments, customer behaviors, regulatory requirements, etc. ITT methodology is applied to a variety of fields needed to create medical product and solutions for healthcare providers, e.g., mechatronics and medical electronics components, medical devices, generating and validating requirements, workflow optimization, operational efficiency, business model development, commercialization, etc. (see Fig. 45.5 for an example).

The experiential learning during acquiring projects and the location of project implementation together shape the ITT methodology with novel problem-solving

Fig. 45.3 3-D Model of the healthcare system stakeholders and disease pathways at ITT Lab in Erlangen, Germany. These models are a key element of the ITT labs and its methodology. During the big picture creation phase, these models support visualization of the customer pain-points, identification of the interdependencies, solution best practices, and root cause analysis. Similar models have been built for radiology, labs, hospitals, product life cycles, departments within the hospitals and disease pathways, etc.

approaches in frequent cycles. The long years of R & D projects experience combined with a focus on product and portfolio definition projects and healthcare provider projects have resulted in creating a strong brand value for ITT. The program attracts and develops talents and has been implemented at several prestigious institutions with some adopting it into their curriculum and expertise development programs. In the last 12 months, over 1500 participants from over 50 universities, hospitals, and government bodies took part in the ITT programs focusing on the *future of healthcare*.

The exceptional presence at both healthcare providers and medical device manufacturers make the co-creation of customer-centric healthcare solutions possible with more efficient market delivery. This also results in the exceptionally fast implementation of new locations and a continuous innovation value streams between the locations.

The high adaptability of ITT, trusted partnerships and brand perception were observed as the key success factors during the COVID-19 pandemic. In the beginning of the lockdown, most of the demand on engineering and co-creation projects needed to be addressed on the basis of existing knowledge within ITT and virtual engagements with partner institutions. However, in few months this proved to be a very efficient way of working and showed a record growth of 40% in overall project volumes, interactions, and number of invitations to set up new ITT locations.

The ITT business model enables the activity locations to be self-sustainable: The beneficiary stakeholders, mainly the host institutions contribute resources for the

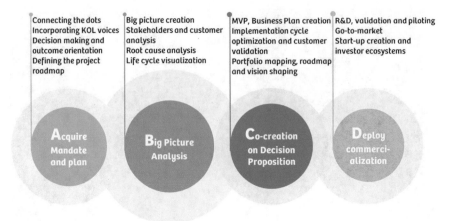

Fig. 45.4 The Innovation Think Tank Methodology—enabling innovation across product life cycle with four steps. Acquire mandate as name signifies highlights the importance of defining the right project scope for maximizing impact. It comprises connecting the dots between various events, outcomes of the previous projects, key opinion leader voices, decision making and outcome orientation resulting in achieving project contacts that includes already a project roadmap defined by the team. Big picture creation highlights the importance of holistic view and includes observations of the customers' environments and visualization of the lifecycle (Fig. 45.3). A decision proposition could be an MVP, a prototype or a business plan and results in portfolio mapping, roadmap and vision shaping and acquisition of further contracts. Deploying commercialization step could happen within R & D teams of the host organization or at a supplier or a start-up created out of the organization. The early engagement in investor ecosystem is provided with stakeholders

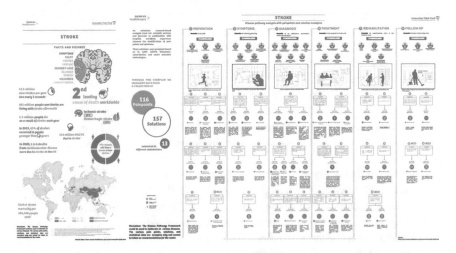

Fig. 45.5 Stroke Disease pathways have been created along with facts and figures and futuristic solution architecture along for other 22 diseases by ITT teams [2]. The diseases have been chosen who have the highest economic burden and mortality rate identified by the World Health Organization. ITT regional co-creation and labs support the adaptation of the disease pathways on global trends and challenges

infrastructure, the ITT location customers pay for (1) the usage and (2) products, generating a recurring revenue.

45.5 Conclusion

The global ITT infrastructure with its portfolio provides opportunities for healthcare system stakeholders to engage in a sustainable way. The ITT methodology is robust to the type of projects and applicable across the innovation lifecycle. The holistic view on the stakeholders and real-time customer insights results in incorporating into the product requirements and shortening time to market. Moreover, ITT supports institutions to adopt an innovation methodology that will allow them to intrinsically identify and approach challenges in their healthcare systems. This will develop a mindset in healthcare workers and institutions that enables them to grow into the contemporary future of healthcare.

Take-Home Messages
- There is a growing need for self-sustaining innovation infrastructures on healthcare providers and medical device manufacturers to meet their customer-centric requirements. Innovation Think Tank (ITT) addresses this with its global footprint and innovative framework methodology.
- Oversimplification of the innovation process and structures, without considering the customer expectations and importance of persistent efforts to gain the trust of the stakeholders has been observed as a limitation for their successful implementations.
- The ITT methodology is independent of the type of projects and is applicable across the innovation lifecycles and industries.

References

1. Enabling healthcare institutions to establish innovative and sustainable infrastructures, P-10-15, 2021. Healthcare guide - partnerships and perspectives of Arab-German cooperation. https://ghorfa.de/wp-content/uploads/Health_Guide_2021_web.pdf
2. Innovation Think Tank, Siemens Healthineers. https://www.siemens-healthineers.com/en-sa/careers/innovation-think-tank
3. Innovation Think Tank, University of South Carolina. https://ittlab.cec.sc.edu/
4. Innovation Think Tank, Imperial College London (2021) https://www.imperial.ac.uk/bioengineering/whats-on/events/siemens-innovation-think-tank/
5. Innovation Think Tank, Baskent University, Turkey (2021) http://itt.baskent.edu.tr/#
6. Innovation Think Tank, Acibadem, Turkey (2021) https://biyotasarim.acibadem.edu.tr/?page_id=20
7. Fakeeh University Hospital partners with Siemens Healthineers to drive greater efficiency and innovation in healthcare delivery (2021) https://fuh.care/media-center/fakeeh-university-hospital-partners-with-siemens-healthineers/

8. Innovation Think Tank, Era's Lucknow Medical College and Hospital. https://www.erauniversity.in/innovationthinktankeu/index.html

9. Mahmeen M, Haider S, Friebe M, Pech M (2021) Mapping and deep analysis of hospital radiology department to identify workflow challenges and their potential digital solutions

10. Newrzella SR, Franklin DW, Haider S (2021) 5-Dimension cross-industry digital twin applications model and analysis of digital twin classification terms and models. IEEE Access 9:131306–131321. https://doi.org/10.1109/ACCESS.2021.3115055

11. Fritzsche H, Barbazzeni B, Mahmeen M, Haider S, Friebe M (2021) A structured pathway toward disruption: a novel HealthTec innovation design curriculum with entrepreneurship in mind. Front Public Health. https://doi.org/10.3389/fpubh.2021.715768

12. Schneider H, Haider S. Method and system to optimize and automate clinical workflow, US Patent 7,895,055

13. Haider S, Popescu S. Imaging modality and method for operating an imaging modality, US Patent App. 4/624,772

14. Deinlein A, Dippl T, Dornberger S, Fuhrmann M, Haider S, Nanke R. Positioning of a mobile x-ray detector, US Patent 10,111,642

15. Distler F, Haider S, Mueller HJ. Rotor with a backplane bus having electrical connection elements to make electrical contact with electrical components in a medical apparatus, as well as rotating unit and medical apparatus with such a rotor, US Patent 9,888,886

Prof. Sultan Haider is the Founder and Global head of Innovation Think Tank (ITT) at Siemens Healthineers, which he established in 2005. His inspiring vision of innovation culture formed ITT to become a global infrastructure of 85 activity locations (Innovation Labs, ITT certification programs, ITT Incubation Centers etc.) in Germany, China, UK, India, USA, UAE, Turkey, Canada, Australia, Egypt, Saudi Arabia, Portugal, Switzerland, Brazil, Jordan, and South Africa. Prof. Haider has generated more than 440 inventions which resulted in more than 150 patent filings. Under his leadership, ITT teams have worked on over 2500 technology, strategy, and product definition projects worldwide. He is a Principal Key Expert at Siemens Healthineers (SHS), a title awarded to him by the SHS Managing Board in 2008 for his outstanding innovation track record. Furthermore, Prof. Haider has been awarded honorary directorships, professorships and has developed innovation infrastructures and implemented innovation management certification programs for top institutions.

The Power of a Collaborative Ecosystem: Introducing the Edison™ Accelerator

46

Jan Beger and Mathias Goyen

Abstract

The opportunities for healthcare organizations with a truly intelligent connected digital enterprise are significant. We at GE Healthcare are accelerating the transformation of healthcare, but we cannot do it alone. We strongly believe in the power of an integrated digital ecosystem leveraging new and legacy technologies with open innovation to enable healthcare resilience and organizational growth.

Launched together with our partner Wayra UK, Edison™ Accelerator is our start-up acceleration and healthcare provider collaboration program. It allows start-ups to learn from our Healthcare and IT expertise and enhance the value proposition of their business ideas. It provides Healthcare Artificial Intelligence (AI) start-ups with what they are lacking: Access to healthcare data, validated problem statements, access to HCPs & clinical mentoring and a chance to run pilots and proof-of-concepts.

Keywords

Open innovation · Co-creation · Collaboration · Artificial intelligence · Start-up · Corporate accelerator

J. Beger · M. Goyen (✉)
GE Healthcare, Solingen, Germany
e-mail: jan.beger@ge.com; mathias.goyen@ge.com

What Will You Learn in This Chapter?
- How a large OEM is engaged in Digital Health Start-Up support
- That meaningful solutions for the health transformation can only be created in a multi-stakeholder environment
- That fostering open innovation is the way forward

46.1 Introduction

No single actor can solve today's healthcare challenges alone. Innovation needs to be far more dynamic, nimble, and collaborative than in the past. It is all about working across silos and collaborating across the healthcare ecosystem, including start-ups, technology providers, research centers, hospitals, clinicians, and patients. Many start-up companies that are creating amazing innovations are barely known. As a result, these small but nimble "makers" often see resources dry up due to a lack of backing by a bigger player from the industry, and from the financial and investment world who can support them logistically, legally, financially, and commercially. And when these start-ups vanish, so do their ideas, their creativity, and their innovations.

GE Healthcare is a leading innovator enabling precision health by integrating clinical care and data across the patient journey via innovative products, digital tools and solutions that simplify decision-making, accelerate care delivery, and improve patient outcomes.

The opportunities for healthcare organizations with a truly intelligent connected digital enterprise are significant. At GE Healthcare we are accelerating the transformation of healthcare, but we cannot do it alone. We strongly believe in the power of a collaborative ecosystem to foster open innovation and the digital transformation of healthcare.

Launched together with our partner Wayra UK, Edison™ Accelerator is our start-up acceleration and healthcare provider collaboration program.

It allows start-ups to learn from our Healthcare and IT expertise and enhance the value proposition of their business ideas. The program focuses on innovation integration and adoption, by launching solutions for validated problem statements and integrating innovative technologies or business models.

It provides Healthcare AI start-ups with what they are lacking: Validated use cases of importance for healthcare providers, access to healthcare data, access to healthcare professionals and clinical mentoring, joint pilots between start-up and healthcare provider partners. In addition, the program also provides support in customer discovery and product validation, mentoring for trials and regulatory clearance preparation, as well as support in adapting, integrating, and launching the solution on GE Healthcare's Edison™ marketplace. The start-ups can tap into

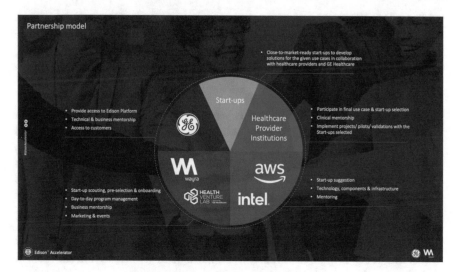

Fig. 46.1   Overview of Edison™ Accelerator program model

GE Healthcare's extensive global network, which includes thousands of sales professionals and distribution partners in 160 countries.

Random acts of innovation rarely pay off. The program is co-designed with healthcare providers to make it easier for digital health start-ups to bring new and innovative solutions to market quickly and cost-effectively. To deliver true value, the effort must clearly align with all stakeholders business strategies. Healthcare provider institutions know where the pain points are that need to be solved together. The Edison™ Accelerator enables healthcare provider partners to work with subject matter experts on real-world problems and to share needs and experiences with other domain experts. They are exposed to agile software development and potentially receive a differentiating digital solution with clear benefits to patients, clinicians, managers, financials, and their brand.

Additionally, the program offers an opportunity to explore how to accelerate the adoption and integration of AI products into existing workflows.

Other program partners are technology partners. They provide technology, components, and/or infrastructure to enable co-creation. They also contribute to mentoring and marketing activities.

Accelerator partners support with start-up scouting, start-up scouting, pre-selection, and onboarding. Accelerator partners provide business mentorship and support the Edison™ Accelerator with marketing and events.

All these stakeholders, also shown in Fig. 46.1, are united within a single, connected ecosystem to deliver real bottom-line impact and ensure better outcomes for patients.

46.2   Case Study

The 2021 edition of the Edison™ Accelerator was a 12-month-long program, including 6 months of preparation (see Fig. 46.2). It consisted of 3 workstreams to address the following use cases:

- Accelerating Imaging AI adoption at scale: How to orchestrate Imaging AI to achieve broader adoption due to seamless integration, more efficiency and productivity?
- Eliminating waste in the oncology care pathway with AI & Digital Technologies: How do we use AI and digital technologies together to improve access to diagnostic services, referral quality, patient prioritization, and automation of operational processes, i.e., scheduling in cancer diagnostic pathways?
- Democratizing cancer care: Putting the healthcare consumer front and central: How can conversational AI platforms, NLP technology and digital PROMS/PREMS tools help to analyze and visualize patient data to inform clinical decision-making, treatment options, and service operations.

GE Healthcare and Wayra UK received more than 350 applications to the 2021 program (see Fig. 46.3). Six start-ups were selected, of which five finished the program at the End of 2021.

By purpose, GE Healthcare and its program partners wanted to work with a small number of start-up companies to be able to fully focus their efforts to make them successful.

The six start-ups that have been part of the Edison™ Accelerator in 2021 were:

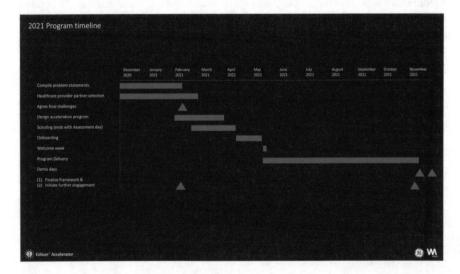

Fig. 46.2 2021 Edison™ Accelerator program timeline

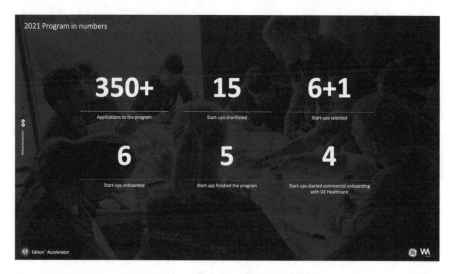

Fig. 46.3 2021 Edison™ Accelerator program in numbers

- Lucida Medical (Cambridge, UK): Lucida uses machine learning and radiogenomics to help identify cancer from MRI and clinical data. Lucida Medical aims to make cancer screening accurate, accessible, cost-effective, and quick.
- Radiobotics (Copenhagen, Denmark): Radiobotics are automating measurements and detections in X-rays to streamline the reading of features in musculoskeletal X-rays.
- SPRYT (London, UK): SPRYT is a smart scheduling solution that improves uptake/coverage rates for screening services as well as reducing the appointment no-show problem. No show rates are a significant problem in diagnostic cancer services.
- My Clinical Outcomes (London, UK): My Clinical Outcomes integrates remote, longitudinal Patient-Reported Outcome Measures into routine clinical practice to enable smarter individual decisions and better overall value care pathways.
- Legit.Health (Bilbao, Spain): Legit Health is a clinical data and communication tool that helps next-generation dermatologists improve diagnosis, score severity and monitor the evolution of wounds, and chronic and malignant skin lesions. This AI-powered technology helps clinicians and patients alike to improve diagnosis.
- Vinehealth (London, UK): Vinehealth combines behavioral science and AI to provide highly personalized patient support that improves the quality of life and survival of cancer patients.

Healthcare provider partners have been: NHS East Midlands Imaging Network, Alliance Medical, Ribera Salud and Manchester University NHS Foundation Trust.

Katherine Boylan, Head of Innovation at Manchester University NHS Foundation Trust summarized: "Collaboration and co-development are key for delivering digital transformation and getting the data-driven technologies that we need into healthcare. The value for us as a healthcare provider for participating in the Edison™ Accelerator is that it allowed us to work with subject matter experts on a real-world problem that helps us and leverages the commercial, strategic, and technical minds of GE Healthcare's Edison™ platform. We benefit from the expertise, we built our own expertise, we built our collaborative network, that we may otherwise not have been part of." [1].

Dr. Peter Strouhal, Medical Director UK at Alliance Medical stated: "Current cancer diagnostic pathways remain laborious, lengthy and sometimes a multi-faceted process with many potential bottlenecks despite recent innovations. With increasing post Covid-19 pressures, any such bottlenecks could delay diagnostic investigations and hence adversely affect patient outcomes. Alliance's focus on service excellence and innovative imaging technologies allows us to create value for patients, clinicians, service providers and other stakeholders while improving patient experiences, clinical outcomes and yet still driving efficiencies to reduce costs. Thus, we are delighted to partner with GE Healthcare through the Edison™ Accelerator as a validation partner to help deliver a healthier world with more precise and efficient care."

Tania Menéndez Hevia, Digital Transformation Officer at Ribera Salud stated: "We believe that collaboration between corporations and start-ups is essential to boost innovation in healthcare. Through the Edison™ Accelerator, we can share needs and experiences with other European stakeholders and create an ecosystem of cooperation and joint development."

With four of the participating start-ups, GE Healthcare has initiated commercial integration efforts. The healthcare provider partners continue partnership with those start-ups.

46.3   Conclusion

We strongly believe that the future of innovation will be about working across silos and collaborating across the healthcare ecosystem, including start-ups, technology providers, research centers, hospitals, clinicians, and patients.

The results of the first edition have shown that the Edison™ Accelerator focusing on adopting innovations, launching solutions for validated problem statements with the active participation from healthcare provider partners is a great means to foster open innovation and the digital transformation of healthcare.

Take-Home Messages
- No single actor can solve today's healthcare challenges alone.
- Random acts of innovation rarely pay off. For any initiative to deliver true value, the effort must clearly align with all stakeholders business strategies.

(continued)

- Strong relationships with healthcare provider institutions and their active involvement is crucial to achieve true market fit.
- Such a program requires clear objective definition, solid preparation and enough time and capacity for serious due diligence during the scouting process.
- At the end, it is about the people involved. If you find the right individuals across all the stakeholders that are passionate, that are interested to drive the digital transformation forward you can achieve a lot of good things.

Reference

1. https://www.gehealthcare.co.uk/article/one-year-of-edison™-accelerator

Jan Beger, As Senior Director of the Digital Ecosystem, Jan's role is to drive GE Healthcare Digital Europe's Ecosystem Strategy and together with his team develop cooperation with partner companies, indirect channels, start-up centers and networks. Jan is the head of GE Healthcare's EdisonTM Accelerator in the EMEA region.

With over 15 years of experience in healthcare informatics, the medical imaging domain and management, Jan is a seasoned leader with vast experience in leading strategic initiatives and influencing strategic decisions that ultimately drive revenue and contribute to enterprise value.

He brings a real passion for the digital transformation of healthcare, understanding healthcare provider needs, improving healthcare workflows and user experience.

Jan lives outside of Leipzig, Germany, with his wife and their two daughters and founded www.gr4ai.academy, a non-profit to create awareness and educate children around Artificial Intelligence.

Mathias Goyen is currently the Chief Medical Officer EMEA for GE Healthcare. Mathias began his career as a diagnostic Radiologist in Essen/Germany. He was appointed Professor of Diagnostic Radiology at the University of Hamburg/Germany in 2010. Mathias' previous experience also includes 5 years as Managing Director of UKE Consult and Management GmbH, a subsidiary of the University Medical Center Hamburg where he was responsible for the overall consulting business of the University Medical Center Hamburg. Mathias holds a medical degree (MD) from the University of Bochum, Germany. He has been secretary-general of the German Chinese Society of Medicine, Berlin, from 2005 to 2019.

Jörg Traub

Abstract

Innovation networks are goal- and topic-oriented associations of several legally independent companies and organizations. Practice shows that these objectives can vary greatly depending on the industry. The objective of medical technology networks is to identify the trends and the challenges of the industry in a structured manner and provide practical solutions to master them. To this end, the various players are brought together to bundle the knowledge of diverse experts and make it available to the network for the benefit of all partners in the network. The association and the bundled access to experts from different sub-sectors of the healthcare industry is the basis for a successful translation from idea to market acceptance. This includes early evaluation of market opportunities, development of a strategy for implementation through the experience of the stakeholder in the network, and consistent implementation and execution. The diverse players highlight emerging trends, the needs, and the challenges, offers solutions and deep understanding of these, and provide insights into the status quo of the market environment. Thus, the strategy for translating an idea, as well as its implementation and market penetration is of higher success being part of a network.

Keywords

Network · Ecosystem · Cluster · Innovation · Forum Medtech Pharma · Market Access Challenge · Innovation

J. Traub (✉)
Forum Medtech Pharma & Healthcare Bayern Innovativ, Munich, Germany

What Will You Learn in This Chapter?
- *Examples of Health Innovation networks.*
- *Access to domain experts is a key for successful translation.*
- *Becoming a member of such a network is a good idea for health innovators.*

47.1   Introduction

The core element of innovation networks is the broadest possible knowledge by the experts in the different phases of the innovation system and thereby a deep, collective understanding of the healthcare system [1].

Network organizations can be active both regionally and globally; they can organize themselves into unions, associations, or institutions with the goal of jointly mastering the complex implementation from idea to market penetration. This can be, for example, a horizontal integration, i.e., from large corporations to the smallest companies and startups in a subject area, or a vertical integration with complementary expertise and fields of specialization along an application or technology domain.

The focus of such networks can then be either application- or technology-driven. In the healthcare industry, networks on applications are, for example, medical societies, e.g., in orthopedics or radiology, while networks of technology focus around a scientific-technical discipline, e.g., artificial intelligence or robotics.

Innovation networks that are less specialized in individual topics can also cover the entire innovation process in healthcare, often with alliances to other application and technology-oriented networks. In network organizations, different actors with similar interests or goals exchange ideas and initiate joint activities in the form of events, strategy-supporting measures, formulation of demands to policymakers, joint projects, or network meetings. As a rule, such communities organize themselves to strengthen economic performance and society.

In regionally focused networks, the focus is on the development of the location to develop it sustainably and to secure and expand its status. The activities of the members of a network can be multi-layered: From loose exchange to joint projects and cooperation, using and utilizing common resources, e.g., for production or export.

In addition to the network organizations for innovations, there are also other active networks, e.g., through alumni associations of universities, which enable a trusting exchange not related to applications or technology. These networks have been an integral part of our society for several centuries and to date have rarely received scientific historical attention [2].

In medical technology, there are many networks with a regional and global focus. In this chapter, we will focus on three exemplary health tech innovation networks from southern Germany and therefore make no claim of completeness. There are other networks and clusters in the field of medical technology in Germany as well as

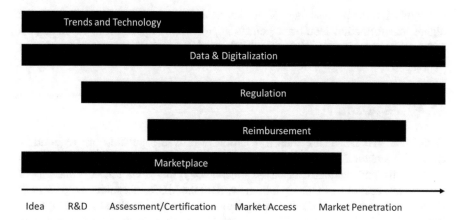

Fig. 47.1 The challenges that are coordinated by networks in the health tech market from the idea to the market success are manifold

in Austria, Switzerland, other European countries, and the rest of the world, each of which is active at the state or regional community level.

Specifically, the challenges in the healthcare innovation process can be categorized into different areas, five of them their application in the innovation value chain are shown in Figure 47.1.

47.2 Trends and Technology

Health tech is an area with a particularly high proportion of research and development. New medical products, improvements to medical devices, but also research and development of new processes and procedural techniques have a particularly high share. Likewise, university and non-university research is particularly broadly positioned in biotechnology and medical technology, as well as in diagnostics and digital health. The particular importance of technological progress is also reflected in the steadily growing proportion of patent applications. At the European Patent Office, medical technology was the sector with the highest number of applications in 2020, while the pharmaceutical and biotechnology sectors recorded the greatest growth compared to the previous year (EPO2020).

47.3 Data and Digitalization

The healthcare industry is evolving from eminence-based medicine to evidence-based medicine using data and models. Often referred to as Artificial Intelligence, the automatic interpretation of data using heuristic approaches or learning-based machine learning algorithms will change medicine in the future. One example describing the dimensions of digitalization in a medical discipline, surgery, highlights that the process must be accompanied by experts [3]. The complexity of digital transformation requires a holistic approach through network organizations with various experts, among others, with the goal that digitalization provides support and quality assurance in surgery.

47.4 Regulation

Throughout the healthcare industry, strong efforts in regulation are required to ensure the technical safety of solutions for use on patients and to comply with all laws, norms, and standards. In medical technology, this is with the CE marking of the products according to the legal framework of the MDR—Regulation (2017/745). Compliance with the standards and regulations requires specialized knowledge and experience, which, especially for new solution, often can only be mapped with external experts and a network.

47.5 Reimbursement

Reimbursement in hospitals and care facilities is complex and it is essential to choose the right strategy to establish a product successfully and economically in the market. On the one hand, investments such as hospital financing or financing of care facilities are possible, on the other hand, case-related flat rates. As a rule, both ways of financing require proof of improved treatment and care quality demonstrated through clinical studies. An application for a public health insurance number in Germany that allows reimbursement in the outpatient sector is a long and tedious process, but essential, as there is a permission proviso (Section 135 (1) SGB V), i.e., a service may only be billed if it has been positively evaluated through the process at the G-BA (see Figure 47.2). Precondition of this is the successful conducting of the necessary studies with a positive patient outcome [4].

47.6 Marketplace

In the healthcare industry, the market environment of products is changing more and more in the direction of solution and process-oriented approaches. The future products and services are not an individual product, but the result of diagnosis or therapy. Digitalization and sensor technology in particular are enabling telemedical procedures that allow diagnosis, therapy, and also subsequent care within the home environment.

Different networks provide services for their member and partners to bundle the activities and overcome the individual hurdles of the market change. Three example networks are:

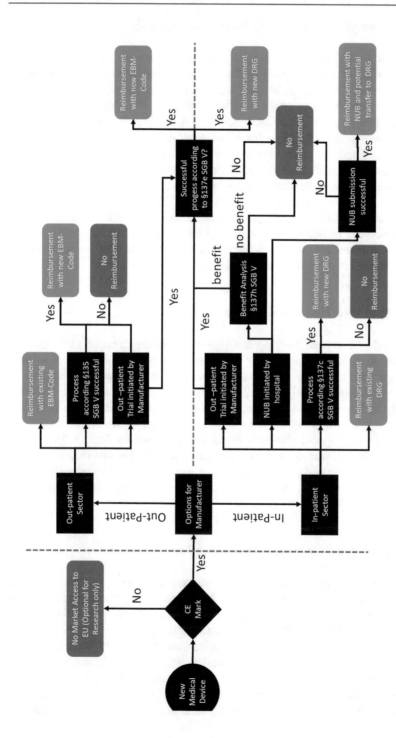

Fig. 47.2 The Process of getting reimbursement of a medical product in Germany is complex and there are different pathways for inpatient DRG codes and out-patient EBM codes. Experts through networks are required to master this complex process. Source: own graphic adopted from [4].

47.6.1 Example 1: Forum MedTech Pharma e.V.

Forum MedTech Pharma e.V. is a nationally active network in Germany for the implementation of innovations in the healthcare industry with a focus on medical technology that also cooperates with international network organizations beyond national borders. The non-profit association is particularly characterized by its interdisciplinary and cross-sectoral perspective, which is reflected both in its broad thematic areas and in the diversity of its membership structure. The network's mission is to support all players in the sector in successfully establishing innovations in healthcare through its networking offerings. Topics taken up by the network are always based on currently identified opportunities and challenges of the industry, which are developed together with its members. The networking tools include, among other things, the conception and organization of specialist events and expert circles, individual consulting and support for members, the provision of knowledge carriers from a pool of experts, an information hub for public relations and industry-specific statements, as well as participation in project consortia [5].

47.6.2 Example 2: Medical Valley EMN e.V.

Medical Valley European Metropolitan Region Nuremberg-Erlangen (MV-EMN) is a international leading regional cluster in northern Bavaria in the field of medical technology and healthcare. Regional partners and urban developers as well as clinical care facilities, highly specialized research institutions, and numerous internationally leading as well as emerging companies and startups are active here. With the aim of jointly finding solutions to the challenges of healthcare today and tomorrow, these cooperate closely with globally renowned institutions. This exceptional regionally focused density, in combination with the international market and competitive position of individual players, provides conditions that enable rapid development of products, processes, and services and thus innovation in the healthcare industry [6].

47.6.3 Example 3: Medical Mountains GmbH

MedicalMountains GmbH is a network for medical technology and brings together players to drive innovation through knowledge and technology transfer, qualification offerings and cooperation. In this way, Medical Mountains sustainably strengthens the top position of the medical technology location in and around the region in Tuttlingen, but also that of Germany as a business location. Medical Mountains GmbH represents the concerns of the medical technology industry on a regional and national level with European visibility and advocates for sensible framework conditions. Through position papers and regular dialogue, the voice of each individual partner is given more weight vis-à-vis politicians, authorities, and associations.

Similarly, Medical Mountains GmbH supports the network in the face of new challenges by feeding in important information and developing pragmatic approaches to solutions for the players. By focusing on the current and relevant content of medical technology development, each individual network partner is strengthened in its work [7].

47.6.4 MedTech Cluster Alliance D-A-CH: Fusion of Leading Clusters in Germany, Austria, and Switzerland

The Medtech Cluster Alliance D-A-CH is an association of cluster organizations, networks and initiatives of the medical technology industry from the DACH region, i.e., from Germany, Austria, and Switzerland. The alliance currently consists of ten German, five Austrian, and one Swiss partner and thus jointly develops an innovation and economic region with several thousand medical technology or research partners, as well as numerous clinical institutions. The central focus of the alliance is the global networking need of companies, universities, research institutions, healthcare providers and investors from the healthcare sector. The partner strengthens the member's position thus globally with the focus on innovation in the German, Austria, and Switzerland region and makes its value chains more resistant to future crises. In addition, common topics such as position papers are to have a better reach with the joint efforts in the association to find its way and consideration in politics and decision-making authorities. Required steps can be moderated in order to strengthen the innovation dynamics and economic power of the industry in the region [8].

47.7 Conclusion

Innovation networks are a trusting association of organizations in a subject area with the aim of bringing solutions, products, and services from the idea to the market. This is an important instrument, especially in market segments with a high degree of complexity, for determining the probability of success on the market at an early stage and for providing expert support on the path to implementation. In the healthcare industry, especially in medical technology and biotechnology, there are already established innovation networks that have been stimulating, initiating, and supporting innovation projects for more than 20 years. Recently also the largest network in Europe for innovation in healthcare, the Forum Medtech Pharma e.V. joint forces with all other leading regional innovation networks in health technology to remain globally competitive providing the best service for the network. Being part of such an ecosystem is key for success in complex markets such as healthcare.

Take-Home Messages

- Innovation networks are trusted ecosystems with complementary partners.
- Complex solutions and markets need experts for implementation, market introduction, and scaling for sustainable success.
- Forum Medtech Pharma e.V. is the largest innovation network for health technology in Europe with around 500 members.
- Cooperation and partnership of networks is required to be globally recognized and successful—the leading innovation clusters in health tech/medtech in Germany, Austria, and Switzerland formed an alliance as "MedTech Cluster Alliance D-A-CH."
- Startups can greatly benefit from these networks to get access in the different domains and phases throughout the creation, startup, market access, and scale-up.
- Startups also actively contribute to the network with new methods, products, solutions, and approaches. The dynamic of them will actively implement exponential innovation into the ecosystem.
- The joint activities of established players and new players will make the ecosystem persistent and sustainable, adopting to change with a great reach and grounded base.

References

1. Schier, H. (2020). Dynamische Innovationsnetzwerke als Erfolgsfaktor. In *Innovationen und Innovationsmanagement im Gesundheitswesen*. Springer.
2. Ferguson, N. (2018). *„Türme und Plätze: Netzwerke, Hierarchien und der Kampf um die globale Macht"*. Propyläen Verlag.
3. Wilhelm, K., Ostler, S., & Meyer, F. *Digitalisierung in der Chirurgie, Was Chirurgen darüber denken und was sie wissen – Ergebnisse einer Onlineumfrage*; Der Chirurg, Ausgabe 01/2020.
4. Winkler O (2020) Market Access von Medizinprodukten. In: Tunder R (ed) Market Access Management für Pharma- und Medizinprodukte. Springer Gabler, Wiesbaden
5. Internetseite des Forum Medtech Pharma e.V. https://www.medtech-pharma.de/. Aufgerufen am 21.12.2021.
6. Medical Valley Europäische Metropolregion Nürnberg-Erlangen. https://www.medical-valley-emn.de/. Aufgerufen am 21.12.2021.
7. Medical Mountain GmbH. https://medicalmountains.de/. Aufgerufen am 21.12.2021.
8. Pressemeldung der Medtech Cluster Allianz – weitergeDACHt. https://bit.ly/3p2w3SS. Aufgerufen am 21.12.2021.

Dr. Jörg Traub received his PhD in Medical Computer Science. He has more than 15 years of experience in the medical device start-up industry and was as founder and CEO responsible from the idea generation, through the implementation, certification, internationalization, and multiple financing round all the way to the exit of the medical technology company and university spin-off SurgicEye. Following this, he was business development director and coordinating the strategic marketing and sales activities at a smart medical robotics solution company, responsible also for strategic corporate partnerships.

Since 2020 he is director of healthcare sector at Bayern Innovativ and is Managing Director of Forum Medtech Pharma e.V., responsible for innovation incubation, infrastructure projects in healthcare innovation and networking.

Leadership-Compass: "Networking and Synapting Empowered by Health Captains"

Henri Michael von Blanquet

Abstract

Platform business for Healthcare 4.0 needs a new Leadershipstrategy based on building sustainable strategic and scaleable synapting Network- and Alliance Ecosystems. In this article we show based on the Neuoleadership Experience of THE HEALTH CAPTAINS CLUB invented by the platform founder a new strategic approach focused on the Convergence of the Dunbar Number, Tipping Point Number, and Medici Effect Knowledge toward a holistic concept to scale together value for the patient to connect the health sciences with the healthcare innovation cycle: "from bench to bedside and bedside to bench."

Keywords

Neuroleadership · Strategic networking · Alliance building · Scaling creativity by synapting · Convergence · Platform business

What Will You Learn in This Chapter?
- Dunbar Numbers (150, 50, 15, 5), Tipping Point Number 150, Medici Effect
- Their significance and relation to building a convergence platform for Health Leaders and Innovators
- Value for patients through a new strategic approach

H. M. von Blanquet (✉)
THE HEALTH CAPTAINS CLUB, Nieblum on Föhr Island, Germany
e-mail: president@healthcaptains.club

© The Author(s), under exclusive license to Springer Nature Switzerland AG 2022
M. Friebe (ed.), *Novel Innovation Design for the Future of Health*,
https://doi.org/10.1007/978-3-031-08191-0_48

The ability to network is an essential part of the leadership skills of **"Health Captains."** Too many executives mistakenly see networking as a freestyle. The quality and ability of **networking** and **synapting** (*linking separate network islands and scale from smaller networks to larger network units*) is a leadership competence, i.e., a core competency that cannot be delegated away by a manager and leader, but is directly connected to his own personality and defines his visibility.

A striking example of the appreciation of the quality of results of successful networking by a leader is the Chemistry Nobel Laureate (1954) and Nobel Peace Prize Winner (1962) Linus Pauling, who did not attribute his creative success to his luck or his immense intellect, but to his diverse contacts: "The best method of having a good idea is to have a lot of ideas."

48.1   Conclusion

The ability to multi-professional and multi-sectoral "silo-free" networks and the professional management of interdisciplinary networks is increasingly becoming a central component of the leadership in the health industry, health sciences, and health policy, because medical-operational or medical-scientific stand-alone solutions are no longer competitive.

The net message is: "*The Only-I-Myself and Me-Company*" has had its days. The Era of "Silo-Medicine" is coming to its end toward Systemmedicine.

Today's reality is, in order to be successful in the long term, the own strategic network that has been built up and maintained over many years (see Table 48.1) has become an alternative without alternative. Building a personal leadership network is not primarily a question of talent but above all a question of your own will to do it.

48.2   The Health Captains Club Platform I

The Health Captains Club Platform was formed with the goal to create such a valuable and transferable starting point for all key-stakeholders to create *the transdisciplinary and international HEALTH 4.0 & ONE HEALTH Ecosystem* for today and future generations of leaders in the spirit of next-generation leadership and next-generation mentorship empowered by its Health Captains Ecosystem:

> Relationships are all there is. Everything in the universe only exists because it is in relationship to everything else. Nothing exists in isolation. We have to stop pretending we are individuals that can go it alone—Margaret Wheatley

Neuroleadership: Dunbar's Key Numbers in Networks are connected to the Tipping Point and creativity for innovation is scaleable by the Medici Effect in Cores of 15, Think-Tanks of 15, and Brainpools of 50.

Table 48.1   The four Qualities of networks

Network	Operative networks	Personal networks	Strategic networks	Strategic alliances
Purpose	Efficient task completion, network maintenance of the necessary knowledge and functions	Growth through expansion of personal and professional environment; Access to useful information and contacts	Finding future priorities and challenges in connection with the organization of mutual support by other stakeholders	Belonging to an international, multi-professional and multi-stakeholder network alliance, which strategically extends your own network radius into a higher dimension that is no longer personally controllable and connects existing networks to form a strategic "network alliance"
Localization and temporary orientation	Focus: "in-house" experts	Mainly external contacts. Orientation toward current and possible future interests	Internal and external contacts—focus on strategic future topics	Nationwide to global orientation through the interaction of several networks under one synchronized "alliance roof"
Teammates and recruiting	Key contacts are based on the operational task and the "organization chart"—this makes it clear who is relevant for an operational network	Key contacts are mostly a matter of discretion and it is not always clear who will be relevant now or later. Guarantors are often a prerequisite for membership.	Key contacts primarily follow the strategic requirements and the environment of your own organization, specific affiliations are a matter of discretion and it is not always clear who is relevant. Memberships	Key contacts are a matter of discretion as well as of the complex strategic solution considerations. The recruitment is done strategically by networking existing organizations and people. Platforms for

(continued)

Table 48.1 (continued)

Network	Operative networks	Personal networks	Strategic networks	Strategic alliances
			are often linked to a guarantee system.	"Elder Statesman"
Network attributes and behaviors	Depth: building strong working relationships	Width: Outstretched hand to contacts that can make the difference	Leverage: linking the internal and external worlds; Broadening horizons and expanding your own radius of action— network basis for creativity and problem solving	Scalability: supra-regional to global— typically to master "large-scale business" for new technologies of formerly separate industries (convergence) and to solve complex global problems
Network examples	"Not public"— hardly any external connections	E.g., Sports clubs, student associations, non-for-profit organizations, orders, alumni clubs	E.g., Business clubs with national and international partner clubs, chambers of commerce, think tanks	E.g., M8 Alliance of the World Health Summit Berlin, World Economic Forum Davos, Kenup Foundation, Malteser, BIOCOM, HIMSS
Examples	"Not public"	www.dcada.de www.suevia.de www. seenotretter.de www. johanniterorden.de www.rotary.org	www. amcham.de www.industrie-club.de www. uebersee-club.de www. healthcaptains. club www. bbug.de	

Managers who are not leadership personalities often believe that they have already adapted successful networking to their actions, but still operate exclusively at the operational and private network level. Successful leaders, on the other hand, maintain strategic networks and try to belong to network alliances

48.2.1 The Dunbar Metrics (SQ)

Investigations by Robin Dunbar, head of the Institute of Anthropology at Oxford University, on our social network behavior and our limited number of capacity to maintain networks, resulted in metrics that clearly show our neurophysiological neocortex capacities for relationships in creativity networks (SQ: **Social Intelligence Hypothesis**—*Understanding and Managing Relationships*).

The number of people with whom we are at best able to maintain lasting and valuable contacts is approx. **150** people, i.e., this is a cognitively limited number of people with whom an individual can fully maintain social mindful relationships. Within this relationship size, the talented networker and not only knows all names and life stories—above all, he also knows all cross-connections and dependencies within this network of knowledge and expertise.

Dunbar has also found that the other group sizes we can control roughly correspond to a "rule of three": a closer circle of friends can be made up of up to **50** personalities (Brainpoolof 50), while an "inner circle" supporting our activities can make up around **15** people (Think-Tank of 15).

Dunbar measures the most intimate and trustworthy group of friends with only **5** friends or family members (Core of 5). With the help of the social key figures for networks of individuals created by Robin Dunbar, we can carry out a self-assessment of our personal cognitive network capacity, which makes each of us authentic and qualitatively more controllable to build and maintain our own networking engagement.

Based on the cognitive ability for lifelong learning known from neuroleadership, we can expand our individual Dunbar parameters through active networking over time. This is a classic neurophysiological adaptation process.

The Dunbar metrics help us analyze our own networks to manage them quantitatively and qualitatively.

48.2.2 Social Networks

They often consist only of clusters (network islands) that have no connection (synapses) to each other. Such unconnected clusters can be traced back to two deficient principles:

- Principle of Self-Similarity: The "principle of self-similarity" says: When you make contacts, you tend primarily to surround yourself with the people whose experience, training, and worldview are similar.
- The Principle of Proximity stands in the same way for the variety of networks: It means that you primarily include people in your networks with whom you already spend most of your time. Employees with similar training work in the same departments, and people from similar backgrounds tend to live in the same area.

If you give in to your natural inclinations and build your networks according to the principle of neighborhood and self-similarity, then you only generate echoes and miss the actual purpose of network diversity: multi-professional and multi-cultural international diversity and broadening horizons through your own multi-stakeholder contact world.

48.2.3 Real Strategic Network and Alliance System

Such a network and alliance system and its added value is only created by interconnecting initially separate network clusters using a connector, i.e., a mediator who is personally anchored in these clusters and has the ability to switch between these clusters.

Unconnected clusters without active mediators (synaptists) do not generate new information for the members of the individual clusters, but only echoes. Synaptists act as an active link between previously unrelated groups. A talented synaptist gives every member of each group access to other parts of the network and only by connecting and synchronizing the synapses of the separate network clusters can visible added value be created to solve complex tasks. **Synapting** is the real purpose of any strategic network. Neurophysiologically, our brain itself makes exactly the same synchronized synaptic interconnection of different anatomically separated brain areas for processing complex sensory impressions or in the genesis of our own complex creative world of thoughts. It is therefore not sufficient to operate networks formally, but an effective synaptic network is only created by the synaptic interconnection of the before separated clusters. Mediators with the ability to interconnect several networks to form strategic alliances are called **Supra-Synpatic-Leaders**.

48.2.4 Added Value from Networks

In order to create a network that is rich in social capital, you should maintain active relationships with influential synaptists and these for your benefit and the benefit of others support network members personally in their network work. If you are a synaptist yourself, then you can team up with other synaptists and perhaps form a network alliance together.

Networks provide access to confidential and private information—only the trust in each other in the network of relationships determines the outcome quality. With a one-sided, short-term participant role you will never become a valuable and respected member of a network and therefore cannot build lasting trust—only if you give by yourself unselfishly and regularly input into your own network you will get something worthwhile back yourself. Good networking is determined by your own attitude and your network loyalty and can only be done by yourself, e.g., especially by making unselfish new contacts for your own contacts. This increases your trust capital in the network and then you can access your network later with another authentication if you need it yourself. The added value of networks is created by your own investment in the network and takes time. In the long term, those who do not cooperate only in times of emergency win.

A strategically sustainable network setup is time-consuming and active work. Net message: "Just do it".

48.2.5 Neuroleadership-Experience: Dunbar Number 150 + Tipping Point Number 150

Already a small, particularly valuable multi-professional and multi-stakeholder network of approx. 150 people (**Dunbar Number**), who are rich in their network of synaptists and charismatists, can already trigger a tipping point by themselves through the power of these multipliers and their external connections and effects in other comparable and different networks.

According to Malcom Gladwell, the **Tipping Point** goes back to *the law of the few* and is the moment when an idea, a trend, a fashion, or a social behavior crosses a threshold, tilts, and spreads like an unstoppable wave. The tipping point is the moment of critical mass that is capable of triggering such a wave and Gladwell was able to show that this already corresponds to approx. **150 people in a network** who can create the conditions for this.

48.3   The Health Captains Club Platform II

How we scale Creativity and Strategy—from "Cores of 5," to "Think-Tanks of 15" toward "Brainpools of 50" and "Twin-Chapters of 150."

The COREs of 5 are the beating hearts of THE HEALTH CAPTAINS CLUB: Following the **Neuroleadership-Experience** we start to take action at THE HEALTH CAPTAINS CLUB with a **CORE of 5**. The CORE Team's built THINK-TANKs of 15 and BRAINPOOLs of 50 as Sounding Boards for their support- and feedback knowledge network.

The COREs of 5 is the creativity and innovation nucleus for all action plans, panel program committees of THE HEALTH CAPTAINS CLUB.

We try to operate the cores in such a way that the participants only partially know each other beforehand, if possible, in order to promote an optimum of creativity and new thinking about patient benefit and the further development of health systems toward sustainability. The COREs are all the starting teams to explore the 360° innovation spectrum for sustainable HEALTH 4.0 and ONE HEALTH.

Twin-Chapters: *To scale success stories* we connect out of the Membership two international Metropolregions as Health Sciences Clusters together with local partners starting with the Leadership of one Twin-Chapter-CORE of 5—These Twin-Chapter-CORE's are building the Chapter Strategy, Action plans, coordinating White Papers-Processes, orchestrating Policy Papers out of BRAINPOOL's and other Publications or Conferencepanels to built and run a Twin Chapter of 150, e.g., THE HEALTH CAPTAINS CLUB started with Berlin—New York and Munich–Brussels in 2021, than actually Colonia–Rome in 2022 and next step we are planning to start Miami–Basel in synchronization with the Art Basel Tradition since 2002.

Our Health Captains act as sovereigns and the CORE teams navigate themselves independently. Everything happens on the basis of intrinsic motivation to improve the existing conditions through personal commitment to give a living example of leadership together.

48.3.1 The Super-Convergence

The convergence of biology and technology causes a long-lasting major global innovation wave for new basic innovations for a new industrial age that replaces our outgoing information age (sixth Kondratieff Cycle).

This was already recognized by one of the greatest protagonists of the digital era, Steve Jobs, who said, "I think the greatest innovations of the twenty-first century will be at the intersection of biology and engineering. A new era is beginning." [1].

The simultaneity of the mutually disruptively reinforcing new technologies (see list below) of the digital-molecular transformation of medicine requires the "super-convergence" of our health systems toward precision medicine and scaleable systemmedicine [2]:

- Wireless sensors
- Genomics, Transcriptomics, Proteomics, Metabolomics, Microbiomics, Epigenomics
- Digital imaging
- Digital information and knowledge systems
- Mobile connectivity and bandwidth
- Internet technologies
- Social Media Networks
- Constantly increasing performance of computers and smartphones
- Modern database systems with big data performance
- Cloud Computing
- Artificial intelligence
- Telemedicine
- Robotics
- Nanomedicine
- Regenerative Medicine
- Precision Medicine and Precision Science

Through the interaction of all these technologies and the use of around 8 billion mobile phones and smartphones today by individuals, who are becoming better and better informed, the medicine of the future will be decisively controlled by the patient: "The patients will be the CEO's of their own Health" [3].

Platform economies will therefore play a central role in the process of developing Medicine 4.0 when it comes to connecting patients and doctors with health sciences and the health economy.

Platforms connect these different stakeholders with each other on the basis of standardized, scalable processes. While social platforms in medicine are conceivable in themselves, such as patient networks for diseases or as diabetes or movements against STDs, aggregation platforms and mobilizing platforms remain the most relevant for systems medicine.

An aggregating platform can facilitate the exchange of information, introduce the right stakeholders to the relevant resources and, as a business model, offer additional services as an added value of the platform. A mobilizing platform can bind groups of researchers together and advance ideas and projects like precision medicine. In this way, algorithms and artificial intelligence can connect suitable cases with each other.

Groups of patients can find each other, doctors can exchange information with each other and the healthcare industry can offer its products and services on the platform. The model could also solve the initial problems of systems medicine through added value.

From an economic point of view, a platform can take on the costs of data maintenance and other IT services and manage the effort, while interactions of companies and institutions with the platform generate profits. This would reduce the average costs for the stakeholders and result in improved and cost-effective healthcare.

Looking at the development of platforms, these are increasingly turning into dynamic "learning platforms." This would allow the newly accumulated knowledge of stakeholders to be brought to all stakeholders in the healthcare sector and to inform them of new findings. In this way, the findings of systems medicine could be quickly distributed to all those willing to learn via the platform economy, shorten the publication processes, and apply these findings more quickly in research and practice.

However, the principle of the platform business comes from a profit-oriented environment of the market and they make their money through services, the evaluation and handling of sensitive patient data. Access to a platform is often free of charge, they charge fees for transactions or charge a kind of subscription fee. Depending on the type of platform, how do they generate their healthcare profits?

The most well-known and discussed threat is that data from systems medicine is bought by third parties and possibly misused. Is it ethical to trust a platform with patient data for many years and is it secure when it is all in one place? Who uses and sees the data and will it be used after my recovery for research and analysis purposes or to accumulate commercial profits? Side effects of systems medicine must also be taken into account in order to enable an ethical approach to the platform economy, since the operators would have to adhere to the same rules as medicine so that they do not act negligently. Trust in a platform that is not controlled by a neutral source must be built and maintained for a long time. Data security will always play an important role in this sector; otherwise, individuals may refuse to permanently interact with platforms in the health sector. On the other hand, if the restrictions are too great for the platforms, this system may no longer be worthwhile.

48.3.2 Platforms for Solving the Integration and Interpretation Tasks of Medical Data

Precision medicine remains a task of integrating and interpreting large patient data that is not yet consistently standardized and of the patient biology with integrity, which can be solved by medical-technological platforms in the future:

- The sequencing of a human genome alone generates around 200 gigabytes of data that have to be interpreted individually.
- Every patient has an individual molecular tumor disease, the same applies to metabolic, immunological - basically to all "molecular diseases."
- Countless drug combinations, co-medications
- Complex co-morbidities
- Complex rare diseases
- The non-responder problem
- Adverse drug reactions and side effects
- Patient-safe in-silico pharmagenomics with the optimal matching from thousands of approved drugs optimized for each individual case
- Clinical Study Navigator—the study comes to the patient, not the patient to the study.

In several database layers, derived from the data-technical structure of the global GIS systems, such as Google Maps (Fig. 48.1), the digitized Precision Medicine data must be offset against each other [1].

Precision medicine seeks to leverage advances in knowledge of genetic factors and biological disease mechanisms, combined with unique considerations about patient care needs, to make healthcare safer and more effective. Due to these contributions to improving the quality of care, precision medicine is a key strategy

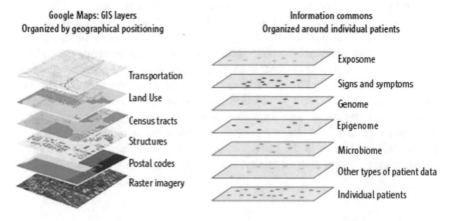

Fig. 48.1 The standardized database layers are the foundation for global GPS navigation today and in the future for international multi-omics precision medicine [3–5]

of any healthcare reform and is the systems medicine implementation of the globally recognized "value-based healthcare" approach. Precision Medicine establishes a new sustainable age for medicine and health.

48.4   Summary

The Transformation and Implementation of 360° Next Generation Healthcare needs high-quality international, multi-professionals, and multi-stakeholder networks and alliance systems based on a deeper knowledge of Neuroleadership. THE HEALTH CAPTAINS CLUB is a next-generation non-for-profit platform model to scale Creative Minds and to navigate New Leadership toward Smart Health powered by Smart Healthcare Business created by Health Captains in togetherness. All new innovations and technologies have to be value-based and scaleable and therefore need to be developed in hubs that are powered by also scaleable strategic networks like THE HEALTH CAPTAINS CLUB. You are welcome on Board to explore sustainable health and to navigate toward One Health together.

> **Take-Home Messages**
> - Networking and Synapting is a Leadershipfactor for Health Captains and real knowledge about the Neuoleaderexperience is a Key-Success-Factor for Innovators and Entrepreneurs [1].
> - Strategic Network- and Alliancebuilding cannot be delegated.
> - THE HEALTH CAPTAINS CLUB is a value-based non-for-profit platform to scale leadership and innovative minds for sustainable solutions for Health 4.0 by creating Tipping Points and Medici Effects together to empower "Profit with Non-Profit."
> - The Superconvergence in the Transformation of Medicine needs "silofree" Platforms to connect and synchronize Healthcare Providers, Health Industry and Health Sciences from Academia to create new Business-Modells with the continuum of innovations cycles by a "from bench to industry and bedside *and* bedside to bench and industry" as scaleable value-chain.

References

1. von Blanquet HM (2017) Networking und Synapting als vornehme Führungsaufgabe. In: Debation JF et al (eds) Krankenhausmanagement Strategien. Methoden Medizinisch Wissenschaftliche Verlagsgesellschaft Berlin, Konzepte

2. Topol EJ (2013) The Creative Destruction of Medicine—How the digital revolution will create better healthcare. Basic Books New York
3. von Blanquet HM (2019) Creating solutions for value-based precision medicine. In: Pieper U, Steidel A, Werner J (eds) XPOMET 360-degree next generation healthcare. Medizinisch Wissenschaftliche Verlagsgesellschaft Berlin
4. National Research Council (2011) Towards Precision Medicine: Building a knowledge network for biomedical research and a new taxonomy of disease. National Academies Press Washington, DC. URL: www.nap.edu/ catalog/13284/toward-precision-medicine-building-a-knowledge-net-work-for-biomedical-research
5. Topol EJ (2014) Individualized Medicine from Prewomb to Tomb. Cell 157(1):241–253

Henri Michael von Blanquet , MD. "At the latest in the face of the global corona pandemic, we have to completely restructure medicine, health sciences, health industry and the health systems worldwide towards sustainability. It is about human life, the life of our families and friends, our employees and colleagues, the life of those entrusted to us and it is about our life and our nature". Dr. von Blanquet is the inventor and President of THE HEALTH CAPTAINS CLUB and the PRECISION MEDICINE ALLI-ANCE. To combine and scale value-based "Profit- with Non--Profit" Innovations and Business, he is, since 2022, the founder of the business-hub "HEALTHCAPTAINS+COMPANY 360° Next Generation Healthcare"—headquartered in Berlin and on Föhr Island at the Westcoast of North Europe.

Part XI

Purpose Launchpad Health: Toolset Templates and Principles

PLH Templates and Principles

Michael Friebe

Abstract

The tools and canvases that are recommended for the 8 segments of the Purpose Launchpad Health in the Exploration and Evaluation phase are listed here, including short explanations on how to use them. Additionally, the course information are including and in some cases also a link to available online guidance tools or documents. All canvases and tools presented can be freely used and downloaded for private use. Please respect the individual licenses and the source/author information. The chapter starts listing the INNOVATION PRINCIPLES of the book one more time. Please read them occasionally! After that—starting with the PURPOSE—all recommended tools are presented. The tools are however only as good as the willingness of the user to actually go beyond hypothesis formulation. Validation and learning, followed by adaptation and formulation of new hypothesis are the basis for success. And please embrace failure at least as much as a confirmed validation. Good luck disrupting and creating a Future of Health that we all want and deserve.

Keywords

Business model canvas · Value proposition canvas · Exponential canvas · Team canvas · Purpose launchpad · Minimal viable prototype · Experimentation · Exploration · Evaluation · Future of health · Health disruption

M. Friebe (✉)
AGH University of Science and Technology, Krakow, Poland

Otto-von-Guericke University, Magdeburg, Germany

IDTM GmbH, Recklinghausen, Germany

FOM University of Applied Science, Center for Innovation and Business Development, Essen, Germany
e-mail: info@friebelab.org

What Will You Learn in This Chapter?
- The key tools and canvases for each of the 8 segments
- How these canvases link together
- The key innovation principles of the book and the importance of continuous learning in the innovation process

49.1   Introduction

This chapter presents all the discussed and suggested PURPOSE LAUNCHPAD HEALTH templates for the 8 segments, as well as short instructions on how to use them.

The Templates and the PLH process as explained and used here follow the official guide authored by Francisco Palao [1] with the adaptations for the health segment [2].

For the health-related activities you should follow the innovation principles as listed below, especially the Lean Startup [3, 4] process of building (a prototype, concept, process - low-fidelity demonstrator); measuring the effect in combination with a customer interview; collect, analyze, and evaluate the collected data to re-adjust or (in)validate the hypotheses.

The goal is to deeply understand the problem space that you are trying to innovate for. When moving from an EXPLORATION/DISCOVERY to an EVALUATION/ VALIDATION phase your prototypes will become more tangible and real (high-fidelity) assess and your learning will be more incremental.

Please never be shy to revisit old hypotheses or discard the ones that you believed were "carved in stone" to replace them with new learnings.

The templates as presented should help you understand the problem, test and fine-tune your idea with respect to a future-oriented viable offering and the needed connection to abundance, which you can demonstrate, validate and take advantage of when you combine Value Proposition Canvas, Business Model Canvas and the internal and external attributes of the Exponential Canvas.

There are many other templates and canvases that you can—and should—use, but that are not listed here.

We do believe that the ones that we suggest are sufficient to gain a thorough understanding of the problem space, subsequently support you in drafting need statements and solution ideas and to help you come up with an impactful future purpose– and product–market fit. But depending on your specific need you may require more in-depth experiments, information, or dedicated tools. We are sure that you are way beyond the initial EXPLORATION phase and possibly even far in your validation before you see the need for them. By that time you are deeply involved in your project and certainly have found them yourself.

The tools provided here are also great for initial and advanced Health Innovation lectures at universities or for project definitions and validations at Healthtech organizations

49.1.1 Core Innovation Principles of this Book

- *Use agile and iterative approaches to validate hypotheses and assumptions*
- *Think 10x whenever possible and feasible.*
- *Experimentation is key.*
- *Put yourself often in the position of the core user/patient and try to see the situation from their point of view—EMPATHIZE.*
- *Do not start building before defining and evaluating the problem that you are trying to address.*
- *If you ask the right questions you will find the answers!*
- *Embrace failure or invalidating ideas and with that focus on LEARNING.*
- *Write everything down for a learning history and to be able to go back.*
- *Come up with many solution ideas—do not limit yourself to the first one that you fell in love with.*
- *Use Minimal Viable Prototypes (MVP) for validation checks, customer experiments.*
- *Evaluate alternative solutions based on their DESIRABILITY, FEASIBILITY, and VIABILITY.*
- *Have a Massive Transformative Purpose, longer-term vision, a shorter-term mission, and core values.*

49.1.2 Core Principles of the Purpose Launchpad

1. Purpose over problem, and problem over solution
2. Exploration outcomes over optimization
3. Customers' insights over market research
4. Abundance over scarcity
5. Meaningful incomes over investment
6. Mindset over processes and tools
7. Validated learnings before building
8. Qualitative metrics over quantitative metrics
9. Purpose-driven synergies over competition
10. Long-term impact over short-term profit

These principles are essential for the use of the PURPOSE LAUNCHPAD HEALTH tools (Fig. 49.1) and are therefore listed one more time as Fig. 49.2.

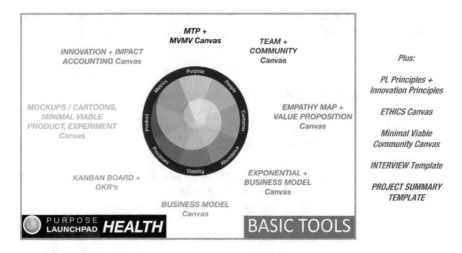

Fig. 49.1 PURPOSE LAUNCHPAD HEALTH and all the recommended templates and tools for the EXPLORATION and EVALUATION phases

PURPOSE LAUNCHPAD PRINCIPLES

1. Purpose over problem, and problem over solution.
2. Exploration outcomes over optimization.
3. Customers' insights over market research.
4. Abundance over scarcity.
5. Meaningful incomes over investment.
6. Mindset over processes and tools.
7. Validated learnings before building.
8. Qualitative metrics over quantitative metrics.
9. Purpose-driven synergies over competition.
10. Long-term impact over short-term profit.

CORE INNOVATION PRINCIPLES of this book

- Use agile and iterative approaches to validate hypotheses and assumptions
- Think 10x whenever possible and feasible
- Experimentation is key
- Put yourself often in the position of the core user / patient and try to see the situation from their point of view - EMPATHIZE
- Do not start building before defining and evaluating the problem that you are trying to address
- If you ask the right questions you will find the answers!
- Embrace failure or invalidating ideas and with that focus on LEARNING
- Write everything down for a learning history and to be able to go back
- Come up with many solution ideas - do not limit yourself to the first one that you fell in love with
- Use Minimal Viable Prototypes (MVP) for validation checks, customer experiments,
- Evaluate alternative solutions based on their DESIRABILITY, FEASIBILITY, and VIABILITY
- Have a Massive Transformative Purpose, longer-term vision, a shorter term mission and core values

Fig. 49.2 PURPOSE LAUNCHPAD principles and the core innovation principles of the FUTURE HEALTH INNOVATION book

49.2 PLH Segment Tools

To develop the PURPOSE of your organization use the checklist for a MASSIVE TRANSFORMATIVE PURPOSE as presented in Fig. 49.3. Your company should also develop a MOONSHOT, VISION, MISSION, and VALUES.

The PEOPLE segment is incredibly helpful to align the goals of the team members with the goals and purpose of the entity you are trying to set up. And, you can also see what expertise is missing in your team (Fig. 49.4). A good starting

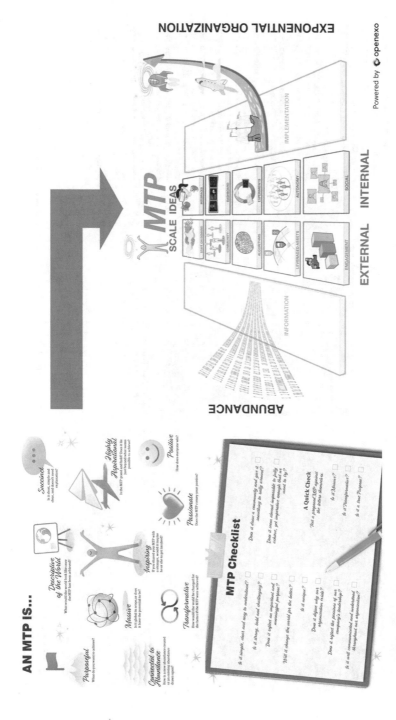

Fig. 49.3 PURPOSE: Massive Transformative Purpose (MTP)—general rules and as main attribute in the Exponential Canvas. Originally published under CC-BY-SA license with permission. Source: www.openexo.com

Fig. 49.4 PEOPLE: Team Canvas and instructions by Alex Ivanov and Mitya Voloshchuk—adapted from https://theteamcanvas.com

point for checking your individual entrepreneurial DNA can be obtained by doing the bosidna.com test. Highly recommendable and insightful. The PEOPLE segment does not only cover the direct team, but also the community around the problem. The Minimal Viable Community Canvas is a great tool to find out more about the community and how one can leverage the power of the community (Fig. 49.5).

The CUSTOMER segment deals of course with the customer of your product solution idea. But who really is the customer? Do we really understand the customer's needs and how they think/act/.../talk? The EMPATHY canvas just like any other canvas is a list of hypothesis/assumptions that needs to be validated. But to put yourself into the position of the customer is especially important for the HEALTH spec. Do we really understand the problems and needs of the PATIENTS or the CLINICIANS or the SUPPORT STAFF? DO we really understand the needs of the person and individual in a certain region or with a certain health problem? Figure 49.6 shows this very important canvas and together with the famous Value Proposition Canvas (VPC—Fig. 49.7) it can be the only tool that you need to understand the customer needs, pains, gains, and gain a deep understanding of the associated problems. Solutions are easily defined then! The highest priorities/values of each of the VPC can then be extracted and put together to formulate an ELEVA-TOR PITCH sentence that describes our PRODUCT/SERVICE idea, what PROB-LEM it will solve, and how it compares to the current solutions (Fig. 49.8). Very powerful, when you want to explain your entire concept in a very short time. And let us face it, if you need more than 1 min to explain your cool idea you may have lost an opportunity to convince someone already.

The ABUNDANCE segment mainly uses the Business Model Canvas (BMC—Fig. 49.9), that is also the key element for the next segment, as well as the EXPONENTIAL CANVAS (Fig. 49.10) with its 5 internal attributes—abbreviated IDEAS—and the 5 external ones—abbreviated SCALE—together with the MTP (see segment PURPOSE). This canvas will help you to connect your business plan (BMC) with abundance attributes and with that help you identify exponential scaling effects. Very powerful as well, but only if you use these two canvases together.

The VIABILITY segment deals with the business model of your idea and the question on whether you can define a sustainable operation. The BMC is possibly the most powerful tool of all and the one that stimulates and forces agile and iterative processes defining relevant Minimal viable Prototypes and customer experiments. The Value Proposition segment in the center together with the customer segment comes from the VPC. The BMC is the collection point for all the learning insights and will give you the confidence that the idea you are developing is worth to invest your future in.

You need PROCESSES to actually understand the problem and define initial solutions for it, but more importantly to constantly improve them and question them. The PLH methodology is such a process, but Objective and Key result (Fig. 49.11) and the Kanban board (Fig. 49.12) are great tools to actually manage the progress and define clear goals to be tested and validated. And, the most important Innovation process is shown in Fig. 49.13 using the VPC and Hypothesis validation together

Fig. 49.5 **PEOPLE**: Minimal Viable Community Canvas. Originally published under CC-BY-SA license with permission. Available at https://community-canvas.org— authors Fabian Pfortmüller, Nico Luchsinger, Sascha Mobartz

Developing your customer and creating the right value proposition means finding out who is (and who isn't) your customer now and in the future and evaluating their problems.

The goal is to define an early adopter who would buy right away and be enthusiastic about the product, service, or customer experience.

Based on seven segments with relevant questions, an Empathy Map creates a shared understanding of user needs and aid in decision making.

Completed Empathy Map provides necessary input for the Value Proposition Canvas.

Source:
Dave Gray, Updated Empathy Map Canvas (https://gamestorming.com/update-to-the-empathy-map/) — how to use it: https://medium.com/the-xplane-collection/updated-empathy-map-canvas-46df22df3c8a

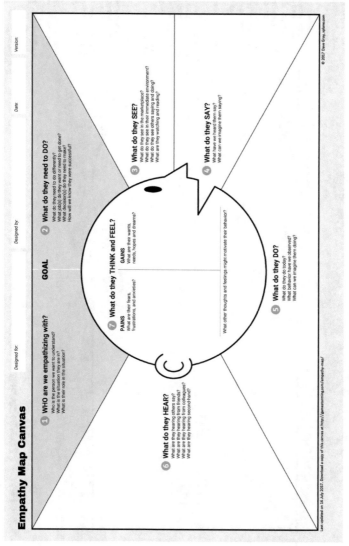

Fig. 49.6 CUSTOMER: Empathy Canvas and segment explanation—developed by Dave Gray—courtesy xplane.com/Dave Gray

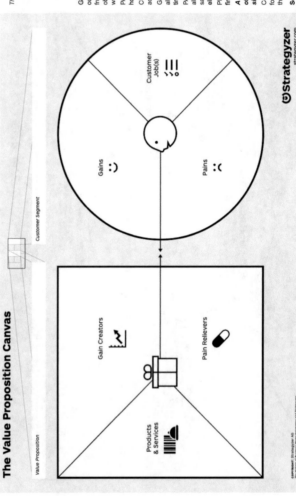

The Value Proposition Canvas

Value Proposition

Customer Segment

Gain Creators

Products & Services

Pain Relievers

Customer Job(s)

Gains :)

Pains :(

Strategyzer
strategyzer.com

*Success **doesn't come from understanding the customer.** It comes from a deep understanding of the job the customer is trying to get done.*
Clayton Christensen

The Value Proposition Canvas includes two building blocks:

- Customer profile for each customer segment to identify your customer's major Jobs-to-be-done, the pains they face when trying to accomplish their jobs-to-be-done, and the gains they perceive by getting their jobs done.
- Value proposition to define the most important components of your offering, how you relieve pain and create gains for your customers.

GAINS - gains are motivational and emotionally driven – outcomes. Pains are process and function-driven – obstacles/ frustrations. Gains are harder than pains to find. You can observe a pain, an obstacle, but understanding why someone wants to achieve a goal, a job, requires you to dig a bit deeper.

PAINS - Pains often go unnoticed. People just accept that this how things are done – **how we do things.**

CUSTOMER JOBS - describes what the customer has to accomplish.

GAIN CREATOR - describes how your products and services alleviate the pains you noted earlier (create savings - money, **time, efforts, make things easier,** ...)

PAIN RELIEVER - describe how your products and services alleviate the pains you noted earlier (could they produce savings, eliminate customer fears, put an end to difficulties, **eliminate barriers,** ...)

PRODUCTS AND SERVICES - tangible, intangible, digital, financial, ...

A product-market-fit is achieved, when the products offered as part of the value proposition address the most significant pains and gains from the customer profile.

Completed Value Proposition Model provides necessary input for the VALUE PROPOSITION and CUSTOMER segments of the Business Model Canvas.

Source:
Alexander Osterwalder et al.: Value Proposition Design: How to Create Products and Services Customers Want.
For a great explanation and guide on how to use and fill the VPC - see https://web.eecs.umich.edu/~sugih/courses/eecs441/w18/03-VPC.pdf

Fig. 49.7 CUSTOMER/ABUNDANCE: Value Proposition Canvas and segment explanation. Courtesy www.strategyzer.com

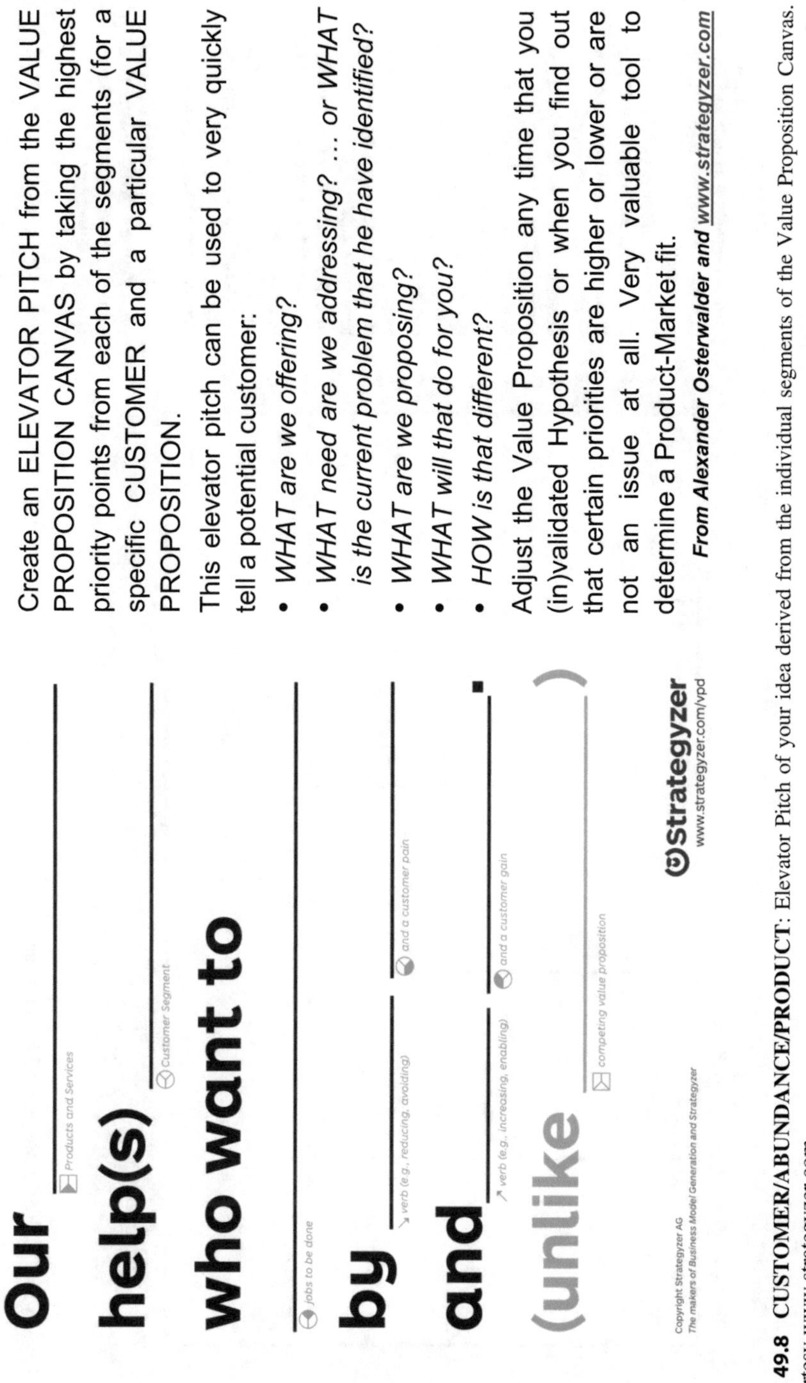

Create an ELEVATOR PITCH from the VALUE PROPOSITION CANVAS by taking the highest priority points from each of the segments (for a specific CUSTOMER and a particular VALUE PROPOSITION.

This elevator pitch can be used to very quickly tell a potential customer:

- *WHAT are we offering?*
- *WHAT need are we addressing? … or WHAT is the current problem that he have identified?*
- *WHAT are we proposing?*
- *WHAT will that do for you?*
- *HOW is that different?*

Adjust the Value Proposition any time that you (in)validated Hypothesis or when you find out that certain priorities are higher or lower or are not an issue at all. Very valuable tool to determine a Product-Market fit.

From Alexander Osterwalder and www.strategyzer.com

Our ___ [Products and Services]

help(s) ___ [Customer Segment]

who want to ___ [jobs to be done]

by ___ [verb (e.g. reducing, avoiding)] ___ [and a customer pain]

and ___ [verb (e.g. increasing, enabling)] ___ [and a customer gain]

(unlike ___ [competing value proposition]

Strategyzer
www.strategyzer.com/vpd

Copyright Strategyzer AG
The makers of Business Model Generation and Strategyzer
www.strategyzer.com

Fig. 49.8 CUSTOMER/ABUNDANCE/PRODUCT: Elevator Pitch of your idea derived from the individual segments of the Value Proposition Canvas. Courtesy www.strategyzer.com

Business Model Canvas consist of nine building blocks representing different fundamental elements of a business

- Customer Segments: Who are the customers?
- Value Propositions: What is compelling about your proposition?
- Channels: How are these propositions promoted, sold and delivered?
- Customer Relationships: How do you interact with the customer?
- Revenue Streams: How does the business earn revenue?
- Key Activities: What uniquely strategic things do you do to deliver the proposition?
- Key Resources: What unique strategic assets must the business have to compete?
- Key Partnerships: What can the company not do so it can focus on its Key Activities?
- Cost Structure: What are the business' major cost drivers associated with operating your business model

Source: Alexander Osterwalder and Yves Pigneur: Business Model Generation: A Handbook for Visionaries, Game Changers, and Challengers

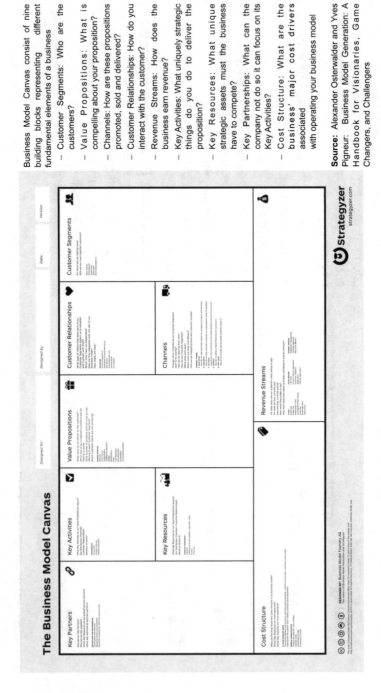

Fig. 49.9 VIABILITY: Business Model Canvas and segment explanation. Originally published under CC-BY-SA license with permission

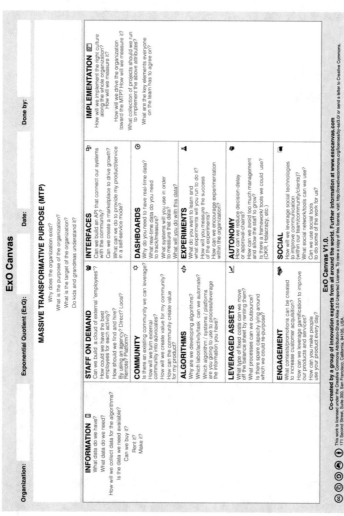

Exponential Organization (ExO) is a purpose-driven entity that leverages a set of 11 common organizational attributes that allow to tap into and manage abundance to scale exponentially.

- **Massive Transformative Purpose**, the higher aspirational purpose of the organization, is the only ExO Attribute out of the eleven that is mandatory.
- **SCALE** (left side), a group of five external attributes is focused on connecting and keeping abundance.
- **IDEAS** (right side), a group of five internal attributes is focused on managing abundance to grow

We recommend having an MTP plus at least four of the ten attributes in order to create an ExO. https://openexo.com

Fig. 49.10 **ABUNDANCE**: Exponential Canvas and segment explanation. Originally published under CC-BY-SA license with permission www.openexo.com

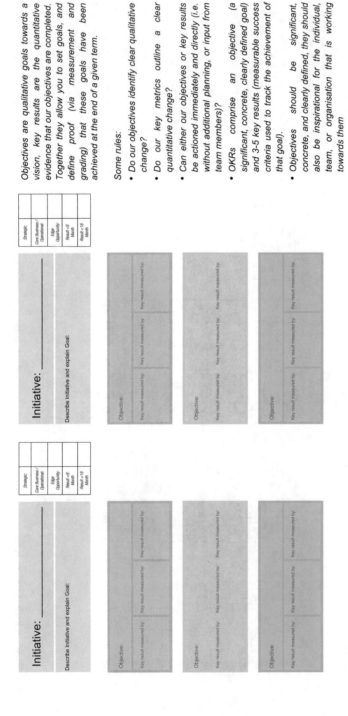

Fig. 49.11 PROCESS: Objective and Key Results Canvas and some rules and explanations

Kanban boards visually depict work at various stages of a process using cards to represent work items and columns to represent each stage of the process.

Cards are moved from left to right to show progress and to help coordinate teams performing the work.

A kanban board may be divided into horizontal "swimlanes" representing different kinds of work or different teams performing the work.

There are plenty of KANBAN board and Project Management Tools available like TRELLO, used by the EASYJECTOR and BODYTUNE Teams

Fig. 49.12 PROCESS: a general Kanban board layout and some examples using a digital board like www.trello.com

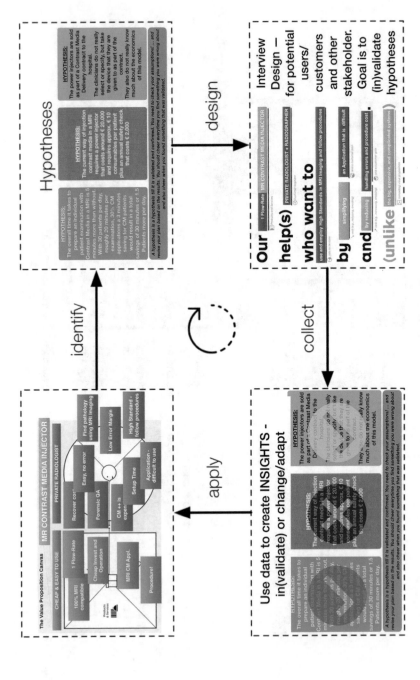

Fig. 49.13 **PROCESS/PRODUCT**: Agile and iterative Lean StartUp oriented LEARNING process—Identify—Design—Collect—Apply

with customer interviews and data collection for a LEARNING process. If I would list the top three tools you should follow then this would be one of them.

For the PRODUCT segment, several low fidelity (Fig. 49.14) and high fidelity (Fig. 49.15) minimal viable prototype ideas are listed. You do not always have to build something tangible. To get insights—especially in the EXPLORATION phase—it is easier and often better to create stories or draw something. This can already give you a good understanding about the DESIRABILITY of a solution idea. In an EVALUATION phase, you also need to show and demonstrate FEASIBILITY, which means can it actually be build, and do we believe—based on rudimentary working prototypes—that it can actually do what it is supposed to do.

Brings us to the last segment, the METRICS. Figure 49.16 shows the TEST card to be used to define the experiments and the key measures/metrics to define a (in)validated experiment result. The LEARNING card lets you reflect on the previously defined hypothesis and the observations that may change your product definitions, value propositions, customer definitions, ... Fig. 49.17 shows the process of testing HYPOTHESIS and the LEARNINGS in form of progress from the validation work. This can and should be used in combination with the LEARNING and TEST CARDS, as well as with the INTERVIEW TEMPLATE presented in Fig. 49.18. Other Innovation METRICS you can use are (but define them based on your specific needs):

- *Number of customer interviews*
- *Number of experiments*
- *Number of (in)validated hypotheses*
- *Number of Prototypes build*
- *Number of patents filed*
- *Number of scientific papers written (but please patent first!)*
- *Number of users on LinkedIn*

All the points can be summarized in the template provided in Fig. 49.19. You can see it as a short idea and business summary in one page. Sufficient detail (and hopefully validated details), but focusing on the key aspects. This is also something you can use for your first discussions with potential supporters.

And finally—little bit outside the main segments. The ETHICS canvas in Fig. 49.20. ETHICS will be a very important element and ties the needs of individuals/groups/behaviors/product performances and other things together. Especially important for any Health and Patient related Innovation and therefore an important tool that should be used after completion of the EXPLORATION phase.

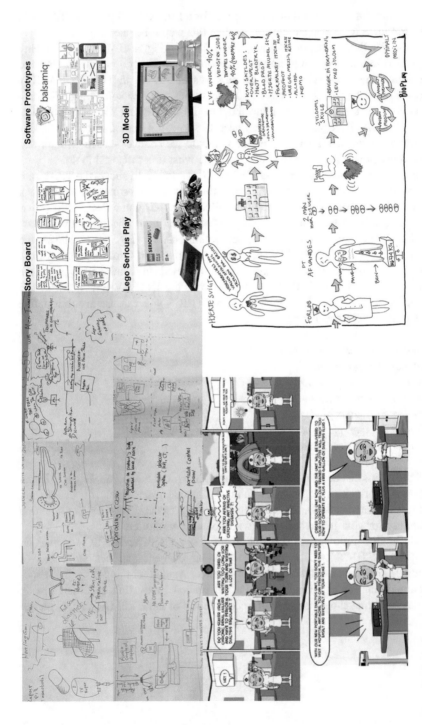

Fig. 49.14 PRODUCT: Prototyping—low-fidelity for the EXPLORATION/DISCOVERY phase—drawings/cartoon stories and story boards, "fake" software, Lego, 3D printed prototypes, patient stories

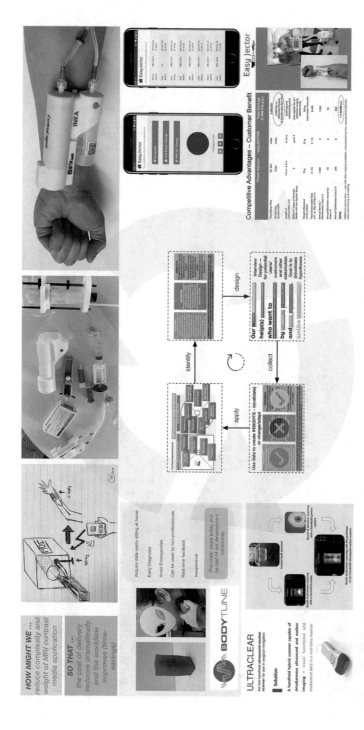

Fig. 49.15 PRODUCT: Prototyping—some examples of more tangible higher fidelity products, actual functional prototypes, one-pager with explanations. Use these prototypes to continuously collect data, apply them to the product and business idea and formulate new hypotheses to be validated through additional interviews ("customer experiments")

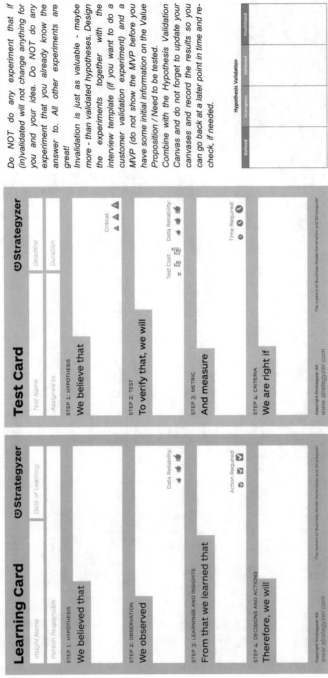

Fig. 49.16 PRODUCT/PROCESS/CUSTOMER: TEST and LEARNING cards—courtesy STRATEGYZER.COM

The presented Innovation Progress Board shows the process of testing HYPOTHESIS and the LEARNINGS in form of progress from the validation work.

This can and should be used in combination with the LEARNING and TEST CARDS, as well as with the INTERVIEW TEMPLATE presented in Figure 18. Create a test BACKLOG for validating the key Business HYPOTHESIS („Experimentation").

Additionally define Metrics you want to track and that are relevant for your initiative.

For the PLH approach it may not be necessary to actually use a dedicated canvas, but define and track metrics like:

- number of customer interviews
- Number of experiments
- Number of (in)validated hypotheses
- Number of Prototypes build
- Number of patents filed
- Number of scientific papers written (but please patent first!)
- Number of users on LinkedIn
- ...

Use in combination TEST and LEARNING cards and the OKR's / KPI's

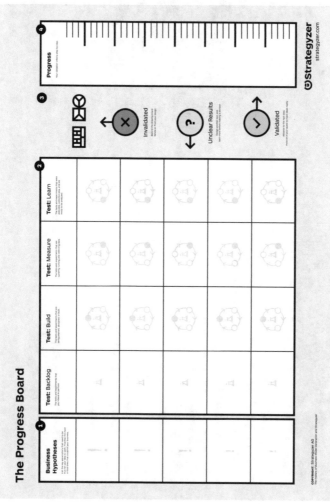

Fig. 49.17 METRICS/PROCESS: Innovation Accounting and Metrics Tracking—courtesy STRATEGZYER.COM

INTERVIEW TEMPLATE

Example for customer discovery interview

Introduction

- Hello, my name is xxx and I'd like to interview you in order to learn about a new initiative we have focused on [rough idea or purpose]. Please, just provide honest feedback.

Segmentation (level of detail is optional, ask or note down in interview list

- What's your name? What's your email? What's your role in the organization?

Customer's problems

- Tell me a story about the last time you experienced (THE PROBLEM)?
- What are the key problems and challenges you see around [topic of the idea]?
- Why was that hard / annoying / difficult? (dig deeper). Try to learn as much as possible about the underlying reasons.

Problems (hypotheses)

- We believe one of the key problems people / organizations have is to [hypotheses]. What do you think about it?

Customer's solutions (optional)

- How do you solve (PROBLEM) now? What's the ACTION they take, if any? And why is that not so great or a perfect solution?

Solutions (our hypotheses - if you are ready to share)

Key hypotheses from Value Proposition Canvas and Business Model Canvas to derive and to rank according to importance and amount of evidence (**hypothesis list**).

Interviews (based on an interview template) with 10+ persons per customer segment to receive fast feedback from markets and to test value proposition and business model.

Summarize any insights that inform future assumptions and experiments.

Summarize your interviews

What is the main problem this segment has AND is trying to solve?

What are the most common current solutions used by customers to solve that problem?

What is not great about those solutions where you can do better?

What did the interviewed person said about the hypotheses. Are they confirmed or refuted?

Fig. 49.18 **PRODUCT/CUSTOMER:** Interview Template for Experimentation. Everything is initially an assumption/not validated hypothesis

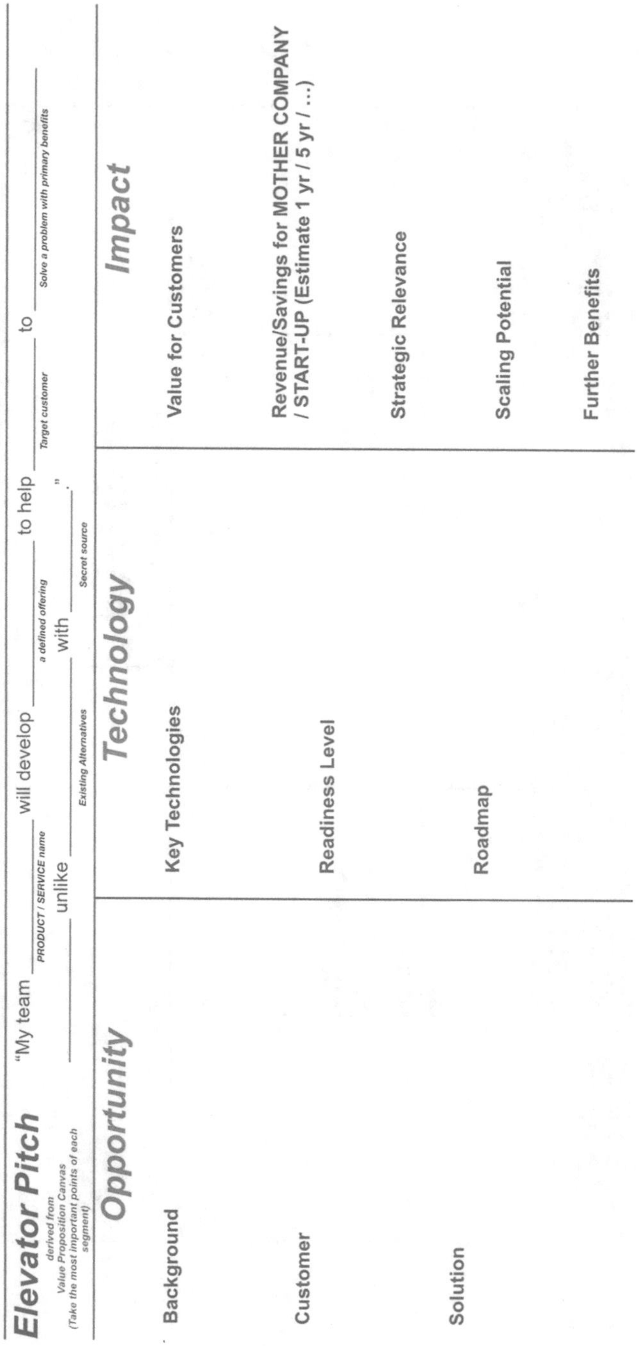

Your Idea Template

Name of Initiative:

Team and Support
Lead:
Members:
Sponsor / Mentor :

Elevator Pitch

"My team _____ will develop _____ to help _____ to _____
                    PRODUCT / SERVICE name   *a defined offering*   *Target customer*
derived from
Value Proposition Canvas   unlike _____ with _____ "   _____
(Take the most important points of each   *Existing Alternatives*   *Secret source*   *Solve a problem with primary benefits*
segment)

Opportunity

Background

Customer

Solution

Technology

Key Technologies

Readiness Level

Roadmap

Impact

Value for Customers

Revenue/Savings for MOTHER COMPANY
/ START-UP (Estimate 1 yr / 5 yr / ...)

Strategic Relevance

Scaling Potential

Further Benefits

Fig. 49.19 PRODUCT/CUSTOMER/VIABILITY: Summary Idea Template

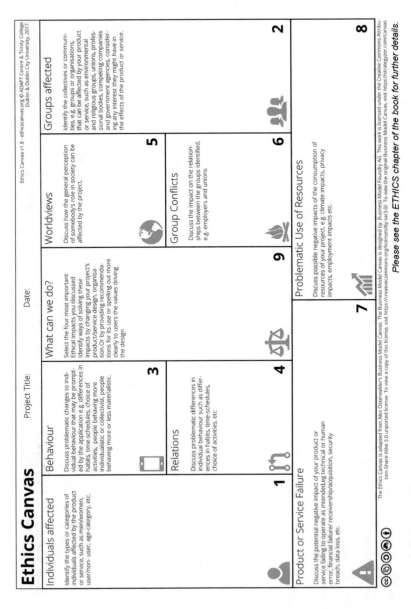

Ethics Canvas Project Title: Date:

Ethics Canvas v1.8 · ethicscanvas.org © ADAPT Centre & Trinity College Dublin & Dublin City University, 2017.

Individuals affected

Identify the types or categories of individuals affected by the product or service, such as men/women, user/non- user, age-category, etc.

1

Behaviour

Discuss problematic changes to individual behaviour that may be prompted by the application e.g. differences in habits, time-schedules, choice of activities, people behaving more individualistic or collectivist, people behaving more or less materialistic.

3

Relations

Discuss problematic differences in individual behaviour such as differences in habits, time-schedules, choice of activities, etc.

4

What can we do?

Select the four most important Ethical impacts you discussed. Identify ways of solving these impacts by changing your project's product/service design, organisation.Or by providing recommendations for its use or spelling out more clearly to users the values driving the design

Worldviews

Discuss how the general perception of somebody's role in society can be affected by the project.

5

Group Conflicts

Discuss the impact on the relationships between the groups identified, e.g. employers and unions

9

Groups affected

Identify the collectives or communities, e.g. groups or organisations, that can be affected by your product or service, such as environmental and religious groups, unions, professional bodies, competing companies and government agencies, considering any interest they might have in the effects of the product or service.

2

Problematic Use of Resources

Discuss possible negative impacts of the consumption of resources of your project, e.g. climate impacts, privacy impacts, employment impacts etc.

7

6

8

Product or Service Failure

Discuss the potential negative impact of your product or service failing to operate as intended, e.g technical or human error, financial failure / receivership/acquisition, security breach, data loss, etc.

https://www.ethicscanvas.org/

Fig. 49.20 PRODUCT/CUSTOMER: Ethics Canvas. Originally published under CC-BY-SA license with permission. Canvas can be downloaded from https://www.ethicscanvas.org/

49.3 Remark with Respect to the Use of the Tools

The permission to use the presented templates has been obtained for this book. If no author or webpage is listed then the tools were developed or created by the Editor/ Authors. Most can be freely used under Creative Commons license 3.0 or 4.0 BY NC SA, details see https://creativecommons.org/licenses/by-nc-sa/4.0/. Please use the proper referencing when using them. The Editor and Authors of this book have obtained permission from the authors to use the tools in this book, which does not mean that you can freely use it in a similar book. For any other use outside the boundaries of the license you need to contact the authors directly and obtain permission.

The ones without the CC license are copyright Michael Friebe, but can be freely used for anything provided that the author and chapter are properly referenced.

Good luck disrupting and creating a Future of Health that we all want and deserve.

Take-Home Messages
- VPV, BMC, and Exponential Canvas work together.
- Without a process that embraces failure and focuses on learning you will not learn the real problem space.
- Use Minimal Viable Prototypes - low-fidelity ones initially and only after the first successful exploration validations steps tangible ones.
- Learning and willingness to apply learnings are essential.
- What cannot be measured cannot be improved . . . true at least most of the time!
- Create an Elevator Pitch to summarize your Idea, the Problem you are trying to address, and the need for that Solution.
- The tools as presented and if used in an agile way will lead to a problem definition and meaningful solution ideas.

References

1. Purpose Launchpad Guide, and Assessment Information, www.purposelaunchpad.com, viewed 30 Jan 2022
2. Friebe M, Fritzsche H, Heryan K (2022) The PLH - Purpose Launchpad Health - Meta-Methodology to Explore Problems and Evaluate Solutions for Biomedical Engineering Impact Creation. IEEE EMBC 2022, Glasgow, Conference Paper
3. Ries E (2011) The Lean Startup: How Today's Entrepreneurs Use Continuous Innovation to Create Radically Successful Businesses. ISBN-13: 978-0307887894
4. Blank S, Dorf B (2020) The startup owner's manual: the step-by-step guide for building a great company. Wiley. ISBN-13: 978-1119690689

Michael Friebe received the B.Sc. degree in electrical engineering, the M.Sc. degree in technology management from Golden Gate University, San Francisco, and the Ph.D. degree in medical physics in Germany. He spent five years in San Francisco, as a Research and Design Engineer with an MRI and ultrasound device manufacturer. He is a German citizen with expertise in diagnostic imaging and image-guided therapies, as a/an Founder/Innovator/CEO/Investor and a Scientist. He is also a Research Fellow with the Technical University of Munich, Munich; an Adjunct Professor with the Queensland University of Technology, Brisbane; and a honorary Professor of image-guided therapies with Otto von Guericke University, Magdeburg, Germany. Since 2022, he has been a Professor of biomedical engineering innovation with the AGH University of Science and Technology, Krakow, Poland. He is a listed inventor of more than 80 patents and has authored over 200 papers, and was part of over 35 Medical Technology Start-Ups. He is a Board Member of four medical technology start-up companies and an investment partner of a MedTec-fund. From 2016 to 2018, he was a Distinguished Lecturer of the IEEE EMBC teaching innovation generation and MedTec entrepreneurship. He is also a coach and trainer for OpenExO, and a master Launchpad mentor for the Purpose Alliance.